Robust Statistics

WILEY SERIES IN PROBABILITY AND STATISTICS

ESTABLISHED BY WALTER A. SHEWHART AND SAMUEL S. WILKS

The *Wiley Series in Probability and Statistics* is well established and authoritative. It covers many topics of current research interest in both pure and applied statistics and probability theory. Written by leading statisticians and institutions, the titles span both state-of-the-art developments in the field and classical methods.

Reflecting the wide range of current research in statistics, the series encompasses applied, methodological and theoretical statistics, ranging from applications and new techniques made possible by advances in computerized practice to rigorous treatment of theoretical approaches. This series provides essential and invaluable reading for all statisticians, whether in academia, industry, government, or research.

A complete list of titles in this series can be found at http://www.wiley.com/go/wsps

Robust Statistics
Theory and Methods (with R)

Second Edition

Ricardo A. Maronna
Consultant Professor, National University of La Plata, Argentina

R. Douglas Martin
Departments of Applied Mathematics and Statistics,
University of Washington, USA

Victor J. Yohai
Department of Mathematics, University of Buenos Aires, and
CONICET, Argentina

Matías Salibián-Barrera
Department of Statistics, The University of British Columbia, Canada

WILEY

This edition first published 2019

Edition History
John Wiley & Sons (1e, 2006)

Registered Offices
John Wiley & Sons, Inc., 111 River Street, Hoboken, NJ 07030, USA
John Wiley & Sons Ltd, The Atrium, Southern Gate, Chichester, West Sussex, PO19 8SQ, UK

Editorial Office
9600 Garsington Road, Oxford, OX4 2DQ, UK

For details of our global editorial offices, customer services, and more information about Wiley products visit us at www.wiley.com.

Wiley also publishes its books in a variety of electronic formats and by print-on-demand. Some content that appears in standard print versions of this book may not be available in other formats.

Library of Congress Cataloging-in-Publication Data

Names: Maronna, Ricardo A., author. | Martin, R. Douglas, author. |
Yohai, Victor J., author. | Salibián-Barrera, Matías, author.
Title: Robust statistics : theory and methods (with R) / Ricardo A. Maronna,
Consultant Professor, National University of La Plata, Argentina, R. Douglas Martin,
Departments of Applied Mathematics and Statistics, University of Washington, USA,
Victor J. Yohai, Department of Mathematics, University of Buenos Aires, and CONICET,
Argentina, Matías Salibián-Barrera, Department of Statistics, The University of
British Columbia, Canada.
Description: Second edition. | Hoboken, NJ : WIley, 2019. | Series: Wiley
series in probability and statistics | Includes bibliographical references
and index. |
Identifiers: LCCN 2018033202 (print) | LCCN 2018037714 (ebook) | ISBN
9781119214670 (Adobe PDF) | ISBN 9781119214663 (ePub) | ISBN 9781119214687
(hardcover)
Subjects: LCSH: Robust statistics.
Classification: LCC QA276 (ebook) | LCC QA276 .M336 2019 (print) | DDC
519.5–dc23
LC record available at https://lccn.loc.gov/2018033202

Cover design by Wiley
Cover image: Courtesy of Ricardo A. Maronna

Set in 10/12pt TimesLTStd by SPi Global, Chennai, India
Printed in Singapore by C.O.S. Printers Pte Ltd

10 9 8 7 6 5 4 3 2 1

To Susana, Jean, Julia, Livia, Paula,
Verónica, Lucas, Dante and Sofia

and

with recognition and appreciation of the foundations laid by the founding fathers of robust statistics: John Tukey, Peter Huber and Frank Hampel.

Contents

Note: sections marked with an asterisk can be skipped on first reading.

Preface

It has now been eleven years since the publication of the first edition of *Robust Statistics: Theory and Methods* in 2006. Since that time, there have been two developments prompting the need for a second edition. The first development is that since 2006 a number of new results in the theory and methods of robust statistics have been developed and published, in particular by the book's authors. The second development is that the S-PLUS software has been superseded by the open source package R, so our original of the S-PLUS robust statistics package became outdated. Thus, for this second edition, we have created a new R-based package called RobStatTM, and in that package and at the publisher's web site we provide scripts for computing all the examples in the book.

We will now discuss the main research advances included in this second edition.

Finite-sample robustness

Asymptotically normal robust estimators have tuning constants that allow users to control their normal distribution variance efficiency, in a trade-off with robustness toward fat-tailed non-normal distributions. The resulting finite-sample performance in terms of mean-squared error (MSE), which takes into account bias as well as variance, can be considerably worse than implied by the asymptotic performance. This second edition contains useful new results concerning the finite-sample MSE performance of robust linear regression and robust covariance estimators. These are briefly described below.

Linear regression estimators

A loss function with optimality properties is introduced in Section 5.8.1, and it is shown that its use gives much better results than the popular bisquare function, in both efficiency and robustness.

Section 5.9.3 focuses on finite-sample efficiency and robustness and introduces a new "distance-constrained maximum-likelihood" (DCML) estimator. The DCML estimator is shown to provide the best trade-off between finite-sample robustness and normal distribution efficiency, in comparison with an MM estimator that is asymptotically 85% efficient, and an adaptive estimator, described in Section 5.9.2, that is asymptotically fully efficient for normal distributions.

Multivariate location and scatter

A number of proposed robust covariance matrix estimators were discussed in the first edition, and some comments about the choice of estimator were made. In this second edition, the new Section 6.10 "Choosing a location/scatter estimator" replaces the previous Section 6.8, and this new section provides new recommendations for choosing a robust covariance matrix estimator, based on extensive finite-sample performance simulation studies.

Fast and reliable starting points for initial estimators

The standard starting point for computing initial S-estimators for linear regression and covariance matrix estimators is based on a subsampling algorithm. Subsampling algorithms have two disadvantages: the first is that their computation time increases exponentially with the number of variables. The second disadvantage of the subsampling method is that the method is stochastic, which means that different final S-estimators and MM-estimators can occur when the computation is repeated.

Linear regression

Section 5.7.4 describes a deterministic algorithm due to Peña and Yohai (1999) for obtaining a starting point for robust regression. Since this algorithm is deterministic, it always yields the same final MM-estimator. This is particularly important in some applications, for example in financial risk calculations. Furthermore, it is shown in Section 5.7.6 that the Peña–Yohai starting-value algorithm is much faster than the subsampling method, and has smaller maximum MSE than the subsampling algorithm, sometimes substantially so.

Multivariate location and scatter

Subsampling methods have also been used to get starting values for robust estimators of location and dispersion (scatter), but they have a similar difficulty as in linear regression, namely that they will be too slow when the number of variables is large. Fortunately, there is an improved algorithm for computing starting values due to Peña and Prieto (2007), which makes use of finding projection directions of maximum and minimum kurtosis plus a set of random directions obtained by a "stratified sampling" procedure. This method, which is referred to as the KSD method, is described in Section 6.9.2. While the KSD method is still stochastic in nature, it provides fast reliable starting values, and is more stable than ordinary subsampling, as is discussed in Sections 6.10.2 and 6.10.3.

Robust regularized regression

The use of penalized regression estimators to obtain good results for high-dimensional but sparse predictor variables has been a hot topic in the "machine learning" literature over the last decade or so. These estimators add L_1 and L_2 penalties to the least squares objective function; the leading estimators of this type are Lasso regression, Least Angle Regression, and Elastic Net regression, among others. A new section on robust regularized regression describes how to extend robust linear model regression to obtain robust versions of the above non-robust least-squares-based regularized regression estimators.

Multivariate location and scatter estimation with missing data

Section 6.12 provides a method for solving the problem of robust estimation of scatter and location with missing data. The method contains two main components. The first is the introduction of a generalized S-estimator of scatter and location that depends on Mahalanobis distances for the non-missing data in each observation. The second component is a weighted version of the well-known expectation-maximization (EM) algorithm for missing data.

Robust estimation with independent outliers in variables

The Tukey–Huber outlier-generating family of distribution models has been a commonly accepted standard model for robust statistics research and associated empirical studies for independent and identically distributed data. In the case of multivariate data, the Tukey–Huber model describes the distribution of the rows, or "cases", of a data matrix whose columns represent variables; outliers generated by this model are known as "case outliers". However, there are important problems where outliers occur independently across cells – that is, across variables – in each row of a data matrix. For example with portfolios of stock returns, where the columns represent different stocks and the rows represent observations at different times, outlier returns in different stocks (representing idiosyncratic risk) occur independently across stocks; that is, across cells/variables.

Section 6.13 discusses an important and relatively recent model for generating independent outliers across cells (across variables), called the *independent contamination* (IC) model. It turns out that estimators that have good robustness properties under the Tukey–Huber model have very poor robustness properties for the IC model. For example, estimators that have high breakdown points under the Tukey–Huber model can have very low breakdown points under the IC model. This section surveys the current state of research on robust methods for IC models, and on robust methods for simultaneously dealing with outliers from both Tukey–Huber and IC models. The problem of obtaining robust estimators that work well for both Tukey–Huber and IC models is an important ongoing area of research.

Mixed linear models

Section 6.15 discusses robust methods for mixed linear models. Two primary methods are discussed, the first of which is an S-estimator method that has good robustness properties for Tukey–Huber model case-wise outliers, but does not perform well for cell-wise independent outliers. The second method is designed to do well for both types of outliers, and achieves a breakdown point of 50% for Tukey–Huber models and 29% for IC models.

Generalized linear models

New material on a family of robust estimators has been added to the chapter on generalized linear models (GLMs). These estimators are based on using M-estimators after a variance-stabilizing transformation has been applied to the response variable.

Regularized robust estimators of the inverse covariance matrix

In Chapter 6, on multivariate analysis, Section 6.14 looks at regularizing robust estimators of inverse covariance matrices in situations where the ratio of the number of variables to the number of cases is closer to or larger than one.

A note on software and book web site

The section on "Recommendations and software"at the end of each chapter indicates the procedures recommended by the authors and the R functions that implement them. These functions are located in several libraries, in particular the R package

`RobStatTM`, which was especially developed for this book. All are available in the CRAN network (https://cran.r-project.org).

The R scripts and datasets that enable the reader to reproduce the book's examples are available at the book's web site at www.wiley.com/go/maronna/robust where each dataset has the same name as the respective script. The scripts and data sets are also directly available in `RobStatTM`. The book web site also contains an errata document.

Preface to the First Edition

Why robust statistics are needed

All statistical methods rely explicitly or implicitly on a number of assumptions. These assumptions generally aim at formalizing what the statistician knows or conjectures about the data analysis or statistical modeling problem he or she is faced with, while at the same time aim at making the resulting model manageable from the theoretical and computational points of view. However it is generally understood that the resulting formal models are simplifications of reality and that their validity is at best approximate. The most widely used model formalization is the assumption that the observed data has a *normal* (Gaussian) distribution. This assumption has been present in statistics for two centuries, and has been the framework for all the classical methods in regression, analysis of variance, and multivariate analysis. There have been attempts to justify the assumption of normality with theoretical arguments, such as the central limit theorem. These attempts however are easily proven wrong. The main justification for assuming a normal distribution is that it gives an approximate representation to many real data sets, and at the same time is theoretically quite convenient because it allows one to derive explicit formulae for optimal statistical methods such as maximum likelihood and likelihood ratio tests, as well as the sampling distribution of inference quantities such as t-statistics. We refer to such methods as *classical* statistical methods, and note that they rely on the assumption that normality holds *exactly*. The classical statistics are by modern computing standards quite easy to compute. Unfortunately theoretical and computational convenience does not always deliver an adequate tool for the practice of statistics and data analysis, as we shall see throughout this book.

It often happens in practice that an assumed normal distribution model (e.g., a location model or a linear regression model with normal errors) holds approximately in that it describes the majority of observations, but some observations follow a different pattern or no pattern at all. In the case when the randomness in the model is assigned to observational errors – as in astronomy which was the first instance of the use of the least squares method – the reality is that while the behavior of many sets of data appeared rather normal, this held only approximately with the main discrepancy being that a small proportion of observations were quite atypical by virtue of being far from the bulk of the data. Behavior of this type is common across the entire spectrum of data analysis and statistical modeling applications. Such atypical data are called *outliers*, and even a single outlier can have a large distorting influence on a classical statistical method that is optimal under the assumption of normality or linearity. The kind of "approximately" normal distribution that gives rise to outliers is one that has a normal shape in the central region, but has tails that are heavier or "fatter" than those of a normal distribution.

One might naively expect that if such approximate normality holds, then the results of using a normal distribution theory would also hold approximately. This is unfortunately not the case. If the data are assumed to be normally distributed but their actual distribution has heavy tails, then estimates based on the maximum likelihood principle not only cease to be "best" but may have unacceptably low statistical efficiency (unnecessarily large variance) if the tails are symmetric and may have very large bias if the tails are asymmetric. Furthermore, for the classical tests their level may be quite unreliable and their power quite low, and for the classical confidence intervals their confidence level may be quite unreliable and their expected confidence interval length may be quite large.

The robust approach to statistical modeling and data analysis aims at deriving methods that produce reliable parameter estimates and associated tests and confidence intervals, not only when the data follow a given distribution exactly, but also when this happens only approximately in the sense just described. While the emphasis of this book is on approximately normal distributions, the approach works as well for other distributions that are close to a nominal model, e.g., approximate gamma distributions for asymmetric data. A more informal data-oriented characterization of robust methods is that they fit the bulk of the data well: if the data contain no outliers the robust method gives approximately the same results as the classical method, while if a small proportion of outliers are present the robust method gives approximately the same results as the classical method applied to the "typical" data. As a consequence of fitting the bulk of the data well, robust methods provide a very reliable method of detecting outliers, even in high-dimensional multivariate situations.

We note that one approach to dealing with outliers is the *diagnostic* approach. Diagnostics are statistics generally based on classical estimates, that aim at giving numerical or graphical clues for the detection of data departures from the assumed model. There is a considerable literature on outlier diagnostics, and a good outlier diagnostic is clearly better than doing nothing. However, these methods present two drawbacks. One is that they are in general not as reliable for detecting outliers as examining departures from a robust fit to the data. The other is that, once suspicious observations have been flagged, the actions to be taken with them remain the analyst's personal decision, and thus there is no objective way to establish the properties of the result of the overall procedure.

Robust methods have a long history that can be traced back at least to the end of the 19th century with Simon Newcomb (see Stigler, 1973). But its first great steps forward occurred in the 60s and the early 70s with the fundamental work of John Tukey (1960, 1962), Peter Huber (1964, 1967) and Frank Hampel (1971, 1974). The applicability of the new robust methods proposed by these researchers was made possible by the increased speed and accessibility of computers. In the last four decades the field of robust statistics has experienced substantial growth as a research area, as evidenced by a large number of published articles. Influential books have been written by Huber (1981), Hampel *et al.* (1986), Rousseeuw and Leroy (1987) and Staudte and Sheather (1990). The research efforts of the current book's authors, many of which are reflected in the various chapters, were stimulated by the early foundation results, as well as work by many other contributors to the field, and the emerging computational opportunities for delivering robust methods to users.

The above body of work has begun to have some impact outside the domain of robustness specialists, and there appears to be a generally increased awareness of the dangers posed by atypical data values and of the unreliability of exact model assumptions. Outlier detection methods are nowadays discussed in many textbooks on classical statistical methods, and implemented in several software packages. Furthermore by now several commercial statistical software packages offer some robust methods, with the Robust Library offering in S-PLUS being the currently most complete and user friendly. In spite of the increased awareness of the impact outliers can have on classical statistical methods and the availability of some commercial software, robust methods remain largely unused and even unknown by most of the communities of applied statisticians, data analysts, and scientists that might benefit from their use. It is our hope that this book will help rectify this unfortunate situation.

Purpose of the book

This book was written to stimulate the routine use of robust methods as a powerful tool to increase the reliability and accuracy of statistical modeling and data analysis. To quote John Tukey (1975a), who used the terms *robust* and *resistant* somewhat interchangeably:

> It is perfectly proper to use both classical and robust/resistant methods routinely, and only worry when they differ enough to matter. But when they differ, you should think hard.

For each statistical model such as location, scale, linear regression, etc., there exist several if not many robust methods, and each method has several variants which an applied statistician, scientist or data analyst must choose from. To select the most appropriate method for each model it is important to understand how the robust methods work, and their pros and cons. The book aims at enabling the reader to select and use the most adequate robust method for each model, and at the same time to understand the theory behind the method; i.e., not only the "how" but also the "why". Thus for each of the models treated in this book we provide:

- conceptual and statistical theory explanations of the main issues;
- the leading methods proposed to date and their motivations;
- a comparison of the properties of the methods;
- computational algorithms, and S-PLUS implementations of the different approaches;
- recommendations of preferred robust methods, based on what we take to be reasonable trade-offs between estimator theoretical justification and performance, transparency to users, and computational costs.

Intended audience

The intended audience of this book consists of the following groups of individuals among the broad spectrum of data analysts, applied statisticians and scientists:

- those who will be quite willing to apply robust methods to their problems once they are aware of the methods, supporting theory and software implementations;
- instructors who want to teach a graduate level course on robust statistics;
- graduate students wishing to learn about robust statistics;
- graduate students and faculty who wish to pursue research on robust statistics and will use the book as background study.

General prerequisites are basic courses in probability, calculus and linear algebra, statistics and familiarity with linear regression at the level of Weisberg (1985), Montgomery, Peck and Vining (2001), and Seber and Lee (2003). Previous knowledge of multivariate analysis, generalized linear models and time series are required for Chapters 6, 7 and 8, respectively.

Organization of the Book

There are many different approaches for each model in robustness, resulting in a huge volume of research and applications publications (though perhaps shorter on the latter than we might like). Doing justice to all of them would require an encyclopedic work that would not necessarily be very effective for our goal. Instead we concentrate on the methods we consider most sound according to our knowledge and experience.

Chapter 1 is a data-oriented motivation chapter. Chapter 2 introduces the main methods in the context of location and scale estimation; in particular we concentrate on the so-called M-estimates that will play a major role throughout the book.

Chapter 3 discusses methods for the evaluation of the robustness of model parameter estimates, and derives "optimal" estimates based on robustness criteria. Chapter 4 deals with linear regression for the case where the predictors contain no outliers, typically because they are fixed non-random values, including for example fixed balanced designs. Chapter 5 treats linear regression with general random predictors which may contain outliers in the form of so-called "leverage" points. Chapter 6 treats robust estimation of multivariate location and dispersion, and robust principal components. Chapter 7 deals with logistic regression and generalized linear models. Chapter 8 deals with robust estimation of time series models, with a main focus on AR and ARIMA. Chapter 9 contains a more detailed treatment of the iterative algorithms for the numerical computing of M-estimates. Chapter 10 develops the asymptotic theory of some robust estimates, and contains proofs of several results stated in the text. Chapter 11 is an appendix containing descriptions of most data sets used in the book. Chapter 12 contains detailed instructions on the use of robust procedures written in S-PLUS.

All methods are introduced with the help of examples with real data, The problems at the end of each chapter consist of both theoretical derivations and analysis of other real data sets.

How to read this book

Each chapter can be read at two levels. The main part of the chapter explains the models to be tackled and the robust methods to be used, comparing their advantages and shortcomings through examples and avoiding technicalities as much as possible. Readers whose main interest is in applications should read enough of each chapter to understand which is the currently preferred method, and the reasons it is preferred. The theoretically oriented reader can find proofs and other mathematical details in appendices and in Chapter 9 and Chapter 10. Sections marked with an asterisk may be skipped at first reading.

Computing

A great advantage of classical methods is that they require only computational procedures based on well-established numerical linear algebra methods which are generally quite fast algorithms. On the other hand computing robust estimates requires solving highly nonlinear optimization problems that typically involve a dramatic increase in computational complexity and running time. Most current robust methods would be unthinkable without the power of today's standard personal computers. Fortunately computers continue getting faster, have larger memory and are cheaper, which is good for the future of robust statistics.

Since the behavior of a robust procedure may depend crucially on the algorithm used, the book devotes considerable attention to algorithmic details for all the methods proposed. At the same time in order that robust statistics be widely accepted by a wide range of users, the methods need to be readily available in commercial software. Robust methods have been implemented in several available commercial statistical packages, including S-PLUS and SAS. In addition many robust procedures have been implemented in the public-domain language R, which is similar to S. References for free software for robust methods are given at the end of Chapter 11. We have focused on S-PLUS because it offers the widest range of methods, and because the methods are accessible from a user-friendly menu and dialog user interface as well as from the command line.

For each method in the book, instructions are given on how to compute it using S-PLUS in Chapter 11. For each example, the book gives the reference to the respective dataset and the S-PLUS code that allow the reader to reproduce the example. Datasets and codes are to be found in the book's web site http://www.wiley.com/go/robuststatistics. This site will also contain corrections to any errata we subsequently discover, and clarifying comments and suggestions as needed. The authors will appreciate any feedback from readers that will result in posting additional helpful material on the web site.

S-PLUS software download

A time-limited version of S-PLUS for Windows software, that expires after 150 days, is being provided by Insightful for this book. To download and install the S-PLUS software, follow the instructions at http://www.insightful.com/support/splusbooks/robstats.

To access the web page, the reader must provide a password. The password is the web registration key provided with this book as a sticker on the inside back cover. In order to activate S-PLUS for Windows the reader must use the web registration key.

Acknowledgements

The authors thank Elena Martínez, Ana Bianco, Mahmut Kutlukaya, Débora Chan, Isabella Locatelli and Chris Green for their helpful comments. Special thanks are due to Ana Julia Villar, who detected hosts of errors and also contributed part of the computer code.

This book could not have been written without the incredible patience of our wives and children for the many hours devoted to the writing of this book and of our associated research over the years that has led to the writing of the books. Untold thanks to Susana, Livia, Jean and Julia.

One of the authors (Martin) wishes to acknowledge his fond memory of, and deep indebtedness to John Tukey for introducing him to robustness and arranging a consulting arrangement with Bell Labs, Murray Hill, that lasted for ten years, and without which he would not be writing this book and without which S-PLUS would not exist

Authors' note on publication of the Second Edition

Some of the text of the First Edition Preface above – in particular the description of the chapter contents, and the information about the S-PLUS package – has now been superseded, but is included here for completeness.

About the Companion Website

The companion website for this book is at

www.wiley.com/go/maronna/robust

The website includes:

Scripts and Datasets

Scan this QR code to visit the companion website.

1

Introduction

1.1 Classical and robust approaches to statistics

This introductory chapter is an informal overview of the main issues to be treated in detail in the rest of the book. Its main aim is to present a collection of examples that illustrate the following facts:

- Data collected in a broad range of applications frequently contain one or more atypical observations, known as *outliers*; that is, observations that are well-separated from the majority or "bulk" of the data, or in some way deviate from the general pattern of the data.
- Classical estimates, such as the sample mean, the sample variance, sample covariances and correlations, or the least-squares fit of a regression model, can be adversely influenced by outliers, even by a single one, and therefore often fail to provide good fits to the bulk of the data.
- There exist *robust* parameter estimates that provide a good fit to the bulk of the data when the data contains outliers, as well as when the data is free of them. A direct benefit of a good fit to the bulk of data is the reliable detection of outliers, particularly in the case of multivariate data.

In Chapter 3 we shall provide some formal probability-based concepts and definitions of robust statistics. Meanwhile, it is important to be aware of the following performance distinctions between classical and robust statistics at the outset. Classical statistical inference quantities such as confidence intervals, t-statistics and p-values, R^2 values and model selection criteria in regression can be adversely influenced by the presence of even one outlier in the data. In contrast, appropriately constructed

Robust Statistics: Theory and Methods (with R), Second Edition.
Ricardo A. Maronna, R. Douglas Martin, Victor J. Yohai and Matías Salibián-Barrera.

Companion website: www.wiley.com/go/maronna/robust

robust versions of those inference quantities are little influenced by outliers. Point estimate predictions and their confidence intervals based on classical statistics can be spoiled by outliers, while predictive models fitted using robust statistics do not suffer from this disadvantage.

It would, however, be misleading to always think of outliers as "bad" data. They may well contain unexpected, but relevant information. According to Kandel (1991, p. 110):

> The discovery of the ozone hole was announced in 1985 by a British team working on the ground with "conventional" instruments and examining its observations in detail. Only later, after reexamining the data transmitted by the TOMS instrument on NASA's Nimbus 7 satellite, was it found that the hole had been forming for several years. Why had nobody noticed it? The reason was simple: the systems processing the TOMS data, designed in accordance with predictions derived from models, which in turn were established on the basis of what was thought to be "reasonable", had rejected the very ("excessively") low values observed above the Antarctic during the Southern spring. As far as the program was concerned, there must have been an operating defect in the instrument.

In the next sections we present examples of classical and robust estimates of the mean, standard deviation, correlation and linear regression for data containing outliers. Except in Section 1.2, we do not describe the robust estimates in any detail, and return to their definitions in later chapters.

1.2 Mean and standard deviation

Let $\mathbf{x} = (x_1, x_2, ..., x_n)$ be a set of observed values. The sample mean $\bar{x}$ and sample standard deviation (SD) s are defined by

$$\bar{x} = \frac{1}{n}\sum_{i=1}^{n} x_i, \quad s^2 = \frac{1}{n-1}\sum_{i=1}^{n}(x_i - \bar{x})^2. \tag{1.1}$$

The sample mean is just the arithmetic average of the data, and as such one might expect that would provide a good estimate of the *center* or *location* of the data. Likewise, one might expect that the sample SD would provide a good estimate of the *dispersion* of the data. Now we shall see how much influence a single outlier can have on these classical estimates.

Example 1.1 *Consider the following 24 determinations of the copper content in wholemeal flour (in parts per million), sorted in ascending order (Analytical Methods Committee, 1989):*

2.20	2.20	2.40	2.40	2.50	2.70	2.80	2.90
3.03	3.03	3.10	3.37	3.40	3.40	3.40	3.50
3.60	3.70	3.70	3.70	3.70	3.77	5.28	28.95

The value 28.95 immediately stands out from the rest of the values and would be considered an outlier by almost anyone. One might conjecture that this

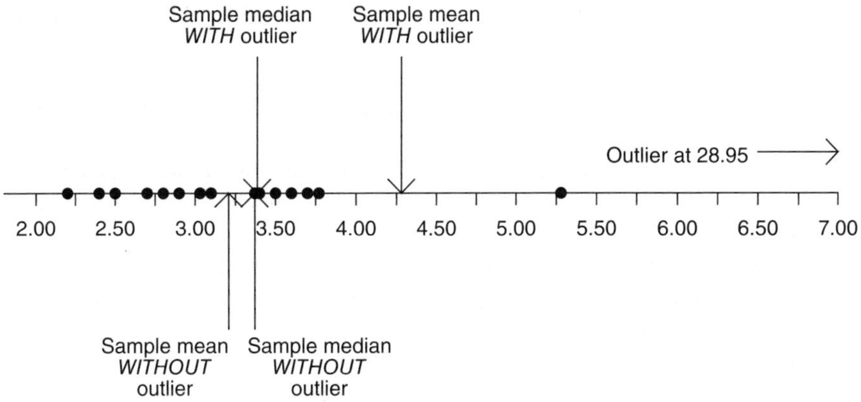

Figure 1.1 Copper content of flour data with sample mean and sample median estimates

inordinately large value was caused by a misplaced decimal point with respect to a "true" value of 2.895. In any event, it is a highly influential outlier, as we now demonstrate.

The values of the sample mean and SD for the above dataset are $\bar{x} = 4.28$ and $s = 5.30$, respectively. Since $\bar{x} = 4.28$ is larger than all but two of the data values, it is not among the bulk of the observations and as such does not represent a good estimate of the center of the data. If one deletes the suspicious value of 28.95, then the values of the sample mean and sample SD are changed to $\bar{x} = 3.21$ and $s = 0.69$. Now the sample mean does provide a good estimate of the center of the data, as is clearly shown in Figure 1.1, and the SD is over seven times smaller than it was with the outlier present. See the leftmost upward pointing arrow and the rightmost downward-pointing arrow in Figure 1.1.

Let us consider how much influence a single outlier can have on the sample mean and sample SD. For example, suppose that the value 28.95 is replaced by an arbitrary value x for the 24th observation, x_{24}. It is clear from the definition of the sample mean that by varying x from $-\infty$ to $+\infty$ the value of the sample mean changes from $-\infty$ to $+\infty$. It is an easy exercise to verify that as x ranges from $-\infty$ to $+\infty$, the sample SD ranges from some positive value smaller than that based on the first 23 observations to $+\infty$. Thus we can say that a single outlier has an *unbounded influence* on these two classical statistics.

An outlier may have a serious adverse influence on confidence intervals. For the flour data, the classical interval based on the t-distribution with confidence level 0.95 is (2.05, 6.51); after removing the outlier, the interval is (2.91, 3.51). The impact of the single outlier has been to considerably lengthen the interval in an asymmetric way.

This example suggests that a simple way to handle outliers is to detect them and remove them from the dataset. There are many methods for detecting outliers (see, for example, Barnett and Lewis, 1998). Deleting an outlier, although better than doing nothing, still poses a number of problems:

- When is deletion justified? Deletion requires a subjective decision. When is an observation "outlying enough" to be deleted?
- The user or the author of the data may think that "an observation is an observation" (i.e., observations should speak for themselves) and hence feel uneasy about deleting them
- Since there is generally some uncertainty as to whether an observation is really atypical, there is a risk of deleting "good" observations, which would result in underestimating data variability
- Since the results depend on the user's subjective decisions, it is difficult to determine the statistical behavior of the complete procedure.

We are thus lead to another approach: why use the sample mean and SD? Maybe are there other better possibilities?

One very old method for estimating the "middle" of the data is to use the sample *median*. Any number t with a value such that the numbers of observations on both sides of it are equal is called a *median* of the dataset: t is a median of the data set $\mathbf{x} = (x_1, \ldots, x_n)$, and will be denoted by

$$t = \text{Med}(\mathbf{x}), \ \textit{if} \ \#\{x_i > t\} = \#\{x_i < t\},$$

where $\#\{A\}$ denotes the number of elements of the set A. It is convenient to define the sample median in terms of the *order statistics* $(x_{(1)}, x_{(2)}, \ldots, x_{(n)})$, obtained by sorting the observations $\mathbf{x} = (x_1, \ldots., x_n)$ in increasing order so that

$$x_{(1)} \leq \ldots \leq x_{(n)}. \tag{1.2}$$

If n is odd, then $n = 2m - 1$ for some integer m, and in that case $\text{Med}(\mathbf{x}) = x_{(m)}$. If n is even, then $n = 2m$ for some integer m, and then any value between $x_{(m)}$ and $x_{(m+1)}$ satisfies the definition of a sample median, and it is customary to take

$$\text{Med}(\mathbf{x}) = \frac{x_{(m)} + x_{(m+1)}}{2}.$$

However, in some cases (e.g. in Section 4.5.1) it may be more convenient to choose $x_{(m)}$ or $x_{(m+1)}$ ("low" and "high" medians, respectively).

The mean and the median are approximately equal if the sample is symmetrically distributed about its center, but not necessarily otherwise. In our example, the median of the whole sample is 3.38, while the median without the largest value is 3.37, showing that the median is not much affected by the presence of this value. See the locations of the sample median with and without the outlier present in Figure 1.1 above. Notice that for this sample, the value of the sample median with the outlier present is relatively close to the sample mean value of 3.21 with the outlier deleted.

Suppose again that the value 28.95 is replaced by an arbitrary value x for the 24th observation $x_{(24)}$. It is clear from the definition of the sample median that when x ranges from $-\infty$ to $+\infty$ the value of the sample median does not change from $-\infty$ to $+\infty$ as was the case for the sample mean. Instead, when x goes to $-\infty$ the sample median undergoes the small change from 3.38 to 3.23 (the latter being the average of $x_{(11)} = 3.10$ and $x_{(12)} = 3.37$ in the original dataset); when x goes to $+\infty$ the sample median goes to the value 3.38 given above for the original data. Since the sample median fits the bulk of the data well, with or without the outlier, and is not much influenced it, it is a good robust alternative to the sample mean.

Likewise, one robust alternative to the SD is the *median absolute deviation about the median* (MAD), defined as

$$\text{MAD}(\mathbf{x}) = \text{MAD}(x_1, x_2, ..., x_n) = \text{Med}\{|\mathbf{x} - \text{Med}(\mathbf{x})|\}.$$

This estimator uses the sample median twice, first to get an estimate of the center of the data in order to form the set of absolute residuals about the sample median, $\{|\mathbf{x} - \text{Med}(\mathbf{x})|\}$, and then to compute the sample median of these absolute residuals. To make the MAD comparable to the SD, we define the *normalized MAD* (MADN) as

$$\text{MADN}(\mathbf{x}) = \frac{\text{MAD}(\mathbf{x})}{0.6745}.$$

The reason for this definition is that 0.6745 is the MAD of a standard normal random variable, and hence a $\text{N}(\mu, \sigma^2)$ variable has $\text{MADN} = \sigma$.

For the above dataset, one gets MADN = 0.53, as compared with $s = 5.30$. Deleting the large outlier yields MADN = 0.50, as compared to the somewhat higher sample SD value of $s = 0.69$. The MAD is clearly not influenced very much by the presence of a large outlier, and as such provides a good robust alternative to the sample SD.

So why not always use the median and MAD? An informal explanation is that if the data contain no outliers, these estimates have a statistical performance that is poorer than that of the classical estimates $\overline{x}$ and s. The ideal solution would be to have "the best of both worlds": estimates that behave like the classical ones when the data contain no outliers, but are insensitive to outliers otherwise. This is the data-oriented idea of robust estimation. A more formal notion of robust estimation based on statistical models, which will be discussed in the following chapters, is that the statistician always has a statistical model in mind (explicitly or implicitly) when analyzing data, for example a model based on a normal distribution or some other idealized parametric model such as an exponential distribution. The classical estimates are in some sense "optimal" when the data are exactly distributed according to the assumed model, but can be very suboptimal when the distribution of the data differs from the assumed model by a "small" amount. Robust estimates on the other hand maintain approximately optimal performance, not just under the assumed model, but under "small" perturbations of it too.

1.3 The "three sigma edit" rule

A traditional measure of the outlyingness of an observation x_i with respect to a sample, is the ratio between its distance to the sample mean and the sample SD:

$$t_i = \frac{x_i - \overline{x}}{s}, \tag{1.3}$$

Observations with $|t_i| > 3$ are traditionally deemed suspicious (the "three-sigma rule"), based on the fact that they would be "very unlikely" under normality, since $\mathrm{P}(|x| \geq 3) = 0.003$ for a random variable x with a standard normal distribution. The largest observation in the flour data has $t_i = 4.65$, and so is suspicious. Traditional "three-sigma edit" rules result in either discarding observations for which $|t_i| > 3$, or adjusting them to one of the values $\overline{x} \pm 3s$, whichever is nearer. Despite its long tradition, this rule has some drawbacks that deserve to be taken into account:

- In a very large sample of "good" data, some observations will be declared suspicious and be altered. More precisely, in a large normal sample, about three observations out of 1000 will have $|t_i| > 3$. For this reason, normal Q–Q plots are more reliable for detecting outliers (see example below).
- In very small samples the rule is ineffective: it can be shown that

$$|t_i| < \frac{n-1}{\sqrt{n}}$$

  for all possible data sample values, and hence if $n \leq 10$ then always $|t_i| < 3$. The proof is left to the reader (Problem 1.3).
- When there are several outliers, their effects may interact in such a way that some or all of them remain unnoticed (an effect called *masking*), as the following example shows.

Example 1.2 *The following data (Stigler 1977) are 20 determinations of the time (in microseconds) needed for the light to travel a distance of 7442 m. The actual times are the table values* $\times\, 0.001 + 24.8$.

28	26	33	24	34	−44	27	16	40	−2
29	22	24	21	25	30	23	29	31	19

The normal Q–Q plot in Figure 1.2 reveals the two lowest observations (−44 and −2) as suspicious. Their respective t_is are −3.73 and −1.35, and so the value of $|t_i|$ for the observation −2 does not indicate that it is an outlier. The reason

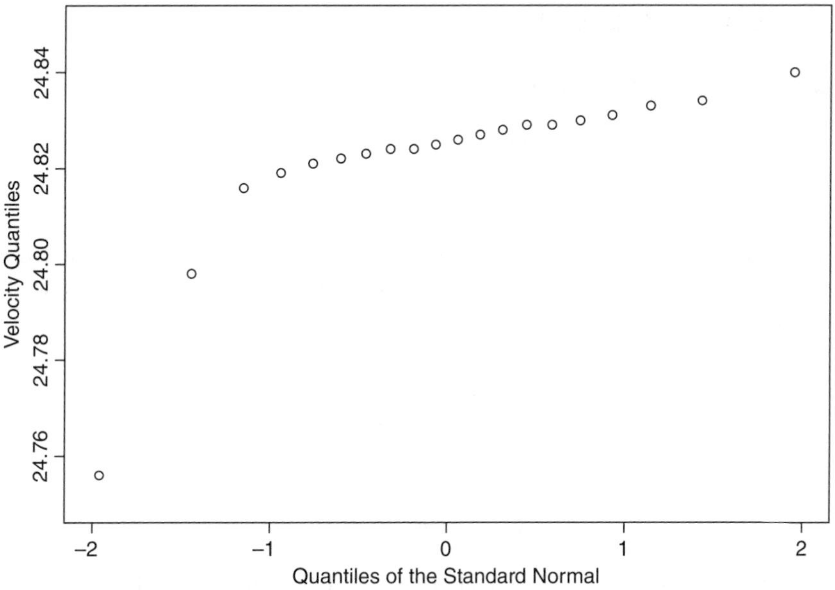

Figure 1.2 Velocity of light: Q–Q plot of observed times

that −2 has such a small $|t_i|$ value is that both observations pull $\bar{x}$ to the left and inflate s; it is said that the value −44 "masks" the value −2.

To avoid this drawback it is better to replace $\bar{x}$ and s in (1.3) by robust location and dispersion measures. A robust version of (1.3) can be defined by replacing the sample mean and SD by the median and MADN, respectively:

$$t'_i = \frac{x_i - \text{Med}(\mathbf{x})}{\text{MADN}(\mathbf{x})}. \tag{1.4}$$

The t_is for the two leftmost observations are now −11.73 and −4.64, and hence the three-sigma edit rule, with t' instead of t, pinpoints both as suspicious. This suggests that even if we only want to detect outliers – rather than to estimate parameters – detection procedures based on robust estimates are more reliable.

A simple robust location estimate could be defined by deleting all observations with $|t'_i|$ larger than a given value, and taking the average of the rest. While this procedure is better than the three-sigma edit rule based on t, it will be seen in Chapter 3 that the estimates proposed in this book handle the data more smoothly, and can be tuned to have certain desirable robustness properties that this procedure lacks.

1.4 Linear regression

1.4.1 Straight-line regression

First consider fitting a straight line regression model to the dataset $\{(x_i, y_i) : i = 1, ., n\}$

$$y_i = \alpha + x_i\beta + u_i, \quad i = 1, \ldots, n$$

where x_i and y_i are the predictor and response variable values, respectively, and u_i are random errors. The time-honored classical way of fitting this model is to estimate the parameters α and β with the least-squares (LS) estimates

$$\hat{\beta} = \frac{\sum_{i=1}^{n}(x_i - \overline{x})(y_i - \overline{y})}{\sum_{i=1}^{n}(x_i - \overline{x})^2}$$

$$\hat{\alpha} = \overline{y} - \overline{x}\hat{\beta}$$

As an example of how influential two outliers can be on these estimates, Figure 1.3 plots the earnings per share (EPS) versus time each year for a company with the stock exchange ticker symbol IVENSYS, along with the straight-line fits of the LS estimate and of a robust regression estimate (called an MM-estimate) that has desirable theoretical properties (to be described in detail in Chapter 5).

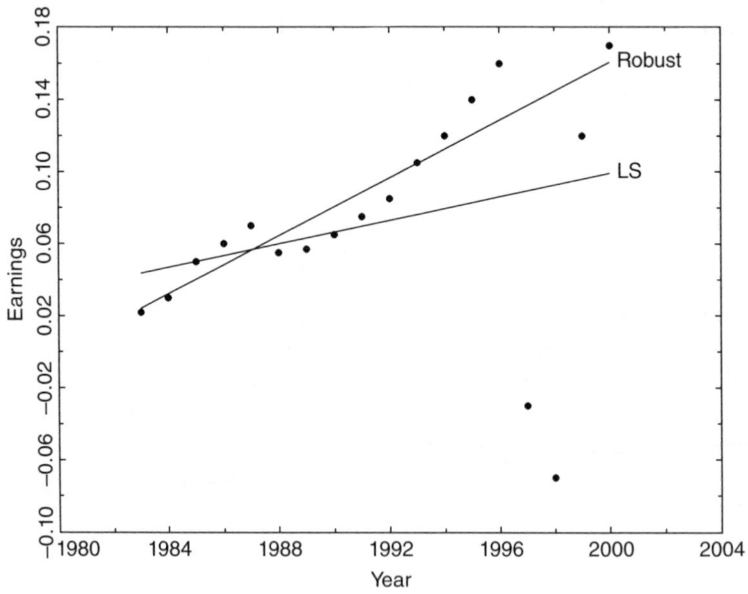

Figure 1.3 EPS data with robust and LS fits

The two unusually low EPS values in 1997 and 1998 cause the LS line to fit the data very poorly, and one would not expect the line to provide a good prediction of EPS in 2001. By way of contrast, the robust line fits the bulk of the data well, and should provide a reasonable prediction of EPS in 2001.

The above EPS example was brought to one of the author's attention by an analyst in the corporate finance department of a well-known large company. The analyst was required to produce a prediction of next year's EPS for several hundred companies, and at first he used the LS fit for this purpose. But then he noticed a number of firms for which the data contained outliers that distorted the LS parameter estimates, resulting in a very poor fit and a poor prediction of next year's EPS. Once he discovered the robust estimate, and found that it gave him essentially the same results as the LS estimate when the data contained no outliers, while at the same time providing a better fit and prediction than LS when outliers were present, he began routinely using the robust estimate for his task.

It is important to note that automatically flagging large differences between a classical estimate (in this case LS) and a robust estimate provides a useful diagnostic alert that outliers may be influencing the LS result.

1.4.2 Multiple linear regression

Now consider fitting a multiple linear regression model

$$y_i = \sum_{j=1}^{p} x_{ij}\beta_j + u_i, \quad i = 1, \dots, n$$

where the response variable values are y_i, and there are p predictor variables x_{ij}, $j = 1, \dots, p$, and p regression coefficients β_j. Not surprisingly, outliers can also have an adverse influence on the LS estimate $\widehat{\boldsymbol{\beta}}$ for this general linear model, a fact which is illustrated by the following example that appears in Hubert and Rousseeuw (1997).

Example 1.3 *The response variable values y_i are the rates of unemployment in various geographical regions around Hannover, Germany, and the predictor variables x_{ij}, $j = 1, \dots, p$ are as follows:*

- *PA: percentage engaged in production activities*
- *GPA: growth in PA*
- *HS: percentage engaged in higher services*
- *GHS: growth in HS*
- *Region: geographical region around Hannover (21 regions)*
- *Period: time period (three periods: 1979–82, 1983–88, 1989–92)*

Note that the categorical variables Region and Period require 20 and 2 parameters respectively, so that, including an intercept, the model has 27 parameters,

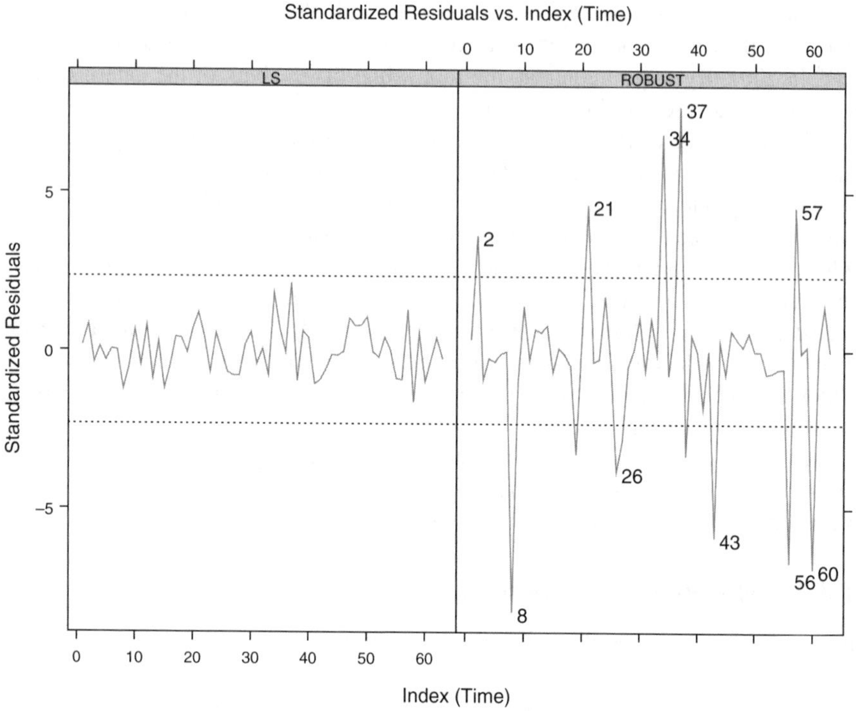

Figure 1.4 Standardized residuals for LS and robust fits

and the number of response observations is 63, one for each region and period. Figures 1.4 and 1.5 show the results of LS and robust fitting in a manner that facilitates easy comparison of the results. The robust fitting is done by a special "M-estimate" that has desirable theoretical properties, and is described in detail in Section 5.7.5.

For a set of estimated parameters $(\widehat{\beta}_1, \ldots, \widehat{\beta}_p)$, with fitted values $\widehat{y}_i = \sum_{j=1}^{p} x_{ij}\widehat{\beta}_j$, residuals $\widehat{u}_i = y_i - \widehat{y}_i$ and residuals dispersion estimate $\widehat{\sigma}$, Figure 1.4 shows the standardized residuals $\tilde{u}_i = \widehat{u}_i/\tilde{\sigma}$ plotted versus the observations' index values i. Standardized residuals that fall outside the horizontal dashed lines at ± 2.33, which occurs with probability 0.02, are declared suspicious. The display for the LS fit does not reveal any outliers, while that for the robust fit clearly reveals 10 to 12 outliers among 63 observations. This is because the robust regression has found a linear relationship that fits the majority of the data points well, and consequently is able to reliably identify the outliers. The LS estimate instead attempts to fit all data points and so is heavily influenced by the outliers. The fact that all of the LS standardized residuals lie inside the horizontal dashed lines is because the outliers have inflated

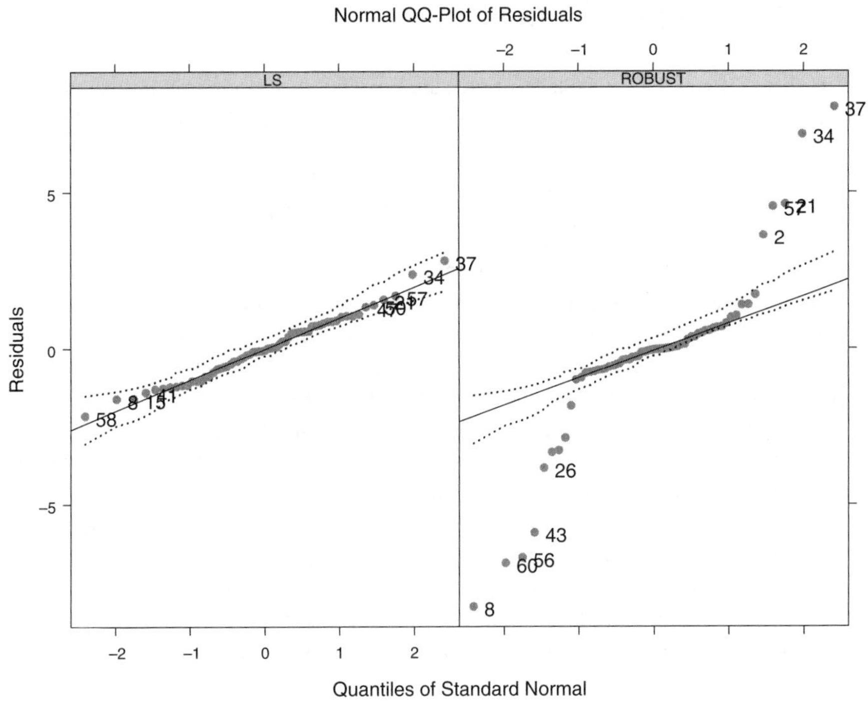

Figure 1.5 Normal Q–Q plots for (left) LS and (right) robust fits

the value of $\tilde{\sigma}$ computed in the classical way based on the sum of squared residuals, while a robust estimate $\tilde{\sigma}$ used for the robust regression is not much influenced by the outliers.

Figure 1.5 shows normal Q–Q plots of the residuals for the LS and robust fits, with light dotted lines showing the 95% simulated pointwise confidence regions to allow an assessment of whether or not there are significant outliers and potential nonnormality. These plots may be interpreted as follows. If the data fall along the straight line (which itself is fitted by a robust method) with no points outside the 95% confidence region, then one is moderately sure that the data are normally distributed.

Performing only the LS fit, and therefore looking only at the normal Q–Q plot in the left-hand plot in Figure 1.5, would lead to the conclusion that the residuals are indeed quite normally distributed, with no outliers. The normal Q–Q plot of residuals for the robust fit in the right-hand panel of Figure 1.5 clearly shows that such a conclusion is wrong. This plot shows that the bulk of the residuals are indeed quite normally distributed, as evidenced by the compact linear behavior in the middle of the plot. At the same time, it clearly reveals the outliers that were evident in the plot of standardized residuals (Figure 1.4).

1.5 Correlation coefficients

Let $\{(x_i, y_i)\}$, $i = 1, \ldots, n$, be a bivariate sample. The most popular measure of association between the x_i and the y_i is the sample correlation coefficient, defined as

$$\hat{\rho} = \frac{\sum_{i=1}^{n}(x_i - \overline{x})(y_i - \overline{y})}{\left(\sum_{i=1}^{n}(x_i - \overline{x})^2\right)^{1/2}\left(\sum_{i=1}^{n}(y_i - \overline{y})^2\right)^{1/2}}$$

where $\overline{x}$ and $\overline{y}$ are the sample means of the x_i and y_i.

The sample correlation coefficient is highly sensitive to the presence of outliers. Figure 1.6 shows a scatterplot of the increase (gain) in numbers of telephones versus the annual change in new housing starts, for a period of 15 years in a geographical region within New York City in the 1960s and 1970s, in coded units.

There are two outliers in this bivariate (two-dimensional) dataset that are clearly separated from the rest of the data. It is important to notice that these two outliers are not one-dimensional outliers; they are not even the largest or smallest values in any of the two coordinates. This observation illustrates an extremely important point: two-dimensional outliers cannot be reliably detected by examining the values of bivariate data one-dimensionally; that is, one variable at a time.

The value of the sample correlation coefficient for the complete gain data is $\hat{\rho} = 0.44$, and deleting the two outliers yields $\hat{\rho} = 0.91$, which is quite a large difference and in the range of what an experienced user might expect for the dataset with the two outliers removed. The dataset with the two outliers deleted can be seen

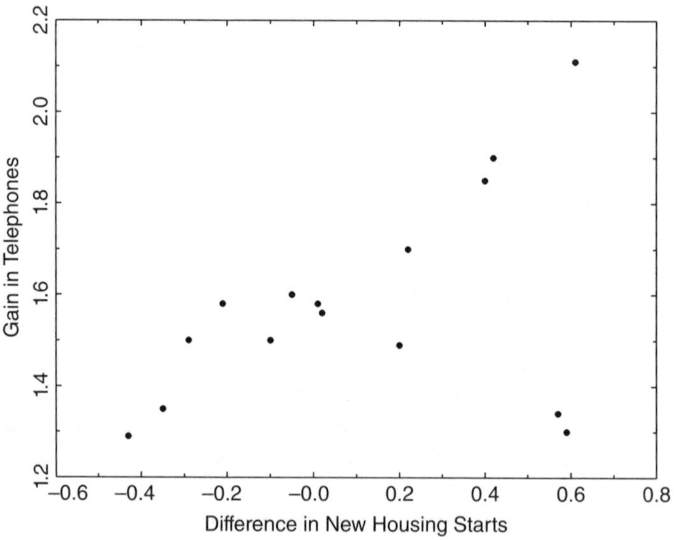

Figure 1.6 Increase in numbers of telephones versus difference in new housing starts

as roughly elliptical, with a major axis sloping up and to the right and the minor axis sloping up and to the left With this picture in mind one can see that the two outliers lie in the minor axis direction, though offset somewhat from the minor axis. The impact of the outliers is to decrease the value of the sample correlation coefficient by the considerable amount of 0.44 from the value of 0.91 it has with the two outliers deleted. This illustrates a general biasing effect of outliers on the sample correlation coefficient: outliers that lie along the minor axis direction of data that is otherwise positively correlated negatively influence the sample correlation coefficient. Similarly, outliers that lie along the minor axis direction of data that is otherwise negatively correlated will increase the sample correlation coefficient. Outliers that lie along a major axis direction of the rest of the data will increase the absolute value of the sample correlation coefficient, making it more positive if the bulk of the data is positively correlated.

If one uses a robust correlation coefficient estimate it will not make much difference whether the outliers in the main-gain data are present or deleted. Using a good robust method $\widehat{\rho}_{Rob}$ for estimating covariances and correlations on the main-gain data yields $\widehat{\rho}_{Rob} = 0.85$ for the entire dataset and $\widehat{\rho}_{Rob} = 0.90$ with the two outliers deleted. For the robust correlation coefficient, the change due to deleting the outlier is only 0.05, compared to 0.47 for the classical estimate. A detailed description of robust correlation and covariance estimates is provided in Chapter 6.

When there are more than two variables, examining all pairwise scatterplots for outliers is hopeless unless the number of variables is relatively small. But even looking at all scatterplots or applying a robust correlation estimate to all pairs does not suffice, for in the same way that there are bivariate outliers that do not stand out in any univariate representation, there may be multivariate outliers that heavily influence the correlations and do not stand out in any bivariate scatterplot. Robust methods deal with this problem by estimating all the correlations simultaneously, in such a manner that points far away from the bulk of the data are automatically downweighted. Chapter 6 considers these methods in detail.

1.6 Other parametric models

We do not want to leave the reader with the impression that robust estimation is only concerned with outliers in the context of an assumed normal distribution model. Outliers can cause problems in fitting other simple parametric distributions such as an exponential, Weibull or gamma distribution, where the classical approach is to use a nonrobust maximum likelihood estimate (MLE) for the assumed model. In these cases one needs robust alternatives to the MLE in order to obtain a good fit to the bulk of the data.

For example, the exponential distribution with density

$$f(x; \lambda) = \frac{1}{\lambda} e^{-x/\lambda}, \quad x \geq 0$$

is widely used to model random inter-arrival and failure times, and it also arises in the context of times-series spectral analysis (see Section 8.14). It is easily shown that the parameter λ is the expected value of the random variable x – in other words, $\lambda = \mathrm{E}(x)$ – and that the sample mean is the MLE. We already know from the previous discussion that the sample mean lacks robustness and can be greatly influenced by outliers. In this case the data are nonnegative so one is only concerned about large positive outliers that cause the value of the sample mean to be inflated in a positive direction. So we need a robust alternative to the sample mean, and one naturally considers use of the sample median $\mathrm{Med}(\mathbf{x})$. It turns out that the sample median is an *inconsistent* estimate of λ: it does not approach λ when the sample size increases, and hence a correction is needed. It is an easy calculation to check that the median of the exponential distribution has value $\lambda \log 2$, where log stands for natural logarithm, and so one can use $\mathrm{Med}(\mathbf{x})/\log 2$ as a simple robust estimate of λ that is consistent with the assumed model. This estimate turns out to have desirable robustness properties, as described in Problem 3.15.

The methods of robustly fitting Weibull and gamma distributions are much more complicated than the above use of the adjusted median for the exponential distribution. We present one important application of robust fitting a gamma distribution due to Marazzi *et al.* (1998). The gamma distribution has density

$$f(x; \alpha, \sigma) = \frac{1}{\Gamma(\alpha)\sigma^{\alpha}} x^{\alpha-1} e^{-x/\sigma}, \quad x \geq 0$$

and the mean of this distribution is known to be $\mathrm{E}(x) = \alpha\sigma$. The problem has to do with estimating the length of stay (LOS) of 315 patients in a hospital. The mean LOS is a quantity of considerable economic importance, and some patients whose hospital stays are much longer than those of the majority of the patients adversely influence the MLE fit of the gamma distribution. The MLE values turn out to be $\widehat{\alpha}_{MLE} = 0.93$ and $\widehat{\sigma}_{MLE} = 8.50$, while the robust estimates are $\widehat{\alpha}_{Rob} = 1.39$ and $\widehat{\sigma}_{Rob} = 3.64$, and the resulting mean LOS estimates are $\widehat{\mu}_{MLE} = 7.87$ and $\widehat{\mu}_{Rob} = 4.97$. Some patients with unusually long LOS values contribute to an inflated estimate of the mean LOS for the majority of the patients. A more complete picture is obtained through the figures below.

Figure 1.7 shows a histogram of the data along with the MLE and robust gamma density fit to the LOS data. The MLE underestimates the density for small values of LOS and overestimates the density for large values of LOS, thereby resulting in a larger MLE estimate of the mean LOS, while the robust estimate provides a better overall fit and a mean LOS that better describes the majority of the patients. Figure 1.8 shows a gamma Q–Q plot based on the robustly fitted gamma distribution. This plot reveals that the bulk of the data is well fitted by the robust method, while approximately 30 of the largest values of LOS appear to come from a sub-population of the patients characterized by longer LOS values. This is best modeled separately using another distribution, possibly another gamma distribution with different values of the parameters α and σ.

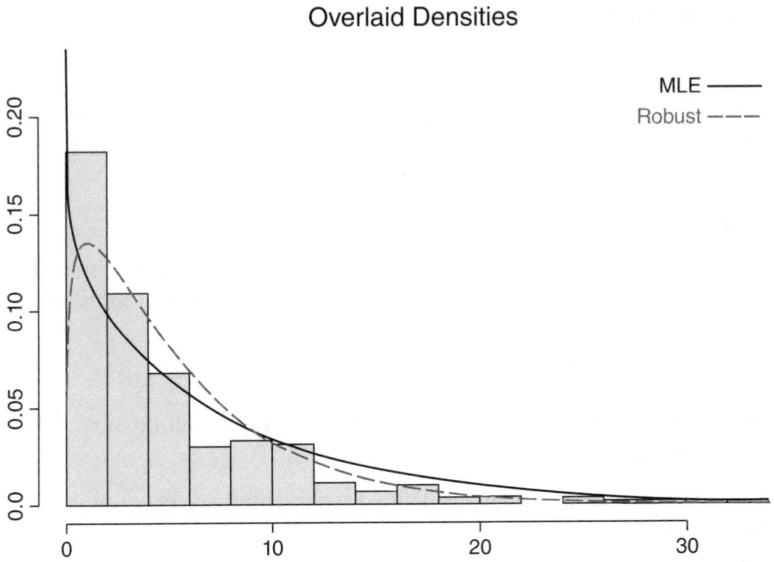

Figure 1.7 MLE and robust fits of a gamma distribution to LOS data

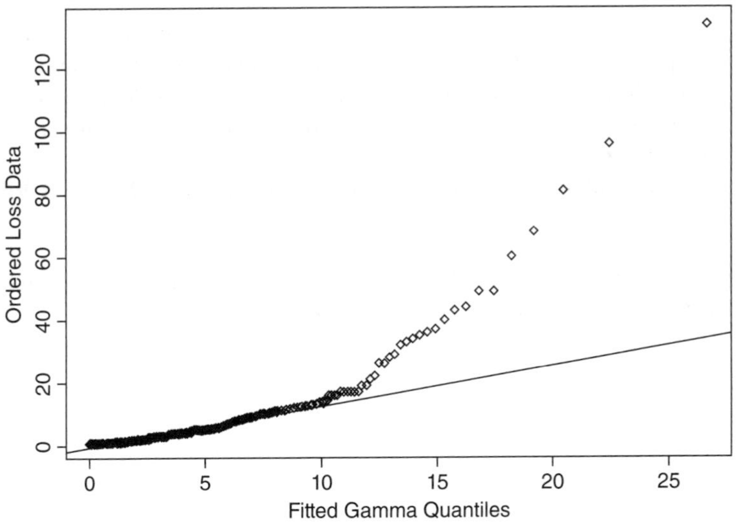

Figure 1.8 Fitted gamma QQ-plot of LOS data

1.7 Problems

1.1. Show that if a value x_0 is added to a sample $\mathbf{x}=\{x_1,...,x_n\}$, when x_0 ranges from $-\infty$ to $+\infty$, the standard deviation of the enlarged sample ranges between a value smaller than SD($\mathbf{x}$) and infinity.

1.2. Consider the situation of the former problem.

(a) Show that if n is even, the maximum change in the sample median when x_0 ranges from $-\infty$ to $+\infty$ is the distance from Med($\mathbf{x}$) to the next order statistic the farthest from Med($\mathbf{x}$).

(b) What is the maximum change if n is odd?

1.3. Show for t_i defined in (1.3) that $|t_i| < (n-1)/\sqrt{n}$ for all possible datasets of size n, and hence for all datasets $|t_i| < 3$ if $n \leq 10$.

1.4. The interquartile range (IQR) is defined as the difference between the third and the first quartiles.

(a) Calculate the IQR of the N(μ, σ^2) distribution.

(b) Consider the sample interquartile range

$$\text{IQR}(\mathbf{x}) = \text{IQR}(x_1, x_2, ..., x_n) = x_{(\lfloor 3n/4 \rfloor)} - x_{(\lfloor n/4 \rfloor)}$$

as a measure of dispersion. It is known that sample quantiles tend to the respective distribution quantiles if these are unique. Based on this fact, determine the constant c such that the normalized interquartile range IQRN($\mathbf{x}$) = IQR ($\mathbf{x}$)$/c$ is a consistent estimate of σ when the data has a N(μ, σ^2) distribution.

(c) Can you think of a reason why you would prefer MADN($\mathbf{x}$) to IQRN($\mathbf{x}$) as a robust estimate of dispersion?

1.5. Show that the median of the exponential distribution is $\lambda \log 2$, and hence Med($\mathbf{x}$)$/\log 2$ is a consistent estimate of λ.

2

Location and Scale

2.1 The location model

For a systematic treatment of the situations considered in Chapter 1, we need to represent them by probability-based statistical models. We assume that the outcome x_i of each observation depends on the "true value" μ of the unknown parameter (in Example 1.1, the copper content of the whole flour batch) and also on some random error process. The simplest assumption is that the error acts additively:

$$x_i = \mu + u_i \ (i = 1, \ldots, n) \tag{2.1}$$

where the errors $u_1, \ldots, u_n$ are random variables. This is called the *location model.*

If the observations are independent replications of the same experiment under equal conditions, it may be assumed that

- $u_1, \ldots, u_n$ have the same distribution function F_0
- $u_1, \ldots, u_n$ are independent.

It follows that $x_1, \ldots, x_n$ are independent, with common distribution function

$$F(x) = F_0(x - \mu) \tag{2.2}$$

and we say that the x_i are *i.i.d.* – independent and identically distributed – random variables.

The assumption that there are no systematic errors can be formalized as follows:

- u_i and $-u_i$ have the same distribution, and consequently $F_0(x) = 1 - F_0(-x)$.

Robust Statistics: Theory and Methods (with R), Second Edition.
Ricardo A. Maronna, R. Douglas Martin, Victor J. Yohai and Matías Salibián-Barrera.

Companion website: www.wiley.com/go/maronna/robust

An *estimator* $\hat{\mu}$ is a function of the observations: $\hat{\mu} = \hat{\mu}(x_1, \ldots, x_n) = \hat{\mu}(\mathbf{x})$ (in some cases, the numeric value of an estimator for a particular sample will be called an *estimate*). We are looking for estimators such that in some sense $\hat{\mu} \approx \mu$ with high probability.

One way to measure the approximation is with the *mean squared error* (MSE):

$$\mathrm{MSE}(\hat{\mu}) = \mathrm{E}(\hat{\mu} - \mu)^2 \tag{2.3}$$

(other measures will be developed later). The MSE can be decomposed as

$$\mathrm{MSE}(\hat{\mu}) = \mathrm{Var}(\hat{\mu}) + \mathrm{Bias}(\hat{\mu})^2,$$

with

$$\mathrm{Bias}(\hat{\mu}) = \mathrm{E}\hat{\mu} - \mu,$$

where "E" stands for the expectation. Note that if $\hat{\mu}$ is the sample mean and c is any constant, then

$$\hat{\mu}(x_1 + c, \ldots, x_n + c) = \hat{\mu}(x_1, \ldots, x_n) + c \tag{2.4}$$

and

$$\hat{\mu}(cx_1, \ldots, cx_n) = c\hat{\mu}(x_1, \ldots, x_n). \tag{2.5}$$

The same holds for the median. These properties are called respectively *shift* (or location) and *scale equivariance* of $\hat{\mu}$. They imply that, for instance, if we express our data in degrees Celsius instead of Fahrenheit, the estimator will automatically adapt to the change of units.

A traditional way to represent "well-behaved" data – data without outliers – is to assume F_0 is normal with mean 0 and unknown variance σ^2, which implies

$$F = \mathcal{D}(x_i) = \mathrm{N}(\mu, \sigma^2),$$

where $\mathcal{D}(x)$ denotes the distribution of the random variable x, and $\mathrm{N}(\mu, \upsilon)$ is the normal distribution with mean μ and variance υ. Classical methods assume that F belongs to an *exactly* known parametric family of distributions. If the data were *exactly* normal, the mean would be an "optimal" estimator – the maximum likelihood estimator (MLE) (see next section) – and minimizes the MSE among unbiased estimators, and also among equivariant ones (Bickel and Doksum, 2001; Lehmann and Casella, 1998). But data are seldom so well behaved.

Figure 2.1 shows the normal Q–Q plots of the observations in Example 1.1. We see that the *bulk* of the data may be described by a normal distribution, but not the whole of it. The same feature can be observed in the Q–Q plot of Figure 1.2. In this sense, we may speak of F as being only *approximately normal*, with normality failing at the tails. We may thus state our initial goal as: looking for estimators that are almost as good as the mean when F is exactly normal, but that are also "good" in some sense when F is only approximately normal.

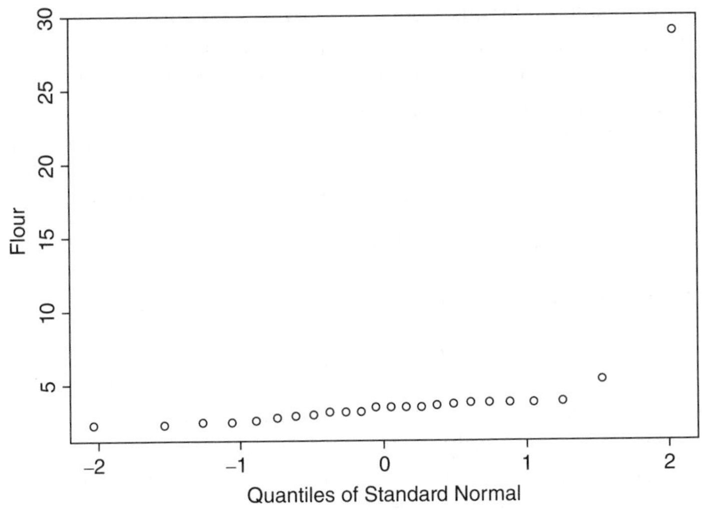

Figure 2.1 Q–Q plot of the flour data

At this point it may seem natural to think that an adequate procedure could be to test the hypothesis that the data are normal; if it is not rejected, we use the mean, otherwise, the median; or, better still, fit a distribution to the data, and then use the MLE for the fitted one. But this has the drawback that very large sample sizes are needed to distinguish the true distribution, especially since here it is the *tails* – precisely the regions with fewer data – that are most influential.

2.2 Formalizing departures from normality

To formalize the idea of approximate normality, we may imagine that a proportion $1 - \epsilon$ of the observations is generated by the normal model, while a proportion ϵ is generated by an unknown mechanism. For instance, repeated measurements are made of something. These measurements are correct 95% of the time, but 5% of the time the apparatus fails or the experimenter makes an incorrect transcription. This may be described by supposing that:

$$F = (1 - \epsilon)G + \epsilon H, \tag{2.6}$$

where $G = \mathrm{N}(\mu, \sigma^2)$ and H may be any distribution; for instance, another normal with a larger variance and a possibly different mean. This is called a *contaminated normal distribution*. This model of contamination is called the *Tukey–Huber* model, after Tukey (1960), who gave an early example of the use of these distributions to show the dramatic lack of robustness of the SD, and Huber (1964), who derived the first optimality results under this model. In general, F is called a *mixture* of G and H, and is called a *normal mixture* when both G and H are normal.

To justify (2.6), let A be the event "the apparatus fails", which has $P(A) = \varepsilon$, and A' its complement. We are assuming that our observation x has distribution G conditional on A' and H conditional on A. Then by the total probability rule:

$$\begin{aligned} F(t) &= \mathrm{P}(x \le t) = \mathrm{P}(x \le t|A')P(A') + \mathrm{P}(x \le t|A)P(A) \\ &= G(t)(1-\varepsilon) + H(t)\varepsilon. \end{aligned}$$

If G and H have densities g and h, respectively, then F has density

$$f = (1-\varepsilon)g + \varepsilon h. \tag{2.7}$$

It must be emphasized that – as in the ozone layer example of Section 1.1 – atypical values are not necessarily due to erroneous measurements: they simply reflect an unknown change in the measurement conditions in the case of physical measurements, or more generally the behavior of a sub-population of the data.An important example of the latter is that normal mixture distributions have been found to often provide quite useful models for stock market returns; that is, the relative change in price from one time period to the next, with the mixture components corresponding to different volatility regimes of the returns.

Another model for outliers are so-called *heavy-tailed* or *fat-tailed* distributions, where the density tails tend to zero more slowly than in the normal density tails. An example is the so-called *Cauchy* distribution, with density

$$f(x) = \frac{1}{\pi(1+x^2)}. \tag{2.8}$$

It is bell shaped like the normal, but its mean does not exist. It is a particular case of the family of *Student* (or t) densities with $\nu > 0$ degrees of freedom, given by

$$f_\nu(x) = c_\nu \left(1 + \frac{x^2}{\nu}\right)^{-(\nu+1)/2} \tag{2.9}$$

where c_υ is a constant:

$$c_\upsilon = \frac{\Gamma((\upsilon+1)/2)}{\sqrt{\upsilon\pi}\,\Gamma(\upsilon/2)},$$

where Γ is the gamma function. This family contains all degrees of heavy-tailedness. When $\nu \to \infty, f_\nu$ tends to the standard normal density; for $\nu = 1$ we have the Cauchy distribution.

Figure 2.2 shows the densities of $\mathrm{N}(0,1)$, the Student distribution with four degrees of freedom, and the contaminated distribution (2.7) with $g = \mathrm{N}(0,1), h = \mathrm{N}(0,100)$ and $\varepsilon = 0.10$, denoted by N, T4 and CN respectively. To make comparisons more clear, the three distributions are normalized to have the same interquartile range.

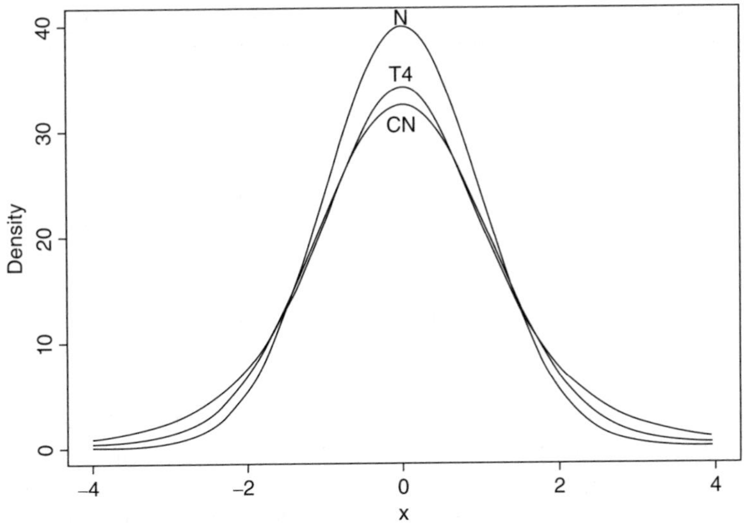

Figure 2.2 Standard normal (N), Student (T4), and contaminated normal (CN) densities, scaled to equal interquartile range

If $F_0 = \mathrm{N}(0, \sigma^2)$ in (2.2), then $\overline{x}$ is $\mathrm{N}(\mu, \sigma^2/n)$. As we shall see later, the sample median is approximately $\mathrm{N}(\mu, 1.57\sigma^2/n)$, so the sample median has a 57% increase in variance relative to the sample mean. We say that the median has a *low efficiency* in the normal distribution.

On the other hand, assume that 95% of our observations are well-behaved, represented by $G = \mathrm{N}(\mu, 1)$, but that 5% of the times the measuring system gives an erratic result, represented by a normal distribution with the same mean but a 10-fold increase in the standard deviation. We thus have the model (2.6) with $\epsilon = 0.05$ and $H = \mathrm{N}(\mu, 100)$. In general, under the model

$$F = (1 - \varepsilon)\mathrm{N}(\mu, 1) + \varepsilon\mathrm{N}(\mu, \tau^2) \tag{2.10}$$

we have (see (2.88), (2.27) and Problem 2.3)

$$\mathrm{Var}(\overline{x}) = \frac{(1-\varepsilon) + \varepsilon\tau^2}{n}, \quad \mathrm{Var}(\mathrm{Med}(\mathbf{x})) \approx \frac{\pi}{2n(1 - \varepsilon + \varepsilon/\tau)^2}. \tag{2.11}$$

Note that Var(Med($\mathbf{x}$)) above means "the *theoretical* variance of the *sample* median of $\mathbf{x}$". It follows that for $\epsilon = 0.05$ and $H = \mathrm{N}(\mu, 100)$, the variance of $\overline{x}$ increases to 5.95, while that of the median is only 1.72. The gain in robustness of using the median is paid for by an increase in variance ("a loss in efficiency") in the normal distribution.

Table 2.1 shows the approximations for large n of n times the variances of the mean and median for different values of τ. It is seen that the former increases rapidly with τ, while the latter stabilizes.

Table 2.1 Variances ($\times n$) of sample mean and median for large n

ε	0.05		0.10	
τ	nVar($\overline{x}$)	nVar(Med)	nVar($\overline{x}$)	nVar(Med)
3	1.40	1.68	1.80	1.80
4	1.75	1.70	2.50	1.84
5	2.20	1.70	3.40	1,86
6	2.75	1.71	4.50	1.87
10	5.95	1.72	10.90	1.90
20	20.9	1.73	40.90	1.92

In the next sections we shall develop estimators that combine the low variance of the mean in the normal with the robustness of the median under contamination. For introductory purposes we will deal only with *symmetric* distributions. The distribution of the variable x is symmetric about μ if $x - \mu$ and $\mu - x$ have the same distribution. If x has a density f, symmetry about μ is equivalent to $f(\mu + x) = f(\mu - x)$. Symmetry implies that $\text{Med}(x) = \mu$, and if the expectation exists, also that $\text{E}x = \mu$. Therefore, if the data have a symmetric distribution, there is no bias and only the variability is at issue. In Chapter 3, general contamination will be addressed.

Two early and somewhat primitive ways to obtain robust estimators were based on *deleting* and *truncating* atypical data. Assume that we define an interval $[a, b]$ (depending on the data) containing supposedly "typical" observations, such as $a = \overline{x} - 2s$, $b = \overline{x} + 2s$. Deletion means using a modified sample, obtained by omitting all points outside $[a, b]$. Truncation means replacing all $x_i < a$ by a and all $x_i > b$ by b, and not altering the other points. In other words, atypical values are swapped for the nearest typical ones. Naive uses of these ideas are not necessarily good, but some of the methods we shall study are elaborate versions of them.

2.3 M-estimators of location

We shall now develop a general family of estimators that contains the mean and the median as special cases.

2.3.1 Generalizing maximum likelihood

Consider again the location model (2.1). Assume that F_0, the distribution function of u_i, has a density $f_0 = F_0'$. The joint density of the observations (the *likelihood function*) is

$$L(x_1, \ldots, x_n; \mu) = \prod_{i=1}^{n} f_0(x_i - \mu)$$

The *maximum likelihood estimator* (MLE) of μ is the value $\widehat{\mu}$ – depending on $x_1, \ldots, x_n$ – that maximizes $L(x_1, \ldots, x_n; \mu)$:

$$\widehat{\mu} = \widehat{\mu}(x_1, \ldots, x_n) = \arg\max_{\mu} \; L(x_1, \ldots, x_n; \mu) \tag{2.12}$$

where "arg max" stands for "the value maximizing".

If we knew F_0 exactly, the MLE would be "optimal" in the sense of attaining the lowest possible asymptotic variance among a "reasonable" class of estimators (see Section 10.8). But since we know F_0 only approximately, our goal will be to find estimators that are "nearly optimal" for both of the following situations:

(A) when F_0 is exactly normal
(B) when F_0 is approximately normal (say, contaminated normal).

If f_0 is everywhere positive, since the logarithm is an increasing function, (2.12) can be written as

$$\widehat{\mu} = \arg\min_{\mu} \sum_{i=1}^{n} \rho(x_i - \mu) \tag{2.13}$$

where

$$\rho = -\log f_0. \tag{2.14}$$

If $F_0 = \mathrm{N}(0, 1)$, then

$$f_0(x) = \frac{1}{\sqrt{2\pi}} \, e^{-x^2/2} \tag{2.15}$$

and apart from a constant, $\rho(x) = x^2/2$. Hence (2.13) is equivalent to

$$\widehat{\mu} = \arg\min_{\mu} \sum_{i=1}^{n} (x_i - \mu)^2. \tag{2.16}$$

If F_0 is the double exponential distribution

$$f_0(x) = \frac{1}{2} e^{-|x|} \tag{2.17}$$

then $\rho(x) = |x|$, and (2.13) is equivalent to

$$\widehat{\mu} = \arg\min_{\mu} \sum_{i=1}^{n} |x_i - \mu|. \tag{2.18}$$

We shall see below that the solutions to (2.16) and (2.18) are the sample mean and median, respectively.

If ρ is differentiable, differentiating (2.13) with respect to μ yields

$$\sum_{i=1}^{n} \psi(x_i - \widehat{\mu}) = 0 \tag{2.19}$$

with $\psi = \rho'$. If ψ is discontinuous, solutions to (2.19) might not exist, and in this case we shall interpret (2.19) to mean that the left-hand side changes sign at μ. Note that if f_0 is symmetric, then ρ is even and hence ψ is odd.

If $\rho(x) = x^2/2$, then $\psi(x) = x$, and (2.19) becomes

$$\sum_{i=1}^{n}(x_i - \widehat{\mu}) = 0$$

which has $\widehat{\mu} = \overline{x}$ as solution.

For $\rho(x) = |x|$, it will be shown that any median of x is a solution of (2.18). In fact, the derivative of $\rho(x)$ exists for $x \neq 0$, and is given by the sign function: $\psi(x) = \text{sgn}(x)$, where

$$\text{sgn}(x) = \begin{cases} -1 & \text{if} \quad x < 0 \\ 0 & \text{if} \quad x = 0 \\ 1 & \text{if} \quad x > 0. \end{cases} \tag{2.20}$$

Since the function to be minimized in (2.18) is continuous, it suffices to find the values of μ where its derivative changes sign. Note that

$$\text{sgn}(x) = \text{I}(x > 0) - \text{I}(x < 0) \tag{2.21}$$

where I(.) stands for the *indicator function*; that is,

$$\text{I}(x > 0) = \begin{cases} 1 & \text{if} \quad x > 0 \\ 0 & \text{if} \quad x \leq 0. \end{cases}$$

Applying (2.21) to (2.19) yields

$$\begin{aligned}\sum_{i=1}^{n} \text{sgn}(x_i - \mu) &= \sum_{i=1}^{n} (\text{I}(x_i - \mu > 0) - \text{I}(x_i - \mu < 0)) \\ &= \#(x_i > \mu) - \#(x_i < \mu) = 0\end{aligned}$$

and hence $\#(x_i > \mu) = \#(x_i < \mu)$, which implies that μ is any sample median.

From now on, the average of a dataset $\mathbf{z} = \{z_1, \ldots, z_n\}$ will be denoted by ave($\mathbf{z}$), or by $\text{ave}_i(z_i)$ when necessary; that is,

$$\text{ave}(\mathbf{z}) = \text{ave}_i(z_i) = \frac{1}{n}\sum_{i=1}^{n} z_i,$$

and its median by Med($\mathbf{z}$) or $\text{Med}_i(z_i)$. If c is a constant, $\mathbf{z} + c$ and $c\mathbf{z}$ will denote the data sets $(z_1 + c, \ldots, z_n + c)$ and $(cz_1, \ldots, cz_n)$. If x is a random variable with distribution F, the mean and median of a function $g(x)$ will be denoted by $\text{E}_F g(x)$ and $\text{Med}_F g(x)$, dropping the subscript F when there is no ambiguity.

Given a function ρ, an *M-estimator of location* is a solution of (2.13). We shall henceforth study estimators of this form, which need not be MLEs for any distribution. The function ρ will be chosen in order to ensure goals (A) and (B) above.

Assume ψ is monotone nondecreasing, with $\psi(-\infty) < 0 < \psi(\infty)$. Then it can be proved (see Theorem 10.1) that (2.19) – and hence (2.13) – always has a solution. If ψ is continuous and increasing, the solution is unique; otherwise the set of solutions is either a point or an interval (throughout this book, we shall call any function g *increasing* (*nondecreasing*) if $a < b$ implies $g(a) < g(b)$ ($g(a) \leq g(b)$)). More details on uniqueness are given in Section 10.1.

It is easy to show that M-estimators are shift equivariant, as defined in (2.4) (Problem 2.5). The mean and median are scale equivariant, but this does not hold in general for M-estimators in their present form. This drawback will be overcome in Section 2.7.

2.3.2 The distribution of M-estimators

In order to evaluate the performance of M-estimators, it is necessary to calculate their distributions. Except for the mean and the median (see (10.60)), there are no explicit expressions for the distribution of M-estimators in finite sample sizes, but approximations can be found and a heuristic derivation is given in Section 2.10.2 (a rigorous treatment is given in Section 10.3).

Assume ψ is increasing. For a given distribution F, define $\mu_0 = \mu_0(F)$ as the solution of

$$\mathrm{E}_F\psi(x - \mu_0) = 0. \tag{2.22}$$

For the sample mean, $\psi(x) = x$, and (2.22) implies $\mu_0 = \mathrm{E}x$; that is, the population mean. For the sample median, (2.21) and (2.22) yield

$$\mathrm{P}(x > \mu_0) - \mathrm{P}(x < \mu_0) = 2F(\mu_0) - 1 = 0$$

which implies $F(\mu_0) = 1/2$, which corresponds to $\mu_0 = \mathrm{Med}(x)$; that is, a population median. In general if F is symmetric, then μ_0 coincides with the center of symmetry (Problem 2.6).

It can be shown (see Section 2.10.2) that when $n \to \infty$,

$$\widehat{\mu} \to_p \mu_0 \tag{2.23}$$

where "$\to_p$" stands for "tends in probability" and μ_0 is defined in (2.22) – we say that $\widehat{\mu}$ is "*consistent* for μ_0" – and the distribution of $\widehat{\mu}$ is approximately

$$\mathrm{N}\left(\mu_0, \frac{v}{n}\right) \text{ with } v = \frac{\mathrm{E}_F(\psi(x-\mu_0)^2)}{(\mathrm{E}_F\psi'(x-\mu_0))^2}. \tag{2.24}$$

Note that under model (2.2) v does not depend on μ_0; that is,

$$v = \frac{\mathrm{E}_{F_0}(\psi(x)^2)}{(\mathrm{E}_{F_0}\psi'(x))^2}. \tag{2.25}$$

If the distribution of an estimator $\hat{\mu}$ is approximately $\mathrm{N}(\mu_0, v/n)$ for large n, we say that $\hat{\mu}$ is *asymptotically normal*, with asymptotic value μ_0 and asymptotic variance v. The *asymptotic efficiency* of $\hat{\mu}$ is the ratio

$$\mathrm{Eff}(\hat{\mu}) = \frac{v_0}{v}, \tag{2.26}$$

where v_0 is the asymptotic variance of the MLE, and measures how near $\hat{\mu}$ is to the optimum. The expression for v in (2.24) is called the *asymptotic variance* of $\hat{\mu}$.

To understand the meaning of efficiency, consider two estimators with asymptotic variances v_1 and v_2. Since their distributions are approximately normal with variances v_1/n and v_2/n, if for example $v_1 = 3v_2$ then the first estimator requires three times as many observations to attain the same variance as the second.

For the sample mean, $\psi' \equiv 1$ and hence $v = \mathrm{Var}(x)$. For the sample median, the numerator of v is one. Here ψ' does not exist, but if x has a density f, it is shown in Section 10.3 that the denominator is $2f(\mu_0)$, and hence

$$v = \frac{1}{4f(\mu_0)^2}. \tag{2.27}$$

Thus for $F = \mathrm{N}(0, 1)$ we have

$$v = \frac{2\pi}{4} = 1.571.$$

It will be seen that a type of ρ- and ψ-functions with important properties is the family of *Huber functions*, plotted in Figure 2.3:

$$\rho_k(x) = \begin{cases} x^2 & \text{if} \quad |x| \le k \\ 2k|x| - k^2 & \text{if} \quad |x| > k \end{cases} \tag{2.28}$$

with derivative $2\psi_k(x)$, where

$$\psi_k(x) = \begin{cases} x & \text{if} \quad |x| \le k \\ \mathrm{sgn}(x)k & \text{if} \quad |x| > k. \end{cases} \tag{2.29}$$

It is seen that ρ_k is quadratic in a central region, but increases only linearly to infinity. The M-estimators corresponding to the limit cases $k \to \infty$ and $k \to 0$ are the mean and the median, and we define $\psi_0(x)$ as $\mathrm{sgn}(x)$.

The value of k is chosen in order to ensure a given asymptotic variance – hence a given asymptotic efficiency – for the normal distribution. Table 2.2 gives the asymptotic variances of the estimator at model (2.6) with $G = \mathrm{N}(0, 1)$ and $H = \mathrm{N}(0, 10)$, for different values of k.

Here we see the trade-off between robustness and efficiency: when $k = 1.4$, the variance of the M-estimator for the normal is only 4.7% larger than that of $\overline{x}$ (which corresponds to $k = \infty$) and much smaller than that of the median (which corresponds to $k = 0$), while for contaminated normals it is clearly smaller than both.

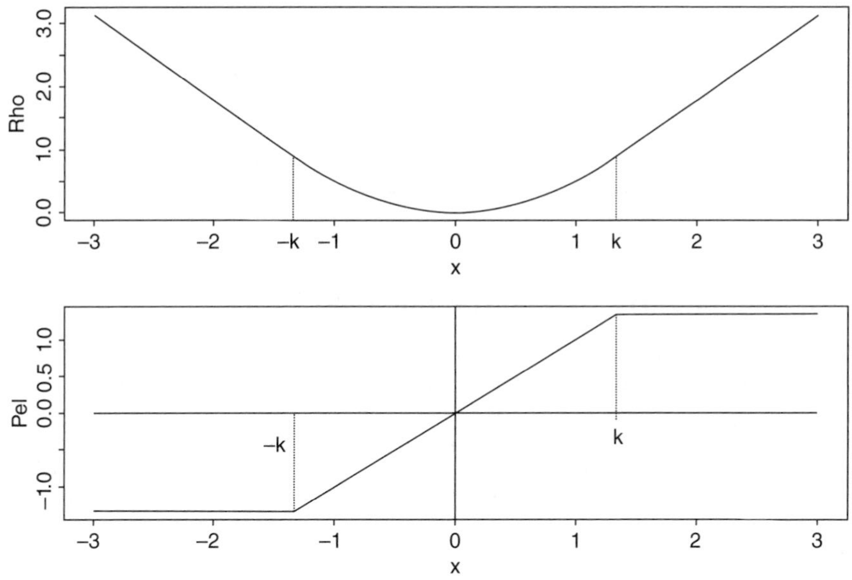

Figure 2.3 Huber ρ- and ψ-functions

Table 2.2 Asymptotic variances of Huber M-estimator

k	$\varepsilon = 0$	$\varepsilon = 0.05$	$\varepsilon = 0.10$
0	1.571	1.722	1.897
0.7	1.187	1.332	1.501
1.0	1.107	1.263	1.443
1.4	1.047	1.227	1.439
1.7	1.023	1.233	1.479
2.0	1.010	1.259	1.550
∞	1.000	5.950	10.900

Huber's ψ is one of the few cases where the asymptotic variance at the normal distribution can be calculated analytically. Since $\psi_k'(x) = \mathrm{I}(|x| \leq k)$, the denominator of (2.24) is $(\Phi(k) - \Phi(-k))^2$. The reader can verify that the numerator is

$$\mathrm{E}_\Phi \psi_k(x)^2 = 2[k^2(1 - \Phi(k)) + \Phi(k) - 0.5 - k\varphi(k)] \tag{2.30}$$

where φ and Φ are the standard normal density and distribution function, respectively (Problem 2.7). In Table 2.3 we give the values of k yielding prescribed asymptotic variances υ for the standard normal. The last row gives values of the quantity $\alpha = 1 - \Phi(k)$, which will play a role in Section 2.4.

Table 2.3 Asymptotic variances for Huber's psi-function

k	0.66	1.03	1.37
v	1.20	1.10	1.05
α	0.25	0.15	0.085

2.3.3 An intuitive view of M-estimators

A location M-estimator can be seen as a weighted mean. In most cases of interest, $\psi(0) = 0$ and $\psi'(0)$ exists, so that ψ is approximately linear at the origin. Let

$$W(x) = \begin{cases} \psi(x)/x & \text{if} \quad x \neq 0 \\ \psi'(0) & \text{if} \quad x = 0. \end{cases} \tag{2.31}$$

Then (2.19) can be written as

$$\sum_{i=1}^{n} W(x_i - \widehat{\mu})(x_i - \widehat{\mu}) = 0,$$

or equivalently

$$\widehat{\mu} = \frac{\sum_{i=1}^{n} w_i x_i}{\sum_{i=1}^{n} w_i}, \text{ with } w_i = W(x_i - \widehat{\mu}), \tag{2.32}$$

which expresses the estimator as a weighted mean. Since in general $W(x)$ is a non-increasing function of $|x|$, outlying observations will receive smaller weights. Note that although (2.32) looks like an explicit expression for $\widehat{\mu}$, actually the weights on the right-hand side depend also on $\widehat{\mu}$. Besides its intuitive value, this representation of the estimator will be useful for its numeric computation in Section 2.8. The weight function corresponding to Huber's ψ is

$$W_k(x) = \min\left\{1, \frac{k}{|x|}\right\} \tag{2.33}$$

which is plotted in the upper panel of Figure 2.4.

Another intuitive way to interpret an M-estimator is to rewrite (2.19) as

$$\widehat{\mu} = \widehat{\mu} + \frac{1}{n}\sum_{i=1}^{n} \psi(x_i - \widehat{\mu}) = \frac{1}{n}\sum_{i=1}^{n} \zeta(x_i, \widehat{\mu}), \tag{2.34}$$

where

$$\zeta(x, \mu) = \mu + \psi(x - \mu), \tag{2.35}$$

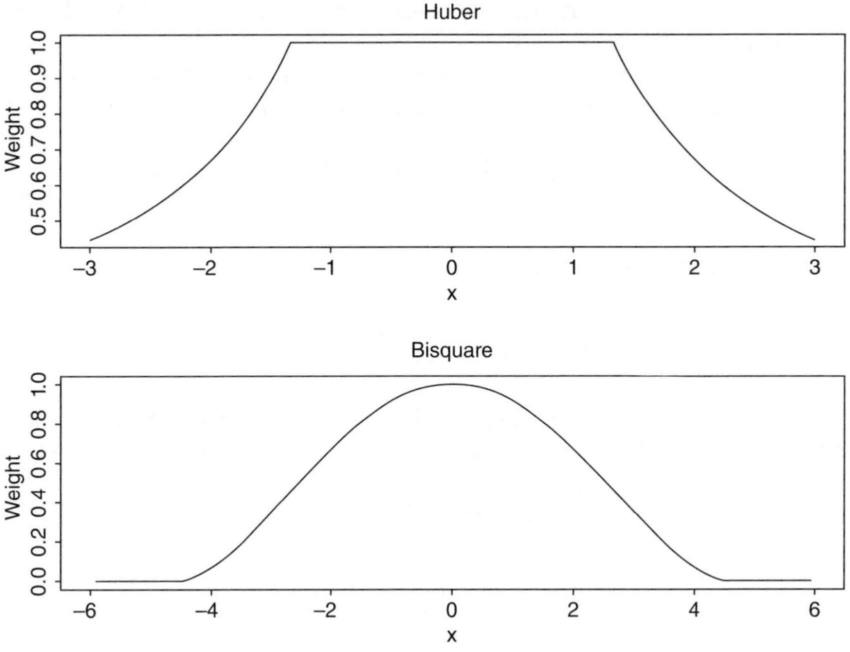

Figure 2.4 Huber and bisquare weight functions

which for the Huber function takes the form

$$\zeta(x,\mu) = \begin{cases} \mu - k & \text{if} \quad x < \mu - k \\ x & \text{if} \quad \mu - k \le x \le \mu + k \\ \mu + k & \text{if} \quad x > \mu + k. \end{cases} \tag{2.36}$$

In other words, $\hat{\mu}$ may be viewed as an average of the modified observations $\zeta(x_i, \hat{\mu})$ (called "pseudo-observations"): observations in the bulk of the data remain unchanged, while those too large or too small are truncated as described at the end of Section 2.1 (note that here the truncation interval depends on the data).

2.3.4 Redescending M-estimators

It is easy to show (Problem 2.15) that the MLE for the Student family of densities (2.9) has the ψ-function

$$\psi(x) = \frac{x}{x^2 + \nu}, \tag{2.37}$$

which tends to zero when $x \to \infty$. This suggests that for symmetric heavy-tailed distributions, it is better to use "redescending" ψs that tend to zero at infinity. This implies that for large x, the respective ρ-function increases more slowly than Huber's ρ (2.28), which is linear for $x > k$.

We will later discuss the advantages of using a *bounded* ρ. A popular choice of ρ- and ψ-functions is the *bisquare* (also called *biweight*) family of functions:

$$\rho(x) = \begin{cases} 1 - [1 - (x/k)^2]^3 & \text{if} \quad |x| \le k \\ 1 & \text{if} \quad |x| > k \end{cases} \tag{2.38}$$

with derivative $\rho'(x) = 6\psi(x)/k^2$ where

$$\psi(x) = x\left[1 - \left(\frac{x}{k}\right)^2\right]^2 \mathrm{I}(|x| \le k). \tag{2.39}$$

These functions are displayed in Figure 2.5. Note that ψ is everywhere differentiable and it vanishes outside $[-k, k]$. M-estimators with ψ vanishing outside an interval are not MLEs for any distribution (Problem 2.12).

The weight function (2.31) for this family is

$$W(x) = \left[1 - \left(\frac{x}{k}\right)^2\right]^2 \mathrm{I}(|x| \le k)$$

and is plotted in Figure 2.4.

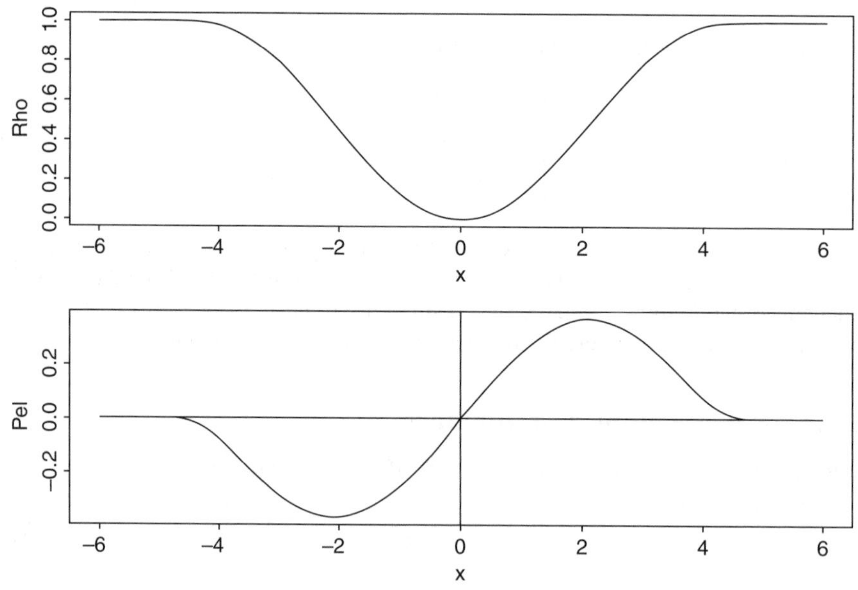

Figure 2.5 ρ- and ψ-functions for the bisquare estimator

Table 2.4 Values of k for prescribed efficiencies of bisquare estimator

Efficiency	0.80	0.85	0.90	0.95
k	3.14	3.44	3.88	4.68

If ρ is everywhere differentiable and ψ is monotonic, then the forms (2.13) and (2.19) are equivalent. If ψ is redescending, some solutions of (2.19) – usually called "bad solutions" – may not correspond to the absolute minimum of the criterion, which defines the M-estimator.

Estimators defined as solutions of (2.19) with monotone ψ will be called "monotone M-estimators" for short, while those defined by (2.13) when ψ is not monotone will be called "redescending M-estimators". Numerical computing of redescending location estimators is essentially no more difficult than for monotone estimators (Section 2.8.1). It will be seen in Section 3.4 that redescending estimators offer an increase in robustness when there are large outliers.

The values of k for prescribed efficiencies (2.26) of the bisquare estimator are given in Table 2.4. If ρ has a nondecreasing derivative, it can be shown (Feller, 1971) that for all x, y

$$\rho(\alpha x + (1-\alpha)y) \leq \alpha\rho(x) + (1-\alpha)\rho(y) \; \forall \; \alpha \in [0,1]. \tag{2.40}$$

Functions verifying (2.40) are referred to as *convex*.

We state the following definitions for later reference.

Definition 2.1 *Unless stated otherwise, a ρ-function will denote a function ρ such that:*

R1 *$\rho(x)$ is a nondecreasing function of $|x|$*
R2 *$\rho(0) = 0$*
R3 *$\rho(x)$ is increasing for $x > 0$ such that $\rho(x) < \rho(\infty)$*
R4 *if ρ is bounded, it is also assumed that $\rho(\infty) = 1$.*

Definition 2.2 *A ψ-function will denote a function ψ that is the derivative of a ρ-function, which implies in particular that*

Ψ1 *ψ is odd and $\psi(x) \geq 0$ for $x \geq 0$.*

2.4 Trimmed and Winsorized means

Another approach to robust estimation of location would be to discard a proportion of the largest and smallest values. More precisely, let $\alpha \in [0, 1/2)$ and $m = [n\alpha]$ where

[.] stands for the integer part, and define the *α-trimmed mean* as

$$\bar{x}_\alpha = \frac{1}{n-2m} \sum_{i=m+1}^{n-m} x_{(i)},$$

where $x_{(i)}$ denotes the order statistics (1.2).

The reader may think that we are again suppressing observations. Note, however, that no subjective choice has been made: the result is actually a function of *all* observations (even of those that have not been included in the sum).

The limit cases $\alpha = 0$ and $\alpha \to 0.5$ correspond to the sample mean and median, respectively. For the data of Example 1.1, the α-trimmed means with $\alpha = 0.10$ and 0.25 are, respectively, 3.20 and 3.27. Deleting the largest observation changes them to 3.17 and 3.22, respectively.

The exact distribution of trimmed means is intractable. Its large-sample approximation is more complicated than that of M-estimators, and will be described in Section 10.7. It can be proved that for large n the distribution under model (2.1) is approximately normal, and for symmetrically distributed u_i, the asymptotic distribution is $\mathcal{D}(\hat{\mu}) \approx \mathrm{N}(\mu, v/n)$, where the asymptotic variance v is that of an M-estimator with Huber's function ψ_k, where k is the $(1-\alpha)$-quantile of u:

$$v = \frac{\mathrm{E}[\psi_k(u)]^2}{(1-2\alpha)^2}. \tag{2.41}$$

The values of α yielding prescribed asymptotic variances at the standard normal are given at the bottom of Table 2.3. Note that the asymptotic efficiency of $\bar{x}_{0.25}$ is 0.83, even though we seem to be "throwing away" 50% of the observations. Note also that the asymptotic variance of a trimmed mean is not a trimmed variance. This would be so if the numerator of (2.41) were

$$\mathrm{E}([(x-\mu)\mathrm{I}(|x-\mu|) \le k])^2.$$

An idea similar to the trimmed mean is the *α-Winsorized mean* (named after the biostatistician Charles P. Winsor), defined as

$$\tilde{x}_\alpha = \frac{1}{n}\left(m x_{(m)} + m x_{(n-m+1)} + \sum_{i=m+1}^{n-m} x_{(i)} \right),$$

where m and the $x_{(i)}$ are as above. That is, extreme values, instead of being deleted as in the trimmed mean, are shifted towards the bulk of the data.

It can be shown that for large n, $\tilde{x}_\alpha$ is approximately normal (Bickel 1965). If the distribution F of u_i is symmetric and has a density f, then $\tilde{x}_\alpha$ is approximately $\mathrm{N}(\mu, v/n)$. with

$$v = 2\alpha\left(u_{1-\alpha} + \frac{\alpha}{f(u_{1-\alpha})}\right)^2 + \mathrm{E}u^2\mathrm{I}(u_\alpha \le u \le u_{1-\alpha}), \tag{2.42}$$

where $u_{1-\alpha}$ is the $(1-\alpha)$-quantile of F.

A more general class of estimators, called *L-estimators*, is defined as linear combinations of order statistics:

$$\widehat{\mu} = \sum_{i=1}^{n} a_i x_{(i)}, \tag{2.43}$$

where the a_is are given constants. For α-trimmed means,

$$a_i = \frac{1}{n-2m} \mathrm{I}(m+1 \leq i \leq n-m), \tag{2.44}$$

and for α-Winsorized means

$$a_i = \frac{1}{n}(m\mathrm{I}(i=m) + m\mathrm{I}(i=n-m+1) + \mathrm{I}(m+1 \leq i \leq n-m)). \tag{2.45}$$

It is easy to show (Problem 2.10) that if the coefficients of an L-estimator satisfy the conditions

$$a_i \geq 0, \quad \sum_{i=1}^{n} a_i = 1, \quad a_i = a_{n-i+1}, \tag{2.46}$$

then the estimator is shift and scale equivariant, and also fulfills the natural conditions

C1 If $x_i \geq 0$ for all i, then $\widehat{\mu} \geq 0$
C2 If $x_i = c$ for all i, then $\widehat{\mu} = c$
C3 $\widehat{\mu}(-x) = -\widehat{\mu}(x)$.

2.5 M-estimators of scale

In this section we discuss a situation that, while not especially important in itself, will play an important auxiliary role in the development of estimators for location, regression and multivariate analysis. Consider observations x_i satisfying the *multiplicative* model

$$x_i = \sigma u_i \tag{2.47}$$

where the u_i are i.i.d with density f_0 and $\sigma > 0$ is the unknown parameter. The distributions of the x_i constitute a *scale family*, with density

$$\frac{1}{\sigma} f_0\left(\frac{x}{\sigma}\right).$$

Examples are the exponential family, with $f_0(x) = \exp(-x)\mathrm{I}(x > 0)$, and the normal scale family $\mathrm{N}(0, \sigma^2)$, with f_0 given by (2.15).

The MLE of σ in (2.47) is

$$\widehat{\sigma} = \arg\max_{\sigma} \frac{1}{\sigma^n} \prod_{i=1}^{n} f_0\left(\frac{x_i}{\sigma}\right).$$

Taking logs and differentiating with respect to σ yields

$$\frac{1}{n}\sum_{i=1}^{n}\rho\left(\frac{x_i}{\widehat{\sigma}}\right)=1 \tag{2.48}$$

where $\rho(t)=t\psi(t)$, with $\psi=-f_0'/f_0$. If f_0 is N(0, 1) then $\rho(t)=t^2$, which yields $\widehat{\sigma}=\sqrt{\text{ave}(\mathbf{x}^2)}$ (the *root mean square*, RMS); if f is double-exponential defined in (2.17), then $\rho(t)=|t|$, which yields $\widehat{\sigma}=\text{ave}(|\mathbf{x}|)$. Note that if f_0 is even, so is ρ, and this implies that $\widehat{\sigma}$ depends only on the absolute values of the x_i.

In general, any estimator satisfying an equation of the form

$$\frac{1}{n}\sum_{i=1}^{n}\rho\left(\frac{x_i}{\widehat{\sigma}}\right)=\delta, \tag{2.49}$$

where ρ is a ρ-function and δ is a positive constant, will be called an *M-estimator of scale*.Note that in order for (2.49) to have a solution we must have $0<\delta<\rho(\infty)$. Hence if ρ is bounded it will be assumed without loss of generality that

$$\rho(\infty)=1,\quad \delta\in(0,1).$$

In the rarely occurring event that $\#(x_i=0)>n(1-\delta)$ should happen, then (2.49) has no solution. In this case it is natural to define $\widehat{\sigma}(\mathbf{x})=0$. It is easy to verify that scale M-estimators are equivariant, in the sense that $\widehat{\sigma}(c\mathbf{x})=c\widehat{\sigma}(\mathbf{x})$ for any $c>0$, and if ρ is even then

$$\widehat{\sigma}(c\mathbf{x})=|c|\widehat{\sigma}(\mathbf{x})$$

for any c. For large n, the sequence of estimators (2.49) converges to the solution of

$$\mathrm{E}\rho\left(\frac{x}{\sigma}\right)=\delta \tag{2.50}$$

if it is unique (Section 10.2); see Problem 10.6.

The reader can verify that the scale MLE for the Student distribution is equivalent to

$$\rho(t)=\frac{t^2}{t^2+v}\quad\text{and}\quad\delta=\frac{1}{v+1}. \tag{2.51}$$

A frequently used scale estimator is the *bisquare scale*, where ρ is given by (2.38) with $k=1$; that is,

$$\rho(x)=\min\{1-(1-x^2)^3,1\} \tag{2.52}$$

and $\delta=0.5$. It is easy to verify that (2.51) and (2.52) satisfy the conditions for a ρ-function in Definition 2.1.

When ρ is the step function

$$\rho(t)=\mathrm{I}(|t|>c), \tag{2.53}$$

where c is a positive constant, and $\delta = 0.5$, we have $\widehat{\sigma} = \mathrm{Med}(|\mathbf{x}|)/c$. The argument in Problem 2.12 shows that it is not the scale MLE for any distribution.

Most often we shall use a ρ that is quadratic near the origin – that is, $\rho'(0) = 0$ and $\rho''(0) > 0$ – and in such cases an M-scale estimator can be represented as a weighted RMS estimator. We define the weight function as

$$W(x) = \begin{cases} \rho(x)/x^2 & \text{if} \quad x \neq 0 \\ \rho''(0) & \text{if} \quad x = 0 \end{cases} \tag{2.54}$$

and then (2.49) is equivalent to

$$\widehat{\sigma}^2 = \frac{1}{n\delta} \sum_{i=1}^{n} W\left(\frac{x_i}{\widehat{\sigma}}\right) x_i^2. \tag{2.55}$$

It follows that $\widehat{\sigma}$ can be seen as a weighted RMS estimator. For the Student MLE

$$W(x) = \frac{1}{\nu + x^2}, \tag{2.56}$$

and for the bisquare scale

$$W(x) = \min\{3 - 3x^2 + x^4, 1/x^2\}. \tag{2.57}$$

It is seen that larger values of x receive smaller weights.

Note that using $\rho(x/c)$ instead of $\rho(x)$ in (2.49) yields $\widehat{\sigma}/c$ instead of $\widehat{\sigma}$. This can be used to normalize $\widehat{\sigma}$ to have a given asymptotic value, as will be done at the end of Section 2.6. If we want $\widehat{\sigma}$ to coincide asymptotically with $\mathrm{SD}(x)$ when x is normal, then (recalling (2.50)) we have to take c as the solution of $\mathrm{E}\rho(x/c) = \delta$ with $x \sim \mathrm{N}(0, 1)$, which can be obtained numerically. For the bisquare scale, the solution is $c = 1.56$.

Although scale M-estimators play an auxiliary role here, their importance will be seen in Chapters 5 and 6.

2.6 Dispersion estimators

The traditional way to measure the variability of a dataset $\mathbf{x}$ is with the standard deviation (SD)

$$\mathrm{SD}(\mathbf{x}) = \left[\frac{1}{n-1} \sum_{i=1}^{n} (x_i - \overline{x})^2\right]^{1/2}.$$

For any constant c the SD satisfies the *shift invariance* and *scale equivariance* conditions

$$\mathrm{SD}(\mathbf{x} + c) = \mathrm{SD}(\mathbf{x}), \qquad \mathrm{SD}(c\mathbf{x}) = |c|\, \mathrm{SD}(\mathbf{x}). \tag{2.58}$$

Any statistic satisfying (2.58) will be called a *dispersion (or scatter) estimator*.

In Example 1.1 we observed the lack of robustness of the standard deviation, and we now consider possible robust alternatives. One alternative estimator proposed in the past is the *mean absolute deviation* (MD):

$$\mathrm{MD}(\mathbf{x}) = \frac{1}{n}\sum_{i=1}^{n}|x_i - \overline{x}| \tag{2.59}$$

which is also sensitive to outliers, although less so than the SD (Tukey, 1960). In the flour example, the MDs with and without the largest observation are, respectively, 2.14 and 0.52: still a large difference.

Both the SD and MD are defined by first centering the data by subtracting $\overline{x}$ (which ensures shift invariance) and then taking a measure of "largeness" of the absolute values. A robust alternative is to subtract the median instead of the mean, and then take the median of the absolute values, which yields the MAD estimator introduced in the previous chapter:

$$\mathrm{MAD}(\mathbf{x}) = \mathrm{Med}(|\mathbf{x} - \mathrm{Med}(\mathbf{x})|) \tag{2.60}$$

which clearly satisfies (2.58). For the flour data with and without the largest observation, the MADs are 0.35 and 0.34, respectively.

In the same way as (2.59) and (2.60), we define the mean and the median absolute deviations of a random variable x as

$$\mathrm{MD}(x) = \mathrm{E}|x - \mathrm{E}x| \tag{2.61}$$

and

$$\mathrm{MAD}(x) = \mathrm{Med}(|x - \mathrm{Med}(x)|), \tag{2.62}$$

respectively.

Two other well-known dispersion estimators are the *range*, defined as $\max(x) - \min(x) = x_{(n)} - x_{(1)}$, and the sample *interquartile range*

$$\mathrm{IQR}(x) = x_{(n-m+1)} - x_{(m)}$$

where $m = [n/4]$. Both are based on order statistics; the former is clearly very sensitive to outliers, while the latter is not.

Note that if $x \sim \mathrm{N}(\mu, \sigma^2)$ (where "$\sim$" stands for "is distributed as") then $\mathrm{SD}(x) = \sigma$ by definition, while $\mathrm{MD}(x)$, $\mathrm{MAD}(x)$ and $\mathrm{IQR}(x)$ are constant multiples of σ:

$$\mathrm{MD}(x) = c_1\sigma,\ \mathrm{MAD}(x) = c_2\sigma,\ \mathrm{IQR}(x) = 2c_2\sigma,$$

where

$$c_1 = 2\varphi(0) \text{ and } c_2 = \Phi^{-1}(0.75)$$

(Problem 2.11). Hence if we want a dispersion estimator that "measures the same thing" as the SD for the normal, we should normalize the MAD by dividing it by $c_2 \approx 0.675$. The "normalized MAD" (MADN) is thus

$$\mathrm{MADN}(x) = \frac{\mathrm{MAD}(x)}{0.675}. \tag{2.63}$$

Likewise, we should normalize the MD and the IQR by dividing them by c_1 and by $2c_2$ respectively.

Observe that for the flour data (which was found to be approximately normal) MADN = 0.53, which is not far from the standard deviation of the data without the outlier: 0.69.

The first step in computing the SD, MD and MAD is "centering" the data; that is, subtracting a location estimator from the data values. The first two are not robust, and the third has a low efficiency. An estimator that combines robustness and efficiency is the following: first compute a location M-estimator $\hat{\mu}$, and then apply a scale M-estimator $\hat{\sigma}$ to the centered data $x_i - \hat{\mu}$. Here $\hat{\sigma}$ should be normalized as described at the end of Section 2.6. We shall call this $\hat{\sigma}$ an *M-dispersion estimator*.

Note that the IQR does not use centering. A dispersion estimator that does not require centering and is more robust than the IQR (in a sense to be defined in the next chapter) was proposed by Croux and Rousseeuw (1992) and Rousseeuw and Croux (1993). The estimator, which they call Q_n, is based only on the differences between data values. Let $m = n/2$. Call $d_{(1)} \leq \ldots \leq d_{(m)}$ the ordered values of the m differences $d_{ij} = x_{(i)} - x_{(j)}$ with $i > j$. Then the estimator is defined as

$$Q_n = d_{(k)}, \quad k = \binom{[n/2]+1}{2}, \tag{2.64}$$

where [.] denotes the integer part. Since $k \approx m/4$, Q_n is approximately the first quartile of the d_{ij}s. It is easy to verify that, for any k, Q_n is shift invariant and scale equivariant. It can be shown that, for the normal, Q_n has an efficiency of 0.82, and the estimator $2.222Q_n$ is consistent for the SD.

Martin and Zamar (1993b) studied another dispersion estimator that does not require centering and has interesting robustness properties (Problem 2.16b).

2.7 M-estimators of location with unknown dispersion

Estimators defined by (2.13) are not scale equivariant.For example, if all the x_i in (2.13) are divided by 10, it does not follow that the respective solution μ is divided by 10. Sections 2.7.1 and 2.7.2 deal with approaches to define scale equivariant estimators. To make this clear, assume we want to estimate μ in model (2.1) where F is given by the mixture (2.6) with $G = \mathrm{N}(\mu, \sigma^2)$. *If* σ were known, it would be natural to divide (2.1) by σ to reduce the problem to the case $\sigma = 1$, which implies estimating μ by

$$\hat{\mu} = \arg\min_{\mu} \sum_{i=1}^{n} \rho\left(\frac{x_i - \mu}{\sigma}\right).$$

It is easy to verify that, as in (2.24), for large n the approximate distribution of $\hat{\mu}$ is $\mathrm{N}(\mu, v/n)$, where

$$v = \sigma^2 \frac{\mathrm{E}\psi((x-\mu)/\sigma)^2}{(\mathrm{E}\psi'((x-\mu)/\sigma))^2}. \tag{2.65}$$

2.7.1 Previous estimation of dispersion

To obtain scale equivariant M-estimators of location, an intuitive approach is to use

$$\hat{\mu} = \arg\min_{\mu} \sum_{i=1}^{n} \rho\left(\frac{x_i - \mu}{\hat{\sigma}}\right), \tag{2.66}$$

where $\hat{\sigma}$ is a previously computed dispersion estimator. It is easy to verify that $\hat{\mu}$ is indeed scale equivariant. Since $\hat{\sigma}$ does not depend on μ, (2.66) implies that $\hat{\mu}$ is a solution of

$$\sum_{i=1}^{n} \psi\left(\frac{x_i - \hat{\mu}}{\hat{\sigma}}\right) = 0. \tag{2.67}$$

It is intuitive that $\hat{\sigma}$ must itself be robust. In Example 1.2, using (2.66) with bisquare ψ with $k = 4.68$, and $\hat{\sigma} = \text{MADN}(\mathbf{x})$, yields $\hat{\mu} = 25.56$; using $\hat{\sigma} = \text{SD}(\mathbf{x})$ instead gives $\hat{\mu} = 25.12$. Now add to the dataset three copies of the lowest value, −44. The results change to 26.42 and 17.19. The reason for this change is that the outliers "inflate" the SD, and hence the location estimator attributes to them too much weight.

Note that since k is chosen in order to ensure a given efficiency for the unit normal, if we want $\hat{\mu}$ to attain the same efficiency for any normal, $\hat{\sigma}$ must "estimate the SD at the normal", in the sense that if the data are $\text{N}(\mu, \sigma^2)$, then when $n \to \infty$, $\hat{\sigma}$ tends in probability to σ. This is why we use the normalized median absolute deviation MADN described previously, rather than the un-normalized version MAD.

If a number $m > n/2$ of data values are concentrated at a single value x_0, we have $\text{MAD}(\mathbf{x}) = 0$, and hence the estimator is not defined. In this case we define $\hat{\mu} = x_0 = \text{Med}(\mathbf{x})$. Besides being intuitively plausible, this definition can be justified by a limit argument. Let the n data values be different, and let m of them tend to x_0. Then it is not difficult to show that, in the limit, the solution of (2.66) is x_0.

It can be proved that if F is symmetric, then when n is large, $\hat{\mu}$ behaves as if $\hat{\sigma}$ were constant, in the following sense: if $\hat{\sigma}$ tends in probability to σ, then the distribution of $\hat{\mu}$ is approximately normal with variance (2.65) (for asymmetric F the asymptotic variance is more complicated; see Section 10.6). Therefore the efficiency of $\hat{\mu}$ does not depend on that of $\hat{\sigma}$. In Chapter 3 it will be seen, however, that its robustness does depend on that of $\hat{\sigma}$.

2.7.2 Simultaneous M-estimators of location and dispersion

An alternative approach is to consider a *location–dispersion* model with two unknown parameters

$$x_i = \mu + \sigma u_i \tag{2.68}$$

where u_i has density f_0, and hence x_i has density

$$f(x) = \frac{1}{\sigma} f_0\left(\frac{x - \mu}{\sigma}\right). \tag{2.69}$$

In this case, σ is the scale parameter of the random variables σu_i, but it is a dispersion parameter for the x_i.

We now derive the simultaneous MLE of μ and σ in model (2.69):

$$(\hat{\mu}, \hat{\sigma}) = \arg\max_{\mu,\sigma} \frac{1}{\sigma^n} \prod_{i=1}^{n} f_0\left(\frac{x_i - \mu}{\sigma}\right)$$

which can be written as

$$(\hat{\mu}, \hat{\sigma}) = \arg\min_{\mu,\sigma} \left\{ \frac{1}{n} \sum_{i=1}^{n} \rho_0\left(\frac{x_i - \mu}{\sigma}\right) + \log \sigma \right\} \tag{2.70}$$

with $\rho_0 = -\log f_0$. The main point of interest here is μ, while σ is a "nuisance parameter".

Proceeding as in the derivations of (2.19) and (2.49) it follows that the MLEs satisfy the system of equations

$$\sum_{i=1}^{n} \psi\left(\frac{x_i - \hat{\mu}}{\hat{\sigma}}\right) = 0 \tag{2.71}$$

$$\frac{1}{n} \sum_{i=1}^{n} \rho_{\text{scale}}\left(\frac{x_i - \hat{\mu}}{\hat{\sigma}}\right) = \delta, \tag{2.72}$$

where

$$\psi(x) = -\rho_0', \ \rho_{\text{scale}}(x) = x\psi(x), \ \delta = 1. \tag{2.73}$$

The reason for notation "ρ_{scale}" is that in all instances considered in this book, ρ_{scale} is a ρ-function in the sense of Definition 2.1; this characteristic is exploited in Section 5.4.1. The notation will be used whenever it is necessary to distinguish this ρ_{scale}, used for scale, from the ρ in (2.14), used for location; otherwise, we shall write just ρ.

We shall deal in general with simultaneous estimators $(\hat{\mu}, \hat{\sigma})$ defined as solutions of systems of equations of the form (2.71)–(2.72), which need not correspond to the MLE for any distribution. It can be proved (see Section 10.5) that for large n the distributions of $\hat{\mu}$ and $\hat{\sigma}$ are approximately normal. If F is symmetric then $\mathcal{D}(\hat{\mu}) \approx \text{N}(\mu, v/n)$, with v given by (2.65), where μ and σ are the solutions of the system

$$\text{E}\,\psi\left(\frac{x - \hat{\mu}}{\sigma}\right) = 0 \tag{2.74}$$

$$\text{E}\,\rho_{\text{scale}}\left(\frac{x - \mu}{\hat{\sigma}}\right) = \delta. \tag{2.75}$$

We may choose Huber's or the bisquare function for ψ. A very robust choice for ρ_{scale} is (2.53), with $c = 0.675$ to make it consistent with the SD for the normal, which yields

$$\hat{\sigma} = \frac{1}{0.675} \text{Med}(|\mathbf{x} - \hat{\mu}|). \tag{2.76}$$

Although this looks similar to using the previously computed MADN, it will be seen in Chapter 6 that the latter yields more robust results.

In general, estimation with a previously computed dispersion is more robust than simultaneous estimation. However, simultaneous estimation will be useful in more general situations, as will be seen in Chapter 6.

2.8 Numerical computing of M-estimators

There are several methods available for computing M-estimators of location and/or scale. In principle one could use any of the general methods for equation solving such as the Newton–Raphson algorithm, but methods based on derivatives may be unsafe with the types of ρ- and ψ-functions that yield good robustness properties (see Chapter 9). Here we shall describe a computational method called *iterative reweighting*, which takes special advantage of the characteristics of the problem.

2.8.1 Location with previously-computed dispersion estimation

For the solution of the robust location estimation optimization problem (2.66), the weighted average expression (2.32) suggests an iterative procedure. Start with a robust dispersion estimator $\widehat{\sigma}$ (for instance, the MADN) and some initial estimator $\widehat{\mu}_0$ (for instance, the sample median). Given $\widehat{\mu}_k$, compute

$$w_{k,i} = W\left(\frac{x_i - \widehat{\mu}_k}{\widehat{\sigma}}\right) \quad (i = 1, \ldots, n) \tag{2.77}$$

where W is the function in (2.31) and let

$$\widehat{\mu}_{k+1} = \frac{\sum_{i=1}^{n} w_{k,i}x_i}{\sum_{i=1}^{n} w_{k,i}}. \tag{2.78}$$

Results to be proved in Section 9.1 imply that if $W(x)$ is bounded and nonincreasing for $x > 0$, then the sequence $\widehat{\mu}_k$ converges to a solution of (2.66). The algorithm, which requires a stopping rule based on a tolerance parameter ε, is thus:

1. Compute $\widehat{\sigma} = \text{MADN}(x)$ and $\mu_0 = \text{Med}(\mathbf{x})$.
2. For $k = 0, 1, 2, \ldots$, compute the weights (2.77) and then $\widehat{\mu}_{k+1}$ in (2.78).
3. Stop when $|\widehat{\mu}_{k+1} - \widehat{\mu}_k| < \varepsilon\widehat{\sigma}$.

If ψ is increasing the solution is unique, and the starting point $\widehat{\mu}_0$ influences only the number of iterations. If ψ is redescending then $\widehat{\mu}_0$ must be robust in order to insure convergence to a "good" solution. Choosing $\widehat{\mu}_0 = \text{Med}(\mathbf{x})$ suffices for this purpose.

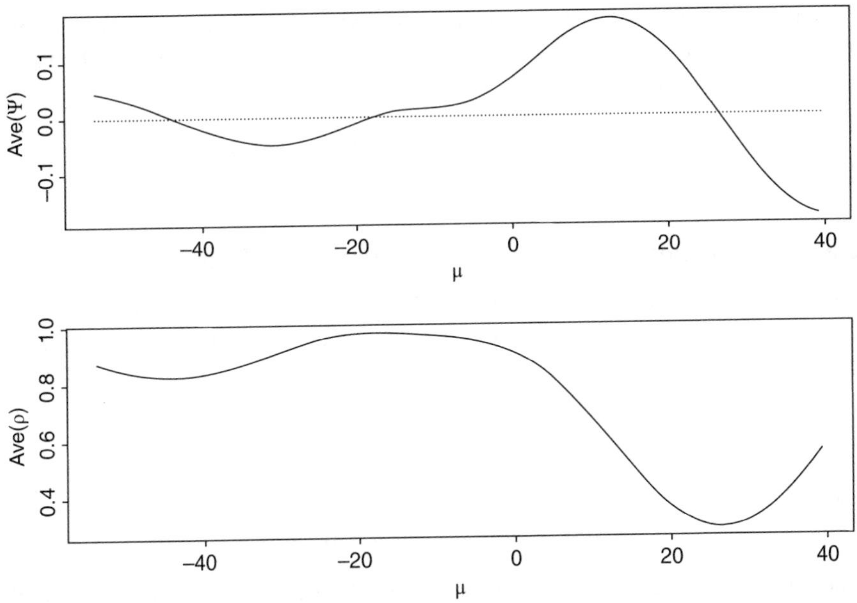

Figure 2.6 Averages of $\psi(x-\mu)$ and $\rho(x-\mu)$ as a function of μ

Figure 2.6 shows the averages of $\psi((\mathbf{x}-\mu)/\hat{\sigma})$ and of $\rho((\mathbf{x}-\mu)/\hat{\sigma})$ as a function of μ, where ψ and ρ correspond to the bisquare estimator with efficiency 0.95, and $\hat{\sigma}$ = MADN, for the data of Example 1.2, to which three extra values of the outlier −44 were added. Three roots of the estimating equation (2.67) are apparent; one corresponds to the absolute minimum of (2.66) while the other two correspond to a relative minimum and a relative maximum. This effect occurs also with the original data, but is less visible.

2.8.2 Scale estimators

For solving (2.49), the expression (2.55) suggests an iterative procedure. Start with some $\hat{\sigma}_0$, for instance, the normalized MAD (MADN). Given $\hat{\sigma}_k$ compute

$$w_{k,i} = W\left(\frac{x_i}{\hat{\sigma}_k}\right) \quad (i = 1, \ldots, n) \tag{2.79}$$

where W is the weight function in (2.54) and let

$$\hat{\sigma}_{k+1} = \sqrt{\frac{1}{n\delta}\sum_{i=1}^{n} w_{k,i}\, x_i^2}. \tag{2.80}$$

Then if $W(x)$ is bounded, even, continuous and nonincreasing for $x > 0$, the sequence σ_N converges to a solution of (2.55) and hence of (2.49) (for a proof see Section 9.4). The algorithm is thus:

1. For $k = 0, 1, 2, \ldots$, compute the weights (2.79) and then $\widehat{\sigma}_{k+1}$ in (2.80).
2. Stop when $|\widehat{\sigma}_{k+1}/\widehat{\sigma}_k - 1| < \varepsilon$.

2.8.3 Simultaneous estimation of location and dispersion

The procedure for solving the system (2.71)–(2.72) is a combination of the ones described in Sections 2.8.1 and 2.8.2. Compute starting values $\widehat{\mu}_0, \widehat{\sigma}_0$, and, given $\widehat{\mu}_k, \widehat{\sigma}_k$, compute for $i = 1, \ldots, n$

$$r_{k,i} = \frac{x_i - \widehat{\mu}_k}{\widehat{\sigma}_k}$$

and

$$w_{1k,i} = W_1(r_{k,i}),\ w_{2k,i} = W_2(r_{k,i})$$

where W_1 is the weight function W in (2.31) and W_2 is the W in (2.54) corresponding to ρ_{scale}. Then at the kth iteration

$$\widehat{\mu}_{k+1} = \frac{\sum_{i=1}^n w_{1k,i} x_i}{\sum_{i=1}^n w_{1k,i}},\ \widehat{\sigma}_{k+1}^2 = \frac{\widehat{\sigma}_k^2}{n\delta} \sum_{i=1}^n w_{2k,i}\, r_i^2.$$

2.9 Robust confidence intervals and tests

2.9.1 Confidence intervals

Since outliers affect both the sample mean $\overline{x}$ and the sample standard deviation s, confidence intervals for $\mu = \mathrm{E}(x)$ based on normal theory may be unreliable. Outliers may displace $\overline{x}$ and/or "inflate" s, resulting in one or both of the following degradations in performance:

- the true coverage probability may be much lower than the nominal one;
- the coverage probability may be either close to or higher than the nominal one, but at the cost of a loss of precision, in the form of an inflated expected confidence interval length.

We briefly elaborate on these points.

Recall that the usual Student confidence interval, justified by the assumption of a normal distribution for i.i.d. observations, is based on the "t-statistic",

$$T = \frac{\overline{x} - \mu}{s/\sqrt{n}}. \tag{2.81}$$

From this, one gets the usual two-sided confidence intervals for μ with level $1-\alpha$

$$\overline{x} \pm t_{n-1,1-\alpha/2}\frac{s}{\sqrt{n}},$$

where $t_{m,\beta}$ is the β-quantile of the t-distribution with m degrees of freedom.

The simplest situation is when the distribution of the data is symmetric about $\mu = \mathrm{E}x$. Then $\mathrm{E}\overline{x} = \mu$ and the confidence interval is centered. However, heavy tails in the distribution will cause the value of s to be inflated, and hence the interval length will be inflated, possibly by a large amount. Thus, in the case of symmetric heavy-tailed distributions, the price paid for maintaining the target confidence interval error rate α will often be unacceptably long confidence intervals. If the data have a mixture distribution $(1-\varepsilon)\mathrm{N}(\mu,\sigma^2)+\varepsilon H$, where H is not symmetric about μ, then the distribution of the data is not symmetric about μ and $\mathrm{E}\overline{x} \neq \mu$. Then the t confidence interval with purported confidence level $1-\alpha$ will not be centered and will not have the error rate α, and will lack robustness of both level and length. If the data distribution is both heavy tailed and asymmetric, then the t confidence interval can fail to have the target error rate and at the same time have unacceptably large interval lengths. Thus the classic t confidence interval lacks robustness of both error rate (confidence level) and length, and we need confidence intervals with both types of robustness.

Approximate confidence intervals for a parameter of interest can be obtained from the asymptotic distribution of a parameter estimator. Robust confidence intervals that are not much influenced by outliers can be obtained by imitating the form of the classical Student t confidence interval, but replacing the average and SD by robust location and dispersion estimators. Consider the M-estimators $\widehat{\mu}$ in Section 2.7, and recall that if $\mathcal{D}(x)$ is symmetric then for large n the distribution of $\widehat{\mu}$ is approximately $\mathrm{N}(\mu, v/n)$, with v given by (2.65). Since v is unknown, an estimator $\hat{v}$ may be obtained by replacing the expectations in (2.65) by sample averages, and the parameters by their estimators:

$$\hat{v} = \widehat{\sigma}^2\frac{\text{ave}\,[\psi((\mathbf{x}-\widehat{\mu})/\widehat{\sigma})]^2}{(\text{ave}[\psi'((\mathbf{x}-\widehat{\mu})/\widehat{\sigma})])^2}. \tag{2.82}$$

A robust approximate t-statistic ("Studentized M-estimator") is then defined as

$$T = \frac{\widehat{\mu}-\mu}{\sqrt{\hat{v}/n}} \tag{2.83}$$

and its distribution is approximately normal N(0, 1) for large n. Thus a robust approximate interval can then be computed as

$$\widehat{\mu} \pm z_{1-\alpha/2}\sqrt{\frac{\hat{v}}{n}} \tag{2.84}$$

where z_β denotes the β-quantile of N(0, 1).

Table 2.5 Confidence intervals for flour data

Estimator	$\widehat{\mu}$	$\sqrt{\widehat{v}(\widehat{\mu}/n)}$	Interval	
Mean	4.280	1.081	2.161	6.400
Bisquare M	3.144	0.130	2.885	3.404
$\overline{x}_{0.25}$	3.269	0.117	3.039	3.499

A similar procedure can be used for the trimmed mean. Recall that the asymptotic variance of the α-trimmed mean for symmetric F is as shown in (2.41). We can estimate v with

$$\widehat{v} = \frac{1}{n-2m}\left(\sum_{i=m+1}^{n-m}(x_{(i)}-\widehat{\mu})^2 + m(x_{(m)}-\widehat{\mu})^2 + m(x_{(n-m+1)}-\widehat{\mu})^2\right). \tag{2.85}$$

An approximate t-statistic is then defined as (2.83). Note again that the variance of the trimmed mean is not a trimmed variance, but rather a "Winsorized" variance.

Table 2.5 gives for the data of Example 1.1 the location estimators, their estimated asymptotic SDs and the respective confidence intervals with level 0.95. The results were obtained with script **flour**.

2.9.2 Tests

It appears that many applied statisticians have the impression that t -tests are sufficiently "robust", and that they should have no worries when using them. Again, this impression no doubt comes from the fact – a consequence of the central limit theorem – that it suffices for the data to have finite variance for the classical t -statistic (2.81) to be approximately N(0, 1) in large samples. See for example the discussion to this effect in the introductory text by Box *et al.* (1978). This means that in large samples the Type 1 error rate of a level α is in fact α for testing a null hypothesis about the value of μ. However, this fact is misleading, as we now demonstrate.

Recall that the t-test with level α for the null hypothesis $H_0 = \{\mu = \mu_0\}$ rejects H_0 when the t-interval with confidence level $1-\alpha$ does not contain μ_0. According to the discussion in Section 2.9.1 on the behavior of the t-intervals under contamination, we conclude that if the data are symmetric but heavy tailed, the intervals will be longer than necessary, with the consequence that the actual Type 1 error rate may be much smaller than α, but the Type 2 error rate may be too large; that is, the test will have low power. If the contaminated distribution is asymmetric and heavy tailed, both errors may become unacceptably high.

Robust tests can be derived from a "robust t-statistic" (2.83) in the same way as was done with confidence intervals. The tests of level α for the null hypothesis

$\mu = \mu_0$ against the two-sided alternative $\mu \neq \mu_0$ and the one-sided alternative $\mu > \mu_0$ have the rejection regions

$$|\hat{\mu} - \mu_0| > \sqrt{\hat{v}} z_{1-\alpha/2} \text{ and } \hat{\mu} > \mu_0 + \sqrt{\hat{v}} z_{1-\alpha}, \tag{2.86}$$

respectively.

The robust t-like confidence intervals and test are easy to apply. They have, however, some drawbacks when the contamination is asymmetric, because of the bias of the estimator. Procedures that ensure a given probability of coverage or Type 1 error probability for a contaminated parametric model were given by Huber (1965, 1968), Huber-Carol (1970), Rieder (1978, 1981) and Fraiman *et al.* (2001). Yohai and Zamar (2004) developed tests and confidence intervals for the median that are "nonparametric", in the sense that their level is valid for arbitrary distributions. Further references on robust tests will be given in Section 4.7.

2.10 Appendix: proofs and complements

2.10.1 Mixtures

Let the density f be given by

$$f = (1 - \varepsilon)g + \varepsilon h. \tag{2.87}$$

This is called a *mixture* of g and h. If the variable x has density f, and q is any function, then

$$\mathrm{E}q(x) = \int_{-\infty}^{\infty} q(x)f(x)dx = (1 - \varepsilon)\int_{-\infty}^{\infty} q(x)g(x)dx + \varepsilon \int_{-\infty}^{\infty} q(x)h(x)dx.$$

With this expression we can calculate $\mathrm{E}x$; the variance is obtained from

$$\mathrm{Var}(x) = \mathrm{E}(x^2) - (\mathrm{E}x)^2.$$

If $g = \mathrm{N}(0, 1)$ and $h = \mathrm{N}(a, b^2)$ then

$$\mathrm{E}x = \varepsilon a \text{ and } \mathrm{E}x^2 = (1 - \varepsilon) + \varepsilon(a^2 + b^2),$$

and hence

$$\mathrm{Var}(x) = (1 - \varepsilon)(1 + \varepsilon a^2) + \varepsilon b^2. \tag{2.88}$$

Evaluating the performance of robust estimators requires simulating distributions of the form (2.87). This is easily accomplished: generate u with uniform distribution in (0, 1); if $u \geq \varepsilon$, generate x with distribution g, else generate x with distribution h.

2.10.2 Asymptotic normality of M-estimators

In this section we give a heuristic proof of (2.24). To this end we begin with an intuitive proof of (2.23). Define the functions

$$\lambda(s) = \mathrm{E}\psi(x - s), \ \hat{\lambda}_n(s) = \frac{1}{n}\sum_{i=1}^{n} \psi(x_i - s),$$

so that $\hat{\mu}$ and μ_0 verify respectively

$$\hat{\lambda}_n(\hat{\mu}) = 0, \ \lambda(\mu_0) = 0.$$

For each s, the random variables $\psi(x_i - s)$ are i.i.d. with mean $\lambda(s)$, and hence the law of large numbers implies that when $n \to \infty$

$$\hat{\lambda}_n(s) \to_p \lambda(s) \, \forall s.$$

It is intuitive that also the solution of $\hat{\lambda}_n(s) = 0$ should tend to that of $\lambda(s) = 0$. This can in fact be proved rigorously (see Theorem 10.5).

Now we prove (2.24). Taking the Taylor expansion of order 1 of (2.19) as a function of $\hat{\mu}$ about μ_0 yields

$$0 = \sum_{i=1}^{n} \psi(x_i - \mu_0) - (\hat{\mu} - \mu_0) \sum_{i=1}^{n} \psi'(x_i - \mu_0) + o(\hat{\mu} - \mu_0) \tag{2.89}$$

where the last ("second-order") term is such that

$$\lim_{t \to 0} \frac{o(t)}{t} = 0.$$

Dropping the last term in (2.89) yields

$$\sqrt{n}(\hat{\mu} - \mu_0) \approx \frac{A_n}{B_n}, \tag{2.90}$$

with

$$A_n = \sqrt{n}\ \mathrm{ave}(\psi(\mathbf{x} - \mu_0)), \ B_n = \mathrm{ave}(\psi'(\mathbf{x} - \mu_0)).$$

The random variables $\psi(x_i - \mu_0)$ are i.i.d. with mean 0 because of (2.22). The central limit theorem implies that the distribution of A_n tends to $\mathrm{N}(0, a)$ with $a = \mathrm{E}\psi(x - \mu_0)^2$, and the law of large numbers implies that B_n tends in probability to $b = \mathrm{E}\psi'(x - \mu_0)$. Hence by Slutsky's lemma (see Section 2.10.3) A_n/B_n can be replaced for large n by A_n/b, which tends in distribution to $\mathrm{N}(0, a/b^2)$, as stated. A rigorous proof will be given in Theorem 10.7.

Note that we have shown that $\sqrt{n}(\hat{\mu} - \mu_0)$ converges in distribution; this is expressed by saying that "$\hat{\mu}$ has order $n^{-1/2}$ consistency".

2.10.3 Slutsky's lemma

Let u_n and v_n be two sequences of random variables such that u_n tends in probability to a constant u, and the distribution of v_n tends to the distribution of a variable v (abbreviated "$v_n \rightarrow_d v$"). Then

$$u_n + v_n \rightarrow_d u + v \text{ and } u_n v_n \rightarrow_d uv.$$

The proof can be found in Bickel and Doksum (2001, p. 467) or Shao (2003, p. 60).

2.10.4 Quantiles

For $\alpha \in (0, 1)$ and F a continuous and increasing distribution function, the α-quantile of F is the unique value $q(\alpha)$ such that $F(q(\alpha)) = \alpha$. If F is discontinuous, such a value might not exist. For this reason we define $q(\alpha)$ in general as a value where $F(t) - \alpha$ changes sign; that is,

$$\operatorname{sgn}\left\{\lim_{t\uparrow q(\alpha)} (F(t) - \alpha)\right\} \neq \operatorname{sgn}\left\{\lim_{t\downarrow q(\alpha)} (F(t) - \alpha)\right\},$$

where "↑" and "↓" denote the limits from the left and from the right, respectively. It is easy to show that such a value always exists. It is unique if F is increasing. Otherwise, it is not necessarily unique, and hence we may speak of *an* α-quantile.

If x is a random variable with distribution function $F(t) = \mathrm{P}(x \le t)$, $q(\alpha)$ will also be considered as an α -quantile of the variable x, and in this case is denoted by x_α.

If g is a monotonic function, and $y = g(x)$, then

$$g(x_\alpha) = \begin{cases} y_\alpha & \text{if } g \text{ is increasing} \\ y_{1-\alpha} & \text{if } g \text{ is decreasing,} \end{cases} \tag{2.91}$$

in the sense that, for example, if z is *an* α-quantile of x, then z^3 is *an* α-quantile of x^3.

When the α-quantile is not unique, there exists an interval $[a, b)$ such that $F(t) = \alpha$ for $t \in [a, b)$. We may obtain uniqueness by defining $q(\alpha)$ as a – the smallest α-quantile – and then (2.91) remains valid. It seems more symmetric to define it as the midpoint $(a + b)/2$, but then (2.91) ceases to hold.

2.10.5 Alternative algorithms for M-estimators

2.10.5.1 The Newton–Raphson procedure

The Newton–Raphson procedure is a widely used iterative method for the solution of nonlinear equations. To solve the equation $h(t) = 0$, at each iteration, h is "linearized"; that is, replaced by its Taylor expansion of order 1 about the current approximation. Thus, if at iteration m we have the approximation t_m, then the next value t_{m+1} is the solution of

$$h(t_m) + h'(t_m)(t_{m+1} - t_m) = 0.$$

In other words,

$$t_{m+1} = t_m - \frac{h(t_m)}{h'(t_m)}. \tag{2.92}$$

If the procedure converges, the convergence is very fast; but it is not guaranteed to converge. If h' is not bounded away from zero, the denominator in (2.92) may become very small, making the sequence t_m unstable unless the initial value t_0 is very near to the solution.

This happens in the case of a location M-estimator, where we must solve the equation $h(\mu) = 0$ with $h(\mu) = \text{ave}\{\psi(\mathbf{x} - \mu)\}$. Here the iterations are

$$\mu_{m+1} = \mu_m + \frac{\sum_{i=1}^{n} \psi(x_i - \mu_m)}{\sum_{i=1}^{n} \psi'(x_i - \mu_m)}. \tag{2.93}$$

If ψ is bounded, its derivative ψ' tends to zero at infinity, and hence the denominator is not bounded away from zero, which makes the procedure unreliable. For this reason, algorithms based on iterative reweighting are preferable, since these are guaranteed to converge.

However, although the result of iterating the Newton–Raphson process indefinitely may be unreliable, the result of a *single* iteration may be a robust and efficient estimator, if the initial value μ_0 is robust but not necessarily efficient, like the median; see Problem 3.16.

2.10.5.2 Iterative pseudo-observations

The expression (2.34) of an M-estimator as a function of the pseudo-observations (2.35) can be used as the basis for an iterative procedure to compute a location estimator with previous dispersion $\widehat{\sigma}$. Starting with an initial $\widehat{\mu}_0$, define

$$\widehat{\mu}_{m+1} = \frac{1}{n}\sum_{i=1}^{n} \zeta(x_i, \widehat{\mu}_m, \widehat{\sigma}), \tag{2.94}$$

where

$$\zeta(x, \mu, \sigma) = \mu + \sigma\psi\left(\frac{x - \mu}{\sigma}\right). \tag{2.95}$$

It can be shown that μ_m converges under very general conditions to the solution of (2.67) (Huber and Ronchetti, 2009). However, the convergence is much slower than that corresponding to the reweighting procedure.

2.11 Recommendations and software

For location we recommend the bisquare M-estimator with MAD scale, and the confidence intervals defined in (2.84). The function **locScaleM** (library `RobStatTM`) computes the bisquare and Huber estimators, their estimated standard deviations needed for the intervals, and the M-dispersion estimator defined in Section 2.6.

The function **scaleM** (`RobStatTM`) computes the bisquare M-scale defined in Section 2.5.

2.12 Problems

2.1. Show that in a sample of size n from a contaminated distribution (2.6), the number of observations from H is random, with binomial distribution $\mathrm{Bi}(n, \varepsilon)$.

2.2. For the data of Example 1.2, compute the mean and median, the 25% trimmed mean and the M-estimator with previous dispersion and Huber's ψ with $k = 1.37$. Use the latter to derive a 90% confidence interval for the true value.

2.3. Verify (2.11) using (2.27).

2.4. For what values of v does the the Student distribution have moments of order k?

2.5. Show that if μ is a solution of (2.19), then $\mu + c$ is a solution of (2.19) with $x_i + c$ instead of x_i.

2.6. Show that if $x = \mu_0 + u$ where the distribution of u is symmetric about 0, then μ_0 is a solution of (2.22).

2.7. Verify (2.30) [hint: use $\varphi'(x) = -x\varphi(x)$ and integration by parts]. From this, find the values of k which yield variances $1/\alpha$ with $\alpha = 0.90, 0.95$ and 0.99 (by using an equation solver, or just trial and error).

2.8. Compute the α-trimmed means with $\alpha = 0.10$ and 0.25 for the data of Example 1.2

2.9. Show that if ψ is odd, then the M-estimator $\hat{\mu}$ satisfies conditions C1-C2-C3 at the end of Section 2.4.

2.10. Show using (2.46) that L-estimators are shift and scale equivariant [recall that the order statistics of $y_i = -x_i$ are $y_{(i)} = -x_{(n-i+1)}$!] and also fulfill C1-C2-C3 of Section 2.4.

2.11. If $x \sim \mathrm{N}(\mu, \sigma^2)$, calculate MD($x$), MAD($x$) and IQR($x$).

2.12. Show that if $\psi = \rho'$ vanishes identically outside an interval, there is no density verifying (2.14).

2.13. Define the sample α-quantile of $x_1, \ldots, x_n$ – with $\alpha \in (1/n, 1/1/n)$ – as $x_{(k)}$, where k is the smallest integer $\geq n\alpha$ and $x_{(i)}$ are the order statistics (1.2). Let

$$\psi(x) = \alpha \mathrm{I}(x > 0) - (1 - \alpha)\mathrm{I}(x < 0).$$

Show that $\mu = x_{(k)}$ is a solution (not necessarily unique) of (2.19). Use this fact to derive the asymptotic distribution of sample quantiles, assuming that $\mathcal{D}(x_i)$ has a unique α-quantile. Note that this ψ is not odd.

2.14. Show that the M-scale (2.49) with $\rho(t) = \mathrm{I}(|t| > 1)$ is the hth order statistic of the $|x_i|$ with $h = n - [n\delta]$.

2.15. Verify (2.37), (2.51) and (2.56)

2.16. Let $[a, b]$ where a and b depend on the data, be the shortest interval containing at least half of the data.

(a) The *Shorth* ("shortest half") location estimator is defined as the midpoint $\hat{\mu} = (a + b)/2$. Show that $\hat{\mu} = \arg\min_{\mu} \mathrm{Med}(|\mathbf{x} - \mu|)$.
(b) Show that the difference $b - a$ is a dispersion estimator.
(c) For a distribution F, let $[a, b]$ be the shortest interval with probability 0.5. Find this interval for $F = \mathrm{N}(\mu, \sigma^2)$.

2.17. Let $\hat{\mu}$ be a location M-estimator. Show that if the distribution of the x_i is symmetric about μ, so is the distribution of $\hat{\mu}$, and that the same happens with trimmed means.

2.18. Verify numerically that the constant c at the end of Section 2.5 that makes the bisquare scale consistent for the normal is indeed equal to 1.56.

2.19. Show that

(a) if the sequence μ_m in (2.93) converges, then the limit is a solution of (2.19)
(b) if the sequence in (2.94) converges, then the limit is a solution of (2.67).

3

Measuring Robustness

In order to measure the effect of different locations of an outlier on an estimate, consider adding to a sample $\mathbf{x} = (x_1, \ldots, x_n)$ an extra data point x_0 that is allowed to range on the whole real line. We define the *sensitivity curve* of the estimator $\hat{\mu}$ for the sample $\mathbf{x}$ as the difference

$$\hat{\mu}(x_1, \ldots, x_n, x_0) - \hat{\mu}(x_1, \ldots, x_n) \tag{3.1}$$

as a function of the location x_0 of the outlier.

For purposes of plotting and comparing sensitivity curves across sample sizes, it is convenient to use *standardized* sensitivity curves, obtained by multiplying (3.1) by $(n + 1)$ (see also Section 3.1). To make our examples clearer, we use a "sample" formed from standard normal distribution quantiles, instead of a random one. Figure 3.1 plots:

- the standardized sensitivity curves of the median
- the 10% trimmed mean $\overline{x}_{0.10}$
- the 10% Winsorized mean $\widetilde{x}_{0.10}$
- the Huber M-estimator with $k = 1.37$, using both the SD and the MADN as previously computed dispersion estimators
- the bisquare M-estimator with $k = 4.68$ using the MADN as dispersion estimator.

Also included is an M-estimator with particular optimality properties, to be defined in Section 5.8.1.

We can see that all curves are bounded, except the one corresponding to the Huber estimator with SD as dispersion estimator, which grows without bound with x_0. The same unbounded behavior (not shown in the figure) occurs with the

Robust Statistics: Theory and Methods (with R), Second Edition.
Ricardo A. Maronna, R. Douglas Martin, Victor J. Yohai and Matías Salibián-Barrera.

Companion website: www.wiley.com/go/maronna/robust

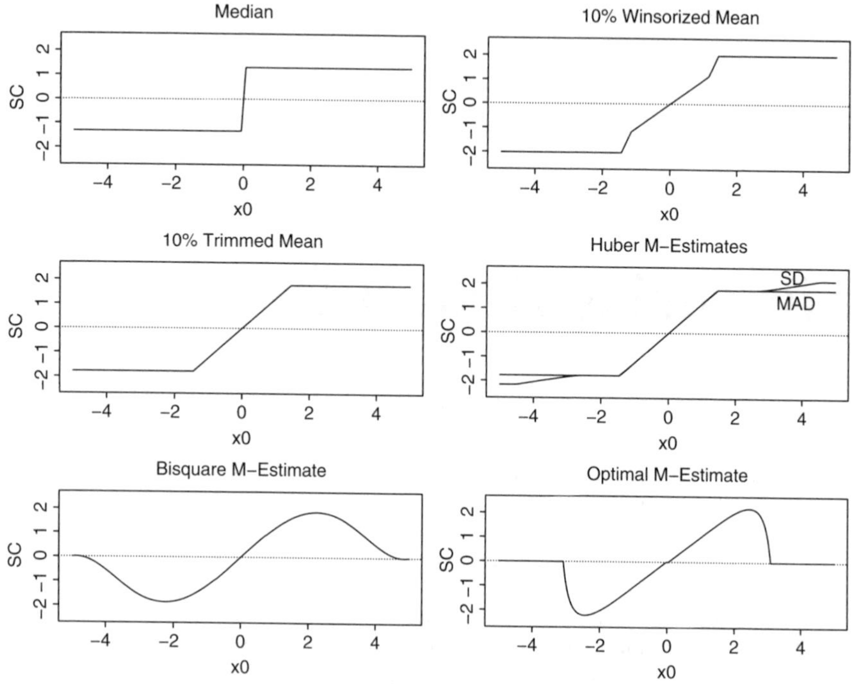

Figure 3.1 Sensitivity curves of location estimators

bisquare estimator with the SD as dispersion estimator. This shows the importance of a robust previous dispersion. All curves are nondecreasing for positive x_0, except the one for the bisquare and "optimal" M-estimators. Roughly speaking, we say that the bisquare M-estimator *rejects* extreme values, while the others do not. The curve for the trimmed mean shows that it does not reject large observations, but just limits their influence. The curve for the median is very steep at the origin.

Figure 3.2 shows the sensitivity curves of the SD along with the normalized MD, MAD and IQR. The SD and MD have unbounded sensitivity curves, while those of the normalized MAD and IQR are bounded.

Imagine now that instead of adding a single point at a variable location, we replace m points by a fixed value $x_0 = 1000$. Table 3.1 shows the resulting "biases"

$$\widehat{\mu}(x_0, x_0, \ldots, x_0, x_{m+1}, .., x_n) - \widehat{\mu}(x_1, .., x_n)$$

as a function of m for the following location estimators:

- the median
- the Huber estimator with $k = 1.37$ and three different dispersions: previously estimated MAD (denoted by MADp), simultaneous MAD ("MADs") and previous SD

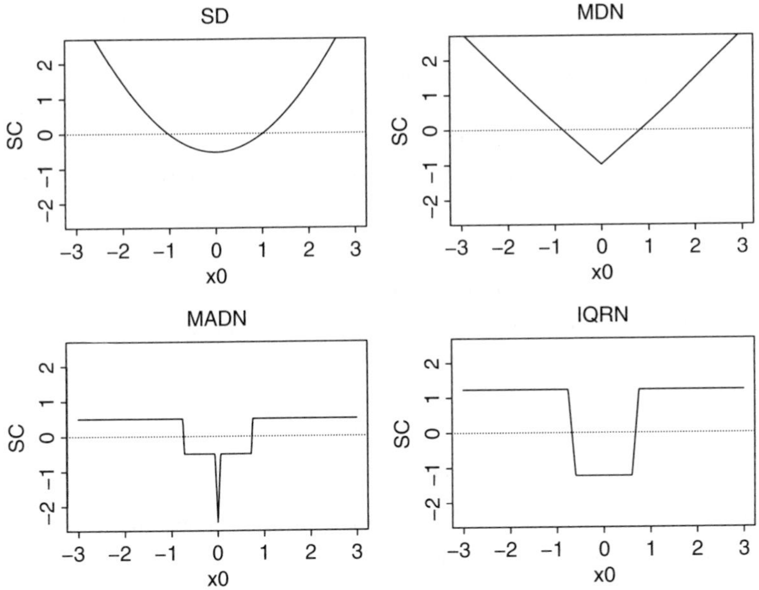

Figure 3.2 Sensitivity curves of dispersion estimators

Table 3.1 The effect of increasing contamination on a sample of size 20

m	Mean	Median	H(MADp)	H(MADs)	H(SD)	$\overline{x}_\alpha$	M-Bisq	MAD	IQR
1	50	0.00	0.03	0.04	16.06	0.04	−0.02	0.12	0.08
2	100	0.01	0.10	0.11	46.78	55.59	0.04	0.22	0.14
4	200	0.21	0.36	0.37	140.5	166.7	0.10	0.46	0.41
5	250	0.34	0.62	0.95	202.9	222.3	0.15	0.56	370.3
7	350	0.48	1.43	42.66	350.0	333.4	0.21	1.29	740.3
9	450	0.76	3.23	450.0	450.0	444.5	0.40	2.16	740.2
10	500	500.5	500.0	500.0	500.0	500.0	500.0	739.3	740.2

- the trimmed mean with $\alpha = 0.085$
- the bisquare estimator.

We also provide the biases for the normalized MAD and IQR dispersion estimators. The choice of k and α was made so that both the Huber estimators and the trimmed mean have the same asymptotic variance for the normal distribution.

The mean deteriorates immediately when $m = 1$, as expected, and since $[\alpha n] = [0.085 \times 20] = 1$ the trimmed mean $\overline{x}_\alpha$ deteriorates when $m = 2$, as could be expected. The H(MADs) deteriorates rapidly, starting at $m = 8$, while H(SD) is already quite bad at $m = 1$. By contrast, the median, H(MADp) and M-Bisq do so only when $m = n/2$, with M-Bisq having smaller bias than H(MADp), and the median (Med) having small biases, comparable to those of the M-Bisq (only slightly higher bias than M-Bisq at $m = 4, 5, 7, 9$).

To formalize these notions, it will be easier to study the behavior of estimators when the sample size tends to infinity ("asymptotic behavior"). Consider an estimator $\widehat{\theta}_n = \widehat{\theta}_n(x)$ depending on a sample $\mathbf{x} = \{x_1, \ldots, x_n\}$ of size n of i.i.d. variables with distribution F. In all cases of practical interest, there is a value depending on F, $\widehat{\theta}_\infty = \widehat{\theta}_\infty(F)$, such that

$$\widehat{\theta}_n \to_p \widehat{\theta}_\infty(F).$$

$\widehat{\theta}_\infty(F)$ is the *asymptotic value* of the estimator at F.

If $\widehat{\theta}_n = \overline{x}$ (the sample mean) then $\widehat{\theta}_\infty(F) = \mathrm{E}_F x$ (the distribution mean), and if $\widehat{\theta}_n(\mathbf{x}) = \mathrm{Med}(\mathbf{x})$ (the sample median) then $\widehat{\theta}_\infty(F) = F^{-1}(0.5)$ (the distribution median). If $\widehat{\theta}_n$ is a location M-estimator given by (2.19) with ψ monotonic, it was stated in Section 2.10.2 that $\widehat{\theta}_\infty(F)$ is the solution of

$$\mathrm{E}_F \psi(x - \theta) = 0.$$

A proof is given in Theorem 10.5. The same reasoning shows that if $\widehat{\theta}_n$ is a scale M-estimator (2.49), then $\widehat{\theta}_\infty(F)$ is the solution of

$$\mathrm{E}_F \rho\left(\frac{x}{\theta}\right) = \delta.$$

It can also be shown that if $\widehat{\theta}_n$ is a location M-estimator given by (2.13), then $\widehat{\theta}_\infty(F)$ is the solution of

$$\mathrm{E}_F \rho(x - \theta) = \min.$$

Details can be found in Huber and Ronchetti (2009; Sec. 6.2). Asymptotic values also exist for the trimmed mean (Section 10.7).

The typical distribution of data depends on one or more unknown parameters. Thus in the location model (2.2) the data have distribution function $F_\mu(x) = F_0(x - \mu)$, and in the location–dispersion model (2.68) the distribution is $F_\theta(x) = F_0((x - \mu)/\sigma)$ with $\theta = (\mu, \sigma)$. These are called *parametric models*. In the location model we have seen in (2.23) that if the data are symmetric about μ and $\widehat{\mu}$ is an M-estimator, then $\widehat{\mu} \to_p \mu$ and so $\widehat{\mu}_\infty(F_\mu) = \mu$. An estimator $\widehat{\theta}$ of the parameter(s) of a parametric family F_θ will be called *consistent* if

$$\widehat{\theta}_\infty(F_\theta) = \theta. \tag{3.2}$$

Since we assume F to be only approximately known, we are interested in the behavior of $\widehat{\theta}_\infty(F)$ when F ranges over a "neighborhood" of a distribution F_0. There are several ways to characterize neighborhoods. The easiest to deal with are *contamination neighborhoods*:

$$\mathcal{F}(F, \varepsilon) = \{(1 - \varepsilon)F + \varepsilon G : G \in \mathcal{G}\} \tag{3.3}$$

where $\mathcal{G}$ is a suitable set of distributions, often the set of all distributions but in some cases the set of point-mass distributions, where the "point mass" δ_{x_0} is the distribution such that $\mathrm{P}(x = x_0) = 1$.

3.1 The influence function

The *influence function* (IF) of an estimator (Hampel, 1974) is an asymptotic version of its sensitivity curve. It is an approximation to the behavior of $\widehat{\theta}_\infty$ when the sample contains a small fraction ε of identical outliers. It is defined as

$$\text{IF}_{\widehat{\theta}}(x_0, F) = \lim_{\varepsilon \downarrow 0} \frac{\widehat{\theta}_\infty((1-\varepsilon)F + \varepsilon\delta_{x_0}) - \widehat{\theta}_\infty(F)}{\varepsilon} \tag{3.4}$$

$$= \frac{\partial}{\partial \varepsilon} \widehat{\theta}_\infty((1-\varepsilon)F + \varepsilon\delta_0)\Big|_{\varepsilon \downarrow 0}, \tag{3.5}$$

where δ_{x_0} is the point-mass at x_0 and "↓" stands for "limit from the right". If there are p unknown parameters, then $\widehat{\theta}_\infty$ is a p-dimensional vector and so is its IF. Henceforth, the argument of $\widehat{\theta}_\infty(F)$ will be dropped if there is no ambiguity.

The quantity $\widehat{\theta}_\infty((1-\varepsilon)F + \varepsilon\delta_{x_0})$ is the asymptotic value of the estimator when the underlying distribution is F and a fraction ε of outliers is equal to x_0. Thus if ε is small, this value can be approximated by

$$\widehat{\theta}_\infty((1-\varepsilon)F + \varepsilon\delta_{x_0}) \approx \widehat{\theta}_\infty(F) + \varepsilon \text{IF}_{\widehat{\theta}}(x_0, F)$$

and the *bias* $\widehat{\theta}_\infty((1-\varepsilon)F + \varepsilon\delta_{x_0}) - \widehat{\theta}_\infty(F)$ is approximated by $\varepsilon\text{IF}_{\widehat{\theta}}(x_0, F)$.

The IF may be considered as a "limit version" of the sensitivity curve, in the following sense. When we add the new observation x_0 to the sample $x_1, \ldots, x_n$ the fraction of contamination is $1/(n+1)$, and so we define the *standardized sensitivity curve* (SC) as

$$\text{SC}_n(x_0) = \frac{\widehat{\theta}_{n+1}(x_1, \ldots, x_n, x_0) - \widehat{\theta}_n(x_1, \ldots, x_n)}{1/(n+1)},$$

$$= (n+1)\left(\widehat{\theta}_{n+1}(x_1, \ldots, x_n, x_0) - \widehat{\theta}_n(x_1, \ldots, x_n)\right)$$

which is similar to (3.4) with $\varepsilon = 1/(n+1)$. One would expect that if the x_i are i.i.d. with distribution F, then $\text{SC}_n(x_0) \approx \text{IF}(x_0, F)$ for large n. This notion can be made precise. Note that for each x_0, $\text{SC}_n(x_0)$ is a random variable. Croux (1998) has shown that if $\widehat{\theta}$ is a location M-estimator with a bounded and continuous ψ-function, or is a trimmed mean, then for each x_0

$$\text{SC}_n(x_0) \to_{a.s.} \text{IF}_{\widehat{\theta}}(x_0, F), \tag{3.6}$$

where "a.s." denotes convergence with probability 1 ("almost sure" convergence). This result is extended to general M-estimators in Section 10.4. See, however, the remarks in Section 3.1.1.

It will be shown in Section 3.8.1 that for a location M-estimator $\widehat{\mu}$

$$\text{IF}_{\widehat{\mu}}(x_0, F) = \frac{\psi(x_0 - \widehat{\mu}_\infty)}{\text{E}\psi'(x - \widehat{\mu}_\infty)} \tag{3.7}$$

and for a scale M-estimator $\widehat{\sigma}$ (Section 2.5)

$$\mathrm{IF}_{\widehat{\sigma}}(x_0, F) = \widehat{\sigma}_\infty \frac{\rho(x_0/\widehat{\sigma}_\infty) - \delta}{\mathrm{E}(x/\widehat{\sigma}_\infty)\rho'(x/\widehat{\sigma}_\infty)}. \tag{3.8}$$

For the median estimator, the denominator is to be interpreted as in (2.27). The similarity between the IF and the SC of a given estimator can be seen by comparing Figure 3.1 to Figures 2.3 and 2.5. The same thing happens with Figure 3.2.

We see above that the IF of an M-estimator is proportional to its ψ-function (or an offset ρ-function in the case of the scale estimator), and this behavior holds in general for M-estimators. Given a parametric model F_θ, a *general M-estimator* $\widehat{\theta}$ is defined as a solution of

$$\sum_{i=1}^{n} \Psi(x_i, \widehat{\theta}) = 0. \tag{3.9}$$

For location, $\Psi(x, \theta) = \psi(x - \theta)$, and for scale, $\Psi(x, \theta) = \rho(x/\theta) - \delta$. It is shown in Section 10.2 that the asymptotic value $\widehat{\theta}_\infty$ of the estimator at F satisfies

$$\mathrm{E}_F \Psi(x, \widehat{\theta}_\infty) = 0. \tag{3.10}$$

It is shown in Section 3.8.1 that the IF of a general M-estimator is

$$\mathrm{IF}_{\widehat{\theta}}(x_0, F) = -\frac{\Psi(x_0, \widehat{\theta}_\infty)}{B(\widehat{\theta}_\infty, \Psi)} \tag{3.11}$$

where

$$B(\theta, \Psi) = \frac{\partial}{\partial \theta} \mathrm{E}\Psi(x, \theta), \tag{3.12}$$

and thus the IF is proportional to the ψ-function $\Psi(x_0, \widehat{\theta}_\infty)$.

If Ψ is differentiable with respect to θ, and the conditions that allow the interchange of derivative and expectation hold, then

$$B(\theta, \Psi) = \mathrm{E}\dot{\Psi}(x, \theta) \tag{3.13}$$

where

$$\dot{\Psi}(x, \theta) = \frac{\partial \Psi(x, \theta)}{\partial \theta}. \tag{3.14}$$

The proof is given in Section 3.8.1. Then, if $\widehat{\theta}$ is consistent for the parametric family F_θ, (3.11) becomes

$$\mathrm{IF}_{\widehat{\theta}}(x_0, F_\theta) = -\frac{\Psi(x_0, \theta)}{\mathrm{E}_F \dot{\Psi}(x, \theta)}.$$

Consider now an M-estimator $\widehat{\mu}$ of location, with known dispersion σ, where the asymptotic value $\widehat{\mu}_\infty$ satisfies

$$\mathrm{E}_F \psi\left(\frac{x - \widehat{\mu}_\infty}{\sigma}\right) = 0.$$

It is easy to show, by applying (3.7) to the estimator defined by the function $\psi^*(x) = \psi(x/\sigma)$, that the IF of $\widehat{\mu}$ is

$$\mathrm{IF}_{\widehat{\mu}}(x_0, F) = \sigma \frac{\psi((x_0 - \widehat{\mu}_\infty)/\sigma)}{\mathrm{E}_F \psi'((x - \widehat{\mu}_\infty)/\sigma)}. \tag{3.15}$$

Now consider location estimation with a previously computed dispersion estimator $\widehat{\sigma}$ as in (2.66). In this case, the IF is much more complicated than the one above, and depends on the IF of $\widehat{\sigma}$. But it can be proved that if F is symmetric, the IF simplifies to (3.15):

$$\mathrm{IF}_{\widehat{\mu}}(x_0, F) = \widehat{\sigma}_\infty \frac{\psi((x_0 - \widehat{\mu}_\infty)/\widehat{\sigma}_\infty)}{\mathrm{E}_F \psi'((x - \widehat{\mu}_\infty)/\widehat{\sigma}_\infty)}. \tag{3.16}$$

The IF for simultaneous estimation of μ and σ is more complicated, but can be derived from (3.48) in Section 3.6.

It can be shown that the IF of an α-trimmed mean $\widehat{\mu}$ at a symmetric F is proportional to Huber's ψ-function:

$$\mathrm{IF}_{\widehat{\mu}}(x_0, F) = \frac{\psi_k(x - \widehat{\mu}_\infty)}{1 - 2\alpha} \tag{3.17}$$

with $k = F^{-1}(1 - \alpha)$. Hence the trimmed mean and the Huber estimator in the example at the beginning of the chapter not only have the same asymptotic variances, but also the same IF. However, Table 3.1 shows that they have very different degrees of robustness.

Comparing (3.7) to (2.24) and (3.17) to (2.41), one sees that the asymptotic variance v of these M-estimators satisfies

$$v = \mathrm{E}_F \mathrm{IF}(x, F)^2. \tag{3.18}$$

It is shown in Section 3.7 that (3.18) holds for a general class of estimators called Fréchet-differentiable estimators, which includes M-estimators with bounded Ψ. However, the relationship (3.18) does not hold in general. For instance, the Shorth location estimator (the midpoint of the shortest half of the data; see Problem 2.16a) has a null IF (Problem 3.12). At the same time, its rate of consistency is $n^{-1/3}$ rather than the usual rate $n^{-1/2}$. Hence the left-hand side of (3.18) is infinite and the right-hand is zero.

3.1.1 *The convergence of the SC to the IF

The plot in the upper left panel of Figure 3.1 for the Huber estimator using the SD as the previously computed dispersion estimator seems to contradict the convergence of $\mathrm{SC}_n(x_0)$ to $\mathrm{IF}(x_0)$. Note, however, that (3.6) asserts only the convergence for *each* x_0. This means that $\mathrm{SC}_n(x_0)$ will be near $\mathrm{IF}(x_0)$ for a given x_0 when n is sufficiently large, but the value of n will in general depend on x_0; in other words, the convergence will not be *uniform*. Rather than convergence at an isolated point, what matters is being able to compare the influence of outliers at different locations; that is, the behavior of the *whole curve* corresponding to the SC. Both curves will be similar along their whole range only if the convergence is uniform. This does not happen with H(SD).

On the other hand Croux (1998) has shown that when $\widehat{\theta}$ is the median, the distribution of $\mathrm{SC}_n(x_0)$ does not converge in probability to any value, and hence (3.6) does not hold. This would seem to contradict the upper right panel of Figure 3.1. However, the *form* of the curve converges to the correct limit in the sense that for each x_0

$$\frac{\mathrm{SC}_n(x_0)}{\max_x |\mathrm{SC}_n(x)|} \to_{a.s.} \frac{\mathrm{IF}(x_0)}{\max_x |\mathrm{IF}(x)|} = \mathrm{sgn}(x - \mathrm{Med}(x)). \tag{3.19}$$

The proof is left to the reader (Problem 3.2).

3.2 The breakdown point

Table 3.1 showed the effect of replacing several data values by outliers. Roughly speaking, the breakdown point (BP) of an estimator $\widehat{\theta}$ of the parameter θ is the largest amount of contamination (proportion of atypical points) that the data may contain such that $\widehat{\theta}$ still gives some information about θ, that is, about the distribution of the "typical" points.

Let θ range over a set Θ. In order for the estimator $\widehat{\theta}$ to give some information about θ, the contamination should not be able to drive $\widehat{\theta}$ to infinity or to the boundary of Θ when it is not empty. For example, for a scale or dispersion parameter, we have $\Theta = [0, \infty]$, and the estimator should remain bounded, and also bounded away from 0, in the sense that the distance between $\widehat{\theta}$ and 0 should be larger than some positive value.

Definition 3.1 *The asymptotic contamination BP of the estimator $\widehat{\theta}$ at F, denoted by $\varepsilon^*(\widehat{\theta}, F)$, is the largest $\varepsilon^* \in (0, 1)$ such that for $\varepsilon < \varepsilon^*$, $\widehat{\theta}_\infty((1-\varepsilon)F + \varepsilon G)$ remains bounded away from the boundary of Θ for all G.*

The definition means that there exists a bounded and closed set $K \subset \Theta$ such that $K \cap \partial\Theta = \emptyset$ (where $\partial\Theta$ denotes the boundary of Θ) such that

$$\widehat{\theta}_\infty((1-\varepsilon)F + \varepsilon G) \in K \forall \varepsilon < \varepsilon^* \text{and } \forall G. \tag{3.20}$$

It is helpful to extend the definition to the case when the estimator is not uniquely defined, for example when it is the solution of an equation that may have multiple roots. In this case, the boundedness of the estimator means that *all* solutions remain in a bounded set.

The BP for each type of estimator has to be treated separately. Note that it is easy to find estimators with high BP. For instance, the "estimator" identically equal to zero has $\varepsilon^* = 1$! However, for "reasonable" estimators it is intuitively clear that there must be more "typical" than "atypical" points and so $\varepsilon^* \leq 1/2$. Actually, it can be proved (Section 3.8.2) that all shift equivariant location estimators as defined in (2.4) have $\varepsilon^* \leq 1/2$.

3.2.1 Location M-estimators

It will be convenient first to consider the case of monotonic but not necessarily odd ψ. Assume that

$$k_1 = -\psi(-\infty),\ k_2 = \psi(\infty)$$

are finite. Then it is shown in Section 3.8.3 that

$$\varepsilon^* = \frac{\min(k_1, k_2)}{k_1 + k_2}. \tag{3.21}$$

It follows that if ψ is odd, then $k_1 = k_2$ and the bound $\varepsilon^* = 0.5$ is attained. Define

$$\varepsilon_j^* = \frac{k_j}{k_1 + k_2}\ (j = 1, 2). \tag{3.22}$$

Then, (3.21) is equivalent to

$$\varepsilon^* = \min(\varepsilon_1^*, \varepsilon_2^*).$$

The proof of (3.21) shows that ε_1^* and ε_2^* are respectively the BPs to $+\infty$ and to $-\infty$. It can be shown that redescending estimators also attain the bound $\varepsilon^* = 0.5$, but the proof is more involved since one has to deal, not with (2.19), but with the minimization (2.13).

3.2.2 Scale and dispersion estimators

We deal first with scale estimators. Note that while a high proportion of atypical points with large values (outliers) may cause the estimator $\widehat{\sigma}$ to overestimate the true scale, a high proportion of data near zero ("inliers") may result in underestimation of the true scale. Thus it is desirable that the estimator remains bounded away from zero ("implosion") as well as away from infinity ("explosion"). This is equivalent to keeping the *logarithm* of $\widehat{\sigma}$ bounded.

Note that a scale M-estimator with ρ-function ρ may be written as a location M-estimator "in the log scale". Put

$$y = \log|x|,\ \mu = \log\sigma,\ \psi(t) = \rho(e^t) - \delta.$$

Since ρ is even and $\rho(0) = 0$, then

$$\rho\left(\frac{x}{\sigma}\right) - \delta = \rho\left(\frac{|x|}{\sigma}\right) - \delta = \psi(y - \mu),$$

and hence $\widehat{\sigma} = \exp(\widehat{\mu})$, where $\widehat{\mu}$ verifies $\text{ave}(\psi(\mathbf{y} - \widehat{\mu})) = 0$, and hence $\widehat{\mu}$ is a location M-estimator.

If ρ is bounded, we have $\rho(\infty) = 1$ by Definition 2.1. Then the BP $\varepsilon^*$of $\hat{\sigma}$ is given by (3.21) with

$$k_1 = \delta,\ k_2 = 1 - \delta,$$

and so

$$\varepsilon^* = \min(\delta, 1 - \delta). \tag{3.23}$$

Since $\mu \to +\infty$ and $\mu \to -\infty$ are equivalent to $\sigma \to \infty$ and $\sigma \to 0$ respectively, it follows from (3.22) that δ and $1 - \delta$ are, respectively, the BPs for explosion and for implosion.

As for dispersion estimators, it is easy to show that the BPs of the SD, the MAD and the IQR are 0, 1/2 and 1/4, respectively (Problem 3.3). In general, the BP of an equivariant dispersion estimator is ≤ 0.5 (Problem 3.5).

3.2.3 Location with previously-computed dispersion estimator

In Table 3.1 we saw the bad consequences of using an M-estimator $\hat{\mu}$ with the SD as the previously computed dispersion estimator $\hat{\sigma}$. The reason is that the outliers inflate this dispersion estimator, and hence outliers do not appear as such in the "standardized" residuals $(x_i - \hat{\mu})/\hat{\sigma}$. Hence the robustness of $\hat{\sigma}$ is essential for that of $\hat{\mu}$.

For monotone M-estimators with bounded and odd ψ, it can be shown that $\varepsilon^*(\hat{\mu}) = \varepsilon^*(\hat{\sigma})$. Thus, if $\hat{\sigma}$ is the MAD then $\varepsilon^*(\hat{\mu}) = 0.5$, but if $\hat{\sigma}$ is the SD then $\varepsilon^*(\hat{\mu}) = 0$.

Note that (3.16) implies that the location estimators using the SD and the MAD as previous dispersion have the same IF, while at the same time they have quite different BPs. Note that this is an example of an estimator with a bounded IF but a zero BP.

For redescending M-estimators (2.66) with bounded ρ, the situation is more complex. Consider first the case of a fixed σ. It can be shown that $\varepsilon^*(\hat{\mu})$ can be made arbitrarily small by taking σ small enough. This suggests that for the case of an estimator σ, it is not only the BP of $\hat{\sigma}$ that matters but also the size of $\hat{\sigma}$. Let $\hat{\mu}_0$ be an initial estimator with BP = 0.5 (say the median), and let $\hat{\sigma}$ be an M-scale centered at $\hat{\mu}_0$, as defined by

$$\frac{1}{n}\sum_{i=1}^{n} \rho_0\left(\frac{x_i - \hat{\mu}_0}{\hat{\sigma}}\right) = 0.5$$

where ρ_0 is another bounded ρ-function. If $\rho \leq \rho_0$, then $\varepsilon^*(\hat{\mu}) = 0.5$ (a proof is given in Section 3.8.3).

Since the MAD has $\rho_0(x) = \mathrm{I}(x \geq 1)$, it does not fulfill $\rho \leq \rho_0$. In this case the situation is more complicated and the BP will in general depend on the distribution (or on the data in the case of the finite-sample BP introduced below). Huber (1984) calculated the BP for this situation, and it follows from his results that for the bisquare ρ with MAD scale, the BP is 1/2 for all practical purposes. Details are given in Section 3.8.3.2.

3.2.4 Simultaneous estimation

The BP for the estimators in Section 2.7.2 is much more complicated, requiring the solution of a nonlinear system of equations (Huber and Ronchetti, 2009, p. 141). In general, the BP of $\widehat{\mu}$ is less than 0.5. In particular, using Huber's ψ_k with $\widehat{\sigma}$ given by (2.76) yields

$$\varepsilon^* = \min\left(0.5, \frac{0.675}{k+0.675}\right),$$

so that with $k = 1.37$ we have $\varepsilon^* = 0.33$. This is clearly lower than the BP = 0.5, which corresponds to using a previously computed dispersion estimator treated above.

3.2.5 Finite-sample breakdown point

Although the asymptotic BP is an important theoretical concept, it may be more useful to define the notion of BP for a finite sample. Let $\widehat{\theta}_n = \widehat{\theta}_n(x)$ be an estimator defined for samples $\mathbf{x} = \{x_1, \ldots, x_n\}$. The *replacement finite-sample breakdown point* (FBP) of $\widehat{\theta}_n$ at x is the largest proportion $\varepsilon_n^*(\widehat{\theta}_n, \mathbf{x})$ of data points that can be arbitrarily replaced by outliers without $\widehat{\theta}_n$ leaving a set which is bounded, and also bounded away from the boundary of Θ (Donoho and Huber, 1983). More formally, call $\mathcal{X}_m$ the set of all datasets $\mathbf{y}$ of size n having $n - m$ elements in common with $\mathbf{x}$:

$$\mathcal{X}_m = \{\mathbf{y} : \#(\mathbf{y}) = n,\ \#(\mathbf{x} \cap \mathbf{y}) = n - m\}.$$

Then

$$\varepsilon_n^*(\widehat{\theta}_n, \mathbf{x}) = \frac{m^*}{n}, \tag{3.24}$$

where

$$m^* = \max\left\{m \geq 0 : \widehat{\theta}_n(\mathbf{y}) \text{ bounded and also bounded away from } \partial\Theta\ \forall\ \mathbf{y} \in \mathcal{X}_m\right\}. \tag{3.25}$$

In most cases of interest, ε_n^* does not depend on $\mathbf{x}$, and tends to the asymptotic BP when $n \to \infty$. For equivariant location estimators, it is proved in Section 3.8.2 that

$$\varepsilon_n^* \leq \frac{1}{n}\left[\frac{n-1}{2}\right] \tag{3.26}$$

and that this bound is attained by M-estimators with odd and bounded ψ. For the trimmed mean, it is easy to verify that $m^* = [n\alpha]$, so that $\varepsilon_n^* \approx \alpha$ for large n.

Another possibility is the *addition* FBP. Call $\mathcal{X}_m$ the set of all datasets of size $n + m$ containing $\mathbf{x}$:

$$\mathcal{X}_m = \{\mathbf{y} : \#(\mathbf{y}) = n + m,\ \mathbf{x} \subset \mathbf{y}\}.$$

Then

$$\varepsilon_n^{**}(\widehat{\theta}_n, \mathbf{x}) = \frac{m^*}{n+m},$$

where

$$m^* = \max \left\{ m \geq 0 : \widehat{\theta}_{n+m}(\mathbf{y}) \text{ bounded and also bounded away from } \partial\, \Theta\, \forall\, y \in \mathcal{X}_m \right\}.$$

Both ε^* and ε^{**} give similar values for large n, but we prefer the former. The main reason for this is that the definition involves only the estimator for the given n, which makes it easier to generalize this concept to more complex cases, as will be seen in Section 4.6.

3.3 Maximum asymptotic bias

The IF and the BP consider extreme situations of contamination. The first deals with "infinitesimal" values of ε, while the second deals with the largest ε an estimator can tolerate. Note that an estimator having a high BP means that $\widehat{\theta}_\infty(F)$ will remain in a bounded set when F ranges in an ε-neighborhood (3.3) with $\varepsilon \leq \varepsilon^*$, but this set may be very large. What we want to do now is, roughly speaking, to measure the worst behavior of the estimator for each given $\varepsilon < \varepsilon^*$.

We again consider F ranging in the ε-neighborhood

$$\mathcal{F}_{\varepsilon,\theta} = \{(1-\varepsilon)F_\theta + \varepsilon G : G \in \mathcal{G}\}$$

of an assumed parametric distribution F_θ, where $\mathcal{G}$ is a family of distribution functions. Unless otherwise specified, $\mathcal{G}$ will be the family of all distribution functions, but in some cases it will be more convenient to choose a more restricted family such as that of point-mass distributions. The *asymptotic bias* of $\widehat{\theta}$ at any $F \in \mathcal{F}_{\varepsilon,\theta}$ is

$$\mathrm{b}_{\widehat{\theta}}(F,\theta) = \widehat{\theta}_\infty(F) - \theta$$

and the maximum bias (MB) is

$$\mathrm{MB}_{\widehat{\theta}}\,(\varepsilon,\theta) = \max\{|\mathrm{b}_{\widehat{\theta}}(F,\theta)| : F \in \mathcal{F}_{\varepsilon,\theta}\}.$$

In the case that the parameter space is the whole set of real numbers, the relationship between MB and BP is

$$\varepsilon^*(\widehat{\theta},F_\theta) = \sup\{\varepsilon \geq 0 : \mathrm{MB}_{\widehat{\theta}}\,(\varepsilon,\theta) < \infty\}.$$

Note that two estimators may have the same BP but different MBs (Problem 3.11).

The *contamination sensitivity* of $\widehat{\theta}$ at θ is defined as

$$\gamma_c(\widehat{\theta},\theta) = \left[\frac{d}{d\varepsilon}\mathrm{MB}_{\widehat{\theta}}\,(\varepsilon,\theta)\right]_{\varepsilon=0}. \tag{3.27}$$

In the case that $\widehat{\theta}$ is consistent, we have $\widehat{\theta}_\infty(F_\theta) = \theta$ and then $\mathrm{MB}_{\widehat{\theta}}(0,\theta) = b_{\widehat{\theta}}(F_\theta, 0) = 0$. Therefore γ_c gives an approximation to the MB for small ε:

$$\mathrm{MB}_{\widehat{\theta}}(\varepsilon,\theta) \approx \varepsilon\gamma_c(\widehat{\theta},\theta). \tag{3.28}$$

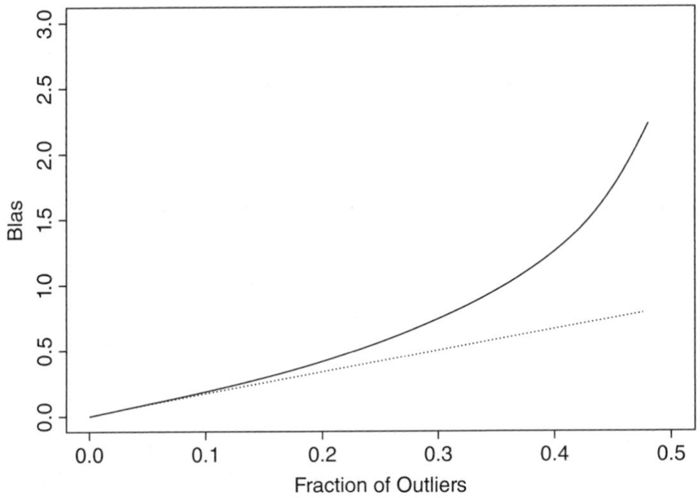

Figure 3.3 Maximum bias of Huber estimator (–) and its linear aproximation (.....) as a function of ε

Note, however, that since $\mathrm{MB}_{\hat{\theta}}(\varepsilon^*, \theta) = \infty$, while the right-hand side of (3.28) always yields a finite result, this approximation will be quite unreliable for sufficiently large values of ε. Figure 3.3 shows $\mathrm{MB}_{\hat{\theta}}(\varepsilon, \theta)$ at $F_\theta = \mathrm{N}(\theta, 1)$ and its approximation (3.28) for the Huber location estimator with $k = 1.37$ (note that the bias does not depend on θ due to the estimator's shift equivariance).

The *gross-error sensitivity* (GES) of $\widehat{\theta}$ at θ is

$$\gamma^*(\widehat{\theta}, \theta) = \max_{x_0} |\mathrm{IF}_{\hat{\theta}}(x_0, F_\theta)|. \tag{3.29}$$

Since $(1-\varepsilon)F_\theta + \varepsilon\delta_{x_0} \in \mathcal{F}_{\varepsilon,\theta}$, we have for all x_0

$$|\widehat{\theta}_\infty((1-\varepsilon)F_\theta + \varepsilon\delta_{x_0}) - \widehat{\theta}_\infty(F_\theta)| \leq \mathrm{MB}_{\hat{\theta}}(\varepsilon, \theta).$$

So dividing by ε and taking the limit we get

$$\gamma^* \leq \gamma_c. \tag{3.30}$$

Equality above holds for M-estimators with bounded ψ-functions, but not in general. For instance, we have seen in Section 3.2.3 that the IF of the Huber estimator with the SD as previous dispersion is bounded, but since $\varepsilon^* = 0$ we have $\mathrm{MB}_{\hat{\theta}}(\varepsilon, \theta) = \infty$ for all $\varepsilon > 0$ and so the right-hand side of (3.30) is infinite.

For location M-estimators $\widehat{\mu}$ with odd ψ and $k = \psi(\infty)$, and assuming a location model $F_\mu(x) = F_0(x - \mu)$, we have

$$\gamma^*(\widehat{\mu}, \mu) = \frac{k}{\mathrm{E}_{F_\mu}\psi'(x - \widehat{\mu}_\infty)} = \frac{k}{\mathrm{E}_{F_0}\psi'(x)} \tag{3.31}$$

so that $\gamma^*(\widehat{\mu}, \mu)$ does not depend on μ.

In general for equivariant estimators, $\mathrm{MB}_{\hat{\theta}}(\varepsilon,\theta)$ does not depend on θ. In particular, the MB for a bounded location M-estimator is as given in Section 3.8.4, where it is shown that the median minimizes the MB for M-estimators in symmetric models.

3.4 Balancing robustness and efficiency

In this section we consider a parametric model F_θ and an estimator $\hat{\theta}$, which is consistent for θ and such that the distribution of $\sqrt{n}(\hat{\theta}_n - \theta)$ under F_θ tends to a normal distribution with mean 0 and variance $v = v(\hat{\theta},\theta)$. This is the most frequent case and contains most of the situations considered in this book.

Under the preceding assumptions, $\hat{\theta}$ has no asymptotic bias and we care only about its variability. Let $v_{\min} = v_{\min}(\theta)$ be the smallest possible asymptotic variance within a "reasonable" class of estimators (for example, equivariant). Under reasonable regularity conditions, $v_{\min}$ is the asymptotic variance of the MLE for the model (Section 10.8). Then the *asymptotic efficiency* of $\hat{\theta}$ at θ is defined as $v_{\min}(\theta)/v(\hat{\theta},\theta)$.

If instead F does not belong to the family F_θ but is in a neighborhood of F_θ, the squared bias will dominate the variance component of MSE for all sufficiently large n. To see this, let $b = \hat{\theta}_\infty(F) - \theta$ and note that in general under F the distribution of $\sqrt{n}(\hat{\theta}_n - \hat{\theta}_\infty)$ tends to normal, with mean 0 and variance v. Then the distribution of $\hat{\theta}_n - \theta$ is approximately $\mathrm{N}(b, v/n)$, so that the variance tends to zero while the bias does not. Thus we must balance the efficiency of $\hat{\theta}$ at the model F_θ with the bias in a neighborhood of it.

We have seen that location M-estimators with a bounded ψ and previously computed dispersion estimator with BP = 1/2 attain the maximum BP of 1/2. To choose among them we must compare their biases for a given efficiency. We consider the Huber and bisquare estimators with previously computed MAD dispersion and efficiency 0.95. Their maximum biases for the model $F_{\varepsilon,\theta} = \{(1-\varepsilon)F_\theta + \varepsilon G : G \in \mathcal{G}\}$ with $F_\theta = \mathrm{N}(0,1)$ and a few values of ε are shown in Table 3.2.

Figure 3.4 shows the respective biases for point contamination at K with $\varepsilon = 0.1$, as a function of the outlier location K. It is seen that although the maximum bias of the bisquare is higher, the difference is very small and its bias remains below that

Table 3.2 Maximum contamination biases of Huber and bisquare location estimators

ε	0.05	0.10	0.20
Huber	0.087	0.184	0.419
Bisquare	0.093	0.197	0.450

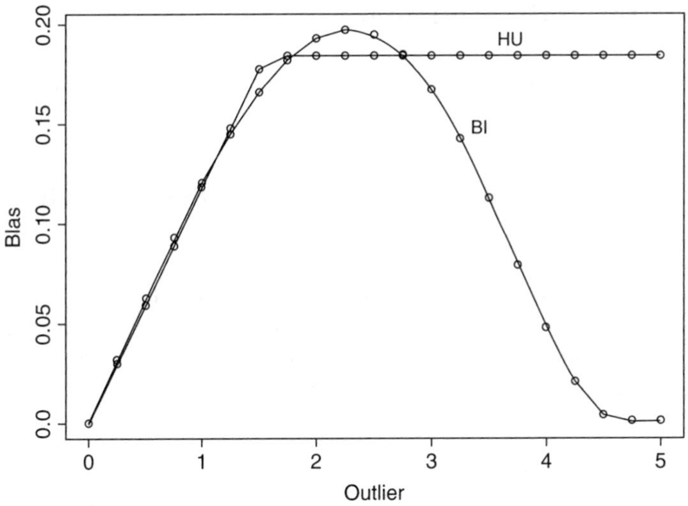

Figure 3.4 Asymptotic biases of Huber and bisquare estimators for 10% contamination as functions of the outlier location K

Table 3.3 Asymptotic efficiencies of three location M-estimators

	Huber	Bisq.	CMLE
Normal	0.95	0.95	0.60
Cauchy	0.57	0.72	1.00

of the Huber estimator for the majority of the values. This shows that, although the maximum bias contains much more information than the BP, it is not informative enough to discriminate among estimators and that one should look at the whole bias behavior when possible.

To study the behavior of the estimators under symmetric heavy-tailed distributions, we computed the asymptotic variances of the Huber and bisquare estimators, and of the Cauchy MLE ("CMLE"), with simultaneous dispersion (Section 2.7.2) for the normal and Cauchy distributions, the latter of which can be considered an extreme case of heavy-tailed behavior. The efficiencies are given in Table 3.3. It is seen that the bisquare estimator yields the best trade-off between the efficiencies for the two distributions.

For all the above reasons we recommend when estimating location the bisquare M-estimator with previously computed MAD.

3.5 *"Optimal" robustness

In this section we consider different ways in which an "optimal" estimator may be defined.

3.5.1 Bias- and variance-optimality of location estimators

3.5.1.1 Minimax bias

If we pay attention only to bias, the quest for an "optimal" location estimator is simple: Huber (1964) has shown that the median has the smallest maximum bias ("minimax bias') among *all* shift equivariant estimators if the underlying distribution is symmetric and unimodal. See Section 3.8.5 for a proof.

3.5.1.2 Minimax variance

Huber (1964) studied location M-estimators in neighborhoods (3.3) of a symmetric F with symmetric contamination (so that there is no bias problem). The dispersion is assumed known. Call $v(\hat{\theta}, H)$ the asymptotic variance of the estimator $\hat{\theta}$ at the distribution H, and

$$v_\varepsilon(\hat{\theta}) = \sup_{H \in \mathcal{F}(F,\varepsilon)} v(\hat{\theta}, H),$$

where $\mathcal{F}(F, \varepsilon)$ is the neighborhood (3.3) with G ranging over all *symmetric* distributions. Assume that F has a density f and that $\psi_0 = -f'/f$ is nondecreasing. Then the M-estimator minimizing $v_\varepsilon(\hat{\theta})$ has

$$\psi(x) = \begin{cases} \psi_0(x) & \text{if} \quad |\psi_0(x)| \le k \\ k \ \mathrm{sgn(x)} & \text{else} \end{cases}$$

where k depends on F and ε. For normal F, this is the Huber ψ_k. Since ψ_0 corresponds to the MLE for f, the result may be described as a truncated MLE.

The same problem with unknown dispersion was considered by Li and Zamar (1991).

3.5.2 Bias optimality of scale and dispersion estimators

The problem of minimax bias scale estimation for positive random variables was considered by Martin and Zamar (1989), who showed that for the case of a nominal exponential distribution the scaled median Med($\mathbf{x}$)/0.693 – as we will see in Problem 3.15, this estimator also minimizes the GES – was an excellent approximation to the minimax bias optimal estimator for a wide range of $\varepsilon < 0.5$. Minimax bias dispersion estimators were treated by Martin and Zamar (1993b) for the case of a nominal normal distribution and two separate families of estimators:

- For simultaneous estimation of location and scale/dispersion with the monotone ψ-function, the minimax bias estimator is well approximated by the MAD for all $\varepsilon < 0.5$, thereby providing a theoretical rationale for an otherwise well-known high-BP estimator.
- For M-estimators of scale with a general location estimator that includes location M-estimators with redescending ψ-functions, the minimax bias estimator is well approximated by the Shorth dispersion estimator (the shortest half of the data, see Problem 2.16b) for a wide range of $\varepsilon < 0.5$. This is an intuitively appealing estimator with BP = 1/2.

3.5.3 The infinitesimal approach

Several criteria have been proposed to define an optimal balance between bias and variance. The treatment can be simplified if ε is assumed to be "very small". Then the maximum bias can be approximated through the gross-error sensitivity (GES) (3.29). We first treat the simpler problem of minimizing the GES. Let F_θ be a parametric family with densities or frequency functions $f_\theta(x)$. Call E_θ the expectation with respect to F_θ; that is, if the random variable $z \sim F_\theta$ and h is any function,

$$\mathrm{E}_\theta h(z) = \begin{cases} \int h(x) f_\theta(x)dx & (z \text{ continuous}) \\ \sum_x h(x) f_\theta(x) & (z \text{ discrete}). \end{cases}$$

We shall deal with general M-estimators $\widehat{\theta}_n$ defined by (3.9), where Ψ is usually called the *score function*. An M-estimator is called *Fisher-consistent* for the family F_θ if:

$$\mathrm{E}_\theta \Psi(x, \theta) = 0. \tag{3.32}$$

In view of (3.10), a Fisher-consistent M-estimator is consistent in the sense of (3.2).

It is shown in Section 10.3 that if $\widehat{\theta}_n$ is Fisher-consistent, then

$$n^{1/2}(\widehat{\theta}_n - \theta) \to_d \mathrm{N}(0, v(\Psi, \theta)),$$

with

$$v(\Psi, \theta) = \frac{A(\theta, \Psi)}{B(\theta, \Psi)^2},$$

where B is defined in (3.12) and

$$A(\theta, \Psi) = \mathrm{E}_\theta(\Psi(x, \theta)^2). \tag{3.33}$$

It follows from (3.11) that the GES of an M-estimator is

$$\gamma^*(\widehat{\theta}, \theta) = \frac{\max_x |\Psi(x, \theta)|}{|B(\theta, \Psi)|}.$$

The MLE is the M-estimator with score function

$$\Psi_0(x,\theta) = -\frac{\dot{f}_\theta(x)}{f_\theta(x)}, \quad \text{with } \dot{f}_\theta(x) = \frac{\partial f_\theta(x)}{\partial \theta}. \tag{3.34}$$

It is shown in Section 10.8 that this estimator is Fisher-consistent; that is,

$$\mathrm{E}_\theta \Psi_0(x,\theta) = 0, \tag{3.35}$$

and has the minimum asymptotic variance among Fisher-consistent M-estimators.

We now consider the problem of minimizing γ^* among M-estimators. To ensure that the estimates consider the correct parameter, we consider only Fisher-consistent estimators.

Call Med_θ the median under F_θ; that is, if $z \sim F_\theta$ and h is any function, then $\mathrm{Med}_\theta(h(z))$ is the value t where

$$\int \mathrm{I}\{h(x) \leq t\} f_\theta(x) dx - 0.5$$

changes sign.

Define

$$M(\theta) = \mathrm{Med}_\theta \Psi_0(x,\theta).$$

It is shown in Section 3.8.6 that the M-estimator $\widetilde{\theta}$ with score function

$$\widetilde{\Psi}(x,\theta) = \mathrm{sgn}(\Psi_0(x,\theta) - M(\theta)) \tag{3.36}$$

is Fisher-consistent and is the M-estimator with smallest γ^* in that class.

This estimator has a clear intuitive interpretation. Recall that the median is a location M-estimator with ψ-function equal to the sign function. Likewise, $\widetilde{\theta}$ is the solution θ of

$$\mathrm{Med}\{\Psi_0(x_1,\theta), \ldots, \Psi_0(x_n,\theta)\} = \mathrm{Med}_\theta \Psi_0(x,\theta). \tag{3.37}$$

Note that, in view of (3.35), the MLE may be written as the solution of

$$\frac{1}{n}\sum_{i=1}^{n} \Psi_0(x_i,\theta) = \mathrm{E}_\theta \Psi_0(x,\theta). \tag{3.38}$$

Hence (3.37) can be seen as a version of (3.38), in which the average on the left-hand side is replaced by the sample median, and the expectation on the right is replaced by the distribution median.

3.5.4 The Hampel approach

Hampel (1974) stated the balance problem between bias and efficiency for general estimators as minimizing the asymptotic variance under a bound on the GES. For a symmetric location model, his result coincides with Huber's. It is remarkable that both approaches coincide at the location problem, and furthermore the result has a high BP.

To simplify the notation, in this section we will write $\gamma^*(\Psi, \theta)$ for the GES $\gamma^*(\hat{\theta}, \theta)$ of an M-estimator $\hat{\theta}$ with score function Ψ. Hampel proposed to choose an M-estimator combining efficiency and robustness by finding Ψ such that, subject to (3.32),

$$v(\Psi, \theta) = \min \text{ with } \gamma^*(\Psi, \theta) \leq G(\theta), \tag{3.39}$$

where $G(\theta)$ is a given bound expressing the desired degree of robustness. It is clear that a higher robustness means a lower $G(\theta)$, but that this implies a higher $v(\Psi, \theta)$. We call this optimization problem Hampel's *direct* problem.

We can also consider a *dual* Hampel problem, in which we look for a function Ψ such that

$$\gamma^*(\Psi, \theta) = \min \text{ with } v(\Psi, \theta) \leq V(\theta), \tag{3.40}$$

with given V. It is easy to see that both problems are equivalent in the following sense: if Ψ^* is optimal for the direct Hampel problem, then it is also optimal for the dual problem with $V(\theta) = v(\Psi^*, \theta)$. Similarly if Ψ^* is optimal for the dual problem, it is also optimal for the direct problem with $G(\theta) = \gamma^*(\Psi^*, \theta)$.

The solution to the direct and dual problems was given by Hampel (1974). The optimal score functions for both problems are of the following form:

$$\Psi^*(x, \theta) = \psi_{k(\theta)}(\Psi_0(x, \theta) - r(\theta)) \tag{3.41}$$

where Ψ_0 is given by (3.34), ψ_k is Huber's ψ-function (2.29), and $r(\theta)$ and $k(\theta)$ are chosen so that that Ψ^* satisfies (3.32). A proof is given in Section 3.8.7. It is seen that the optimal score function is obtained from Ψ_0 by first centering through r and then bounding its absolute value by k. Note that (3.36) is the limit case of (3.41) when $k \to 0$. Note also that for a solution to exist, $G(\theta)$ must be larger than the minimum GES $\gamma^*(\widetilde{\Psi}, \theta)$, and $V(\theta)$ must be larger than the asymptotic variance of the MLE: $v(\Psi_0, \theta)$.

It is not clear which one may be a practical rule for the choice of $G(\theta)$ for the direct Hampel problem. But for the second problem a reasonable criterion is to choose $V(\theta)$ as

$$V(\theta) = \frac{v(\Psi_0, \theta)}{1 - \alpha}, \tag{3.42}$$

where $1 - \alpha$ is the desired asymptotic efficiency of the estimator with respect to the MLE.

Finding k for a given V or G may be complicated. The problem simplifies considerably when F_θ is a location or a scale family, for in these cases the MLE is location (or scale) equivariant. We shall henceforth deal with bounds (3.42). We shall see that k may be chosen as a constant, which can then be found numerically.

For the location model we know from (2.19) that

$$\Psi_0(x, \xi) = \xi_0(x - \theta) \text{ with } \xi_0(x) = -\frac{f_0'(x)}{f_0(x)}, \tag{3.43}$$

Hence $v(\Psi_0, \theta)$ does not depend on θ, and

$$\Psi^*(x, \theta) = \psi_k(\xi_0(x - \theta) - r(\theta)). \tag{3.44}$$

If $k(\theta)$ is constant, then the $r(\theta)$ that fulfills (3.32) is constant too, which implies that $\Psi^*(x, \theta)$ depends only on $x - \theta$, and hence the estimator is location equivariant. This implies that $v(\Psi^*, \theta)$ does not depend on θ either, and depends only on k, which can be found numerically to attain equality in (3.40).

In particular, if f_0 is symmetric, it is easy to show that $r = 0$. When $f_0 = \mathrm{N}(0, 1)$ we obtain the Huber score function.

For a scale model it follows from (2.48) that

$$\Psi_0(x, \xi) = \frac{x}{\theta}\xi_0\left(\frac{x}{\theta}\right) - 1,$$

with ξ_0 as in (3.43). It follows that $v(\Psi_0, \theta)$ is proportional to θ, and that Ψ^* has the form

$$\Psi^*(x, \theta) = \psi_k\left(\frac{x}{\theta}\xi_0\left(\frac{x}{\theta}\right) - r(\theta)\right). \tag{3.45}$$

If k is constant, then the $r(\theta)$ that fulfills (3.32) is proportional to θ, which implies that $\Psi^*(x, \theta)$ depends only on x/θ, and hence the estimator is scale equivariant. This implies that $v(\Psi^*, \theta)$ is also proportional to θ^2, and hence k, which can be found numerically to attain equality in (3.40).

The case of the exponential family is left for the reader in Problem 3.15. Extensions of this approach when there is more than one parameter may be found in Hampel *et al.* (1986).

3.5.5 Balancing bias and variance: the general problem

More realistic results are obtained by working with a positive (not "infinitesimal") ε. Martin and Zamar (1993a) found the location estimator minimizing the asymptotic variance under a given bound on the maximum asymptotic bias for a given $\varepsilon > 0$. Fraiman *et al.* (2001) derived the location estimators minimizing the MSE of a given function of the parameters in an ε-contamination neighborhood. This allowed them to derive "optimal" confidence intervals that retain the asymptotic coverage probability in a neighborhood.

3.6 Multidimensional parameters

We now consider the estimation of p parameters $\theta_1, \ldots, \theta_p$ (e.g., location and dispersion), represented by the vector $\boldsymbol{\theta} = (\theta_1, \ldots, \theta_p)'$. Let $\widehat{\boldsymbol{\theta}}_n$ be an estimator with asymptotic value $\widehat{\boldsymbol{\theta}}_\infty$. Then the asymptotic bias is defined as

$$\mathrm{b}_{\widehat{\theta}}(F, \boldsymbol{\theta}) = \mathrm{disc}(\widehat{\boldsymbol{\theta}}_\infty(F), \boldsymbol{\theta}),$$

where disc(**a**, **b**) is a measure of the discrepancy between the vectors **a** and **b**, which depends on the particular situation. In many cases one may take the Euclidean distance $\| \boldsymbol{a} - \boldsymbol{b} \|$, but in other cases it may be more complex (as in Section 6.7).

We now consider the efficiency. Assume $\widehat{\boldsymbol{\theta}}_n$ is asymptotically normal with covariance matrix $\mathbf{V}$. Let $\widetilde{\boldsymbol{\theta}}_n$ be the MLE, with asymptotic covariance matrix $\mathbf{V}_0$. For $\mathbf{c} \in R^p$ the asymptotic variances of linear combinations $\mathbf{c}'\widehat{\boldsymbol{\theta}}_n$ and $\mathbf{c}'\widetilde{\boldsymbol{\theta}}_n$ are respectively $\mathbf{c}'\mathbf{V}\mathbf{c}$ and $\mathbf{c}'\mathbf{V}_0\mathbf{c}$, and their ratio would yield an efficiency measure for each $\mathbf{c}$. To express them though a single number, we take the worst situation, and define the asymptotic efficiency of $\widehat{\boldsymbol{\theta}}_n$ as

$$\text{eff}(\widehat{\boldsymbol{\theta}}_n) = \min_{\mathbf{c} \neq \mathbf{0}} \frac{\mathbf{c}'\mathbf{V}_0\mathbf{c}}{\mathbf{c}'\mathbf{V}\mathbf{c}}.$$

It is easy to show that

$$\text{eff}(\widehat{\boldsymbol{\theta}}_n) = \lambda_1(\mathbf{V}^{-1}\mathbf{V}_0), \tag{3.46}$$

where $\lambda_1(\mathbf{M})$ denotes the smallest eigenvalue of the matrix $\mathbf{M}$.

In many situations (as in Section 4.4) $\mathbf{V} = a\mathbf{V}_0$, where a is a constant, and then the efficiency is simply $1/a$.

Consider now simultaneous M-estimators of location and dispersion (Section 2.7.2). Here we have two parameters, μ and σ, which satisfy a system of two equations. Put $\boldsymbol{\theta} = (\mu, \sigma)$, and

$$\Psi_1(x, \boldsymbol{\theta}) = \psi\left(\frac{x - \mu}{\sigma}\right) \text{ and } \Psi_2(x, \boldsymbol{\theta}) = \rho_{\text{scale}}\left(\frac{x - \mu}{\sigma}\right) - \delta.$$

Then the estimators satisfy

$$\sum_{i=1}^{n} \boldsymbol{\Psi}(x_i, \widehat{\boldsymbol{\theta}}) = \mathbf{0}, \tag{3.47}$$

with $\boldsymbol{\Psi} = (\Psi_1, \Psi_2)$. Given a parametric model $F_{\boldsymbol{\theta}}$, where $\boldsymbol{\theta}$ is a multidimensional parameter of dimension p, a general M-estimator is defined by (3.46), where $\boldsymbol{\Psi} = (\Psi_1, \ldots, \Psi_p)$.

Then (3.11) can be generalized by showing that the IF of $\widehat{\boldsymbol{\theta}}$ is

$$\text{IF}_{\widehat{\theta}}(x_0, F) = -\mathbf{B}^{-1}\boldsymbol{\Psi}(x_0, \widehat{\boldsymbol{\theta}}_\infty), \tag{3.48}$$

where the matrix $\mathbf{B}$ has elements

$$B_{jk} = \text{E}\left\{ \left. \frac{\partial \Psi_j(x, \boldsymbol{\theta})}{\partial \theta_k} \right|_{\boldsymbol{\theta} = \boldsymbol{\theta}_F} \right\}.$$

M-estimators of multidimensional parameters are further considered in Section 10.5. It can be shown that they are asymptotically normal with asymptotic covariance matrix

$$\mathbf{V} = \mathbf{B}^{-1}(\text{E}\boldsymbol{\Psi}(x_0, \boldsymbol{\theta})\boldsymbol{\Psi}(x_0, \boldsymbol{\theta})')\mathbf{B}^{-1\prime}, \tag{3.49}$$

This therefore verifies the analogue of (3.18):

$$\mathbf{V} = \mathrm{E}\{\mathrm{IF}(x, F)\mathrm{IF}(x, F)'\}. \tag{3.50}$$

The results in this section hold also when the observations x are multidimensional.

3.7 *Estimators as functionals

The mean value may be considered as a "function" that attributes to each distribution F its expectation (when it exists); and the sample mean may be considered as a function attributing to each sample $\{x_1, \ldots, x_n\}$ its average $\overline{x}$. The same can be said of the median. This correspondence between distribution and sample values can be made systematic in the following way. Define the *empirical distribution function* of a sample $\mathbf{x} = \{x_1, \ldots, x_n\}$ as

$$\widehat{F}_{n,\mathbf{x}}(t) = \frac{1}{n}\sum_{i=1}^{n} \mathrm{I}(x_i \le t)$$

(the argument $\mathbf{x}$ will be dropped when there is no ambiguity). Then, for any continuous function g,

$$\mathrm{E}_{\widehat{F}_n} g(x) = \frac{1}{n}\sum_{i=1}^{n} g(x_i).$$

Define a "function" T whose argument is a distribution (a *functional*) as

$$T(F) = \mathrm{E}_F x = \int x dF(x).$$

It follows that $T(\widehat{F}_n) = \overline{x}$. If $\mathbf{x}$ is an i.i.d. sample from F, the law of large numbers implies that $T(\widehat{F}_n) \to_p T(F)$ when $n \to \infty$.

Likewise, define the functional $T(F)$ as the 0.5 quantile of F; if it is not unique, define $T(F)$ as the midpoint of 0.5 quantiles (see Section 2.10.4). Then $T(F) = \mathrm{Med}(x)$ for $x \sim F$, and $T(\widehat{F}_n) = \mathrm{Med}(x_1, \ldots, x_n)$. If $\mathbf{x}$ is a sample from F and $T(F)$ is unique, then $T(\widehat{F}_n) \to_p T(F)$.

More generally, M-estimators can be cast in this framework. For a given Ψ, define the functional $T(F)$ as the solution θ (assumed unique) of

$$\mathrm{E}_F \Psi(x, \theta) = 0. \tag{3.51}$$

Then $T(\widehat{F}_n)$ is a solution of

$$\mathrm{E}_{\widehat{F}_n} \Psi(x, \theta) = \frac{1}{n}\sum_{i=1}^{n} \Psi(x_i, \theta) = 0. \tag{3.52}$$

We see that $T(\widehat{F}_n)$ and $T(F)$ correspond to the M-estimator $\widehat{\theta}_n$ and to its asymptotic value $\widehat{\theta}_\infty(F)$, respectively.

A similar representation can be found for L-estimators. In particular, the α-trimmed mean corresponds to the functional

$$T(F) = \frac{1}{1-2\alpha}\mathrm{E}_F x\mathrm{I}(\alpha \leq F(x) \leq 1-\alpha).$$

Almost all of the estimators considered in this book can be represented as functionals; that is,

$$\widehat{\theta}_n = T(\widehat{F}_n) \tag{3.53}$$

for some functional T. The intuitive idea of robustness is that "modifying a small proportion of observations causes only a small change in the estimator". Thus robustness is related to some form of *continuity*. Hampel (1971) gave this intuitive concept a rigorous mathematical expression. The following is an informal exposition of these ideas; mathematical details and further references can be found in Huber and Ronchetti (2009; Ch. 3).

The concept of continuity requires the definition of a measure of distance $d(F,G)$ between distributions. Some distances (the *Lévy, bounded Lipschitz*, and *Prokhorov* metrics) are adequate to express the intuitive idea of robustness, in the sense that if the sample $\mathbf{y}$ is obtained from the sample $\mathbf{x}$ by

- arbitrarily modifying a small proportion of observations, and/or
- slightly modifying all observations,

then $d(\widehat{F}_{n,\mathbf{x}}, \widehat{F}_{n,\mathbf{y}})$ is "small". Hampel (1971) defined the concept of *qualitative robustness*. A simplified version of his definition is that an estimator corresponding to a functional T is said to be qualitatively robust at F if T is continuous at F according to the metric d; that is, for all ε there exists δ such that $d(F,G) < \delta$ implies $|T(F) - T(G)| < \varepsilon$.

It follows that robust estimators are consistent, in the sense that $T(\widehat{F}_n)$ converges in probability to $T(F)$. To see this, recall that if $\mathbf{x}$ is an i.i.d. sample from F, then the law of large numbers implies that $\widehat{F}_n(t) \to_p F(t)$ for all t. A much stronger result called the Glivenko–Cantelli theorem (Durrett 1996) states that $\widehat{F}_n \to F$ uniformly with probability 1; that is,

$$\mathrm{P}\left(\sup_t |\widehat{F}_n(t) - F(t)| \to 0\right) = 1.$$

It can be shown that this implies $d(\widehat{F}_n, F) \to_p 0$ for the Lévy metric. Moreover, if T is continuous then

$$\widehat{\theta}_\infty = T(F) = T\left(\mathrm{plim}_{n\to\infty}\widehat{F}_n\right) = \mathrm{plim}_{n\to\infty} T(\widehat{F}_n) = \mathrm{plim}_{n\to\infty}\widehat{\theta}_n,$$

where "plim" stands for "limit in probability".

A general definition of BP can be given in this framework. For a given metric, define an ε-neighborhood of F as

$$\mathcal{U}(\varepsilon, F) = \{G : d(F, G) < \varepsilon\},$$

and the maximum bias of T at F as

$$b_\varepsilon = \sup\{|T(G) - T(F)| : G \in \mathcal{U}(\varepsilon, F)\}.$$

For all the metrics considered, we have $d(F, G) < 1$ for all F, G; hence $\mathcal{U}(1, F)$ is the set of all distributions, and $b_1 = \sup\{|T(G) - T(F)| : \text{all } G\}$. Then the BP of T at F is defined as

$$\varepsilon^* = \sup\{\varepsilon : b_\varepsilon < b_1\}.$$

In this context, the IF may be viewed as a derivative. It will help to review some concepts from calculus. Let $h(\mathbf{z})$ be a function of m variables, with $\mathbf{z} = (z_1, ..., z_m) \in R^m$. Then h is *differentiable* at $\mathbf{z}_0$ if there exists a vector $\mathbf{d} = (d_1, .., d_m)$ such that for all $\mathbf{z}$

$$h(\mathbf{z}) - h(\mathbf{z}_0) = \sum_{j=1}^{m} d_j(z_j - z_{0j}) + o(\|\mathbf{z} - \mathbf{z}_0\|), \tag{3.54}$$

where "o" is a function such that $\lim_{t\to 0} o(t)/t = 0$. This means that in a neighborhood of $\mathbf{z}_0$, h can be approximated by a linear function. In fact, if $\mathbf{z}$ is near $\mathbf{z}_0$ we have

$$h(\mathbf{z}) \approx h(\mathbf{z}_0) + L(\mathbf{z} - \mathbf{z}_0),$$

where the linear function L is defined as $L(\mathbf{z}) = \mathbf{d}'\mathbf{z}$. The vector $\mathbf{d}$ is called the *derivative* of h at $\mathbf{z}_0$, which will be denoted by $\mathbf{d} = D(h, \mathbf{z}_0)$.

The *directional derivative* of h at $\mathbf{z}_0$ in the direction $\mathbf{a}$ is defined as

$$D(h, \mathbf{z}_0, \mathbf{a}) = \lim_{t\to 0} \frac{h(\mathbf{z}_0 + t\mathbf{a}) - h(\mathbf{z}_0)}{t}.$$

If h is differentiable, directional derivatives exist for all directions, and it can be shown that

$$D(h, \mathbf{z}_0, \mathbf{a}) = \mathbf{a}'D(h, \mathbf{z}_0).$$

The converse is not true: there are functions for which $D(h, \mathbf{z}_0, \mathbf{a})$ exists for all $\mathbf{a}$, but $D(h, \mathbf{z}_0)$ does not exist.

For an estimator $\widehat{\theta}$ represented as in (3.53), the IF may also be viewed as a directional derivative of T as follows. Since

$$(1 - \varepsilon)F + \varepsilon\delta_{x_0} = F + \varepsilon(\delta_{x_0} - F),$$

we have

$$\text{IF}_{\widehat{\theta}}(x_0, F) = \lim_{\varepsilon\to 0} \frac{1}{\varepsilon}\{T[F + \varepsilon(\delta_{x_0} - F)] - T(F)\},$$

which is the derivative of T in the direction $\delta_{x_0} - F$.

In some cases, the IF may be viewed as a derivative in the stronger sense of (3.54). This means that $T(H) - T(F)$ can be approximated by a linear function of H for *all* H in a neighborhood of F, and not just along each single direction. For a given $\widehat{\theta}$ represented by (3.53) and a given F, put for brevity

$$\xi(x) = \mathrm{IF}_{\widehat{\theta}}(x, F).$$

Then T is *Fréchet-differentiable* if for any distribution H

$$T(H) - T(F) = \mathrm{E}_H \xi(x) + o(d(F, H)). \tag{3.55}$$

The class of Frêchet differentiable estimators contains M-estimators with a bounded score function. Observe that the function

$$H \longrightarrow \mathrm{E}_H \xi(x) = \int \xi(x) dH(x)$$

is linear in H. Putting $H = F$ in (3.55) yields

$$\mathrm{E}_F \xi(x) = 0. \tag{3.56}$$

Some technical definitions are necessary at this point. A sequence z_n of random variables is said to be *bounded in probability* (abbreviated as $z_n = O_p(1)$) if for each ε there exists K such that $\mathrm{P}(|z_n| > K) < \varepsilon$ for all n; in particular, if $z_n \rightarrow_d z$ then $z_n = O_p(1)$. We say that $z_n = O_p(u_n)$ if $z_n/u_n = O_p(1)$, and that $z_n = o_p(u_n)$ if $z_n/u_n \rightarrow_p 0$.

It is known that the distribution of $\sup_t\{\sqrt{n}|\widehat{F}_n(t) - F(t)|\}$ (the so-called Kolmogorov–Smirnov statistic) tends to a distribution (see Feller, 1971), so that $\sup|\widehat{F}_n(t) - F(t)| = O_p(n^{-1/2})$. For the Lévy metric mentioned above, this fact implies that also $d(\widehat{F}_n, F) = O_p(n^{-1/2})$. Then, taking $H = F_n$ in (3.55) yields

$$\begin{aligned}\widehat{\theta}_n - \widehat{\theta}_\infty(F) &= T(F_n) - T(F) = \mathrm{E}_{F_n}\xi(x) + o\left(d\left(\widehat{F}_n, F\right)\right) \\ &= \frac{1}{n}\sum_{i=1}^{n} \xi(x_i) + o_p\left(n^{-1/2}\right).\end{aligned} \tag{3.57}$$

Estimators satisfying (3.57) (called a *linear expansion* of $\widehat{\theta}_\infty$) are asymptotically normal and verify (3.18). In fact, the i.i.d. variables $\xi(x_i)$ have mean 0 (by (3.56)) and variance

$$v = \mathrm{E}_F \xi(x)^2.$$

Hence

$$\sqrt{n}(\widehat{\theta}_n - \widehat{\theta}_\infty) = \frac{1}{\sqrt{n}}\sum_{i=1}^{n} \xi(x_i) + o_p(1),$$

which by the central limit theorem tends to $\mathrm{N}(0, v)$.

For further work in this area, see Fernholz (1983) and Clarke (1983).

3.8 Appendix: Proofs of results

3.8.1 IF of general M-estimators

Assume for simplicity that $\dot{\Psi}$ exists. For a given x_0, put for brevity

$$F_\varepsilon = (1-\varepsilon)F + \varepsilon\delta_{x_0} \text{ and } \theta_\varepsilon = \widehat{\theta}_\infty(F_\varepsilon).$$

Recall that by definition

$$\mathrm{E}_F\Psi(x,\theta_0) = 0. \tag{3.58}$$

Then θ_ε verifies

$$0 = \mathrm{E}_{F_\varepsilon}\Psi(x,\theta_\varepsilon) = (1-\varepsilon)\mathrm{E}_F\Psi(x,\theta_\varepsilon) + \varepsilon\Psi(x_0,\theta_\varepsilon).$$

Differentiating with respect to ε yields

$$-\mathrm{E}_F\Psi(x,\theta_\varepsilon) + (1-\varepsilon)\frac{\partial\theta_\varepsilon}{\partial\varepsilon}\mathrm{E}_F\dot{\Psi}(x,\theta_\varepsilon) + \Psi(x_0,\theta_\varepsilon) + \varepsilon\dot{\Psi}(x_0,\theta_\varepsilon)\frac{\partial\theta_\varepsilon}{\partial\varepsilon} = 0. \tag{3.59}$$

The first term vanishes at $\varepsilon = 0$ by (3.58). Taking $\varepsilon \downarrow 0$ above yields the desired result.

Note that this derivation is heuristic, since it is taken for granted that $\partial\theta_\varepsilon/\partial\varepsilon$ exists and that $\theta_\varepsilon \to \theta_0$. A rigorous proof may be found in Huber and Ronchetti (2009).

The same approach serves to prove (3.48) (Problem 3.9).

3.8.2 Maximum BP of location estimators

It suffices to show that $\varepsilon < \varepsilon^*$ implies $1-\varepsilon > \varepsilon^*$. Let $\varepsilon < \varepsilon^*$. For $t \in R$ define $F_t(x) = F(x-t)$, and let

$$H_t = (1-\varepsilon)F + \varepsilon F_t \in \mathcal{F}_\varepsilon,\; H_t^* = \varepsilon F + (1-\varepsilon)F_{-t} \in \mathcal{F}_{1-\varepsilon},$$

with

$$\mathcal{F}_\varepsilon = \{(1-\varepsilon)F + \varepsilon G : G \in \mathcal{G}\},$$

where $\mathcal{G}$ is the set of all distributions. Note that

$$H_t(x) = H_t^*(x-t). \tag{3.60}$$

The equivariance of $\widehat{\mu}$ and (3.60) imply

$$\widehat{\mu}_\infty(H_t) = \widehat{\mu}_\infty(H_t^*) + t \;\forall\; t.$$

Since $\varepsilon < \varepsilon^*$, $\widehat{\mu}_\infty(H_t)$ remains bounded when $t \to \infty$, and hence $\widehat{\mu}_\infty(H_t^*)$ is unbounded; since $H_t^* \in \mathcal{F}_{1-\varepsilon}$, this implies $1-\varepsilon > \varepsilon^*$.

A similar approach proves (3.26). The details are left to the reader.

3.8.3 BP of location M-estimators

3.8.3.1 Proof of (3.21)

Put for a given G

$$F_\varepsilon = (1-\varepsilon)F + \varepsilon G \text{ and } \mu_\varepsilon = \widehat{\mu}_\infty(F_\varepsilon).$$

Then

$$(1-\varepsilon)\mathrm{E}_F\psi(x-\mu_\varepsilon) + \varepsilon \mathrm{E}_G\psi(x-\mu_\varepsilon) = 0. \tag{3.61}$$

We shall prove first that ε^* is not larger than the right-hand side of (3.21). Let $\varepsilon < \varepsilon^*$. Then for some C, $|\mu_\varepsilon| \leq C$ for all G. Take $G = \delta_{x_0}$, so that

$$(1-\varepsilon)\mathrm{E}_F\psi(x-\mu_\varepsilon) + \varepsilon\psi(x_0-\mu_\varepsilon) = 0. \tag{3.62}$$

Let $x_0 \to \infty$. Since μ_ε is bounded, we have $\psi(x_0 - \mu_\varepsilon) \to k_2$. Since $\psi \geq -k_1$, (3.62) yields

$$0 \geq -k_1(1-\varepsilon) + \varepsilon k_2, \tag{3.63}$$

which implies $\varepsilon \leq k_1/(k_1+k_2)$. Letting $x_0 \to -\infty$ yields likewise $\varepsilon \leq k_2/(k_1+k_2)$.

We shall now prove the opposite inequality. Let $\varepsilon > \varepsilon^*$. Then there exists a sequence G_n such that

$$\mu_{\varepsilon,n} = \widehat{\mu}_\infty((1-\varepsilon)F + \varepsilon G_n)$$

is unbounded. Suppose it contains a subsequence tending to $+\infty$. Then for this subsequence, $x - \mu_{\varepsilon,n} \to -\infty$ for each x, and since $\psi \leq k_2$, (3.61) implies

$$0 \leq (1-\varepsilon)\lim_{n\to\infty} \mathrm{E}_F\psi(x-\mu_{\varepsilon,n}) + \varepsilon k_2,$$

and since the bounded convergence theorem (Section 10.3) implies

$$\lim_{n\to\infty} \mathrm{E}_F\psi(x-\mu_{\varepsilon,n}) = \mathrm{E}_F\left(\lim_{n\to\infty}\psi(x-\mu_{\varepsilon,n})\right)$$

we have

$$0 \leq -k_1(1-\varepsilon) + \varepsilon k_2,$$

This is the opposite inequality to (3.63), from which it follows that $\varepsilon \geq \varepsilon_1^*$ in (3.22). If instead the subsequence tends to $-\infty$, we have $\varepsilon \geq \varepsilon_2^*$. This concludes the proof.

3.8.3.2 Location with previously estimated dispersion

Consider first the case of monotone ψ. Since $\varepsilon < \varepsilon^*(\widehat{\sigma})$ is equivalent to $\widehat{\sigma}$ being bounded away from zero and infinity when the contamination rate is less than ε, the proof is similar to that of Section 3.8.3.1.

Now consider the case of a bounded ρ. Assume $\rho \leq \rho_0$. We shall show that $\varepsilon^* = 0.5$. Let $\varepsilon < 0.5$ and let $\mathbf{y}_N = (y_{N1}, \ldots, y_{Nn})$ be a sequence of data sets having m elements in common with $\mathbf{x}$, with $m \geq n(1-\varepsilon)$. Call $\widehat{\mu}_{0N}$ the initial location estimator,

$\widehat{\sigma}_N$ the previous scale and $\widehat{\mu}_N$ the final location estimator for $\mathbf{y}_N$. Then it follows from the definitions of $\widehat{\mu}_{0N}$, $\widehat{\sigma}_N$ and $\widehat{\mu}_N$ that

$$\frac{1}{n}\sum_{i=1}^{n}\rho\left(\frac{y_{Ni}-\widehat{\mu}_N}{\widehat{\sigma}_N}\right) \leq \frac{1}{n}\sum_{i=1}^{n}\rho\left(\frac{y_{Ni}-\widehat{\mu}_{0N}}{\widehat{\sigma}_N}\right) \leq \frac{1}{n}\sum_{i=1}^{n}\rho_0\left(\frac{y_{Ni}-\widehat{\mu}_{0N}}{\widehat{\sigma}_N}\right) = 0.5. \tag{3.64}$$

Since $\widehat{\mu}_0$ and $\widehat{\sigma}_0$ have BP $= 0.5 > \varepsilon$, $\widehat{\mu}_{0N}$ – and hence $\widehat{\sigma}_N$ – remains bounded for any choice of $\mathbf{y}_N$.

Assume now that there is a sequence $\mathbf{y}_N$ such that $\widehat{\mu}_N \to \infty$. Let $D_N = \{i : y_{Ni} = x_i\}$. Therefore

$$\lim_{N\to\infty}\frac{1}{n}\sum_{i=1}^{n}\rho\left(\frac{y_{Ni}-\widehat{\mu}_N}{\widehat{\sigma}_N}\right) \geq \lim_{N\to\infty}\frac{1}{n}\sum_{i\in D_N}\rho\left(\frac{x_i-\widehat{\mu}_N}{\widehat{\sigma}_N}\right) \geq 1-\varepsilon > 0.5,$$

which contradicts (3.64), and therefore $\widehat{\mu}_N$ must be bounded, which implies $\varepsilon^* \geq 0.5$.

We now deal with the case of bounded ρ when $\rho \leq \rho_0$ does not hold. Consider first the case of fixed σ. Huber (1984) calculated the finite BP for this situation. For the sake of simplicity we treat the asymptotic case with point-mass contamination.

Let

$$\gamma = \mathrm{E}_F\rho\left(\frac{x-\mu_0}{\sigma}\right),$$

where F is the underlying distribution and $\mu_0 = \widehat{\mu}_\infty(F)$:

$$\mu_0 = \arg\min_{\mu}\mathrm{E}_F\rho\left(\frac{x-\mu}{\sigma}\right).$$

It will be shown that

$$\varepsilon^* = \frac{1-\gamma}{2-\gamma}. \tag{3.65}$$

Consider a sequence x_N tending to infinity, and let $F_N = (1-\varepsilon)F + \varepsilon\delta_{x_N}$. For $\mu \in R$

$$A_N(\mu) = \mathrm{E}_{F_N}\rho\left(\frac{x-\mu}{\sigma}\right) = (1-\varepsilon)\mathrm{E}_F\rho\left(\frac{x-\mu}{\sigma}\right) + \varepsilon\rho\left(\frac{x_N-\mu}{\sigma}\right).$$

Let $\varepsilon < \mathrm{BP}(\widehat{\mu})$ first. Then $\mu_N = \widehat{\mu}_\infty(F_N)$ remains bounded when $x_N \to \infty$. By the definition of μ_0,

$$A_N(\mu_N) \geq (1-\varepsilon)\gamma + \varepsilon\rho\left(\frac{x_N-\mu_N}{\sigma}\right).$$

Since $\widehat{\mu}_\infty(F_N)$ minimizes A_N, we have $A_N(\mu_N) \leq A_N(x_N)$, and the latter tends to $1-\varepsilon$. The boundedness of μ_N implies that $x_N - \mu_N \to \infty$, and hence we have in the limit

$$(1-\varepsilon)\gamma + \varepsilon \leq 1-\varepsilon,$$

which is equivalent to $\varepsilon < \varepsilon^*$. The reverse inequality follows likewise. When ρ is the bisquare with efficiency 0.95, $F = \mathrm{N}(0,1)$ and $\sigma = 1$, we have $\varepsilon^* = 0.47$. Note that ε^* is an increasing function of σ.

In the more realistic case that σ is previously estimated, the situation is more complicated; but intuitively it can be seen that the situation is actually more favorable, since the contamination implies a larger σ. The procedure used above can be used to derive numerical bounds for ε^*. For the same ρ and MAD dispersion, it can be shown that $\varepsilon^* > 0.49$ for the normal distribution.

3.8.4 Maximum bias of location M-estimators

Let $F_\mu(x) = F_0(x - \mu)$ where F_0 is symmetric about zero. Let ψ be a nondecreasing and bounded ψ-function and call $k = \psi(\infty)$. The asymptotic value of the estimator is $\hat{\mu}_\infty(F_\mu) = \mu$, and the bias for an arbitrary distribution H is $\hat{\mu}_\infty(H) - \mu$. Define for brevity the function

$$g(b) = \mathrm{E}_{F_0}\psi(x + b),$$

which is odd. It will be assumed that g is increasing. This holds either if ψ is increasing, or if F_0 has positive density everywhere.

Let $\varepsilon < 0.5$. Then it will be shown that the maximum bias is the solution b_ε of the equation

$$g(b) = \frac{k\varepsilon}{1 - \varepsilon}. \tag{3.66}$$

Since the estimator is shift equivariant, it may be assumed without loss of generality that $\mu = 0$. Write, for brevity, $\mu_H = \hat{\mu}_\infty(H)$. For a distribution $H = (1 - \varepsilon)F_0 + \varepsilon G$ (with G arbitrary), μ_H is the solution of

$$(1 - \varepsilon)g(-\mu_H) + \varepsilon \mathrm{E}_G\psi(x - \mu_H) = 0. \tag{3.67}$$

Since $|g(b)| \leq k$, we have for any G

$$(1 - \varepsilon)g(-\mu_H) - \varepsilon k \leq 0 \leq (1 - \varepsilon)g(-\mu_H) + \varepsilon k,$$

which implies

$$-\frac{k\varepsilon}{1 - \varepsilon} \leq g(-\mu_H) \leq \frac{k\varepsilon}{1 - \varepsilon},$$

and hence $|\mu_H| \leq b_\varepsilon$. By letting $G = \delta_{x_0}$ in (3.67) with $x_0 \to \pm\infty$, we see that the bound is attained. This completes the proof.

For the median, $\psi(x) = \mathrm{sgn}(x)$ and $k = 1$, and a simple calculation shows (recalling the symmetry of F_0) that $g(b) = 2F_0(b) - 1$, and therefore

$$b_\varepsilon = F_0^{-1}\left(\frac{1}{2(1 - \varepsilon)}\right). \tag{3.68}$$

To calculate the contamination sensitivity γ_c, put $\dot{b}_\varepsilon = db_\varepsilon/d\varepsilon$, so that $\dot{b}_0 = \gamma_c$. Then differentiating (3.66) yields

$$g'(b_\varepsilon)\dot{b}_\varepsilon = \frac{k}{(1 - \varepsilon)^2},$$

and hence (recalling $b_0 = 0$) $\gamma_c = k/g'(0)$. Since $g'(0) = \mathrm{E}_{H_0}\psi'(x)$, we see that this coincides with (3.31) and hence $\gamma_c = \gamma^*$.

3.8.5 The minimax bias property of the median

Let F_0 have a density $f_0(x)$ which is a nonincreasing function of $|x|$ (a symmetric *unimodal* distribution) . Let b_ε be the maximum asymptotic bias of the median given in (3.68). Let $\widehat{\theta}$ be any location equivariant estimator. It will be shown that the maximum bias of $\widehat{\theta}$ in a neighborhood $\mathcal{F}(F_0, \varepsilon)$, defined in (3.3), is not smaller than b_ε.

Call F_+ the distribution with density

$$f_+(x) = \begin{cases} (1-\varepsilon)f_0(x) & \text{if } x \le b_\varepsilon \\ (1-\varepsilon)f_0(x-2b_\varepsilon) & \text{otherwise.} \end{cases}$$

Then f_+ belongs to $\mathcal{F}(F_0, \varepsilon)$. In fact, it can be written as

$$f_+ = (1-\varepsilon)f_0 + \varepsilon g,$$

with

$$g(x) = \frac{1-\varepsilon}{\varepsilon}(f_0(x-2b_\varepsilon) - f_0(x))\mathrm{I}(x > b_\varepsilon).$$

We must show that g is a density. It is nonnegative, since $x \in (b_\varepsilon, 2b_\varepsilon)$ implies $|x - 2b_\varepsilon| \le |x|$, and the unimodality of f_0 yields $f_0(x-2b_\varepsilon) \ge f_0(x)$; the same thing happens if $x > 2b_\varepsilon$. Moreover, its integral equals one, since by (3.68),

$$\int_{b_\varepsilon}^{\infty}(f_0(x-2b_\varepsilon) - f_0(x))dx = 2F_0(b_\varepsilon) - 1 = \frac{\varepsilon}{1-\varepsilon}.$$

Define

$$F_-(x) = F_+(x + 2b_\varepsilon),$$

which also belongs to $\mathcal{F}(F_0, \varepsilon)$ by the same argument. The equivariance of $\widehat{\theta}$ implies that

$$\widehat{\theta}_\infty(F_+) - \widehat{\theta}_\infty(F_-) = 2b_\varepsilon,$$

and hence $|\widehat{\theta}_\infty(F_+)|$ and $|\widehat{\theta}_\infty(F_-)|$ cannot both be less than b_ε.

3.8.6 Minimizing the GES

To avoid cumbersome technical details, we assume henceforth that $\Psi_0(x, \theta)$ has a continuous distribution for all θ. We prove first that the M-estimator $\widetilde{\theta}$ is Fisher-consistent. In fact, by the definition of the function M,

$$\begin{aligned} \mathrm{E}_\theta\widetilde{\Psi}(x,\theta) &= -\mathrm{P}_\theta(\Psi_0(x,\theta) \le M(\theta)) + \mathrm{P}_\theta(\Psi_0(x,\theta) > M(\theta)) \\ &= -\frac{1}{2} + \frac{1}{2} = 0. \end{aligned}$$

Since $\max_x|\widetilde{\Psi}(x,\theta)| = 1$, the estimator has GES

$$\gamma^*(\widetilde{\Psi}, \theta) = \frac{1}{|B(\theta, \widetilde{\Psi})|}.$$

It will be shown first that for any Fisher-consistent Ψ,

$$B(\theta, \Psi) = \mathrm{E}_\theta \Psi(x, \theta)\Psi_0(x, \theta). \tag{3.69}$$

We give the proof for the continuous case; the discrete case is similar. Condition (3.32) may be written as

$$\mathrm{E}_\theta \Psi(x, \theta) = \int_{-\infty}^{\infty} \Psi(x, \theta) f_\theta(x) dx = 0.$$

Differentiating the above expression with respect to θ yields

$$B(\theta, \Psi) + \int_{-\infty}^{\infty} \Psi(x, \theta)\dot{f}_\theta(x) dx = 0,$$

and (3.33)–(3.34) yield

$$\begin{aligned} B(\theta, \Psi) &= -\int_{-\infty}^{\infty} \Psi(x, \theta)\dot{f}_\theta(x) dx \\ &= \int_{-\infty}^{\infty} \Psi(x, \theta)\Psi_0(x, \theta) f_\theta(x)\, dx = \mathrm{E}_\theta \Psi(x, \theta)\Psi_0(x, \theta), \end{aligned}$$

as stated. Note that $\partial\widetilde{\Psi}/\partial\theta$ does not exist, and hence we must define $B(\theta, \widetilde{\Psi})$ through (3.12) and not (3.13).

Now let $C = \{x : \Psi_0(x, \theta) > M(\theta)\}$, with complement C'. It follows from $\mathrm{I}(C') = 1 - \mathrm{I}(C)$ that

$$\mathrm{P}_\theta(\widetilde{\Psi} = \mathrm{I}(C) - \mathrm{I}(C') = 2\mathrm{I}(C) - 1) = 1.$$

Using (3.35) and (3.36) we have

$$\begin{aligned} B(\theta, \widetilde{\Psi}) &= \mathrm{E}_\theta \widetilde{\Psi}(x, \theta)\Psi_0(x, \theta) \\ &= 2\mathrm{E}_\theta \Psi_0(x, \theta)\mathrm{I}(C) - \mathrm{E}_\theta \Psi_0(x, \theta) = 2\mathrm{E}_\theta \Psi_0(x, \theta)\mathrm{I}(C). \end{aligned}$$

Hence

$$\gamma^*(\widetilde{\Psi}, \theta) = \frac{1}{2|\mathrm{E}_\theta \Psi_0(x, \theta)\mathrm{I}(C)|}. \tag{3.70}$$

Consider a Fisher-consistent Ψ. Then

$$\gamma^*(\Psi, \theta) = \frac{\max_x |\Psi(x, \theta)|}{|B(\theta, \Psi)|}. \tag{3.71}$$

Using (3.32) and (3.69) we have

$$\begin{aligned} B(\theta, \Psi) &= \mathrm{E}_\theta \Psi(x, \theta)\Psi_0(x, \theta) \\ &= \mathrm{E}_\theta \Psi(x, \theta)(\Psi_0(x, \theta) - M(\theta)) \\ &= \mathrm{E}_\theta \Psi(x, \theta)(\Psi_0(x, \theta) - M(\theta))\mathrm{I}(C) \\ &\quad + \mathrm{E}_\theta \Psi(x, \theta)(\Psi_0(x, \theta) - M(\theta))\mathrm{I}(C'). \end{aligned} \tag{3.72}$$

Besides

$$\begin{aligned}&|\mathrm{E}_\theta \Psi(x,\theta)(\Psi_0(x,\theta) - M(\theta))\mathrm{I}(C)|\\&\quad\leq \max_x |\Psi(x,\theta)|\mathrm{E}_\theta(\Psi_0(x,\theta) - M(\theta))\mathrm{I}(C)\\&\quad= \max_x |\Psi(x,\theta)| \left(\mathrm{E}_\theta \Psi_0(x,\theta)\mathrm{I}(C) - \frac{M(\theta)}{2}\right).\end{aligned}\tag{3.73}$$

Similarly

$$\begin{aligned}&|\mathrm{E}_\theta \Psi(x,\theta)(\Psi_0(x,\theta) - M(\theta))\mathrm{I}(C')|\\&\quad\leq -\max_x |\Psi(x,\theta)|\mathrm{E}_\theta(\Psi_0(x,\theta) - M(\theta))\mathrm{I}(C')\\&\quad= \max_x |\Psi(x,\theta)| \left(\mathrm{E}_\theta \Psi_0(x,\theta)\mathrm{I}(C) + \frac{M(\theta)}{2}\right).\end{aligned}\tag{3.74}$$

Therefore, by (3.72)–(3.74) we get

$$|B(\theta,\Psi)| \leq 2\max_x |\Psi(x,\theta)|\mathrm{E}_\theta \Psi_0(x,\theta)\mathrm{I}(C).$$

Therefore, using (3.71), we have

$$\gamma^*(\Psi,\theta) \geq \frac{1}{2|\mathrm{E}_\theta \Psi_0(x,\theta)\mathrm{I}(C)|}.\tag{3.75}$$

And finally (3.70) and (3.75) yield

$$\gamma^*(\widetilde{\Psi},\theta) \leq \gamma^*(\Psi,\theta).$$

The case of a discrete distribution is similar, but the details are much more involved.

3.8.7 Hampel optimality

It will be shown first that estimators with score function (3.41) are optimal for Hampel problems with certain bounds.

Theorem 3.2 *Given $k(\theta)$, the function Ψ^* given by (3.41) and satisfying (3.32) is optimal for the direct Hampel problem with bound*

$$G_k(\theta) = \gamma^*(\Psi^*,\theta) = \frac{k(\theta)}{B(\theta,\Psi^*)},$$

and for the dual Hampel problem with bound

$$V_k(\theta) = \upsilon(\Psi^*,\theta).$$

Proof of Theorem 3.2: We shall show that Ψ^* solves Hampel's direct problem. Observe that Ψ^* satisfies the side condition in (3.39), since by definition $\gamma^*(\Psi^*, \theta) = g(\theta)$. Let now Ψ satisfy (3.32) and

$$\gamma^*(\Psi, \theta) \leq g(\theta). \tag{3.76}$$

We must show that

$$v(\Psi, \theta) \geq v(\Psi^*, \theta). \tag{3.77}$$

We prove (3.77) for a fixed θ. Since for any real number $\lambda \neq 0$, $\lambda\Psi$ defines the same estimator as Ψ, we can assume without loss of generality that

$$B(\theta, \Psi) = B(\theta, \Psi^*), \tag{3.78}$$

and hence

$$\gamma^*(\Psi, \theta) = \frac{\max_x(|\Psi(x, \theta)|)}{|B(\theta, \Psi^*)|}.$$

Then, condition (3.76) becomes

$$\max_x(|\Psi(x, \theta)|) \leq k(\theta) \tag{3.79}$$

and (3.77) becomes $A(\theta, \Psi) \geq A(\theta, \Psi^*)$, so that we have to prove

$$\mathrm{E}_\theta \Psi^2(x, \theta) \geq \mathrm{E}_\theta \Psi^{*2}(x, \theta) \tag{3.80}$$

for any Ψ satisfying (3.79).

Call Ψ_0^c the ML score function centered by r :

$$\Psi_0^c(x, \theta) = \Psi_0(x, \theta) - r(\theta).$$

It follows from (3.69) and (3.32) that

$$\mathrm{E}_\theta \Psi(x, \theta) \Psi_0^c(x, \theta) = B(\theta, \Psi).$$

We now calculate $\mathrm{E}_\theta \Psi^2(x, \theta)$. Recalling (3.78) we have

$$\begin{aligned} \mathrm{E}_\theta \Psi^2(x, \theta) &= \mathrm{E}_\theta\{[\Psi(x, \theta) - \Psi_0^c(x, \theta)] + \Psi_0^c(x, \theta)\}^2 \\ &= \mathrm{E}_\theta(\Psi(x, \theta) - \Psi_0^c(x, \theta))^2 + \mathrm{E}_\theta \Psi_0^c(x, \theta)^2 \\ &\quad + 2\mathrm{E}_\theta \Psi(x, \theta)\Psi_0^c - 2\mathrm{E}_\theta \Psi_0^{c2}(x, \theta) \\ &= \mathrm{E}_\theta(\Psi(x, \theta) - \Psi_0^c(x, \theta))^2 - \mathrm{E}_\theta \Psi_0^c(x, \theta)^2 + 2B(\theta, \Psi^*). \end{aligned} \tag{3.81}$$

Since $\mathrm{E}\Psi_0(x, \theta)^2$ and $B(\theta, \Psi^*)$ do not depend on Ψ, it suffices to prove that putting $\Psi = \Psi^*$ minimizes

$$\mathrm{E}_\theta(\Psi(x, \theta) - \Psi_0^c(x, \theta))^2$$

subject to (3.79). Observe that for any function $\Psi(x,\theta)$ satisfying (3.79) we have

$$|\Psi(x,\theta) - \Psi_0^c(x,\theta)| \geq ||\Psi_0^c(x,\theta)| - k(\theta)|\mathrm{I}\{|\Psi_0^c(x,\theta)| > k(\theta)\},$$

and since

$$|\Psi^*(x,\theta) - \Psi_0^c(x,\theta)| = ||\Psi_0^c(x,\theta)| - k(\theta)|\mathrm{I}\{|\Psi_0^c(x,\theta)| > k(\theta)\},$$

we get

$$|\Psi(x,\theta) - \Psi_0^c(x,\theta)| \geq |\Psi^*(x,\theta) - \Psi_0^c(x,\theta)|.$$

Then

$$\mathrm{E}_\theta(\Psi(x,\theta) - \Psi_0^c(x,\theta))^2 \geq \mathrm{E}_\theta(\Psi^*(x,\theta) - \Psi_0^c(x,\theta))^2,$$

which proves the statement for the direct problem. The dual problem is treated likewise.

The last theorem proves optimality for a certain class of bounds. The next one shows that in fact any feasible bounds can be considered.

Theorem 3.3 *Let*

$$G(\theta) \geq \gamma^*(\widetilde{\Psi},\theta), \quad V(\theta) \geq \upsilon(\Psi_0,\theta) \text{ for all } \theta, \tag{3.82}$$

where Ψ_0 and $\widetilde{\Psi}$ are defined in (3.34) and (3.36) respectively. Then the solutions to both the direct and the dual Hampel problems have the form (3.41) for a suitable function $k(\theta)$.

Proof of Theorem 3.3: We treat the dual problem; the direct problem is dealt with in the same way.

We show first that given any k, there exists r so that $\Psi^*(x,\theta)$ is Fisher-consistent. Let

$$\lambda(r) = \mathrm{E}_\theta \psi_k(\Psi_0(x,\theta) - r).$$

Then λ is continuous, and $\lim_{r\to\pm\infty}\lambda(r) = \mp k$. Hence by the Intermediate Value Theorem, there exists some r such that $\lambda(r) = 0$. In addition, it can be shown that $B(\theta,\Psi^*) \neq 0$. The proof is involved and can be found in Hampel *et al.* (1986).

In view of (3.69):

$$\upsilon(\Psi^*_{(k(\theta))},\theta) = \frac{\mathrm{E}_\theta \Psi^*_{(k(\theta))}(x,\theta)^2}{[\mathrm{E}_\theta \Psi^*_{(k(\theta))}(x,\theta)\Psi_0(x,\theta)]^2}, \tag{3.83}$$

where Ψ^* in (3.41) is written as $\Psi^*_{(k(\theta))}$ to stress its dependence on k. Recall that the limit cases $k \to 0$ and $k \to \infty$ yield $\upsilon(\widetilde{\Psi},\theta)$ (which may be infinite) and $\upsilon(\Psi_0,\theta)$ respectively. Let $V(\theta)$ be given and such that $V(\theta) \geq \upsilon(\Psi_0,\theta)$. Consider a fixed θ. If $V(\theta) \leq \upsilon(\widetilde{\Psi},\theta)$, then there exists a value $k(\theta)$ such that $\upsilon(\Psi^*_{(k(\theta))},\theta) = V(\theta)$. If $V(\theta) > \upsilon(\widetilde{\Psi},\theta)$, then putting $k(\theta) = 0$ (i.e., $\Psi^* = \widetilde{\Psi}$) minimizes $\gamma^*(\Psi^*_{(k(\theta))},\theta)$ and satisfies $\upsilon(\Psi^*_{(k(\theta))},\theta) \leq V(\theta)$.

3.9 Problems

3.1. Verify (3.15).

3.2. Prove (3.19).

3.3. Verify that the breakdown points of the SD, the MAD and the IQR are 0, 1/2 and 1/4, respectively

3.4. Show that the asymptotic BP of the α-trimmed mean is α.

3.5. Show that the BP of equivariant dispersion estimators is ≤ 0.5.

3.6. Show that the asymptotic BP of sample β-quantiles is $\min(\beta, 1-\beta)$ (recall Problem 2.13)

3.7. Prove (3.26).

3.8. Verify (3.46).

3.9. Prove (3.48).

3.10. Prove (3.46).

3.11. Consider the location M-estimator with Huber function ψ_k and the MADN as previously computed dispersion. Recall that it has BP = 1/2 for all k. Show however that for each given $\varepsilon < 0.5$, its maximum bias MB(ε) at a given distribution is an unbounded function of k.

3.12. Let the density $f(x)$ be a decreasing function of $|x|$. Show that the shortest interval covering a given probability is symmetric about 0. Use this result to calculate the IF of the Shorth estimator (Problem 2.16a) for data with distribution f.

3.13. Show that the BP of the estimator Q_n in (2.64) is 0.5. Calculate the BP for the estimator defined as the median of the differences; that is, with $k = m/2$ in (2.64).

3.14. Show the equivalence of the direct and dual Hampel problems (3.39) and (3.40)

3.15. For the exponential family $f_\theta(x) = \mathrm{I}(x \geq 0)\exp(-x/\theta)/\theta$:

(a) Show that the estimator with smallest GES is Med $(\mathbf{x})/\ln 2$.
(b) Find the asymptotic distribution of this estimator and its efficiency with respect to the MLE.
(c) Find the form of the Hampel-optimal estimator for this family.
(d) Write a program to compute the Hampel-optimal estimator with efficiency 0.95.

3.16. Consider the estimator $\widehat{\mu}_1$ defined by the one-step Newton–Raphson procedure defined in Section 2.10.5.1. Assume that the underlying distribution is symmetric about μ, that ψ is odd and differentiable, and that the initial estimator $\widehat{\mu}_0$ is consistent for μ.

(a) Show that $\widehat{\mu}_1$ is consistent for μ.
(b) If ψ is twice differentiable, show that $\widehat{\mu}_1$ has the same influence function as the M-estimator $\widehat{\mu}$ defined by $\text{ave}\{\psi(\mathbf{x}-\mu)\}=0$ (and hence, by (3.18), $\widehat{\mu}_1$ has the same asymptotic variance as $\widehat{\mu}$).
(c) If ψ is bounded and $\psi'(x)>0$ for all x, and the asymptotic BP of $\widehat{\mu}_0$ is 0.5, show that also $\widehat{\mu}_1$ has an asymptotic BP of 0.5.

4

Linear Regression 1

4.1 Introduction

In this chapter we begin the discussion of the estimation of the parameters of linear regression models, which will be pursued in the next chapter. M-estimators for regression are developed in the same way as for location. In this chapter we deal with fixed (nonrandom) predictors. Recall that our estimators of choice for location were redescending M-estimators using the median as starting point and the MAD as dispersion. Redescending estimators will also be our choice for regression. When the predictors are fixed and fulfill certain conditions that are satisfied in particular for analysis of variance models, monotone M-estimators – which are easy to compute – are robust, and can be used as starting points to compute a redescending estimator. When the predictors are random, or when they are fixed but in some sense "unbalanced", monotone estimators cease to be reliable, and the starting points for redescending estimators must be computed otherwise. This problem is considered in the next chapter.

We start with an example that shows the weakness of the least-squares estimator.

Example 4.1 *In an experiment on the speed of learning of rats (Bond, 1979), times were recorded for a rat to go through a shuttlebox in successive attempts. If the time exceeded 5s, the rat received an electric shock for the duration of the next attempt. The data are the number of shocks received and the average time for all attempts between shocks. Tables and figures for this example can be obtained with script* ***shock.R****.*

Robust Statistics: Theory and Methods (with R), Second Edition.
Ricardo A. Maronna, R. Douglas Martin, Victor J. Yohai and Matías Salibián-Barrera.

Companion website: www.wiley.com/go/maronna/robust

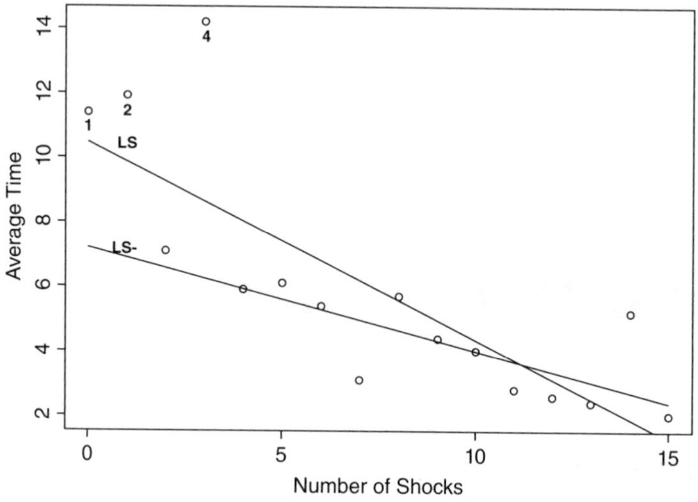

Figure 4.1 Shock data: LS fit with all data and omitting points 1, 2 and 4

Figure 4.1 shows the data and the straight line fitted by least squares (LS) to the linear regression model

$$y_i = \beta_0 + \beta_1 x_i + u_i.$$

The relationship between the variables is seen to be roughly linear, except for the three upper-left points. The LS line does not fit the bulk of the data, being a compromise between those three points and the rest. The figure also shows the LS fit computed without using the three points. It gives a better representation of the majority of the data, while indicating the exceptional character of points 1, 2 and 4.

We aim to develop procedures that give a good fit to the bulk of the data without being perturbed by a small proportion of outliers, and that do not require deciding in advance which observations are outliers.

Table 4.1 gives the estimated parameters for an LS fit, with the complete data and with the three atypical points deleted, and also for two robust estimators (L_1 and bisquare) that will be defined later.

Table 4.1 Regression estimates for rats data

	Intercept	Slope
LS	10.48	−0.61
LS (−1, 2, 4)	7.22	−0.32
L_1	8.22	−0.42
Bisquare M-est.	7.83	−0.41

The LS fit of a straight line consists of finding $\widehat{\beta}_0, \widehat{\beta}_1$ such that the residuals

$$r_i = y_i - (\widehat{\beta}_0 + \widehat{\beta}_1 x_i)$$

satisfy

$$\sum_{i=1}^{n} r_i^2 = \min. \tag{4.1}$$

Recall that in the location case obtained by setting $\beta_1 = 0$, the solution of (2.16) is the sample mean; that is, the LS estimator of location is the average of the data values. Since the median satisfies (2.18), the regression analogue of the median, often called an L_1 estimator (also called the least absolute deviation or LAD estimator), is defined by

$$\sum_{i=1}^{n} |r_i| = \min. \tag{4.2}$$

For our data, the solution of (4.2) is given in Table 4.1, and it can be seen that its slope is smaller than that of the LS estimator; that is, it is less affected by the outliers.

Now consider the more general case of a dataset of n observations $(x_{i1}, \ldots, x_{ip}, y_i)$ where $x_{i1}, \ldots x_{ip}$ are predictor variables (the *predictors* or *independent variables*) and y_i is a response variable (the *response* or *dependent variable*). The data are assumed to follow the *linear model*

$$y_i = \sum_{j=1}^{p} x_{ij}\beta_j + u_i, \quad i = 1, \ldots, n \tag{4.3}$$

where $\beta_1, \ldots, \beta_p$ are unknown parameters to be estimated, and the u_i are random variables (the "errors"). In a designed experiment, the x_{ij} are nonrandom (or *fixed*); that is, determined before the experiment. When the data are observational the x_{ij} are random variables. We sometimes have mixed situations with both fixed and random predictors.

Denoting by $\mathbf{x}_i$ and $\boldsymbol{\beta}$ the p-dimensional column vectors with coordinates $(x_{i1}, ..., x_{ip})$ and $(\beta_1, ..., \beta_p)$ respectively, the model can be more compactly written as

$$y_i = \mathbf{x}_i'\boldsymbol{\beta} + u_i \tag{4.4}$$

where $\mathbf{x}'$ is the transpose of $\mathbf{x}$. In the common case where the model has a constant term, the first coordinate of each $\mathbf{x}_i$ is 1 and the model may be written as

$$y_i = \beta_0 + \underline{\mathbf{x}}_i'\boldsymbol{\beta}_1 + u_i \tag{4.5}$$

where $\underline{\mathbf{x}}_i = (x_{i1}, \ldots, x_{i(p-1)})'$ and $\boldsymbol{\beta}_1$ are in R^{p-1} and

$$\mathbf{x}_i = \begin{pmatrix} 1 \\ \underline{\mathbf{x}}_i \end{pmatrix}, \quad \boldsymbol{\beta} = \begin{pmatrix} \beta_0 \\ \boldsymbol{\beta}_1 \end{pmatrix}. \tag{4.6}$$

Here β_0 is called the *intercept* and the elements of $\boldsymbol{\beta}_1$ are the *slopes*. Call $\mathbf{X}$ the $n \times p$ matrix with elements x_{ij}, and let $\mathbf{y}$ and $\mathbf{u}$ be the vectors with elements y_i and u_i respectively ($i = 1, \ldots, n$). Then the linear model (4.4) may be written

$$\mathbf{y} = \mathbf{X}\boldsymbol{\beta} + \mathbf{u}. \tag{4.7}$$

The *fitted values* $\widehat{y}_i$ and the *residuals* r_i corresponding to a vector $\boldsymbol{\beta}$ are defined respectively as

$$\widehat{y}_i(\boldsymbol{\beta}) = \boldsymbol{x}_i'\boldsymbol{\beta} \text{ and } r_i(\boldsymbol{\beta}) = y_i - \widehat{y}_i(\boldsymbol{\beta}).$$

The dependence of the fitted values and residuals on $\boldsymbol{\beta}$ will be dropped when this does not cause confusion. In order to combine robustness and efficiency, along the lines of Chapter 2, we shall discuss regression M-estimators $\widehat{\boldsymbol{\beta}}$, defined as solutions of equations of the form

$$\sum_{i=1}^{n} \rho\left(\frac{r_i(\widehat{\boldsymbol{\beta}})}{\widehat{\sigma}}\right) = \min. \tag{4.8}$$

Here ρ is a ρ-function (Definition 2.1), and $\widehat{\sigma}$ is an auxiliary scale estimator that is required to make $\widehat{\boldsymbol{\beta}}$ scale equivariant (see (2.5) and (4.16)). The LS estimator and the L_1 estimator correspond respectively to $\rho(t) = t^2$ and $\rho(t) = |t|$. In these two cases $\widehat{\sigma}$ becomes a constant factor outside the summation sign and minimizing (4.8) is equivalent to minimizing $\sum_{i=1}^{n} r_i^2$ or $\sum_{i=1}^{n} |r_i|$, respectively. Thus neither the LS nor the L_1 estimators require a scale estimator.

In a designed experiment, the predictors x_{ij} are fixed. An important special case of fixed predictors is when they represent *categorical* predictors with values of either 0 or 1. The simplest situation is the comparison of several treatments, usually called a *one-way analysis of variance* (or "one-way ANOVA"). Here we have p samples y_{ik} ($i = 1, \ldots, n_k, k = 1, \ldots, p$) and the model

$$y_{ik} = \beta_k + u_{ik} \tag{4.9}$$

where the u_{ik} are i.i.d. Call $\mathbf{1}_m$ the column vector of m ones. Then the matrix $\mathbf{X}$ of predictors is

$$\mathbf{X} = \begin{bmatrix} \mathbf{1}_{n_1} & & & \\ & \mathbf{1}_{n_2} & & \\ & & \ddots & \\ & & & \mathbf{1}_{n_p} \end{bmatrix}$$

with the blank positions filled with zeros. The next level of model complexity is a factorial design with two factors represented by two categorical variables. This is usually called a *two-way analysis of variance*. In this case we have data y_{ijk}, $i = 1, \ldots, I, j = 1, \ldots, J, k = 1, \ldots, K_{ij}$, following an *additive model* usually written in the form

$$y_{ijk} = \mu + \alpha_i + \gamma_j + u_{ijk} \tag{4.10}$$

with "cells" i, j, and K_{ij} observations per cell. Here $p = I + J + 1$ and $\boldsymbol{\beta}$ has coordinates $(\mu, \alpha_1, \ldots, \alpha_I, \gamma_1, \ldots, \gamma_J)$. The rank of $\mathbf{X}$ is $p^* = I + J - 1 < p$ and constraints on the parameters need to be added to make the estimators unique, typically

$$\sum_{i=1}^{I} \alpha_i = \sum_{j=1}^{J} \gamma_j = 0. \tag{4.11}$$

4.2 Review of the least squares method

The LS method was proposed in 1805 by Legendre (for a fascinating account, see Stigler (1986)). The main reason for its immediate and lasting success was that it was the only method of estimation that could be effectively computed before the advent of electronic computers. We shall review the main properties of LS for multiple regression. (See any standard text on regression analysis for further details, for example Weisberg (1985), Draper and Smith (2001), Montgomery *et al.* (2001) or Stapleton (1995).) The LS estimator of $\boldsymbol{\beta}$ is the $\widehat{\boldsymbol{\beta}}$ such that

$$\sum_{i=1}^{n} r_i^2(\widehat{\boldsymbol{\beta}}) = \min. \tag{4.12}$$

Differentiating with respect to $\boldsymbol{\beta}$ yields

$$\sum_{i=1}^{n} r_i(\widehat{\boldsymbol{\beta}})\mathbf{x}_i = \mathbf{0}, \tag{4.13}$$

which is equivalent to the linear equations

$$\mathbf{X}'\mathbf{X}\widehat{\boldsymbol{\beta}} = \mathbf{X}'\mathbf{y}.$$

The above equations are usually called the "normal equations". If the model contains a constant term, it follows from (4.13) that the residuals have zero average.

The matrix of predictors $\mathbf{X}$ is said to have *full rank* if its columns are linearly independent. This is equivalent to

$$\mathbf{Xa} \neq \mathbf{0} \ \forall \ \mathbf{a} \neq \mathbf{0}$$

and also equivalent to the nonsingularity of $\mathbf{X}'\mathbf{X}$. If $\mathbf{X}$ has full rank then the solution of (4.13) is unique and is given by

$$\widehat{\boldsymbol{\beta}}_{LS} = \widehat{\boldsymbol{\beta}}_{LS}(\mathbf{X}, \mathbf{y}) = (\mathbf{X}'\mathbf{X})^{-1}\mathbf{X}'\mathbf{y}. \tag{4.14}$$

If the model contains a constant term, then the first column of $\mathbf{X}$ is identically one, and the full-rank condition implies that no other column is constant. If $\mathbf{X}$ is not of full rank, for example as in (4.10), then we have what is called *collinearity*. When there is collinearity, the parameters are not *identifiable* in the sense that there exist

$\boldsymbol{\beta}_1 \neq \boldsymbol{\beta}_2$ such that $\mathbf{X}\boldsymbol{\beta}_1 = \mathbf{X}\boldsymbol{\beta}_2$, which implies that (4.13) has infinite solutions, all yielding the same fitted values and hence the same residuals.

The LS estimator satisfies

$$\widehat{\boldsymbol{\beta}}_{LS}(\mathbf{X}, \mathbf{y} + \mathbf{X}\boldsymbol{\gamma}) = \widehat{\boldsymbol{\beta}}_{LS}(\mathbf{X}, \mathbf{y}) + \boldsymbol{\gamma} \qquad \text{for all } \boldsymbol{\gamma} \in R^p \tag{4.15}$$

$$\widehat{\boldsymbol{\beta}}_{LS}(\mathbf{X}, \lambda\mathbf{y}) = \lambda\widehat{\boldsymbol{\beta}}_{LS}(\mathbf{X}, \mathbf{y}) \qquad \text{for all } \lambda \in R \tag{4.16}$$

and for all nonsingular $p \times p$-matrices $\mathbf{A}$

$$\widehat{\boldsymbol{\beta}}_{LS}(\mathbf{X}\mathbf{A}, \mathbf{y}) = \mathbf{A}^{-1}\widehat{\boldsymbol{\beta}}_{LS}(\mathbf{X}, \mathbf{y}). \tag{4.17}$$

The properties (4.15), (4.16) and (4.17) are called respectively *regression*, *scale* and *affine equivariance*. These are desirable properties, since they allow us to know how the estimator changes under these transformations of the data. A more precise justification is given in Section 4.9.1.

Assume now that the u_i are i.i.d. with

$$\mathrm{E}u_i = 0 \quad \text{and} \quad \mathrm{Var}(u_i) = \sigma^2$$

and that $\mathbf{X}$ is fixed; that is, nonrandom and of full rank. Under the linear model (4.4) with $\mathbf{X}$ of full rank, $\widehat{\boldsymbol{\beta}}_{LS}$ is unbiased and its mean and covariance matrix are given by

$$\mathrm{E}\widehat{\boldsymbol{\beta}}_{LS} = \boldsymbol{\beta}, \ \mathbf{Var}(\widehat{\boldsymbol{\beta}}_{LS}) = \sigma^2(\mathbf{X}'\mathbf{X})^{-1} \tag{4.18}$$

where henceforth $\mathbf{Var}(\mathbf{y})$ will denote the covariance matrix of the random vector $\mathbf{y}$.

Under model (4.5) we have the decomposition

$$(\mathbf{X}'\mathbf{X})^{-1} = \begin{bmatrix} \frac{1}{n} + \underline{\overline{\mathbf{x}}}'\mathbf{C}^{-1}\underline{\overline{\mathbf{x}}} & -(\mathbf{C}^{-1}\underline{\overline{\mathbf{x}}})' \\ -\mathbf{C}^{-1}\underline{\overline{\mathbf{x}}} & \mathbf{C}^{-1} \end{bmatrix}$$

where

$$\underline{\overline{\mathbf{x}}} = \mathrm{ave}_i(\underline{\mathbf{x}}_i), \ \mathbf{C} = \sum_{i=1}^{n} (\underline{\mathbf{x}}_i - \underline{\overline{\mathbf{x}}})(\underline{\mathbf{x}}_i - \underline{\overline{\mathbf{x}}})' \tag{4.19}$$

and hence

$$\mathbf{Var}(\widehat{\boldsymbol{\beta}}_{1,LS}) = \sigma^2\mathbf{C}^{-1}. \tag{4.20}$$

If $\mathrm{E}u_i \neq 0$, then $\widehat{\boldsymbol{\beta}}_{LS}$ will be biased. However, if the model contains an intercept, the bias will only affect the intercept and *not* the slopes. More precisely, under (4.5)

$$\mathrm{E}\widehat{\boldsymbol{\beta}}_{1,LS} = \boldsymbol{\beta}_1 \tag{4.21}$$

although $\mathrm{E}\widehat{\boldsymbol{\beta}}_{0,LS} \neq \boldsymbol{\beta}_0$ (see Section 4.9.2 for details).

Let p^* be the rank of $\mathbf{X}$ and recall that if $p^* < p$ – that is, if $\mathbf{X}$ is collinear – then $\widehat{\boldsymbol{\beta}}_{LS}$ is not uniquely defined but all solutions of (4.13) yield the same residuals. Then an unbiased estimator of σ^2 is defined by

$$s^2 = \frac{1}{n-p^*}\sum_{i=1}^{n} r_i^2. \tag{4.22}$$

whether or not $\mathbf{X}$ is of full rank.

If the u_i are normal and $\mathbf{X}$ is of full rank, then $\widehat{\boldsymbol{\beta}}_{LS}$ is multivariate normal

$$\widehat{\boldsymbol{\beta}}_{LS} \sim \mathrm{N}_p(\boldsymbol{\beta}, \sigma^2(\mathbf{X}'\mathbf{X})^{-1}), \tag{4.23}$$

where $\mathrm{N}_p(\boldsymbol{\mu}, \boldsymbol{\Sigma})$ denotes the p-variate normal distribution with mean vector $\boldsymbol{\mu}$ and covariance matrix $\boldsymbol{\Sigma}$.

Let γ now be a linear combination of the parameters: $\gamma = \boldsymbol{\beta}'\mathbf{a}$ with $\mathbf{a}$ a constant vector. Then the natural estimator of γ is $\widehat{\gamma} = \widehat{\boldsymbol{\beta}}'\mathbf{a}$, which according to (4.23) is $\mathrm{N}(\gamma, \sigma_\gamma^2)$, with

$$\sigma_\gamma^2 = \sigma^2\mathbf{a}'(\mathbf{X}'\mathbf{X})^{-1}\mathbf{a}.$$

An unbiased estimator of σ_γ^2 is

$$\widehat{\sigma}_\gamma^2 = s^2\mathbf{a}'(\mathbf{X}'\mathbf{X})^{-1}\mathbf{a}. \tag{4.24}$$

Confidence intervals and tests for γ may be obtained from the fact that under normality the "t-statistic"

$$T = \frac{\widehat{\gamma} - \gamma}{\widehat{\sigma}_\gamma} \tag{4.25}$$

has a t-distribution with $n - p^*$ degrees of freedom, where $p^* = \mathrm{rank}(\mathbf{X})$. In particular, a confidence upper bound and a two-sided confidence interval for γ with level $1 - \alpha$ are given by

$$\widehat{\gamma} + \widehat{\sigma}_\gamma t_{n-p^*,1-\alpha} \quad \text{and} \quad (\widehat{\gamma} - \widehat{\sigma}_\gamma t_{n-p^*,1-\alpha/2}, \widehat{\gamma} + \widehat{\sigma}_\gamma t_{n-p^*,1-\alpha/2}) \tag{4.26}$$

where $t_{n.\delta}$ is the δ-quantile of a t-distribution with n degrees of freedom. Similarly, the tests of level α for the null hypothesis $H_0 : \gamma = \gamma_0$ against the two-sided alternative $\gamma \neq \gamma_0$ and the one-sided alternative $\gamma > \gamma_0$ have the rejection regions

$$|\widehat{\gamma} - \gamma_0| > \widehat{\sigma}_\gamma t_{n-p^*,1-\alpha/2} \quad \text{and} \quad \widehat{\gamma} > \gamma_0 + \widehat{\sigma}_\gamma t_{n-p^*,1-\alpha}, \tag{4.27}$$

respectively.

If the u_i are not normal but have a finite variance, then for large n it can be shown using the central limit theorem that $\widehat{\boldsymbol{\beta}}_{LS}$ is approximately normal, with parameters given by (4.18), provided that:

$$\text{none of the } \mathbf{x}_i \text{ is "much larger" than the rest.} \tag{4.28}$$

This condition is formalized in (10.33) in Section 10.10.1. Recall that for large n the quantiles $t_{n,\beta}$ of the t-distribution converge to the quantiles z_β of N(0, 1). For the large-sample theory of the LS estimator, see Stapleton (1995) and Huber and Ronchetti (2009, p. 157).

4.3 Classical methods for outlier detection

The most popular way to deal with regression outliers is to use LS and try to find the influential observations. After they have been identified, a decision must be taken, for example modifying or deleting them and applying LS to the modified data. Many numerical and/or graphical procedures – so-called *regression diagnostics* – are available for detecting influential observations based on an initial LS fit. They include the familiar Q–Q plots of residuals, and plots of residuals against fitted values. See Weisberg (1985), Belsley *et al.* (1980) or Chatterjee and Hadi (1988) for further details on these methods, as well as for proofs of the statements in this section.

The influence of one observation $\mathbf{z}_i = (\mathbf{x}_i, y_i)$ on the LS estimator depends both on y_i being too large or too small compared to instances of y from similar instances of $\mathbf{x}$, and on how "large" $\mathbf{x}_i$ is; that is, how much *leverage* $\mathbf{x}_i$ has. Most popular diagnostics for measuring the influence of $\mathbf{z}_i = (\mathbf{x}_i, y_i)$ are based on comparing the LS estimator based on the full data with LS when $\mathbf{z}_i$ is omitted. Call $\hat{\boldsymbol{\beta}}$ and $\hat{\boldsymbol{\beta}}_{(i)}$ the LS estimates based on the full data and on the data without $\mathbf{z}_i$, and let

$$\hat{\mathbf{y}} = \mathbf{X}\hat{\boldsymbol{\beta}}, \quad \hat{\mathbf{y}}_{(i)} = \mathbf{X}\hat{\boldsymbol{\beta}}_{(i)}$$

where $r_i = r_i(\hat{\boldsymbol{\beta}})$. Note that if $p^* < p$, then $\hat{\boldsymbol{\beta}}_{(i)}$ is not unique, but $\hat{\mathbf{y}}_{(i)}$ is unique. Then the *Cook distance* of $\mathbf{z}_i$ is

$$D_i = \frac{1}{p^* s^2} \|\hat{\mathbf{y}}_{(i)} - \hat{\mathbf{y}}\|^2$$

where $p^* = \text{rank}(\mathbf{X})$ and $\hat{\sigma}$ is the residual standard deviation estimator

$$s^2 = \frac{1}{n - p^*} \sum_{i=1}^{n} r_i^2.$$

Call $\mathbf{H}$ the matrix of the orthogonal projection on the image of $\mathbf{X}$; that is, on the subspace $\{\mathbf{X}\boldsymbol{\beta} : \boldsymbol{\beta} \in R^p\}$. The matrix $\mathbf{H}$ is the so-called "hat matrix" and its diagonal elements $h_1, ..., h_n$ are the *leverages* of $\mathbf{x}_1, ..., \mathbf{x}_n$. If $p^* = p$, then $\mathbf{H}$ fulfills

$$\mathbf{H} = \mathbf{X}(\mathbf{X}'\mathbf{X})^{-1}\mathbf{X}' \quad \text{and} \quad h_i = \mathbf{x}_i'(\mathbf{X}'\mathbf{X})^{-1}\mathbf{x}_i. \tag{4.29}$$

The h_i satisfy

$$\sum_{i=1}^{n} h_i = p^*, \quad h_i \in [0, 1]. \tag{4.30}$$

It can be shown that the Cook distance is easily computed in terms of the h_i:

$$D_i = \frac{r_i^2}{s^2} \frac{h_i}{p^*(1-h_i)^2}. \tag{4.31}$$

It follows from (4.31) that observations with high leverage are more influential than observations with low leverage having the same residuals.

When the regression has an intercept,

$$h_i = \frac{1}{n} + (\underline{\mathbf{x}}_i - \overline{\underline{\mathbf{x}}})'(\mathbf{X}^{*\prime}\mathbf{X}^*)^{-1}(\mathbf{x}_i - \overline{\underline{\mathbf{x}}}) \tag{4.32}$$

where $\overline{\mathbf{x}}$ is the average of the $\mathbf{x}_i$s and $\mathbf{X}^*$ is the $n \times (p-1)$ matrix whose ith row is $(\underline{\mathbf{x}}_i - \overline{\mathbf{x}})'$. In this case h_i is a measure of how far $\mathbf{x}_i$ is from the average value $\overline{\underline{\mathbf{x}}}$.

Calculating h_i does not always require the explicit computation of $\mathbf{H}$. For example, in the case of the two-way design (4.10), it follows from the symmetry of the design that all the h_i are equal, and then (4.30) yields

$$h_i = \frac{p^*}{n} = \frac{I+J-1}{IJ}.$$

While D_i can detect outliers in simple situations, it fails for more complex configurations and may even fail to recognize a single outlier. The reason is that r_i, h_i and s may be greatly influenced by the outlier. It is safer to use statistics based on the "leave-one-out" approach, as follows. The leave-one-out residual $r_{(i)} = y_i - \widehat{\boldsymbol{\beta}}'_{(i)}\mathbf{x}_i$ is known to be expressible as

$$r_{(i)} = \frac{r_i}{1-h_i}. \tag{4.33}$$

and it is shown in the above references that

$$\mathrm{Var}(r_{(i)}) = \frac{\sigma^2}{1-h_i}.$$

An estimator of σ^2 that is free of the influence of $\mathbf{x}_i$ is the quantity $s_{(i)}^2$, which is defined like s^2, but has the ith observation deleted from the sample. It is also shown in the references above that

$$s_{(i)}^2 = \frac{1}{n-p^*-1}\left[(n-p^*)s^2 - \frac{r_i^2}{1-h_i}\right], \tag{4.34}$$

and a Studentized version of $r_{(i)}$ is given by

$$t_{(i)} = \sqrt{1-h_i}\frac{r_{(i)}}{s_{(i)}} = \frac{1}{\sqrt{1-h_i}}\frac{r_i}{s_{(i)}}. \tag{4.35}$$

Under the normal distribution model, $t_{(i)}$ has a t-distribution with $n-1$ degrees of freedom. Then a test of outlyingness with significance level α is to decide that the ith observation is an outlier if $|t_{(i)}| > t_{n-1,(1-\alpha)/2}$. A graphical analysis is provided by the normal Q–Q plot of $t_{(i)}$.

While the above "complete" leave-one-out approach ensures the detection of an isolated outlier, it can still be fooled by the combined action of several outliers, an effect that is referred to as *masking*.

Example 4.2 *The data in Table 4.2 (Scheffé 1959, p. 138) are the yields of grain for eight varieties of oats in five replications of a randomized-block experiment. Tables and figures for this example can be obtained with script* ***oats.R***.

Fitting (4.10) by LS yields residuals with no noticeable structure, and the usual F-tests for row and column effects have highly significant p-values of 0.00002 and 0.001, respectively. To show the effect of outliers on the classical procedure, we have modified five data values. Table 4.3 shows the data with the five altered values in boldface. Figure 4.2 shows the normal Q–Q plot of $t_{(i)}$ for the altered data. Again, nothing suspicious appears. But the p-values of the F-tests are now 0.13 and 0.04, the first of which is quite insignificant and the second of which is barely significant at the liberal 0.05 level. The diagnostics have thus failed to indicate a departure from the model, with serious consequences.

There is a vast literature on regression diagnostics. These procedures are fast, and are much better than naively fitting LS without further care. But they are inferior to robust methods in several senses:

- they may fail in the presence of masking;
- the distribution of the resulting estimator is unknown;
- the variability may be underestimated;
- once an outlier is found further ones may appear, and it is not clear when one should stop.

Table 4.2 Oats data

	Block				
Variety	I	II	III	IV	V
1	296	357	340	331	348
2	402	390	431	340	320
3	437	334	426	320	296
4	303	319	310	260	242
5	469	405	442	487	394
6	345	342	358	300	308
7	324	339	357	352	230
8	488	374	401	338	320

Table 4.3 Modified oats data

	Block				
Variety	I	II	III	IV	V
1	**476**	357	340	331	348
2	402	390	431	340	320
3	437	334	426	320	296
4	303	319	310	260	**382**
5	469	405	442	**287**	394
6	345	342	358	300	308
7	324	339	357	352	**410**
8	**288**	374	401	338	320

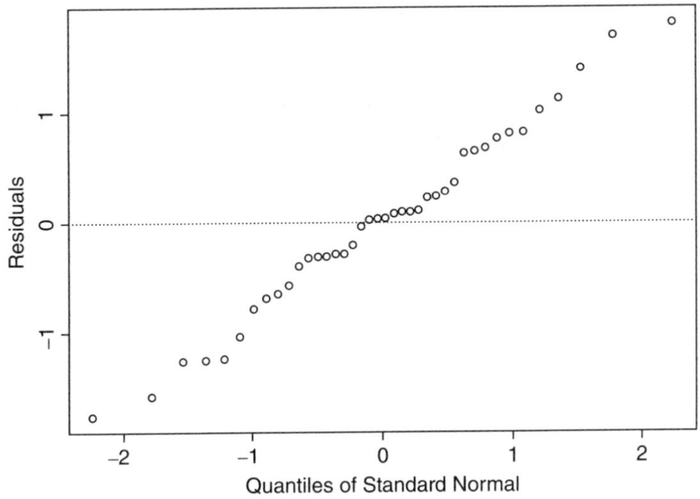

Figure 4.2 Altered oats data: Q–Q plot of LS residuals

4.4 Regression M-estimators

As in Section 2.3 we shall now develop estimators combining robustness and efficiency. Assume model (4.4) with *fixed* **X** where u_i has a density

$$\frac{1}{\sigma}f_0\left(\frac{u}{\sigma}\right),$$

where σ is a scale parameter. For the linear model (4.4) the y_i are independent but not identically distributed, y_i has density

$$\frac{1}{\sigma}f_0\left(\frac{y-\mathbf{x}_i'\boldsymbol{\beta}}{\sigma}\right)$$

and the likelihood function for $\boldsymbol{\beta}$ assuming a fixed value of σ is

$$L(\boldsymbol{\beta}) = \frac{1}{\sigma^n} \prod_{i=1}^{n} f_0 \left(\frac{y_i - \mathbf{x}_i' \boldsymbol{\beta}}{\sigma} \right).$$

Calculating the MLE means maximizing $L(\boldsymbol{\beta})$, which is equivalent to finding $\widehat{\boldsymbol{\beta}}$ such that

$$\frac{1}{n} \sum_{i=1}^{n} \rho_0 \left(\frac{r_i(\widehat{\boldsymbol{\beta}})}{\sigma} \right) + \ln \sigma = \min, \tag{4.36}$$

where $\rho_0 = -\ln f_0$, as in (2.14). We shall deal with estimators defined by (4.36). Continuing to assume σ is known, and differentiating with respect to $\boldsymbol{\beta}$, we have the analogue of the normal equations:

$$\sum_{i=1}^{n} \psi \left(\frac{r_i(\widehat{\boldsymbol{\beta}})}{\sigma} \right) \mathbf{x}_i = \mathbf{0}, \tag{4.37}$$

where $\psi_0 = \rho_0' = -f_0'/f_0$. If f_0 is the standard normal then $\widehat{\boldsymbol{\beta}}$ is the LS estimator (4.12), and if f_0 is the double exponential density then $\widehat{\boldsymbol{\beta}}$ satisfies

$$\sum_{i=1}^{n} \left| r_i(\widehat{\boldsymbol{\beta}}) \right| = \min$$

and $\widehat{\boldsymbol{\beta}}$ is called an L_1 *estimator,* which is the regression equivalent of the median. It is remarkable that this estimator was studied before LS (by Boscovich in 1757 and Laplace in 1799). Differentiating the likelihood function in this case gives

$$\sum_{i=1}^{n} \operatorname{sgn}(r_i(\widehat{\boldsymbol{\beta}}))\mathbf{x}_i = \mathbf{0} \tag{4.38}$$

where "sgn" denotes the sign function (2.20). If the model contains an intercept term, (4.38) implies that the residuals have zero median.

Unlike LS there are in general no explicit expressions for an L_1 estimator. However, there exist very fast algorithms to compute it (Barrodale and Roberts, 1973; Portnoy and Koenker, 1997). We note also that an L_1 estimator $\widehat{\boldsymbol{\beta}}$ may not be unique, and it has the property that at least p residuals are zero (Bloomfield and Staiger, 1983).

We define *regression M-estimators* as solutions $\widehat{\boldsymbol{\beta}}$ to

$$\sum_{i=1}^{n} \rho \left(\frac{r_i(\widehat{\boldsymbol{\beta}})}{\widehat{\sigma}} \right) = \min \tag{4.39}$$

where $\widehat{\sigma}$ is an error scale estimator. Differentiating (4.39) yields the equation

$$\sum_{i=1}^{n} \psi \left(\frac{r_i(\widehat{\boldsymbol{\beta}})}{\widehat{\sigma}} \right) \mathbf{x}_i = \mathbf{0} \tag{4.40}$$

where $\psi = \rho'$. The last equation need not be the estimating equation of an MLE. In most situations considered in this chapter, $\hat{\sigma}$ is computed previously, but it can also be computed simultaneously through a scale M-estimating equation.

It will henceforth be assumed that ρ and ψ are respectively a ρ- and a ψ-function in the sense of Definitions 2.1 and 2.2. The matrix $\mathbf{X}$ will be assumed to have full rank. In the special case where σ is assumed known, the reader may verify that the estimators are regression and affine equivariant (see Problem 4.1). The case of estimated σ is considered in Section 4.4.2.

Solutions to (4.40) with monotone (resp. redescending) ψ are called *monotone* (resp. *redescending) regression M-estimators*. The main advantage of monotone estimators is that all solutions of (4.40) are solutions of (4.39). Furthermore, if ψ is increasing then the solution is unique (see Theorem 10.15). The example in Section 2.8.1 showed that in the case of redescending location estimators, the estimating equation may have "bad" roots. This cannot happen with monotone estimators. On the other hand, we have seen in Section 3.4 that redescending M-estimators of location yield a better trade-off between robustness and efficiency, and the same can be shown to hold in the regression context. Computing redescending estimators requires a starting point, and this will be the main role of monotone estimators. This matter is pursued further in Section 4.4.2.

4.4.1 M-estimators with known scale

Assume model (4.4) with u such that

$$\mathrm{E}\psi\left(\frac{u}{\sigma}\right) = 0 \tag{4.41}$$

which holds in particular if u is symmetric. Then, if (4.28) holds, $\hat{\boldsymbol{\beta}}$ is consistent for $\boldsymbol{\beta}$ in the sense that

$$\hat{\boldsymbol{\beta}} \to_p \boldsymbol{\beta} \tag{4.42}$$

when $n \to \infty$, and furthermore for large n

$$\mathcal{D}(\hat{\boldsymbol{\beta}}) \approx \mathrm{N}_p(\boldsymbol{\beta}, v(\mathbf{X}'\mathbf{X})^{-1}) \tag{4.43}$$

where v is the same as in (2.65):

$$v = \sigma^2 \frac{\mathrm{E}\psi(u/\sigma)^2}{(\mathrm{E}\psi'(u/\sigma))^2}. \tag{4.44}$$

A general proof is given by Yohai and Maronna (1979).

Thus the approximate covariance matrix of an M-estimator differs only by a constant factor from that of the LS estimator. Hence its efficiency for normal u does not depend on $\mathbf{X}$; that is,

$$\mathrm{Eff}(\hat{\boldsymbol{\beta}}) = \frac{\sigma_0^2}{v} \tag{4.45}$$

where υ is given by (4.44) with the expectations computed for $u \sim \mathrm{N}(0, \sigma_0^2)$. It is easy to see that the efficiency does not depend on σ_0.

It is important to note that if we have a model with intercept (4.5), and (4.41) does not hold, then the intercept is asymptotically biased, but the slope estimators are nonetheless consistent (see Section 4.9.2):

$$\widehat{\boldsymbol{\beta}}_1 \to_p \boldsymbol{\beta}_1. \tag{4.46}$$

4.4.2 M-estimators with preliminary scale

When estimating location with an M-estimator in Section 2.7.1, we estimated σ using the MAD. Here, the equivalent procedure is first to compute the L_1 fit and from it obtain the analog of the normalized MAD by taking the median of the nonnull absolute residuals:

$$\widehat{\sigma} = \frac{1}{0.675}\mathrm{Med}_i(|r_i||r_i \neq 0). \tag{4.47}$$

The reason for using only *nonnull* residuals is that since at least p residuals are null, including all residuals when p is large could lead to underestimating σ. Recall that the L_1 estimator does not require estimating a scale.

Write $\widehat{\sigma}$ in (4.47) as $\widehat{\sigma}(\mathbf{X}, \mathbf{y})$. Then, since the L_1 estimator is regression, scale and affine equivariant, it is easy to show that

$$\widehat{\sigma}(\mathbf{X}, \mathbf{y} + \mathbf{X}\boldsymbol{\gamma}) = \widehat{\sigma}(\mathbf{X}, \mathbf{y}),\ \widehat{\sigma}(\mathbf{X}\mathbf{A}, \mathbf{y}) = \widehat{\sigma}(\mathbf{X}, \mathbf{y}),\ \widehat{\sigma}(\mathbf{X}, \lambda\mathbf{y}) = |\lambda|\widehat{\sigma}(\mathbf{X}, \mathbf{y}) \tag{4.48}$$

for all $\boldsymbol{\gamma} \in R^p$, nonsingular $\mathbf{A} \in R^{p\times p}$ and $\lambda \in R$. We say that $\widehat{\sigma}$ is regression and affine *invariant* and scale equivariant.

We then obtain a regression M-estimator by solving (4.39) or (4.40) with $\widehat{\sigma}$ instead of σ. Then (4.48) implies that $\widehat{\boldsymbol{\beta}}$ is regression, affine and scale equivariant (Problem 4.2).

Assume that $\widehat{\sigma} \to_p \sigma$ and that (4.41) holds. Under (4.4) we would expect that for large n the distribution of $\widehat{\boldsymbol{\beta}}$ is approximated by (4.43) and (4.44); that is, that $\widehat{\sigma}$ can be replaced by σ. Since ψ is odd, this holds in general if the distribution of u_i is symmetric. Thus the efficiency of the estimator does not depend on $\mathbf{X}$.

If the model contains an intercept, the approximate distribution result holds for the slopes *without* any requirement on u_i. More precisely, assume model (4.5). Then $\widehat{\boldsymbol{\beta}}_1$ is approximately normal, with mean $\boldsymbol{\beta}_1$ and covariance matrix $\upsilon\mathbf{C}^{-1}$, with υ given by (4.44) and $\mathbf{C}$ defined in (4.19) (see Section 10.10.1 for a heuristic proof).

We can estimate υ in (4.44) as

$$\widehat{\upsilon} = \widehat{\sigma}^2 \frac{\mathrm{ave}_i\{\psi(r_i/\widehat{\sigma})^2\}}{[\mathrm{ave}_i\{\psi'(r_i/\widehat{\sigma})\}]^2}\frac{n}{n-p} \tag{4.49}$$

where the denominator $n - p$ appears for the same reasons as in (4.22). Hence for large n we may treat $\widehat{\boldsymbol{\beta}}$ as approximately normal:

$$\mathcal{D}(\widehat{\boldsymbol{\beta}}) \approx \mathrm{N}_p(\boldsymbol{\beta}, \widehat{\upsilon}(\mathbf{X}'\mathbf{X})^{-1}). \tag{4.50}$$

Thus we can proceed as in (4.24)–(4.27), but replacing s^2 in (4.24) by the estimator $\hat{v}$ above so that $\hat{\sigma}^2_\gamma = \hat{v}\mathbf{a}'(\mathbf{X}'\mathbf{X})^{-1}\mathbf{a}$, to obtain approximate confidence intervals and tests. In the case of intervals and tests for a single coefficient β_i we have

$$\hat{\sigma}^2_{\beta_i} = \hat{v}(\mathbf{X}'\mathbf{X})^{-1}_{ii}$$

where the subscripts ii mean the ith diagonal element of matrix $(\mathbf{X}'\mathbf{X})^{-1}$.

As we have seen in the location case, one important advantage of redescending estimators is that they give null weight to large residuals, which implies the possibility of a high efficiency for both normal and heavy-tailed data. This is also true for regression, since the efficiency depends only on v, which is the same as for location. Therefore our recommended procedure is to use L_1 as a basis for computing $\hat{\sigma}$ and as a starting point for the iterative computing of a bisquare M-estimator.

Example 4.1 (continuation) The slope and intercept values for the bisquare M-estimator with 0.85 efficiency are shown in Table 4.1, along with those of the LS estimator using the full data, the LS estimator computed without the points labeled 1, 2, and 4, and the L_1 estimator. The corresponding fitted lines are shown in Figure 4.3. The results are very similar to the LS estimator computed without the three atypical points.

The estimated standard deviations of the slope are 0.122 for LS and 0.050 for the bisquare M-estimator, and the respective confidence intervals with level 0.95 are (−0.849, −0.371) and (−0.580, −0.384). It is seen that the outliers inflate the confidence interval based on the LS estimator relative to that based on the bisquare M-estimator.

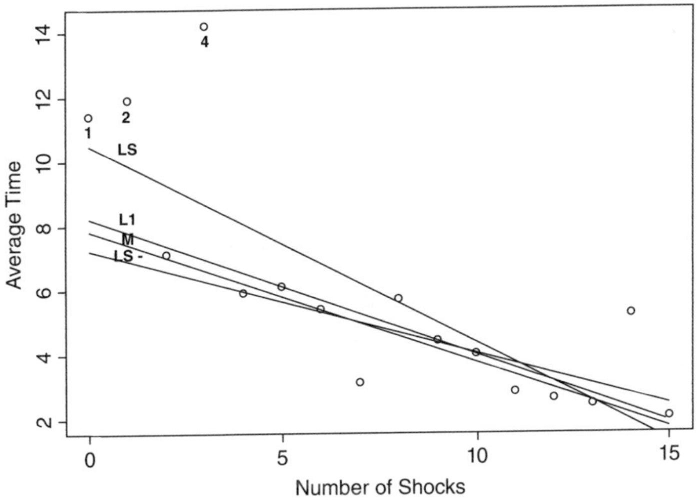

Figure 4.3 Rats data: fits by least squares (LS), L_1, bisquare M-estimator (M) and least squares with outliers omitted (LS–)

Example 4.2 (continuation) Tables and figures for this example are obtained with script **oats.R**. Figure 4.4 shows the residual Q–Q plot based on the bisquare M-estimator and it is seen that the five modified values stand out from the rest.

Table 4.4 gives the p-values of the robust likelihood ratio-type test to be described in Section 4.7.2 for row and column effects. Values are shown for the original and the altered data, together with those of the classical F-test already given.

We see that the M-estimator results for the altered data are quite close to those for the original data. Furthermore, for the altered data the robust test again gives strong evidence of row and column effects.

4.4.3 Simultaneous estimation of regression and scale

Another approach to deal with the estimation of σ is to proceed as in Section 2.7.2, namely to add to the estimating equation (4.40) for $\boldsymbol{\beta}$ an M-estimating equation for σ, resulting in the system

$$\sum_{i=1}^{n} \psi\left(\frac{r_i(\widehat{\boldsymbol{\beta}})}{\widehat{\sigma}}\right)\mathbf{x}_i = \mathbf{0}, \tag{4.51}$$

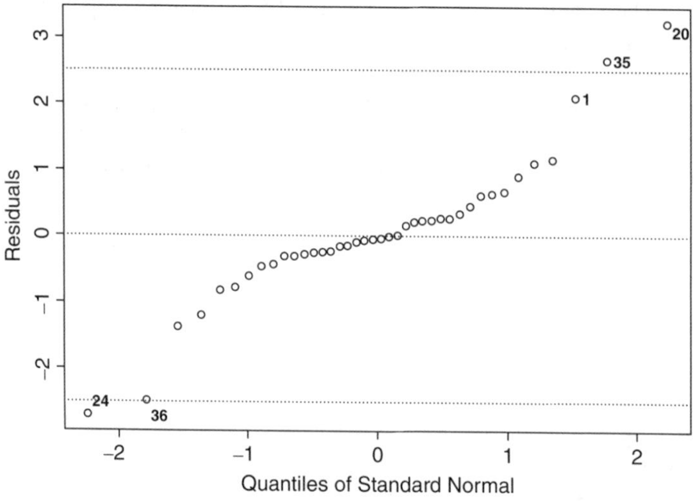

Figure 4.4 Altered oats data: normal Q–Q plot of residuals from M-estimator

Table 4.4 Oats data: p-values of tests

	Rows		Columns	
	F	Robust	F	Robust
Original	1.56×10^{-5}	1.7×10^{-6}	0.001	2.6×10^{-5}
Altered	0.13	4×10^{-5}	0.04	1.7×10^{-4}

$$\frac{1}{n}\sum_{i=1}^{n}\rho_{\text{scale}}\left(\frac{r_i(\boldsymbol{\beta})}{\hat{\sigma}}\right)=\delta, \tag{4.52}$$

where ρ_{scale} is a ρ-function. Note that differentiating (4.36) with respect to $\boldsymbol{\beta}$ and σ yields a system of the form (4.51)–(4.52), with ρ_{scale} given in (2.73). Therefore this class of estimators includes the MLE.

Simultaneous estimators with monotonic ψ are less robust than those of the former Section 4.4.2 (recall Section 3.2.4 for the location case), but they will be used with redescending ψ in another context in Section 5.4.1.

4.5 Numerical computing of monotone M-estimators

4.5.1 The L_1 estimator

As was mentioned above, computing the L_1 estimator requires sophisticated algorithms, such as the one due to Barrodale and Roberts (1973). There are, however, some cases in which this estimator can be computed explicitly. For regression through the origin ($y_i = \beta x_i + u_i$), the reader can verify that $\hat{\beta}$ is a "weighted median" (Problem 4.4). For one-way ANOVA (4.9) we immediately have that $\hat{\beta}_k = \text{Med}_i(y_{ik})$. And for two-way ANOVA with one observation per cell (i.e., (4.10)–(4.11) with $K_{ij} = 1$), there is a simple method that we now describe.

Let $y_{ij} = \mu + \alpha_i + \gamma_j + u_{ij}$. Then differentiating $\sum_i\sum_j|y_{ij} - \mu - \alpha_i - \gamma_j|$ with respect to μ, α_i and γ_j, and recalling that the derivative of $|x|$ is sgn(x), it follows that (4.38) is equivalent to

$$\text{Med}_{i,j}(r_{ij}) = \text{Med}_j(r_{ij}) = \text{Med}_i(r_{ij}) = 0 \quad \text{for all} \quad i,j \tag{4.53}$$

where $r_{ij} = y_{ij} - \hat{\mu} - \hat{\alpha}_i - \hat{\gamma}_j$. These equations suggest an iterative procedure due to Tukey (1977), known as "median polish', which goes as follows (where "$a \leftarrow b$" stands for "replace a by b"):

1. Put $\hat{\alpha}_i = \hat{\gamma}_j = 0$ for $i = 1, ..., I$ and $j = 1, ..., J$, and $\hat{\mu} = 0$, and hence $r_{ij} = y_{ij}$.
2. For $i = 1, ..., I$: let $\delta_i = \text{Med}_j(r_{ij})$. Update $\hat{\alpha}_i \leftarrow \hat{\alpha}_i + \delta_i$ and $r_{ij} \leftarrow r_{ij} - \delta_i$.
3. For $j = 1, ..., J$: let $\delta_j = \text{Med}_i(r_{ij})$. Update $\hat{\gamma}_j \leftarrow \hat{\gamma}_j + \delta_j$ and $r_{ij} \leftarrow r_{ij} - \delta_j$.
4. Repeat steps 2–3 until no more changes take place.
5. Put $a = I^{-1}\sum_i\hat{\alpha}_i$ and $b = J^{-1}\sum_j\hat{\gamma}_j$, and $\hat{\alpha}_i \leftarrow \hat{\alpha}_i - a$, $\hat{\gamma}_j \leftarrow \hat{\gamma}_j - b$, $\hat{\mu} \leftarrow a + b$.

If I or J is even, the median must be understood as the "high" or "low" median (Section 1.2), otherwise the procedure may oscillate indefinitely.

It can be shown (Problem 4.5) that the sum of absolute residuals

$$\sum_{i=1}^{I}\sum_{j=1}^{J}\left|y_{ij} - \hat{\mu} - \hat{\alpha}_i - \hat{\gamma}_j\right|$$

decreases at each step of the algorithm. The result frequently coincides with an L_1 estimator, and is otherwise generally close to it. Sposito (1987) gives conditions under which the median polish coincides with the L_1 estimator.

4.5.2 M-estimators with smooth ψ-function

In the case of a smooth ψ-function, one can solve (4.37) using an iterative reweighting method similar to that of Section 2.8.Define W as in (2.31), and then with σ replaced by $\hat{\sigma}$, the M-estimator equation (4.37) for $\hat{\boldsymbol{\beta}}$ may be written as

$$\sum_{i=1}^{n} w_i r_i \mathbf{x}_i = \sum_{i=1}^{n} w_i \mathbf{x}_i (y_i - \mathbf{x}_i' \hat{\boldsymbol{\beta}}) = \mathbf{0} \tag{4.54}$$

with $w_i = W(r_i/\hat{\sigma})$. These are "weighted normal equations", and if the w_i were known, the equations could be solved by applying LS to $\sqrt{w_i} y_i$ and $\sqrt{w_i}\mathbf{x}_i$. But the w_i are not known and depend upon the data. So the procedure, which depends on a tolerance parameter ε, is

1. Compute an initial L_1 estimator $\hat{\boldsymbol{\beta}}_0$ and compute $\hat{\sigma}$ from (4.47).
2. For $k = 0, 1, 2, \ldots$:
   (a) Given $\hat{\boldsymbol{\beta}}_k$, for $i = 1, \ldots, n$ compute $r_{i,k+1} = y_i - \mathbf{x}_i' \hat{\boldsymbol{\beta}}_k$ and $w_{i,k+1} = W(r_{i,k+1}/\hat{\sigma})$.
   (b) Compute $\hat{\boldsymbol{\beta}}_{k+1}$ by solving

$$\sum_{i=1}^{n} w_{i,k} \mathbf{x}_i (y_i - \mathbf{x}_i' \hat{\boldsymbol{\beta}}) = \mathbf{0}.$$

3. Stop when $\max_i(|r_{i,k} - r_{i,k+1}|)/\hat{\sigma} < \varepsilon$.

This algorithm converges if $W(x)$ is nonincreasing for $x > 0$ (Section 9.1). If ψ is monotone, since the solution is essentially unique, the choice of the starting point influences the number of iterations but not the final result. This procedure is called "iteratively reweighted least squares" (IRWLS).

For simultaneous estimation of $\boldsymbol{\beta}$ and σ, the procedure is the same, except that at each iteration $\hat{\sigma}$ is also updated as in (2.80).

4.6 BP of monotone regression estimators

In this section we discuss the breakdown point of monotone estimators for nonrandom predictors. Assume $\mathbf{X}$ is of full rank so that the estimators are well defined. Since $\mathbf{X}$ is fixed, only $\mathbf{y}$ can be changed, and this requires a modification of the definition of the breakdown point (BP). The FBP for regression with fixed predictors is defined as

$$\varepsilon^* = \frac{m^*}{n},$$

with

$$m^* = \max\{m \geq 0 : \hat{\boldsymbol{\beta}}(\mathbf{X}, \mathbf{y}_m) \text{ bounded } \forall\, \mathbf{y}_m \in \mathcal{Y}_m\}, \tag{4.55}$$

where $\mathcal{Y}_m$ is the set of n–vectors with at least $n - m$ elements in common with $\mathbf{y}$. It is clear that the LS estimator has $\varepsilon^* = 0$.

Let $k^* = k^*(\mathbf{X})$ be the maximum number of $\mathbf{x}_i$ lying on the same subspace of dimension $< p$:

$$k^*(\mathbf{X}) = \max\{\#(\boldsymbol{\beta}'\mathbf{x}_i = 0) : \boldsymbol{\beta} \in R^p,\ \boldsymbol{\beta} \neq \mathbf{0}\} \tag{4.56}$$

where a subspace of dimension 0 is the set $\{0\}$. In the case of simple straight-line regression, k^* is the maximum number of repeated x_i. We have $k^* \geq p - 1$ always. If $k^* = p - 1$ then $\mathbf{X}$ is said to be in *general position*. In the case of a model with intercept (4.6), $\mathbf{X}$ is in general position iff no more than $p - 1$ of the $\mathbf{x}_i$ lie on a hyperplane.

It is shown in Section 4.9.3 that for all regression equivariant estimators

$$\varepsilon^* \leq \varepsilon^*_{\max} := \frac{m^*_{\max}}{n} \tag{4.57}$$

where

$$m^*_{\max} = \left[\frac{n - k^* - 1}{2}\right] \leq \left[\frac{n - p}{2}\right]. \tag{4.58}$$

In the location case, $k^* = 0$ and $m^*_{\max}/n$ becomes (3.26). The FBP of monotone M-estimators is given in Section 4.9.4. For the one-way design (4.9) and the two-way design (4.10), it can be shown that the FBP of monotone M-estimators attains the maximum (4.57) (see Section 4.9.3). In the first case

$$m^* = m^*_{\max} = \left[\frac{\min_j n_j - 1}{2}\right], \tag{4.59}$$

and so if at least half of the elements of the smallest sample are outliers then one of the $\hat{\beta}_j$ is unbounded. In the second case

$$m^* = m^*_{\max} = \left[\frac{\min(I, J) - 1}{2}\right], \tag{4.60}$$

and so if at least half of the elements of a row or column are outliers then at least one of the estimators $\hat{\mu}, \hat{\alpha}_i$ or $\hat{\gamma}_j$ breaks down. It is natural to conjecture that the FBP of monotone M-estimators attains the maximum (4.57) for all $\mathbf{X}$ such that x_{ij} is either 0 or 1, but no general proof is known.

For designs that are not zero–one designs, the FBP of M-estimators will in general be lower than $\varepsilon^*_{\max}$. This may happen even when there are no leverage points. For example, in the case of a uniform design $x_i = i, i = 1, \ldots, n$, for the fitting of a straight line through the origin, we have $k^* = 1$ and hence $\varepsilon^*_{\max} \approx 1/2$, while for large n it can be shown that $\varepsilon^* \approx 0.3$ (see Section 4.9.4). The situation is worse for fitting a polynomial (Problem 4.7). It is even worse when there are leverage points. Consider for instance the design

$$x_i = i \quad \text{for} \quad i = 1, \ldots, 10, \quad x_{11} = 100. \tag{4.61}$$

Then it can be shown that $m^* = 0$ for a linear fit (Problem 4.8). The intuitive reason for this fact is that here the estimator is determined almost solely by y_{11}.

As a consequence, monotone M-estimators can be recommended as initial estimators for zero–one designs, and perhaps also for uniform designs, but not for designs where $\mathbf{X}$ has leverage points. The case of random $\mathbf{X}$ will be examined in the next chapter. The techniques discussed there will also be applicable to fixed designs with leverage points.

4.7 Robust tests for linear hypothesis

Regression M-estimators can be used to obtain robust approximate confidence intervals and tests for a single linear combination of the parameters. Define $\hat{\sigma}_\gamma^2$ as in (4.24), but with s^2 replaced by $\hat{v}$, as defined in (4.49). Then the tests and intervals are of the form (4.26)–(4.27). We shall now extend the theory to inference for several linear combinations of the β_j represented by the vector $\boldsymbol{\gamma} = \mathbf{A}\boldsymbol{\beta}$, where $\mathbf{A}$ is a $q \times p$ matrix of rank q.

4.7.1 Review of the classical theory

To simplify the exposition, it will be assumed that $\mathbf{X}$ has full rank; that is, $p^* = p$, but the results can be shown to hold for general p^*. Assume normally distributed errors and let $\hat{\boldsymbol{\gamma}} = \mathbf{A}\hat{\boldsymbol{\beta}}$, where $\hat{\boldsymbol{\beta}}$ is the LS estimator. Then $\hat{\boldsymbol{\gamma}} \sim \mathrm{N}(\boldsymbol{\gamma}, \boldsymbol{\Sigma}_\gamma)$ where

$$\boldsymbol{\Sigma}_\gamma = \sigma^2 \mathbf{A}(\mathbf{X}'\mathbf{X})^{-1}\mathbf{A}'.$$

An estimator of $\boldsymbol{\Sigma}_\gamma$ is given by

$$\hat{\boldsymbol{\Sigma}}_\gamma = s^2 \mathbf{A}(\mathbf{X}'\mathbf{X})^{-1}\mathbf{A}'. \tag{4.62}$$

It is proved in standard regression textbooks that $(\hat{\boldsymbol{\gamma}} - \boldsymbol{\gamma})'\hat{\boldsymbol{\Sigma}}_\gamma^{-1}(\hat{\boldsymbol{\gamma}} - \boldsymbol{\gamma})'/q$ has an F-distribution with q and $n - p^*$ degrees of freedom, and hence a confidence ellipsoid for $\boldsymbol{\gamma}$ of level $1 - \alpha$ is given by

$$\left\{ \boldsymbol{\gamma} : (\hat{\boldsymbol{\gamma}} - \boldsymbol{\gamma})'\hat{\boldsymbol{\Sigma}}_\gamma^{-1}(\hat{\boldsymbol{\gamma}} - \boldsymbol{\gamma}) \le qF_{q,n-p^*}(1 - \alpha) \right\},$$

where $F_{n_1,n_2}(\delta)$ is the δ-quantile of an F-distribution with n_1 and n_2 degrees of freedom.

We consider testing the linear hypothesis $H_0 : \boldsymbol{\gamma} = \boldsymbol{\gamma}_0$ for a given $\boldsymbol{\gamma}_0$, with level α. The so-called *Wald-type test* rejects H_0 when $\boldsymbol{\gamma}_0$ does not belong to the confidence ellipsoid, and hence has rejection region

$$T > F_{q,n-p^*}(1-\alpha) \tag{4.63}$$

with

$$T = \frac{1}{q}(\widehat{\boldsymbol{\gamma}} - \boldsymbol{\gamma}_0)'\widehat{\boldsymbol{\Sigma}}_{\boldsymbol{\gamma}}^{-1}(\widehat{\boldsymbol{\gamma}} - \boldsymbol{\gamma}_0). \tag{4.64}$$

It is also shown in standard texts, such as Scheffé (1959), that the statistic T can be written in the form

$$T = \frac{(S_R - S)/q}{S/(n-p^*)} \tag{4.65}$$

where

$$S = \sum_{i=1}^{n} r_i^2\left(\widehat{\boldsymbol{\beta}}\right), \quad S_R = \sum_{i=1}^{n} r_i^2\left(\widehat{\boldsymbol{\beta}}_R\right),$$

and where $\widehat{\boldsymbol{\beta}}_R$ is the LS estimator with the restriction $\boldsymbol{\gamma} = \boldsymbol{A}\boldsymbol{\beta} = \boldsymbol{\gamma}_0$. It is also shown that the test based on (4.65) coincides with the likelihood ratio test (LRT). We can also write the test statistic T (4.65) as

$$T = \frac{(S_R^* - S^*)}{q} \tag{4.66}$$

where

$$S^* = \sum_{i=1}^{n}\left(\frac{r_i(\widehat{\boldsymbol{\beta}})}{s}\right)^2, \quad S_R^* = \sum_{i=1}^{n}\left(\frac{r_i(\widehat{\boldsymbol{\beta}}_R)}{s}\right)^2. \tag{4.67}$$

The most common application of these tests is when H_0 is the hypothesis that some of the coefficients β_i are zero. We may assume, without loss of generality, that the hypothesis is

$$H_0 = \{\beta_1 = \beta_2 = \ldots = \beta_q = 0\}$$

which can be written as $H_0 : \boldsymbol{\lambda} = \boldsymbol{A}\boldsymbol{\beta} = \mathbf{0}$ with $\boldsymbol{A} = (\boldsymbol{I}, \mathbf{0})$, where $\mathbf{I}$ is the $q \times q$ identity matrix and $\mathbf{0}$ is a $(p-q) \times p$ matrix with all its elements zero.

When $q = 1$, we have $\gamma = \mathbf{a}'\boldsymbol{\beta}$ with $\mathbf{a} \in R^p$ and then the variance of $\widehat{\gamma}$ is estimated by

$$\widehat{\sigma}_\gamma^2 = \widehat{\sigma}^2\mathbf{a}'(\mathbf{X}'\mathbf{X})^{-1}\mathbf{a}.$$

In this special case the Wald test (4.64) simplifies to

$$T = \left(\frac{\widehat{\gamma} - \gamma_0}{\widehat{\sigma}_\gamma}\right)^2$$

and is equivalent to the two-sided test in (4.27).

When the errors u_i are not normal, but the conditions for the asymptotic normality of $\widehat{\boldsymbol{\beta}}$ given at the end of Section 4.2 hold, the test and confidence regions given in this section will still be approximately valid for large n. For this case, recall that if T has an $F(q,m)$ distribution then when $m \to \infty$, $qT \to_d \chi^2_q$.

4.7.2 Robust tests using M-estimators

Let $\widehat{\boldsymbol{\beta}}$ now be an M-estimator, and let $\widehat{\boldsymbol{\Sigma}}_{\widehat{\beta}} = \widehat{v}(\mathbf{X}'\mathbf{X})^{-1}$ be the estimator of its covariance matrix, with $\widehat{v}$ defined as in (4.49). Let

$$\widehat{\boldsymbol{\gamma}} = \mathbf{A}\widehat{\boldsymbol{\beta}}, \quad \widehat{\boldsymbol{\Sigma}}_{\gamma} = \mathbf{A}\widehat{\boldsymbol{\Sigma}}_{\widehat{\beta}}A' = \widehat{v}\mathbf{A}(\mathbf{X}'\mathbf{X})^{-1}\mathbf{A}'.$$

Then a robust "Wald-type test" is defined by the rejection region

$$\{T_W > F_{q,n-p^*}(1-\alpha)\}$$

with T_W equal to the right-hand side of (4.64), but the classical quantities there are replaced by the above robust estimators $\widehat{\boldsymbol{\gamma}}$ and $\widehat{\boldsymbol{\Sigma}}_{\gamma}$.

Let $\widehat{\boldsymbol{\beta}}_R$ be the M-estimator computed with the restriction that $\boldsymbol{\gamma} = \boldsymbol{\gamma}_0$:

$$\widehat{\boldsymbol{\beta}}_R = \arg\min_{\beta} \left\{ \sum_{i=1}^{n} \rho\left(\frac{r_i(\boldsymbol{\beta})}{\widehat{\sigma}}\right) : \mathbf{A}\boldsymbol{\beta} = \boldsymbol{\gamma}_0 \right\}.$$

A "likelihood ratio-type test" (LRTT) could be defined by the region

$$\{T > F_{q,n-p^*}(1-\alpha)\},$$

with T equal to the right-hand side of (4.66), but where the residuals in (4.67) correspond to an M-estimator $\widehat{\boldsymbol{\beta}}$. But this test would not be robust, since outliers in the observations y_i would result in corresponding residual outliers and hence an overdue influence on the test statistic.

A robust LRTT can instead be defined by the statistic

$$T_L = \sum_{i=1}^{n} \rho\left(\frac{r_i(\widehat{\boldsymbol{\beta}}_R)}{\widehat{\sigma}}\right) - \sum_{i=1}^{n} \rho\left(\frac{r_i(\widehat{\boldsymbol{\beta}})}{\widehat{\sigma}}\right)$$

with a bounded ρ. Let

$$\xi = \frac{\mathrm{E}\psi'(u/\sigma)}{\mathrm{E}\psi(u/\sigma)^2}.$$

Then it can be shown (see Hampel *et al.*, 1986) that under adequate regularity conditions, ξT_L converges in distribution under H_0 to a chi-squared distribution with q degrees of freedom. Since ξ can be estimated by

$$\widehat{\xi} = \frac{\mathrm{ave}_i\left\{\psi'(r_i(\widehat{\boldsymbol{\beta}})/\widehat{\sigma})\right\}}{\mathrm{ave}_i\left\{\psi(r_i(\widehat{\boldsymbol{\beta}})/\widehat{\sigma})^2\right\}},$$

an approximate LRTT for large n has rejection region

$$\hat{\xi} T_L > \chi^2_q(1-\alpha),$$

where $\chi^2_n(\delta)$ denotes the δ-quantile of the chi-squared distribution with n degrees of freedom.

Wald-type tests have the drawback of being based on $\mathbf{X}'\mathbf{X}$, which may affect the robustness of the test when there are high-leverage points. This makes LRTTs preferable. The influence of high-leverage points on inference is discussed further in Section 5.6.

4.8 *Regression quantiles

Let, for $\alpha \in (0, 1)$,

$$\rho_\alpha(x) = \begin{cases} \alpha x & \text{if} \quad x \geq 0 \\ -(1-\alpha)x & \text{if} \quad x < 0. \end{cases}$$

Then it is easy to show (Problem 2.13) that the solution of

$$\sum_{i=1}^{n} \rho_\alpha(y_i - \mu) = \min$$

is the sample α-quantile. In the same way, the solution of

$$\mathrm{E}\rho_\alpha(y - \mu) = \min$$

is an α-quantile of the random variable y.

Koenker and Bassett (1978) extended this concept to regression, defining the *regression* α-quantile as the solution $\hat{\boldsymbol{\beta}}$ of

$$\sum_{i=1}^{n} \rho_\alpha(y_i - \mathbf{x}_i'\hat{\boldsymbol{\beta}}) = \min. \tag{4.68}$$

The case $\alpha = 0.5$ corresponds to the L_1 estimator. Assume the model

$$y_i = \mathbf{x}_i'\boldsymbol{\beta}_\alpha + u_i,$$

where the $\mathbf{x}_i$ are fixed and the α-quantile of u_i is zero; this is equivalent to assuming that the α-quantile of y_i is $\mathbf{x}_i'\boldsymbol{\beta}_\alpha$. Then $\hat{\boldsymbol{\beta}}$ is an estimator of $\boldsymbol{\beta}_\alpha$.

Regression quantiles are especially useful with heteroskedastic data. Assume the usual situation when the model contains a constant term. If the u_i are identically distributed, then the $\boldsymbol{\beta}_\alpha$ for different α differ only in the intercept, and hence regression quantiles do not give much useful information. But if the u_i have different variability, then the $\boldsymbol{\beta}_\alpha$ will also have different slopes.

If the model is correct, one would like to have for $\alpha_1 < \alpha_2$ that $\mathbf{x}_0'\boldsymbol{\beta}_{\alpha_1} < \mathbf{x}_0'\boldsymbol{\beta}_{\alpha_2}$ for all $\mathbf{x}_0$ in the range of the data. But this cannot be mathematically ensured. Although this fact may be taken as an indication of model failure, it is better to ensure it from the start. Methods for avoiding the "crossing" of regression quantiles have been proposed by He (1997) and Zhao (2000).

There is a very large literature on regression quantiles; see Koenker *et al.* (2005) for references.

4.9 Appendix: Proofs and complements

4.9.1 Why equivariance?

In this section we want to explain why equivariance is a desirable property for a regression estimator. Let $\mathbf{y}$ verify the model (4.7). Here $\boldsymbol{\beta}$ is the vector of model parameters. If we put for some vector $\boldsymbol{\gamma}$

$$\mathbf{y}^* = \mathbf{y} + \mathbf{X}\boldsymbol{\gamma}, \tag{4.69}$$

then $\mathbf{y}^* = \mathbf{X}(\boldsymbol{\beta} + \boldsymbol{\gamma}) + \mathbf{u}$, so that $\mathbf{y}^*$ verifies the model with parameter vector

$$\boldsymbol{\beta}^* = \boldsymbol{\beta} + \boldsymbol{\gamma}. \tag{4.70}$$

If $\widehat{\boldsymbol{\beta}} = \widehat{\boldsymbol{\beta}}(\mathbf{X}, \mathbf{y})$ is an estimator, it would be desirable that if the data were transformed according to (4.69), the estimator would also transform according to (4.70); that is, $\widehat{\boldsymbol{\beta}}(\mathbf{X}, \mathbf{y}^*) = \widehat{\boldsymbol{\beta}}(\mathbf{X}, \mathbf{y}) + \boldsymbol{\gamma}$, which corresponds to regression equivariance (4.15).

Likewise, if $\mathbf{X}^* = \mathbf{X}\mathbf{A}$ for some matrix $\mathbf{A}$, then $\mathbf{y}$ verifies the model

$$\mathbf{y} = (\mathbf{X}^*\mathbf{A}^{-1})\boldsymbol{\beta} + \mathbf{u} = X^*(\mathbf{A}^{-1}\boldsymbol{\beta}) + \mathbf{u},$$

which is (4.7) with $\mathbf{X}$ replaced by $\mathbf{X}^*$ and $\boldsymbol{\beta}$ by $\mathbf{A}^{-1}\boldsymbol{\beta}$. Again, it is desirable that estimators transform the same way; that is, $\widehat{\boldsymbol{\beta}}(\mathbf{X}^*, \mathbf{y}) = \mathbf{A}^{-1}\widehat{\boldsymbol{\beta}}(\mathbf{X}, \mathbf{y})$, which corresponds to affine equivariance (4.17). Scale equivariance (4.16) is dealt with in the same manner.

It must be noted that although equivariance is desirable, it must sometimes be sacrificed for other properties, such as a lower prediction error. In particular, the estimators resulting from a procedure for variable selection considered in Section 5.6.2 are neither regression nor affine equivariant. The same thing happens in general with procedures for dealing with a large number of variables, such as ridge regression or least-angle regression (Hastie *et al.*, 2001).

4.9.2 Consistency of estimated slopes under asymmetric errors

We shall first prove (4.21). Let $\alpha = \mathrm{E}u_i$. Then (4.5) may be rewritten as

$$y_i = \beta_0^* + \underline{\mathbf{x}}_i'\boldsymbol{\beta}_1 + u_i^*, \tag{4.71}$$

where

$$u_i^* = u_i - \alpha, \quad \beta_0^* = \beta_0 + \alpha. \tag{4.72}$$

Since $\mathrm{E}u_i^* = 0$, the LS estimator is unbiased for the parameters, which means that $\mathrm{E}(\hat{\boldsymbol{\beta}}_1) = \boldsymbol{\beta}_1$ and $\mathrm{E}(\hat{\beta}_0) = \beta_0^*$, so that only the intercept will be biased.

We now prove (4.46) along the same lines. Let α be such that

$$\mathrm{E}\psi\left(\frac{u_i - \alpha}{\sigma}\right) = 0.$$

Then reexpressing the model as (4.71)–(4.72), since $\mathrm{E}\psi(u_i^*/\sigma) = 0$, we may apply (4.42), and hence

$$\hat{\beta}_0 \to_p \beta_0^*, \quad \hat{\boldsymbol{\beta}}_1 \to_p \boldsymbol{\beta}_1,$$

which implies that the estimator of the slopes is consistent, although that of the intercept may be inconsistent.

4.9.3 Maximum FBP of equivariant estimators

The definition of the FBP in Section 4.6 can be modified to include the case of rank$(\mathbf{X}) < p$. Since in this case there exists $\boldsymbol{\theta} \neq \mathbf{0}$ such that $\mathbf{X}\boldsymbol{\theta} = \mathbf{0}$, (4.56) is modified as

$$k^*(\mathbf{X}) = \max\{\#(\boldsymbol{\theta}'\mathbf{x}_i = 0) : \boldsymbol{\theta} \in R^p, \ \mathbf{X}\boldsymbol{\theta} \neq \mathbf{0}\}. \tag{4.73}$$

If rank$(\mathbf{X}) < p$, there are infinite solutions to the equations, but all of them yield the same fit, $\mathbf{X}\hat{\boldsymbol{\beta}}$. We thus modify (4.55) with the requirement that the *fit* remains bounded:

$$m^* = \max\left\{m \geq 0 : \mathbf{X}\hat{\boldsymbol{\beta}}(\mathbf{X}, \mathbf{y}_m) \text{ bounded } \forall \ \mathbf{y}_m \in \mathcal{Y}_m\right\}.$$

We now prove the bound (4.58). Let $m = m^*_{\max} + 1$. We have to show that $\mathbf{X}\hat{\boldsymbol{\beta}}(\mathbf{X}, \mathbf{y})$ is unbounded for $\mathbf{y} \in \mathcal{Y}_m$. By decomposing into the case of even and odd $n - k^*$, it follows that

$$2m \geq n - k^*. \tag{4.74}$$

In fact, if $n - k^*$ is even, $n - k^* = 2q$, hence

$$m^*_{\max} = \left[\frac{n - k^* - 1}{2}\right] = [q - 0.5] = q - 1,$$

which implies $m = q$ and hence $2m = n - k^*$; the other case follows in a similar way. By the definition of k^*, there exists $\boldsymbol{\theta}$ such that $\mathbf{X}\boldsymbol{\theta} \neq \mathbf{0}$ and $\boldsymbol{\theta}'\mathbf{x}_i = 0$ for a set of size k^*. To simplify notation, we reorder the $\mathbf{x}_i$ so that

$$\boldsymbol{\theta}'\mathbf{x}_i = 0 \quad \text{for} \quad i = 1, \ldots, k^*. \tag{4.75}$$

Let, for some $t \in R$,

$$y_i^* = y_i + t\boldsymbol{\theta}'\mathbf{x}_i \quad \text{for} \quad i = k^* + 1, \ldots, k^* + m \tag{4.76}$$

$$y_i^* = y_i \quad \text{otherwise.} \tag{4.77}$$

Then $\mathbf{y}^* \in \mathcal{Y}_m$. Now let $\mathbf{y}^{**} = \mathbf{y}^* - t\mathbf{X}\boldsymbol{\theta}$. Then $y_i^{**} = y_i$ for $1 \le i \le k^*$ by (4.75), and also for $k^* + 1 \le i \le k^* + m$ by (4.76). Then $\mathbf{y}^{**} \in \mathcal{Y}_m$, since

$$\#(i : y_i^{**} = y_i) \ge k^* + m$$

and $n - (k^* + m) \le m$ by (4.74). Hence the equivariance (4.15) implies that

$$\mathbf{X}\hat{\boldsymbol{\beta}}(\mathbf{X}, \mathbf{y}^*) - \mathbf{X}\hat{\boldsymbol{\beta}}(\mathbf{X}, \mathbf{y}^{**}) = \mathbf{X}(\hat{\boldsymbol{\beta}}(\mathbf{X}, \mathbf{y}^*) - \hat{\boldsymbol{\beta}}(\mathbf{X}, \mathbf{y}^* - t\mathbf{X}\boldsymbol{\theta})) = t\mathbf{X}\boldsymbol{\theta},$$

which is unbounded for $t \in R$, and thus both $\mathbf{X}\hat{\boldsymbol{\beta}}(\mathbf{X}, \mathbf{y}^*)$ and $\mathbf{X}\hat{\boldsymbol{\beta}}(\mathbf{X}, \mathbf{y}^{**})$ cannot be bounded.

4.9.4 The FBP of monotone M-estimators

We now state the FBP of monotone M-estimators, which was derived by Ellis and Morgenthaler (1992) for the L_1 estimator and generalized by Maronna and Yohai (2000).

Let ψ be nondecreasing and bounded. Call Ξ the image of $\boldsymbol{X} : \Xi = \{\mathbf{X}\boldsymbol{\theta} : \boldsymbol{\theta} \in R^p\}$. For each $\boldsymbol{\xi} = (\xi_1, \ldots, \xi_n)' \in R^n$ let $\{i_j : j = 1, \ldots, n\} = \{i_j(\boldsymbol{\xi})\}$ be a permutation that sorts the $|\xi_i|$ in reverse order:

$$|\xi_{i_1}| \ge \ldots \ge |\xi_{i_n}|; \tag{4.78}$$

and let

$$m(\boldsymbol{\xi}) = \min\left\{ m : \sum_{j=1}^{m+1} |\xi_{i_j}| \ge \sum_{j=m+2}^{n} |\xi_{i_j}| \right\}. \tag{4.79}$$

Then it is proved in Maronna and Yohai (1999) that

$$m^* = m^*(\mathbf{X}) = \min\{m(\boldsymbol{\xi}) : \boldsymbol{\xi} \in \Xi,\ \boldsymbol{\xi} \ne \mathbf{0}\}. \tag{4.80}$$

Ellis and Morgenthaler (1992) give a version of this result for the L_1 estimator, and use the ratio of the sums on both sides of the inequality in (4.79) as a measure of leverage.

In the location case we have $x_i \equiv 1$, hence all ξ_i are equal, and the condition in (4.79) is equivalent to $m + 1 \ge n - m + 1$, which yields $m(\boldsymbol{\xi}) = [(n-1)/2]$ as in (3.26).

Consider now fitting a straight line through the origin with a uniform design $x_i = i$ $(i = 1, \ldots, n)$. Then, for all $\boldsymbol{\xi} \ne \mathbf{0}$, ξ_i is proportional to $n - i + 1$, and hence

$$m(\boldsymbol{\xi}) = \min\left\{ m : \sum_{j=1}^{m+1} (n - j + 1) \ge \sum_{j=m+2}^{n} (n - j + 1) \right\}.$$

The condition between braces is equivalent to

$$n(n+1) \geq 2(n-m)(n-m-1),$$

and for large n this is equivalent to $(1-m/n)^2 \leq 1/2$; that is,

$$\frac{m}{n} = 1 - \sqrt{1/2} \approx 0.29.$$

The case of a general straight line is dealt with in a similar way. The proof of (4.59) is not difficult, but that of (4.60) is rather involved (see Maronna and Yohai, 1999).

If $\mathbf{X}$ is uniformly distributed on a p-dimensional spherical surface, it can be proved that $\varepsilon^* \approx \sqrt{0.5/p}$ for large p (Maronna *et al.*, 1979) showing that even a fixed design without leverage points may yield a low BP if p is large.

4.10 Recommendations and software

For linear regression with fixed predictors without leverage points we recommend the bisquare M-estimator starting from L_1 (Section 4.4.2), which can be computed using **lmrobM** (`RobustTM`). **lmrobLinTest** (`RobStatTM`) performs the robust LRT test for linear hypotheses described in Section 4.7.2.

4.11 Problems

4.1. Let $\widehat{\boldsymbol{\beta}}$ be a solution of (4.39) with *fixed* σ. Show that:

(a) if y_i is replaced by $y_i + \mathbf{x}_i'\boldsymbol{\gamma}$, then $\widehat{\boldsymbol{\beta}} + \boldsymbol{\gamma}$ is a solution

(b) if $\mathbf{x}_i$ is replaced by $\mathbf{A}\mathbf{x}_i$, then $\mathbf{A}'^{-1}\widehat{\boldsymbol{\beta}}$ is a solution.

4.2. Let $\widehat{\boldsymbol{\beta}}$ be a solution of (4.39) where $\widehat{\sigma}$ verifies (4.48). Then $\widehat{\boldsymbol{\beta}}$ is regression, affine and scale equivariant.

4.3. Show that the solution $\widehat{\beta}$ of (4.37) is the LS estimator of the regression of y_i^* on $\mathbf{x}_i$, where $y_i^* = \xi(y_i, \mathbf{x}_i'\widehat{\boldsymbol{\beta}}, \widehat{\sigma})$, with ξ being "pseudo-observations", defined as in (2.95). Use this fact to define an iterative procedure to compute a regression M-estimator.

4.4. Show that the L_1 estimator for the model of regression through the origin $y_i = \beta x_i + u_i$ is the median of $z_i = y_i/x_i$, where z_i has probability proportional to $|x_i|$.

4.5. Verify (4.53) and show that, at each step of the median polish algorithm, the sum $\sum_i \sum_j |y_{ij} - \widehat{\mu} - \widehat{\alpha}_i - \widehat{\gamma}_j|$ does not increase.

Table 4.5 Hearing data

	Occupation						
Frequency	I	II	III	IV	V	VI	VII
500	2.1	6.8	8.4	1.4	14.6	7.9	4.8
1000	1.7	8.1	8.4	1.4	12.0	3.7	4.5
2000	14.4	14.8	27.0	30.9	36.5	36.4	31.4
3000	57.4	62.4	37.4	63.3	65.5	65.6	59.8
4000	66.2	81.7	53.3	80.7	79.7	80.8	82.4
6000	75.2	94.0	74.3	87.9	93.3	87.8	80.5
Normal	4.1	10.2	10.7	5.5	18.1	11.4	6.1

I, Professional-managerial; II, farm; III, clerical sales; IV, craftsmen; V, operatives; VI, service; VII, laborers.

4.6. Write computer code for the median polish algorithm and apply it to the original and modified oats data of Example 4.2 and to the data of Problem 4.9.

4.7. Show that, for large n, the FBP given by (4.80) for fitting $y_i = \beta x_i^k + u_i$ with a uniform design of n points is approximately $1 - 0.5^{1/(k+1)}$.

4.8. Show that for the fit of $y_i = \beta x_i + u_i$ with design (4.61), the FBP given by (4.80) is zero.

4.9. Table 4.5 (Roberts and Cohrssen, 1968) gives prevalence rates in percentage terms for men aged 55–64 with hearing levels 16 dB or more above the audiometric zero, at different frequencies (hertz) and for normal speech. The columns classify the data into seven occupational groups: professional-managerial, farm, clerical sales, craftsmen, operatives, service, and laborers. (The dataset is called **hearing**). Fit an additive ANOVA model by LS and robustly. Compare the effect estimations. This data has also been analyzed by Daniel (1978).

5

Linear Regression 2

5.1 Introduction

Chapter 4 concentrated on robust regression estimators for situations where the predictor matrix $\mathbf{X}$ contains no rows $\mathbf{x}_i$ with high leverage, and only the responses $\mathbf{y}$ may contain outliers. In that case a monotone M-estimator is a reliable starting point for computing a robust scale estimator and a redescending M-estimator. But when $\mathbf{X}$ is random, outliers in $\mathbf{X}$ operate as leverage points, and may completely distort the value of a monotone M-estimator when some pairs $(\mathbf{x}_i, y_i)$ are atypical. This chapter will deal with the case of random predictors and one of its focuses is on how to obtain good initial values for redescending M-estimators.

The following example shows the failure of a monotone M-estimator when $\mathbf{X}$ is random and there is a single atypical observation.

Example 5.1 *Smith* et al. *(1984) measured the contents (in parts per million) of 22 chemical elements in 53 samples of rocks in Western Australia. Tables and figures for this example can be obtained with script* ***mineral.R***.

Figure 5.1 plots the zinc (Zn) and the copper (Cu) contents against each other. Observation 15 stands out as clearly atypical. The LS fit is seen to be influenced more by this observation than by the rest. However, the L_1 fit exhibits the same drawback. Neither the LS nor the L_1 fits represent the bulk of the data, since they are "attracted" by observation 15, which has a very large abscissa and too high an ordinate. By contrast, the LS fit omitting observation 15 gives a good fit to the rest of the data. Figures 5.2 and 5.3 show the Q–Q plot and the plot of residuals vs fitted values for

Robust Statistics: Theory and Methods (with R), Second Edition.
Ricardo A. Maronna, R. Douglas Martin, Victor J. Yohai and Matías Salibián-Barrera.

Companion website: www.wiley.com/go/maronna/robust

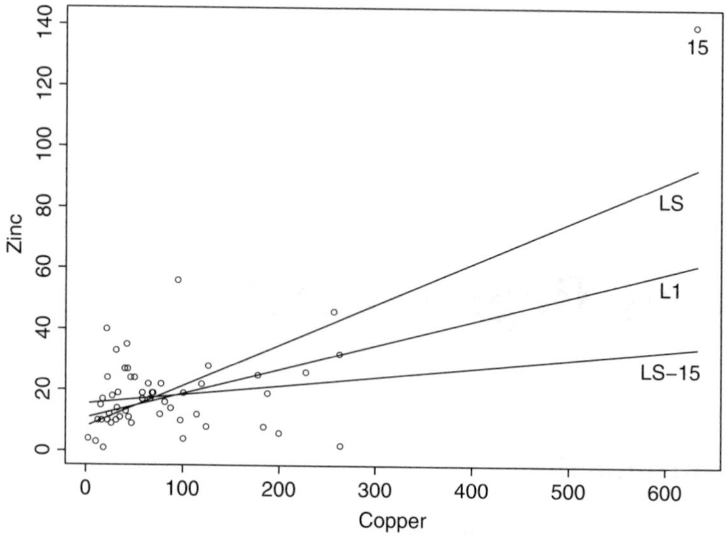

Figure 5.1 Mineral data: fits with LS, L_1, and LS without observation 15

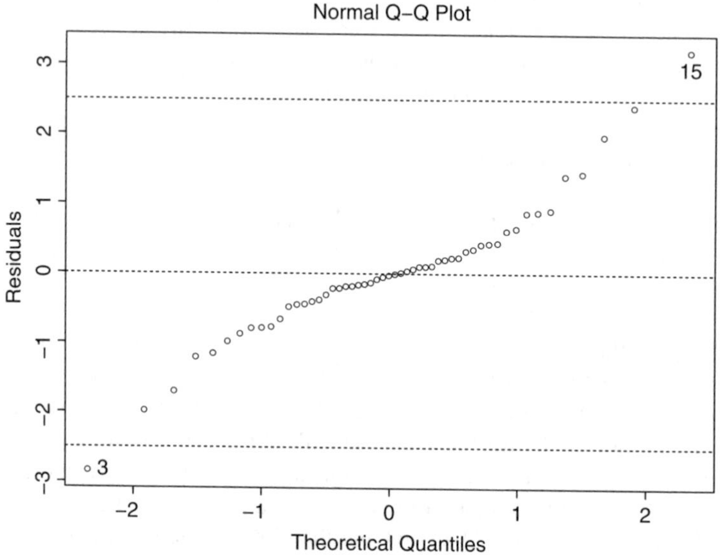

Figure 5.2 Mineral data: Q–Q plot of LS residuals

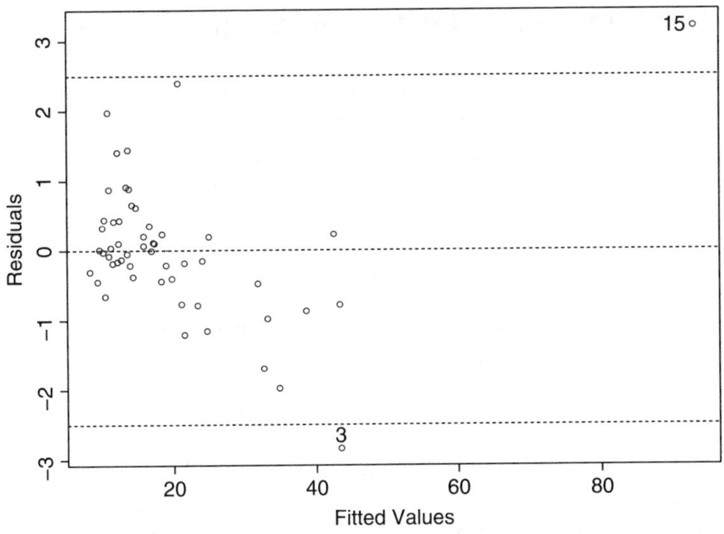

Figure 5.3 Mineral data: LS residuals versus fit

Table 5.1 Regression coefficients for mineral data

	LS	L_1	LS(−15)	Robust
Intercept	7.960	10.412	15.491	12.913
Slope	0.134	0.080	0.030	0.044

the LS estimator. Neither figure reveals the existence of an outlier, as indicated by an exceptionally large residual. However, the second figure shows an approximately linear relationship between residuals and fitted values – except for the point with largest fitted value – and this indicates that the fit is not correct.

Table 5.1 gives the estimated parameters for the LS and L_1 fits, as well as for the LS fit computed without observation 15, and for a redescending regression M-estimator to be described shortly.

The intuitive reason for the failure of the L_1 estimator (and of monotone M-estimators in general) in this situation is that the $\mathbf{x}_i$ outlier dominates the solution to (4.40) in the following sense. If, for some i, $\mathbf{x}_i$ is "much larger than the rest", then in order to make the sum zero, the residual $y_i - \mathbf{x}_i'\hat{\boldsymbol{\beta}}$ must be near zero and hence $\hat{\boldsymbol{\beta}}$ is essentially determined by $(\mathbf{x}_i, y_i)$. This does not happen with the redescending M-estimator.

5.2 The linear model with random predictors

Situations like the one in the previous example occur primarily when $\mathbf{x}_i$ are not fixed, as in designed experiments, but instead are random variables observed together with y_i. We now briefly discuss the properties of a linear model with random $\mathbf{X}$. Our observations are now the i.i.d. $(p+1)$-dimensional random vectors $(\mathbf{x}_i, y_i)$ $(i = 1, \ldots, n)$ satisfying the linear model relation

$$y_i = \mathbf{x}_i'\boldsymbol{\beta} + u_i. \tag{5.1}$$

In the case of fixed $\mathbf{X}$ we assumed that the distribution of u_i does not depend on $\mathbf{x}_i$. The analogous assumption here is that

$$\text{the } u_i \text{ are i.i.d. and independent of the } \mathbf{x}_i. \tag{5.2}$$

The analogue of assuming $\mathbf{X}$ is of full rank is to assume that the distribution of $\mathbf{x}$ is not concentrated on any subspace; that is, $\mathrm{P}(\mathbf{a}'\mathbf{x} = 0) < 1$ for all $\mathbf{a} \neq \mathbf{0}$. This condition implies that the probability that $\mathbf{X}$ has full rank tends to 1 when $n \to \infty$, and holds in particular if the distribution of $\mathbf{x}$ has a density. Then the LS estimator is well defined, and (4.18) holds *conditionally* on $\mathbf{X}$:

$$\mathrm{E}(\widehat{\boldsymbol{\beta}}_{LS}|\mathbf{X}) = \boldsymbol{\beta}, \ \mathbf{Var}(\widehat{\boldsymbol{\beta}}_{LS}|\mathbf{X}) = \sigma^2(\mathbf{X}'\mathbf{X})^{-1},$$

where $\sigma^2 = \mathrm{Var}(u)$,

Also (4.23) holds conditionally: if the u_i are normal then the conditional distribution of $\widehat{\boldsymbol{\beta}}_{LS}$ given $\mathbf{X}$ is multivariate normal. If the u_i are not normal, assume that

$$\mathbf{V}_\mathbf{x} = \mathrm{E}\mathbf{x}\mathbf{x}' \tag{5.3}$$

exists. It can be shown that

$$\mathcal{D}(\widehat{\boldsymbol{\beta}}_{LS}) \approx \mathrm{N}_p\left(\boldsymbol{\beta}, \frac{\mathbf{C}_{\widehat{\beta}}}{n}\right) \tag{5.4}$$

where

$$\mathbf{C}_{\widehat{\beta}} = \sigma^2 \mathbf{V}_\mathbf{x}^{-1} \tag{5.5}$$

is the asymptotic covariance matrix of $\widehat{\boldsymbol{\beta}}$; see Section 10.10.2. The estimation of $\mathbf{C}_{\widehat{\beta}}$ is discussed in Section 5.6.

In the case (4.5) where the model has an intercept term, it follows from (5.5) that the asymptotic covariance matrix of $(\beta_0, \boldsymbol{\beta}_1)$ is

$$\sigma^2 \begin{pmatrix} 1 + \boldsymbol{\mu}_\mathbf{x}'\mathbf{C}_\mathbf{x}^{-1}\boldsymbol{\mu}_\mathbf{x} & \boldsymbol{\mu}_\mathbf{x}' \\ \boldsymbol{\mu}_\mathbf{x} & \mathbf{C}_\mathbf{x}^{-1} \end{pmatrix} \tag{5.6}$$

where

$$\boldsymbol{\mu}_\mathbf{x} = \mathrm{E}\mathbf{x}, \quad \mathbf{C}_\mathbf{x} = \mathbf{Var}(\mathbf{x}).$$

5.3 M-estimators with a bounded ρ-function

Our approach to robust regression estimators where both the $\mathbf{x}_i$ and the y_i may contain outliers is to use an M-estimator $\widehat{\boldsymbol{\beta}}$ defined by

$$\sum_{i=1}^{n} \rho\left(\frac{r_i(\widehat{\boldsymbol{\beta}})}{\widehat{\sigma}}\right) = \min \tag{5.7}$$

with a *bounded* ρ and a high BP preliminary scale $\widehat{\sigma}$. The scale $\widehat{\sigma}$ will be required to fulfill certain requirements discussed in Section 5.5. If ρ has a derivative ψ it follows that $\widehat{\boldsymbol{\beta}}$ solves

$$\sum_{i=1}^{n} \psi\left(\frac{r_i(\widehat{\boldsymbol{\beta}})}{\widehat{\sigma}}\right)\mathbf{x}_i = \mathbf{0}, \tag{5.8}$$

where ψ is redescending (it is easy to verify that a function ρ with a monotonic derivative ψ cannot be bounded). Consequently, the estimating equation (5.8) may have multiple solutions corresponding to multiple *local* minima of the function on the left-hand side of (5.7), and generally only one of them (the "good solution") corresponds to the global minimizer $\widehat{\boldsymbol{\beta}}$ defined by (5.7). We shall see that ρ and $\widehat{\sigma}$ may be chosen in order to attain both a high BP and a high efficiency.

In Section 5.5 we describe a particular computing method for approximating $\widehat{\boldsymbol{\beta}}$ as defined by (5.7). The method is called an *MM-estimator*, and as a demonstration of its use we apply it to the data of Example 5.1. The results, displayed in Figure 5.4, show

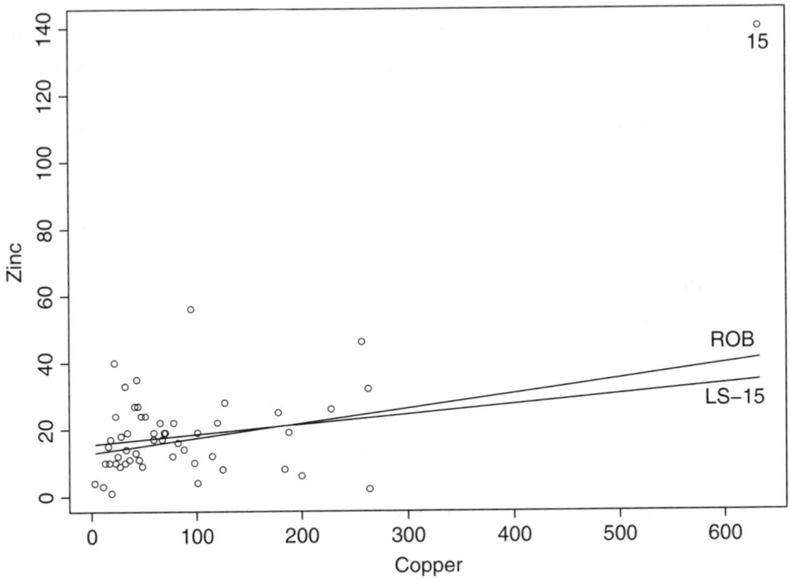

Figure 5.4 Mineral data: fits by MM estimator ("ROB") and by LS without the outlier

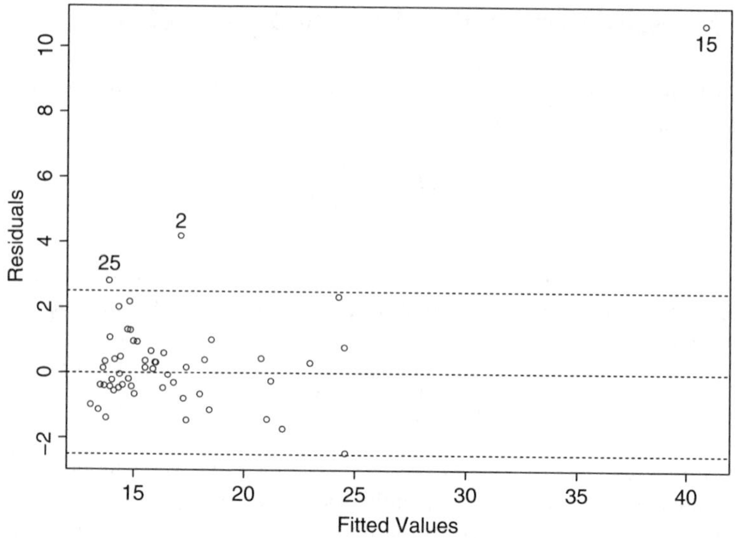

Figure 5.5 Mineral data: residuals versus fitted values of MM-estimator

that the MM-estimator almost coincides with the LS estimator computed with data point 15 deleted. The MM-estimator intercept and slope parameters are now 14.05 and 0.02, respectively, as compared to 7.96 and 0.13 for the LS estimator (recall Table 5.1).

Figure 5.5 shows the residuals plotted against fitted values and Figure 5.6 shows the Q–Q plot of the residuals. The former now lacks the suspicious structure of Figure 5.3 and point 15 is now revealed as a large outlier in the residuals as well as the fit, with a considerably reduced value of fit (roughly 40 instead of more than 90). Moreover, compared to Figure 5.2 the Q–Q plot now clearly reveals point 15 as an outlier. Figure 5.7 compares the sorted absolute values of residuals from the MM-estimator fit and the LS fit, with point 15 omitted for reasons of scale. It is seen that most points lie below the identity diagonal, showing that, except for the outlier, the sorted absolute MM-residuals are smaller than those from the LS estimator, and hence the MM-estimator fits the data better.

5.3.1 Properties of M-estimators with a bounded ρ-function

If $\widehat{\sigma}$ is regression and affine equivariant, as defined in (4.48), then the estimator $\widehat{\beta}$ defined by (5.7) is regression, scale and affine equivariant. We now discuss the breakdown point, influence function and asymptotic normality of such estimators.

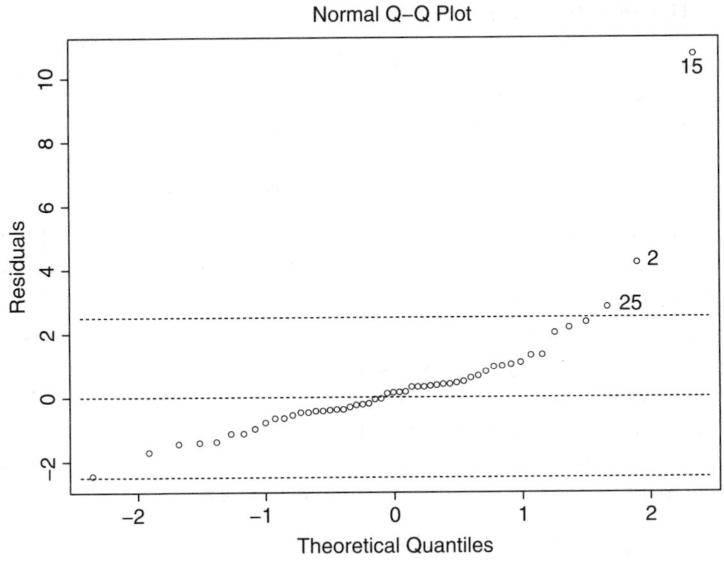

Figure 5.6 Mineral data: Q–Q plot of robust residuals

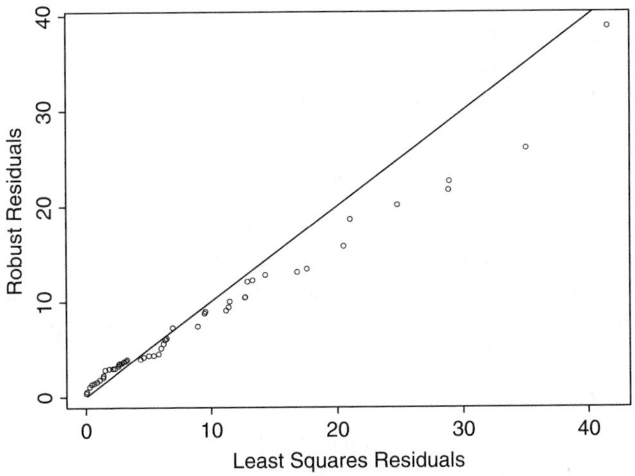

Figure 5.7 Mineral data: sorted absolute values of robust versus LS residuals (point 15 omitted)

5.3.1.1 Breakdown point

We focus on the finite breakdown point (FBP) of $\widehat{\beta}$. Since the $\mathbf{x}$ are now random, we are in the situation of Section 3.2.5. Let $\mathbf{z}_i = (\mathbf{x}_i, y_i)$ and write the estimator as $\widehat{\beta}(\mathbf{Z})$ with $\mathbf{Z} = \{\mathbf{z}_1, ..., \mathbf{z}_n\}$. Then, instead of (4.55), define $\varepsilon^* = m^*/n$, where

$$m^* = \max\{m \geq 0 : \widehat{\beta}(\mathbf{Z}_m) \text{ bounded } \forall\, \mathbf{Z}_m \in \mathcal{Z}_m\}, \tag{5.9}$$

and $\mathcal{Z}_m$ is the set of datasets with at least $n - m$ elements in common with $\mathbf{Z}$. Note that since not only $\mathbf{y}$ but also $\mathbf{X}$ are variable here, the FBP given by (5.9) is less than or equal to that given earlier by (4.55).

It is then easy to show that the FBP of monotone M-estimators is zero (Section 5.13.1). Intuitively, this is due to the fact that a term with a "large" $\mathbf{x}_i$ "dominates" the sum in (5.7). Then the scale used in Section 4.4.2, which is based on the residuals from the L_1 estimator, also has a zero BP.

On the other hand, it can be shown that the maximum FBP of any regression equivariant estimator is again the one given in Section 4.6:

$$\varepsilon^* \leq \varepsilon^*_{\max} =: \frac{1}{n}\left[\frac{n - k^* - 1}{2}\right] \leq \frac{1}{n}\left[\frac{n-p}{2}\right], \tag{5.10}$$

with k^* as in (4.56):

$$k^*(\mathbf{X}) = \max\{\#(\boldsymbol{\theta}'\mathbf{x}_i = 0) : \boldsymbol{\theta} \in R^p,\ \boldsymbol{\theta} \neq \mathbf{0}\} \tag{5.11}$$

The proof is similar to that of Section 4.9.3, and we shall see that this bound is attained by several types of estimator to be defined in this chapter. It can be shown in the same way that the maximum asymptotic BP for regression equivariant estimators is $(1 - \alpha)/2$, where

$$\alpha = \max_{\boldsymbol{\theta} \neq \mathbf{0}} \mathrm{P}(\boldsymbol{\theta}'\mathbf{x} = 0). \tag{5.12}$$

In the previous chapter, our method for developing robust estimators was to generalize the MLE, which leads to M-estimators with unbounded ρ. In the present setting, calculating the MLE again yields (4.36); in particular, LS is the MLE for normal u, for any $\mathbf{x}$. Thus no new class of estimators emerges from the ML approach.

5.3.1.2 Influence function

If the joint distribution F of $(\mathbf{x}, y)$ is given by the model (5.1)–(5.2), then it follows from (3.48) that the influence function (IF) of an M-estimator with known σ under the model is

$$\mathrm{IF}((\mathbf{x}_0, y_0), F) = \frac{\sigma}{b}\psi\left(\frac{y_0 - \mathbf{x}_0'\beta}{\sigma}\right)\mathbf{V}_{\mathbf{x}}^{-1}\mathbf{x}_0 \text{ with } b = \mathrm{E}\psi'\left(\frac{u}{\sigma}\right) \tag{5.13}$$

and with $\mathbf{V}_{\mathbf{x}}$ defined by (5.3). The proof is similar to that of Section 3.8.1. It follows that the IF is unbounded. However, the IFs for the cases of monotone and of redescending ψ are rather different. If ψ is monotone, then the IF tends to infinity for any fixed $\mathbf{x}_0$ if y_0 tends to infinity. If ψ is redescending and is such that $\psi(x) = 0$ for $|x| \geq k$, then the IF will tend to infinity only when $\mathbf{x}_0$ tends to infinity and $|y_0 - \mathbf{x}_0'\boldsymbol{\beta}|/\sigma \leq k$, which means that large outliers have no influence on the estimator.

When σ is unknown and is estimated by $\widehat{\sigma}$, it can be shown that if the distribution of u_i is symmetric, then (5.13) also holds, with σ replaced by the asymptotic value of $\widehat{\sigma}$.

The fact that the IF is unbounded does not necessarily imply that the bias is unbounded for any positive contamination rate ε. In fact, while a monotone ψ implies BP = 0, we shall see in Section 5.5 that with a bounded ρ it is possible to attain a high BP, and hence that the bias is bounded for large values of ε. On the other hand, in Section 5.11.1 we shall define a family of estimators with bounded IF, but such that their BP may be very low for large p. These facts indicate that the IF need not yield a reliable approximation to the bias.

5.3.1.3 Asymptotic normality

Assume that the model (5.1)–(5.2) holds, that $\mathbf{x}$ has finite variances, and that $\widehat{\sigma}$ converges in probability to some σ. Then it can be proved under rather general conditions (see Section 10.10.2 for details) that the estimator $\widehat{\boldsymbol{\beta}}$ defined by (5.7) is consistent and asymptotically normal. More precisely

$$\sqrt{n}(\widehat{\boldsymbol{\beta}} - \boldsymbol{\beta}) \rightarrow_d \mathrm{N}_p(\mathbf{0}, v\mathbf{V}_{\mathbf{x}}^{-1}), \tag{5.14}$$

where $\mathbf{V}_{\mathbf{x}} = \mathrm{E}\mathbf{x}\mathbf{x}'$, and v is as in (4.44):

$$v = \sigma^2 \frac{\mathrm{E}\psi(u/\sigma)^2}{(\mathrm{E}\psi'(u/\sigma))^2}. \tag{5.15}$$

This result implies that as long as $\mathbf{x}$ has finite variances, the efficiency of $\widehat{\boldsymbol{\beta}}$ does not depend on the distribution of $\mathbf{x}$.

The Fisher-consistency (Section 3.5.3) of M-estimators with random predictors is shown in general in Section 10.11.

We have seen in Chapter 4 that a leverage point forces the fit of a monotone M-estimator to pass near the point, and this has a double-edged effect: if the point is a "typical" observation, the fit improves (although the normal approximation to the distribution of the estimator deteriorates); if it is "atypical", the overall fit worsens. The implications of these facts for the case of random $\mathbf{x}$ are as follows. Suppose that $\mathbf{x}$ is heavy tailed so that its variances do not exist. If the model (5.1)–(5.2) holds, then the normal ceases to be a good approximation to the distribution of $\widehat{\boldsymbol{\beta}}$, but at the same time $\widehat{\boldsymbol{\beta}}$ is "closer" to $\boldsymbol{\beta}$ than in the case of "typical" $\mathbf{x}$ (see Section 5.13.2 for details). But if the model does not hold, then $\widehat{\boldsymbol{\beta}}$ may have a higher bias than in the case of "typical" $\mathbf{x}$.

5.4 Estimators based on a robust residual scale

The M-estimators defined in (5.7) require a robust scale estimator $\widehat{\sigma}$, which typically is computed using residuals from a robust regression estimator. In this section, we shall present a family of regression estimators that do not depend on a preliminary residual scale, and thus break this cycle. They can be used to compute the preliminary scale $\widehat{\sigma}$ in (5.7).

Note that the LS and the L_1 estimators minimize the averages of the squared and of the absolute residuals respectively, and therefore they minimize measures of residual largeness that can be seriously influenced by even a single residual outlier. A more robust alternative is to minimize a scale measure of residuals that is insensitive to large values, and one such possibility is the *median* of the absolute residuals. This is the basis of the least median of squares (LMS) estimator, introduced as the first estimator of this kind by Hampel (1975) and by Rousseeuw (1984) who also proposed a computational algorithm. In the location case, the LMS estimator is equivalent to the Shorth estimator, defined as the mid-point of the shortest half of the data (see Problem 2.16a). For fitting a linear model, the LMS estimator has the intuitive property of generating the strip of minimum width that contains half of the observations (Problem 5.9).

Let $\widehat{\sigma} = \widehat{\sigma}(\mathbf{r})$ be a location-invariant and scale-equivariant robust scale estimator based on a vector of residuals

$$\mathbf{r}(\boldsymbol{\beta}) =(r_1(\boldsymbol{\beta}), \ldots, r_n(\boldsymbol{\beta})). \tag{5.16}$$

Then a regression estimator can be defined as

$$\widehat{\boldsymbol{\beta}} = \arg\min_{\boldsymbol{\beta}} \widehat{\sigma}(\mathbf{r}(\boldsymbol{\beta})). \tag{5.17}$$

Such estimators are regression, scale, and affine equivariant (Problem 5.1).

5.4.1 S-estimators

A very important case of (5.17) is when $\widehat{\sigma}(\mathbf{r})$ is a scale M-estimator defined for each $\mathbf{r}$ as the solution to

$$\frac{1}{n}\sum_{i=1}^{n} \rho\left(\frac{r_i}{\widehat{\sigma}}\right) = \delta, \tag{5.18}$$

where ρ is a bounded ρ-function. By (3.23), the asymptotic BP of $\widehat{\sigma}$ is $\min(\delta, 1-\delta)$. The resulting estimator (5.17) is called an S-estimator (Rousseeuw and Yohai, 1984). See Section 5.9.4 for the choice of an initial estimator $\widehat{\boldsymbol{\beta}}_0$ for the MM-estimator.

We now consider the BP of S-estimators. Proofs of all results on the BP are given in Section 5.13.4. The maximum FBP of an S-estimator with a bounded ρ-function is

$$\varepsilon^*_{\max} = \frac{m^*_{\max}}{n}, \tag{5.19}$$

where $m^*_{\max}$ is the same as in (4.58), namely

$$m^*_{\max} = \left[\frac{n-k^*-1}{2}\right], \tag{5.20}$$

where k^* is defined in (5.11). Hence $\varepsilon^*_{\max}$ coincides with the maximum BP for equivariant estimators given in (5.10). This BP is attained by taking any δ of the form

$$\delta = \frac{m^*_{\max}+\gamma}{n} \quad \text{with} \quad \gamma \in (0,1). \tag{5.21}$$

Recall that $k^* \geq p-1$, and if $k^* = p-1$ we say that $\mathbf{X}$ is in general position. When $\mathbf{X}$ is in general position, the maximum FBP is

$$\varepsilon^*_{\max} = \frac{1}{n}\left[\frac{n-p}{2}\right],$$

which is approximately 0.5 for large n. Similarly, the maximum asymptotic BP of a regression S-estimator with a bounded ρ is

$$\varepsilon^* = \frac{1-\alpha}{2}, \tag{5.22}$$

with α defined in (5.12), and thus coincides with the maximum asymptotic BP for equivariant estimators given in Section 5.3.1.1 This maximum is attained by taking $\delta = (1-\alpha)/2$. If $\mathbf{x}$ has a density then $\alpha = 0$, and hence $\delta = 0.5$ yields $\varepsilon^* = 0.5$.

Since the median of absolute values is a scale M-estimator, the LMS estimator may be written as the estimator minimizing the scale $\widehat{\sigma}$ given by (5.18), with $\rho(t) = \mathrm{I}(|t| < 1)$ and $\delta = 0.5$. For a general δ, a solution $\widehat{\sigma}$ of (5.18) is the hth order statistic of $|r_i|$, with $h = n - [n\delta]$ (Problem 2.14). The regression estimator defined by minimizing $\widehat{\sigma}$ is called the *least α-quantile estimator*, with $\alpha = h/n$. Although it has a discontinuous ρ-function, the proof of the preceding results (5.20)–(5.21) can be shown to imply that the maximum BP is again (5.19) and that it can be attained by choosing

$$h = n - m^*_{\max} = \left[\frac{n+k^*+2}{2}\right], \tag{5.23}$$

which is slightly larger than $n/2$. See the end of Section 5.13.4.

We deal now with the efficiency of S-estimators. Since an S-estimator $\widehat{\boldsymbol{\beta}}$ minimizes $\widehat{\sigma} = \widehat{\sigma}(\mathbf{r}(\boldsymbol{\beta}))$ it follows that $\widehat{\boldsymbol{\beta}}$ is also an M-estimator (5.7) in that

$$\sum_{i=1}^{n} \rho\left(\frac{r_i(\widehat{\boldsymbol{\beta}})}{\widehat{\sigma}}\right) \leq \sum_{i=1}^{n} \rho\left(\frac{r_i(\widetilde{\boldsymbol{\beta}})}{\widehat{\sigma}}\right) \quad \text{for all } \widetilde{\boldsymbol{\beta}}, \tag{5.24}$$

where $\widehat{\sigma} = \widehat{\sigma}(\mathbf{r}(\widehat{\boldsymbol{\beta}}))$ is the same in the denominator on both sides of the equation. To see that this is indeed the case, suppose that for some $\widetilde{\boldsymbol{\beta}}$ we had

$$\sum_{i=1}^{n} \rho\left(\frac{r_i(\widetilde{\boldsymbol{\beta}})}{\widehat{\sigma}}\right) < \sum_{i=1}^{n} \rho\left(\frac{r_i(\widehat{\boldsymbol{\beta}})}{\widehat{\sigma}}\right).$$

Then by the monotonicity of ρ there would exist $\widetilde{\sigma} < \widehat{\sigma}$ such that

$$\sum_{i=1}^{n} \rho\left(\frac{r_i(\widetilde{\beta})}{\widetilde{\sigma}}\right) = n\delta,$$

which would contradict the fact that $\widehat{\sigma}$ is the minimum M-scale. If ρ has a derivative ψ, it follows that $\widehat{\boldsymbol{\beta}}$ is also an M-estimator in the sense of (5.8), but with the condition that the scale $\widehat{\sigma} = \widehat{\sigma}(\mathbf{r}(\widehat{\boldsymbol{\beta}}))$ is estimated simultaneously with $\widehat{\boldsymbol{\beta}}$.

Because S-estimators are M-estimators, it follows that the asymptotic distribution of an S-estimator with a smooth ρ under the model (5.1)–(5.2) is given by (4.43)–(4.44); see Davies (1990) and Kim and Pollard (1990) for a rigorous proof. For the LMS estimator, which has a discontinuous ρ, Davies (1990) shows that $\widehat{\boldsymbol{\beta}} - \boldsymbol{\beta}$ has a slow convergence rate of $n^{-1/3}$, while estimators based on a smooth ρ-function have the usual convergence rate $n^{-1/2}$. Thus the LMS estimator is highly inefficient for large n.

Unfortunately S-estimators with a smooth ρ cannot simultaneously have high BP and high efficiency. In particular, it was shown by Hössjer (1992) that an S-estimator with BP = 0.5 has an asymptotic efficiency under normally distributed errors that is not larger than 0.33. In fact, numerical computation shows that, for normal distributions, the efficiency of S-estimators based on the bisquare ρ function is 0.29, which is adequately close to the upper bound.

Since an S-estimator with a differentiable ρ-function satisfies (5.8), its IF is given by (5.13) and hence is unbounded. See, however, the comments on p. 132. Note also that S-estimators are "redescending" in the sense that if some of the y_i are "too large", the estimator is completely unaffected by these observations, and coincides with an M-estimator computed after deleting such outliers. A precise statement is given in Problem 5.10.

Algorithms to compute S-estimators are discussed in Section 5.7.1.

5.4.2 L-estimators of scale and the LTS estimator

An alternative to using an M-scale is to use an L-estimator of scale. Call $|r|_{(1)} \leq .... \leq |r|_{(n)}$ the ordered absolute values of residuals. Then we can define scale estimators as linear combinations of the $|r|_{(i)}$ in one of the two following forms:

$$\widehat{\sigma} = \sum_{i=1}^{n} a_i |r|_{(i)}, \quad \text{or} \quad \widehat{\sigma} = \left(\sum_{i=1}^{n} a_i |r|_{(i)}^2\right)^{1/2},$$

where the a_i are nonnegative constants.

A particular version of the second form is the α−trimmed squares scale where $\alpha \in (0, 1)$, and $n - h = [n\alpha]$ of the largest absolute residuals are trimmed:

$$\widehat{\sigma} = \left(\sum_{i=1}^{h} |r|_{(i)}^2\right)^{1/2}. \tag{5.25}$$

The corresponding regression estimator is called the least trimmed squares (LTS) estimator (Rousseeuw, 1984). The FBP of the LTS estimator depends on h in the same way as that of the LMS estimator, so that for the LTS estimator to attain the maximum BP one must choose h in (5.25) as in (5.23). In particular, when $\mathbf{X}$ is in general position one must choose

$$h = n - \left[\frac{n-p-2}{2}\right] = \left[\frac{n+p+1}{2}\right],$$

which is approximately $n/2$ for large n. The asymptotic behavior of the LTS estimator is more complicated than that of smooth S-estimators. However, it is known that they have the standard convergence rate of $n^{-1/2}$, and it can be shown that the asymptotic efficiency of the LTS estimator for the normal distribution has the exceedingly low value of about 7%; see Rousseeuw and Leroy (1987; p. 180).

5.4.3 τ–estimators

In order to improve the efficiency of regression estimators based on scale estimators Yohai and Zamar (1988) proposed a different scale estimator to be used in (5.17). As before, given a vector of residuals $\mathbf{r} = (r_1, \ldots, r_n)$, let $\widehat{\sigma}(\mathbf{r})$ be a robust M-scale satisfying

$$\frac{1}{n}\sum_{i=1}^{n} \rho_0\left(\frac{r_i}{\widehat{\sigma}}\right) = \delta, \tag{5.26}$$

where ρ_0 is a bounded ρ-function, tuned to obtain the desired BP. Now use another bounded ρ-function ρ_1 to define the τ-scale as

$$\tau(\mathbf{r})^2 = \widehat{\sigma}(\mathbf{r})^2 \frac{1}{n\,\delta_1} \sum_{i=1}^{n} \rho_1\left(\frac{r_i}{\widehat{\sigma}(\mathbf{r})}\right), \tag{5.27}$$

where the constant δ_1 satisfies $E\rho_1(Z) = \delta_1$, with $Z \sim N(0,1)$, which ensures consistency for Gaussian errors. The τ-regression estimator is defined by

$$\widehat{\boldsymbol{\beta}} = \arg\min_{\boldsymbol{\beta}} \tau(\mathbf{r}(\boldsymbol{\beta})). \tag{5.28}$$

Note that although the above definition is in line with (5.17), an important difference is that the ρ-function in (5.27) can be tuned separately from ρ_0 in (5.26) in order to improve the efficiency of the resulting regression estimator. An intuitive motivation is that the LS estimator is a special case of (5.28) when $\rho_1(r) = r^2$, and hence by using an adequate choice of ρ_1 the estimator can be made arbitrarily close to the LS estimator, which will yield an arbitrarily high efficiency for the normal distribution.

Yohai and Zamar (1988) showed that τ-estimators satisfy an M-estimating equation (5.8), where the score function ψ is a linear combination of ρ_0' and ρ_1' with coefficients depending on the data. From this observation it follows that $\widehat{\boldsymbol{\beta}}$ has an asymptotically normal distribution. Its asymptotic efficiency at the normal distribution can be adjusted to be arbitrarily close to 1, by tuning the function ρ_1, just as in the case of MM-estimators. The BP of the τ-estimators is the same as that of an S-estimator based on ρ_0, and so by suitable choice of ρ_0, the estimator can attain the maximum BP for regression estimators.

Efficient algorithms to compute τ-regression estimators were studied by Salibian-Barrera *et al.* (2008a). They showed that the strategy discussed in Section 5.7.3 can be applied successfully to τ-regression estimators. R code implementing this algorithm is publicly available online at https://github.com/msalibian/fast-tau.

Note that the value of the τ-scale estimator corresponding to the regression estimator – the value $\tau(\mathbf{r}(\hat{\boldsymbol{\beta}}))$ in (5.28) – is a robust and efficient residual scale estimator. As such, these estimators can be used to build robust tests for linear hypotheses (see Section 4.7) with good level and power properties. In particular, one can construct ANOVA-type tests of the form (4.65) where the sums of squared residuals are replaced by τ-scale estimators. Salibian-Barrera *et al.* (2016) studied such tests and used a robust bootstrap method (see Section 5.6.1) to estimate the corresponding p-values. R code implementing these tests is publicly available at https://github.com/msalibian/tau-tests.

5.5 MM-estimators

Computing an M-estimator requires finding the absolute minimum of

$$L(\boldsymbol{\beta}) = \sum_{i=1}^{n} \rho\left(\frac{r_i(\boldsymbol{\beta})}{\hat{\sigma}}\right). \tag{5.29}$$

When ρ is bounded (and thus non-convex) this is an exceedingly difficult task, except for the cases when $p = 1$ or 2 where a grid search would work. However, we shall see that it suffices to find a "good" local minimum to achieve both a high BP and high efficiency for a normal distribution. This local minimum will be obtained by starting from a reliable starting point and applying the IRWLS algorithm of Section 4.5.2. This starting point will also be used to compute the robust residual scale $\hat{\sigma}$ required to define the M-estimator, and hence it is necessary that it can be computed without requiring a previous scale.

The L_1 estimator does not require a scale, but we have already seen that it is not a convenient estimator when $\mathbf{X}$ is random. Hence we need an initial estimator that is robust toward any kind of outliers and that does not require a previously computed scale. We choose an S-estimator.

The steps of the proposed procedure are thus:

1. Compute an initial consistent estimator $\hat{\boldsymbol{\beta}}_0$ with high breakdown point but possibly low normal efficiency.
2. Compute a robust scale $\hat{\sigma}$ of the residuals $r_i(\hat{\boldsymbol{\beta}}_0)$.
3. Find a solution $\hat{\boldsymbol{\beta}}$ of (5.7) using an iterative procedure starting at $\hat{\boldsymbol{\beta}}_0$.

We shall demonstrate that in this way we can obtain $\widehat{\boldsymbol{\beta}}$ having both a high BP and a prescribed high efficiency at the normal distribution.

Now we look at the details of the above steps. The robust initial estimator $\widehat{\boldsymbol{\beta}}_0$ must be regression, scale and affine equivariant, which ensures that $\widehat{\boldsymbol{\beta}}$ inherits the same properties. We choose an S-estimator with bisquare scale. We shall use two different functions, ρ and ρ_0, and each of these must be a *bounded* ρ-function in the sense of Definition 2.1 at the end of Section 2.3.4. The scale estimator $\widehat{\sigma}$ must be an M-scale estimator (2.49) given by

$$\frac{1}{n}\sum_{i=1}^{n}\rho_0\left(\frac{r_i}{\widehat{\sigma}}\right) = 0.5. \tag{5.30}$$

By (3.23) the asymptotic BP of $\widehat{\sigma}$ is 0.5. As was seen at the end of Section 2.5, we can always find c_0 such that using $\rho_0(r/c_0)$ ensures that the asymptotic value of σ coincides with the standard deviation when the u_i are normal. For the bisquare scale given by (2.52) this value is $c_0 = 1.56$.

The key result is given by Yohai (1987), who called these estimators *MM-estimators*. Recall that all local minima of $L(\boldsymbol{\beta})$ satisfy (5.8). Let ρ satisfy

$$\rho_0 \geq \rho. \tag{5.31}$$

Yohai (1987) shows that if $\widehat{\boldsymbol{\beta}}$ is such that

$$L(\widehat{\boldsymbol{\beta}}) \leq L(\widehat{\boldsymbol{\beta}}_0), \tag{5.32}$$

then $\widehat{\boldsymbol{\beta}}$ is consistent. It can also be shown – in the same way as the similar result for location in Section 3.2.3 – that its BP is not less than that of $\widehat{\boldsymbol{\beta}}_0$. If, furthermore, $\widehat{\boldsymbol{\beta}}$ is any solution of (5.8), then it has the same efficiency as the global minimum. Thus it is not necessary to find the absolute minimum of (5.7) to ensure a high BP and high efficiency.

The numerical computation of the estimator follows the approach in Section 4.5: starting with $\widehat{\boldsymbol{\beta}}_0$ we use the IRWLS algorithm to obtain a solution of (5.8). It is shown in Section 9.1 that $L(\boldsymbol{\beta})$ given in (5.29) decreases at each iteration, which ensures (5.32).

It remains to choose ρ in order to attain the desired normal efficiency, which is $1/v$, where v is the expression (5.15) computed at the standard normal. Let ρ^* be any bounded ρ-function; for instance the bisquare given by (2.38) with $k = 1$. Let

$$\rho_0(r) = \rho^*\left(\frac{r}{c_0}\right) \quad \text{and} \quad \rho(r) = \rho^*\left(\frac{r}{c_1}\right),$$

where c_0 is chosen for consistency of the scale at the normal. In particular, when ρ^* is the bisquare function, $c_0 = 1.56$. In order that $\rho \leq \rho_0$ we must have $c_1 \geq c_0$: the larger c_1, the higher the efficiency at the normal distribution. The values of c_1 for prescribed efficiencies are the values of k in Table (2.4).

In Section 5.8 we shall demonstrate the basic trade-off between normal efficiency and bias under contamination: the larger the efficiency, the larger the bias. It is therefore important to choose the efficiency so as to maintain reasonable bias control. The results in Section 5.8 show that an efficiency of 0.95 yields too high a bias, and hence it is safer to choose an efficiency of 0.85, which gives a smaller bias while retaining sufficiently high efficiency.

Note that M-estimators, and MM-estimators in particular, have an unbounded IF but a high BP. This seeming contradiction can be resolved by noting that an infinite gross-error sensitivity means only that the maximum bias for ε-contamination, $\mathrm{MB}(\varepsilon)$, is not $O(\varepsilon)$ for small ε, but does not imply that it is infinite! Actually, Yohai and Zamar (1997) have shown that $\mathrm{MB}(\varepsilon) = O(\sqrt{\varepsilon})$ for the estimators considered in this section. This implies that the bias induced by altering a single observation is bounded by $c/\sqrt{n}$ for some constant c, instead of the stronger bound c/n.

Example 5.2 *The next example is based on the "modified wood gravity data". The raw data came from Draper and Smith (1966, p. 227) and were used to determine the influence of anatomical factors on wood specific gravity, with 20 cases, five explanatory variables and an intercept. Rousseeuw and Leroy (1987) modified the data by replacing four observations (4, 6, 8, and 19) by outliers (dataset* ***wood****). The tables and figures for this example can be obtained with script* ***wood.R****.*

Figures 5.8 and 5.9 show the plot of the residuals against fit and the normal Q–Q plot for the LS estimator. No outliers are apparent. Figures 5.10 and 5.11 are the respective plots for the 85% normal efficiency MM-estimator, clearly showing the four outliers 4, 6, 8 and 19. Figure 5.12 plots the ordered absolute residuals

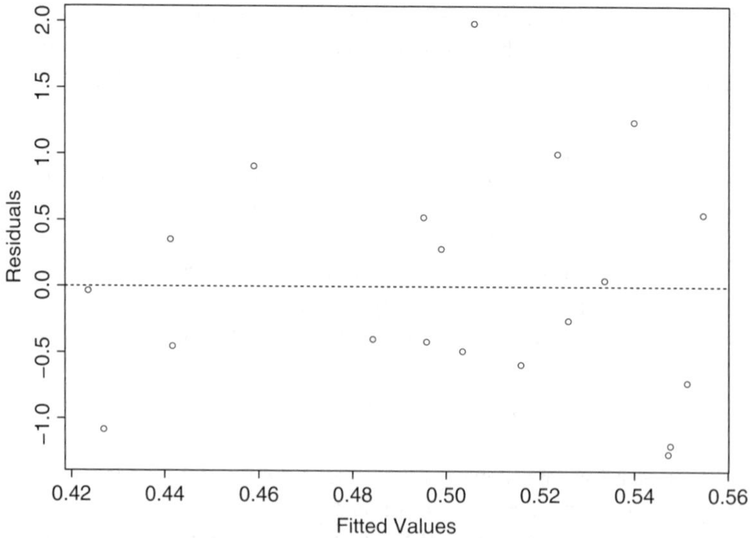

Figure 5.8 Wood gravity data: LS residuals versus fit

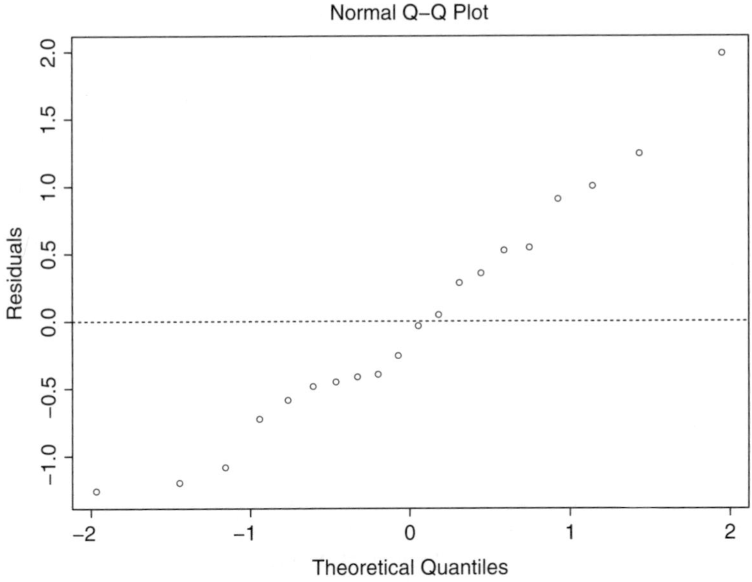

Figure 5.9 Wood gravity data: Q–Q plot of LS residuals

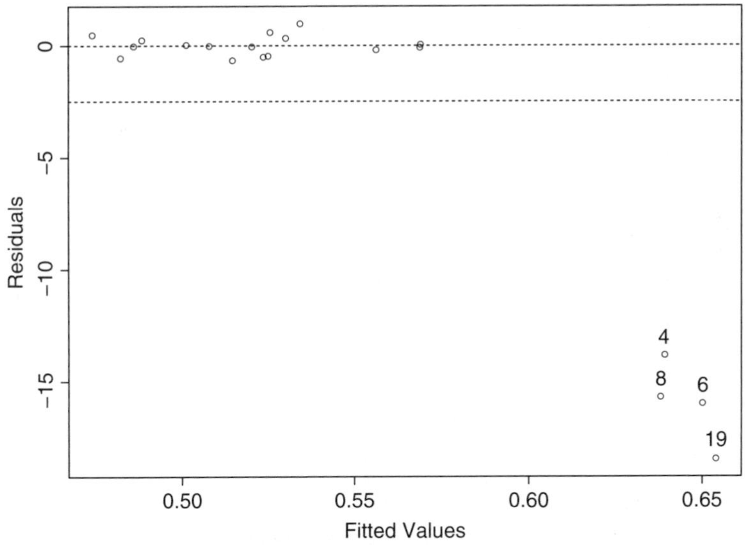

Figure 5.10 Wood gravity data: MM residuals versus fit

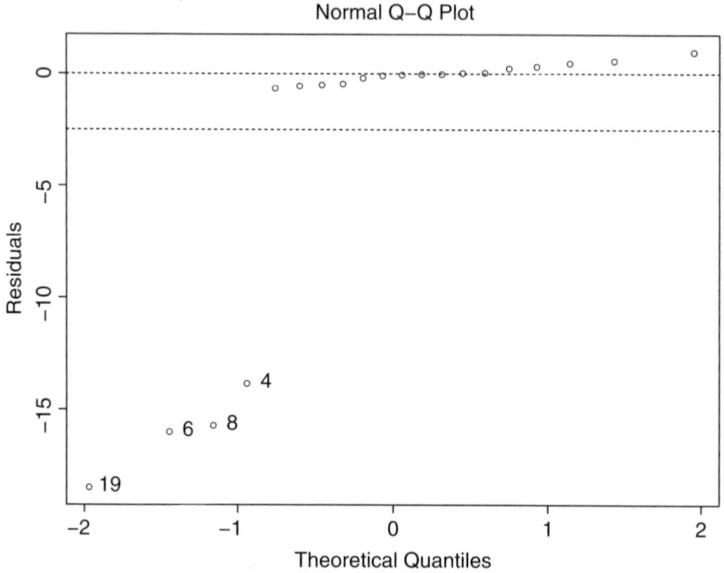

Figure 5.11 Wood gravity data: Q–Q plot of robust residuals

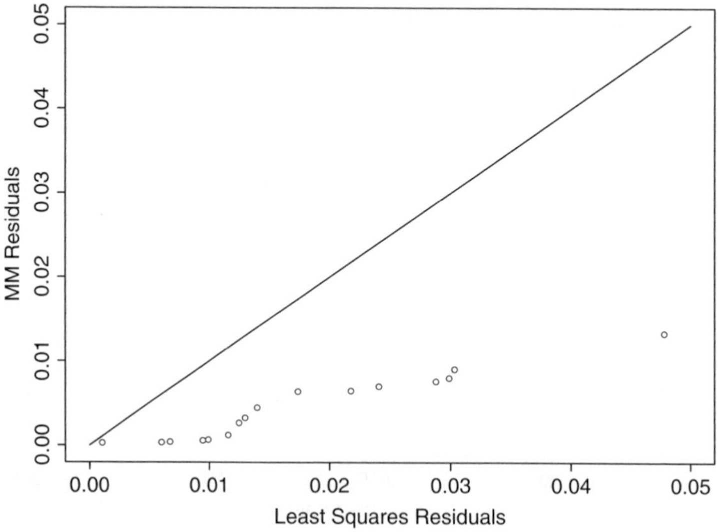

Figure 5.12 Wood gravity data: ordered absolute residuals from MM and from LS (largest residuals omitted)

from LS as the abscissa and those from the MM-estimator as the ordinate, as compared to the identity line; the observations with the four largest absolute residuals from the MM-estimator were omitted for reasons of scale. The plot shows that the MM-residuals are in general smaller than the LS residuals, and hence MM gives a better fit to the bulk of the data.

5.6 Robust inference and variable selection for M-estimators

In general, estimators that fulfill an M-estimating equation like (5.8) are asymptotically normal, and hence approximate confidence intervals and tests can be obtained as in Sections 4.4.2 and 4.7.2. Recall that, according to (5.14), $\widehat{\boldsymbol{\beta}}$ has an approximately normal distribution, with covariance matrix given $vn^{-1}\mathbf{V}_{\mathbf{x}}^{-1}$. For the purposes of inference, $\mathbf{V}_{\mathbf{x}}$ and v can be estimated by

$$\widehat{\mathbf{V}}_{\mathbf{x}} = \frac{1}{n}\sum_{i=1}^{n}\mathbf{x}_i\mathbf{x}_i' = \frac{1}{n}\mathbf{X}'\mathbf{X}, \quad \widehat{v} = \widehat{\sigma}^2\frac{\mathrm{ave}_i\{\psi(r_i/\widehat{\sigma})^2\}}{[\mathrm{ave}_i\{\psi'(r_i/\widehat{\sigma})\}]^2}\frac{n}{n-p}, \tag{5.33}$$

and hence the resulting confidence intervals and tests are the same as those for fixed $\mathbf{X}$.

Actually, this estimator of $\mathbf{V}_{\mathbf{x}}$ has the drawback of not being robust. In fact, just one large $\mathbf{x}_i$ corresponding to an outlying observation with a large residual may have a large distorting influence on $\mathbf{X}'\mathbf{X}$, with diagonal elements typically inflated. Since $\widehat{v}$ is stable with respect to outlier influence, the confidence intervals based on $\widehat{v}(\mathbf{X}'\mathbf{X})^{-1}$ may be too small and hence the coverage probabilities may be much smaller than the nominal.

Yohai *et al.* (1991) proposed a more robust estimator of the matrix $\mathbf{V}_{\mathbf{x}}$, defined as

$$\widetilde{\mathbf{V}}_{\mathbf{x}} = \frac{1}{\sum_{i=1}^{n} w_i}\sum_{i=1}^{n} w_i\mathbf{x}_i\mathbf{x}_i', \tag{5.34}$$

with $w_i = W(r_i/\widehat{\sigma})$, where W is the weight function (2.31). Under the model with n large, the residual r_i is close to the error u_i, and since u_i is independent of $\mathbf{x}_i$ we have, as $n \to \infty$,

$$\frac{1}{n}\sum_{i=1}^{n} w_i\mathbf{x}_i\mathbf{x}_i' \to_p \mathrm{E}W\left(\frac{u}{\sigma}\right)(\mathrm{E}\mathbf{x}\mathbf{x}')$$

and

$$\frac{1}{n}\sum_{i=1}^{n} w_i \to_p \mathrm{E}W\left(\frac{u}{\sigma}\right),$$

and thus

$$\widetilde{\mathbf{V}}_{\mathbf{x}} \to_p \mathrm{E}\mathbf{x}\mathbf{x}'.$$

Assume that $\psi(t) = 0$ if $|t| > k$ for some k, as happens with the bisquare. Then, if $|r_i|/\hat{\sigma} > k$, the weight w_i is zero. If observation i has high leverage (i.e., $\mathbf{x}_i$ is "large") and is an outlier, then, since the estimator is robust, $|r_i|/\hat{\sigma}$ is also "large", and this observation will have null weight and hence will not influence $\widetilde{\mathbf{V}}_\mathbf{x}$. On the other hand, if $\mathbf{x}_i$ is "large" but r_i is small or moderate, then w_i will be nonnull, and $\mathbf{x}_i$ will still have a beneficial influence on $\widetilde{\mathbf{V}}_\mathbf{x}$ by virtue of reducing the variance of $\hat{\boldsymbol{\beta}}$. Hence the advantage of $\widetilde{\mathbf{V}}_\mathbf{x}$ is that it downweights high-leverage observations only when they are outlying. Therefore, we recommend the routine use of $\widetilde{\mathbf{V}}_\mathbf{x}$ instead of $\mathbf{V}_\mathbf{x}$ for all instances of inference, in particular the Wald-type tests defined in Section 4.7.2, which also require estimating the covariance matrix of the estimators.

The following example shows how different the inference can be when using an MM-estimator rather than the LS estimator.

For the straight-line regression of the mineral data in Example 5.1, the slope given by LS and its estimated SD are 0.135 and 0.020 respectively, while the corresponding values for the MM-estimator are 0.044 and 0.021; hence the classical and robust two-sided intervals with level 0.95 are (0.0958, 0.1742) and (0.00284, 0.08516), which are disjoint, showing the influence of the outlier.

5.6.1 Bootstrap robust confidence intervals and tests

Since the confidence intervals and tests for robust estimators are asymptotic, their actual level may be lower than the desired one if n is not large. This occurs especially when the error distribution is very heavy tailed or asymmetric (see the end of Section 10.3). Better results can be obtained using the bootstrap method, which simulates the distribution of the estimator of interest by recomputing the estimator on a large number of new samples randomly drawn from the original data (*bootstrap samples*). It can be shown that, under certain regularity conditions, the empirical distribution of the recomputed estimator converges to the sampling distribution of the estimator of interest. See, for example, Efron and Tibshirani (1993) and Davison and Hinkley (1997) for more details.

While the bootstrap approach has proved successful in many situations, its application to robust estimators presents two main problems. One is that recomputing robust estimators on a large number of bootstrap samples may demand impractical computing times. Another difficulty is that the proportion of outliers in some of the bootstrap samples might be much higher than in the original one, severely affecting the recomputed estimator and the resulting estimated distribution. Salibian-Barrera and Zamar (2002) proposed a bootstrap method that is faster and more robust than the naive application of the bootstrap approach. The main idea is to express the robust estimator $\hat{\boldsymbol{\beta}}$ as the solution to a fixed-point equation

$$\hat{\boldsymbol{\beta}} = \mathbf{g}_n(\hat{\boldsymbol{\beta}}),$$

where the function $\mathbf{g}_n : R^p \to R^p$ generally depends on the sample. Then, given a bootstrap sample, instead of solving the above equation for the bootstrap sample, we compute a one-step approximation:

$$\widehat{\boldsymbol{\beta}}^{1,*} = \mathbf{g}_n^*(\widehat{\boldsymbol{\beta}}),$$

where $\mathbf{g}_n^*$ is the function corresponding to the bootstrap sample, but evaluated on the estimator computed on the original sample. It it is easy to see that the distribution of $\widehat{\boldsymbol{\beta}}^{1,*}$ generally will not estimate that of $\widehat{\boldsymbol{\beta}}$. However, it can be corrected by applying a simple linear transformation. A Taylor expansion of the fixed-point equation above shows that, under certain regularity conditions, the following approximation has the same limiting distribution as the fully bootstrapped estimator:

$$\widehat{\boldsymbol{\beta}} + [\mathbf{I} - \nabla \mathbf{g}_n(\widehat{\boldsymbol{\beta}})]^{-1} \, (\widehat{\boldsymbol{\beta}}^{1,*} - \widehat{\boldsymbol{\beta}}), \tag{5.35}$$

where $\mathbf{I}$ denotes the identity matrix, and $\nabla \mathbf{g}_n$ is the matrix of first derivatives of $\mathbf{g}_n$. Thus, as usual, we can estimate the distribution of (5.35) by evaluating it on many bootstrap samples. Formal consistency proofs for this approach under mild regularity conditions can be found in Salibian-Barrera and Zamar (2002) and Salibian-Barrera *et al.* (2006).

This fast and robust way to estimate the sampling distribution of robust estimators can be used to derive confidence intervals and tests of hypotheses with good robustness properties (see, for example, Salibian-Barrera, 2005; Van Aelst and Willems, 2011; and Salibian-Barrera et al, 2016). It has also been shown to provide a robust alternative to bootstrap-based model selection procedured (Salibian-Barrera and Van Aelst, 2008).

A review of the method with applications to different models and settings can be found in Salibian-Barrera *et al.* (2008b). Code implementing this method for different robust estimators is available at https://github.com/msalibian and http://users.ugent.be/~svaelst/software/.

5.6.2 Variable selection

In many situations, the main purpose of fitting a regression equation is to predict the response variable. If the number of predictor variables is large and the number of observations relatively small, fitting the model using all the predictors will yield poorly estimated coefficients, especially when predictors are highly correlated. More precisely, the variances of the estimated coefficients will be high and therefore the forecasts made with the estimated model will have a large variance too. A common practice to overcome this difficulty is to fit a model using only a subset of variables, selected according to some statistical criterion.

Consider evaluating a model using the mean squared error (MSE) of the forecast. This MSE is composed of the variance plus the squared bias. Deleting some predictors may cause an increase in the bias and a reduction of the variance. Hence the problem of finding the best subset of predictors can be viewed as that of finding the best trade-off between bias and variance. There is a very large literature on the subset selection problem when the LS estimator is used as an estimation procedure; see for example Miller (1990), Seber (1984) and Hastie *et al.* (2001).

Let the sample be $(\mathbf{x}_i, y_i)$, $i = 1, \ldots, n$, where $\mathbf{x}_i = (x_{i1}, \ldots, x_{ip})$. The predictors are assumed to be random, but the case of fixed predictors is treated in a similar manner. For each set $C \subset \{1, 2, \ldots, p\}$, let $q = \#(C)$ and $\mathbf{x}_{iC} = (x_{ij})_{j \in C} \in R^q$. Akaike's (1970) *Final Prediction Error* (FPE) criterion based on the LS estimator is defined as

$$\mathrm{FPE}(C) = \mathrm{E}(y_0 - \mathbf{x}_{0C}'\widehat{\boldsymbol{\beta}}_C)^2. \tag{5.36}$$

where $\widehat{\boldsymbol{\beta}}_C$ is the estimator based on the set C and $(\mathbf{x}_0, y_0)$ have the same joint distribution as $(\mathbf{x}_i, y_i)$ and are independent of the sample. The expectation on the right-hand side of (5.36) is with respect to both $(\mathbf{x}_0, y_0)$ and $\widehat{\boldsymbol{\beta}}_C$. Then, it is shown that an approximately unbiased estimator of FPE is

$$\mathrm{FPE}^*(C) = \frac{1}{n}\sum_{i=1}^{n} r_{iC}^2 \left(1 + 2\frac{q}{n}\right), \tag{5.37}$$

where

$$r_{iC} = y_i - \mathbf{x}_{iC}'\widehat{\boldsymbol{\beta}}_C.$$

The first term of (5.37) evaluates the goodness of the fit when the estimator is $\widehat{\boldsymbol{\beta}}_C$, and the second term penalizes the use of a large number of explanatory variables. The best subset C is chosen as the one minimizing FPE $^*(C)$.

It is clear, however, that a few outliers may distort the value of FPE* (C), so that the choice of the predictors may be determined by a few atypical observations. To robustify FPE, we must note that not only the regression estimator must be robust, but the value of the criterion should not be sensitive to a few residuals.

We shall therefore robustify the FPE criterion by using for $\widehat{\boldsymbol{\beta}}$ a robust M-estimator (5.7) along with a robust error scale estimator $\widehat{\sigma}$. In addition, we shall bound the influence of large residuals by replacing the square in (5.36) with a bounded ρ-function, namely the same ρ as in (5.7). To make the procedure invariant under scale changes, the error must be divided by a scale σ, and to make consistent comparisons among different subsets of the predictor variable, σ must remain the same for all C. Thus the proposed criterion, which will be called the *robust final prediction error* (RFPE), is defined as

$$\mathrm{RFPE}(C) = \mathrm{E}\rho\left(\frac{y_0 - \mathbf{x}_{0C}'\widehat{\boldsymbol{\beta}}_C}{\sigma}\right) \tag{5.38}$$

where σ is the asymptotic value of $\widehat{\sigma}$.

To estimate RFPE for each subset C, we first compute

$$\widehat{\boldsymbol{\beta}}_C = \arg\min_{\boldsymbol{\beta} \in R^q} \sum_{i=1}^{n} \rho\left(\frac{y_i - \mathbf{x}_{iC}'\boldsymbol{\beta}}{\widehat{\sigma}}\right)$$

where the scale estimator $\widehat{\sigma}$ is based on the full set of variables, and define the estimator by

$$\text{RFPE}^*(C) = \frac{1}{n}\sum_{i=1}^{n}\rho\left(\frac{r_{iC}}{\widehat{\sigma}}\right) + \frac{q}{n}\frac{\widehat{A}}{\widehat{B}} \tag{5.39}$$

where

$$\widehat{A} = \frac{1}{n}\sum_{i=1}^{n}\psi\left(\frac{r_{iC}}{\widehat{\sigma}}\right)^2, \quad \widehat{B} = \frac{1}{n}\sum_{i=1}^{n}\psi'\left(\frac{r_{iC}}{\widehat{\sigma}}\right). \tag{5.40}$$

Note that if $\rho(r) = r^2$, then $\psi(r) = 2r$, and the result is equivalent to (5.37) since $\widehat{\sigma}$ cancels out. The criterion (5.39) is justified in Section 5.13.7.

When p is large, finding the optimal subset may be very costly in terms of computation time and therefore strategies to find suboptimal sets can be used. Two problems arise:

- Searching over all subsets may be impractical because of the extremely large number of subsets.
- Each computation of RFPE* requires recomputing a robust estimator $\widehat{\boldsymbol{\beta}}_C$ for each C, which can be very time-consuming when performed a large number of times.

In the case of the LS estimator, there exist very efficient algorithms to compute the classical FPE (5.37) for all subsets (see the references above), and so the first problem above is tractable for the classical approach if p is not too large. But computing a robust estimator $\widehat{\boldsymbol{\beta}}_C$ for all subsets C would be infeasible unless p were small. A simple but frequently effective suboptimal strategy is *stepwise regression*: add or remove one variable at a time ("forward" or "backward" regression), choosing the one whose inclusion or deletion yields the lowest value of the criterion. Various simulation studies indicate that the backward procedure is better. Starting with $C = \{1, \ldots, p\}$, we remove one variable at a time. At step k $(= 1, \ldots, p-1)$, we have a subset C with $\#(C) = p - k + 1$, and the next predictor to be deleted is found by searching over all subsets of C of size $p - k$ to find the one with smallest RFPE*.

The second problem above arises because robust estimators are computationally intensive, the more so when there is a large number of predictors. A simple way to reduce the computational burden is to avoid repeating the subsampling for each subset C by computing $\widehat{\boldsymbol{\beta}}_C$, starting from the approximation given by the weighted LS estimator with weights w_i obtained from the estimator corresponding to the full model.

Example 5.3 *To demonstrate the advantages of using a robust model selection approach based on RFPE*, we shall use a simulated dataset from a known model for which the "correct solution" is clear. The results in this example are obtained with script* ***step.R***.

We generated $n = 50$ observations from the model $y_i = \beta_0 + \mathbf{x}_i'\boldsymbol{\beta}_1 + u_i$ with $\beta_0 = 1$ and $\boldsymbol{\beta}_1' = (1, 1, 1, 0, 0, 0)$, so that $p = 7$. The u_i and x_{ij} are i.i.d. standard

Table 5.2 Variable selection for simulated data

LS		Robust	
Vars.	AIC	Vars.	RFPE*
1 2 3 4 5 6	240.14	1 2 3 4 5 6	6.73
1 3 4 5 6	238.16	1 2 3 5 6	6.54
2 4 5 6	237.33	1 2 3 5	6.35
4 5 6	237.83	1 2 3	6.22
		2 3	7.69

normal. Here, a perfect model selection method would select the variables $\{1, 2, 3\}$. We changed the values of the first six observations for outliers, the values of y by $25 + 5i$ and those of x_4, x_5, x_6 by $i/2$, $1 \leq i \leq 6$. We then applied the backward stepwise procedure using both the RFPE* criterion based on a MM-estimator and the Akaike information criterion (AIC) based on the LS estimator. While the RFPE* criterion gives the correct answer, selecting variables 1, 2 and 3, the AIC criterion selects variables 2, 4, 5 and 6. Both selecting processes are shown in Table 5.2.

Other approaches to robust model selection were given by Qian and Künsch (1998), Ronchetti and Staudte (1994) and Ronchetti *et al.* (1997).

5.7 Algorithms

In this section, we discuss successful strategies to compute the robust regression estimators discussed above. Both classes of estimators presented in this chapter (redescending M-estimators, and those based on minimizing a robust residual scale estimator) can be challenging to calculate because they are defined as the minimum of non-convex objective functions with many variables. In addition, some of these functions are not differentiable. As an illustration, consider an estimator based on a robust scale, as in (5.7.3), for a simple linear regression model through the origin. We simulated $n = 50$ observations from the model $y_i = \beta\, x_i + u_i$, where x_i and u_i are i.i.d. $N(0, 1)$. The true $\beta = 0$, and we added three outliers located at $(x, y) = (10, 20)$. Figure 5.13 shows the loss functions to be minimized when computing the LMS, the LTS estimator with $\alpha = 0.5$, an S-estimator with a bisquare ρ-function, and the LS estimator. The corresponding figure for an MM-estimator with a bisquare ρ function is very similar to that of the S-estimator. We see that the LMS loss function is very jagged, and that all estimators except LS exhibit a local minimum at about 2, which is a "bad solution". The global minima of the loss functions for these four estimators are attained at the values of 0.06, 0.14, 0.07 and 1.72, respectively.

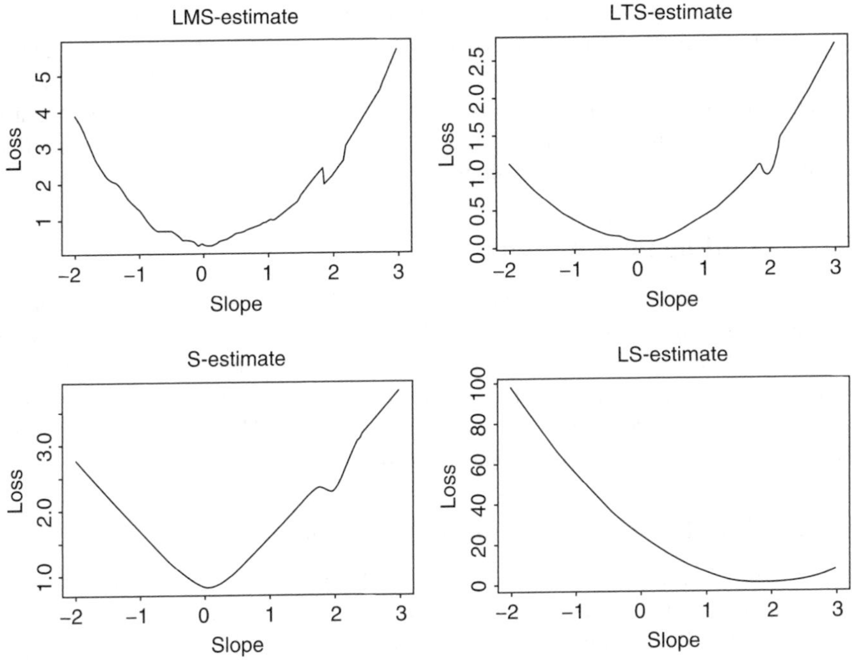

Figure 5.13 Loss functions for different regression estimators

The loss functions for the LMS and LTS estimators are not differentiable, and hence gradient methods cannot be applied to them. Stromberg (1993a,b) gives an exact algorithm for computing the LMS estimator, but the number of operations it requires is of order $\binom{n}{p+1}$, which is only practical for very small values of n and p. For other approaches see Agulló (1997, 2001) and Hawkins (1994). The loss function for the bisquare S-estimator is differentiable, but since gradient methods ensure only the attainment of a local minimum, a "good" starting point is needed.

In Section 5.7.1 we present iterative algorithms that can be shown to improve the corresponding objective function at each step. We will refer to these as "local improvements". The lack of convexity, however, means that usually there are several local minima, and hence these algorithms may not converge to the global minimum of the objective function. A strategy to solve this problem is to start the iterations from a large number of different initial points. Rather than using starting values taken completely at random from the set of possible parameter values, it is generally more efficient to let the data guide the construction of initial values. Below we discuss two such methods: subsampling, and an estimator proposed by Peña and Yohai (1999). The first is widely used and is discussed in Section 5.7.2. Since this method requires using a large number of starting points, one can reduce the overall computing time by identifying, early in the local improvement or "concentration" iterations, which sub-sampling candidates are most promising (see Rousseeuw and van Driessen, 2000; and Salibian-Barrera and Yohai, 2006). This variant of the subsampling algorithm is presented in Section 5.7.3. The use of *Peña–Yohai starting points* has only recently started to receive attention in the literature, but we show in Section 5.7.4 below that it compares very favorably to sub-sampling.

5.7.1 Finding local minima

The iterative reweighted least squares algorithm discussed in Section 4.5.2 can be used as a *local improvement strategy* for M-estimators defined in (5.7). In Section 9.1 we show that the objective function decreases in each iteration, and thus this method leads to a local minimum of (5.7). In what follows, we describe iterative procedures for S- and LTS-estimators that also lead to local minima. These work by decreasing the value of the objective function at each step.

5.7.1.1 Local improvements for S-estimators

Recall that S-estimators can be thought of as M-estimators (see p. 127), and thus a simple approach to find a local minimum that satisfies the first-order conditions (5.8) is to adapt the IRWLS algorithm described in Section 4.5.2 by updating σ at each step. In other words, if $\widehat{\boldsymbol{\beta}}_k$ is the estimator at the kth iteration, the scale estimator $\widehat{\sigma}_k$ is obtained by solving

$$\sum_{i=1}^{n} \rho\left(\frac{r_i(\widehat{\boldsymbol{\beta}}_k)}{\widehat{\sigma}_k}\right) = \delta \tag{5.41}$$

with the method of Section 2.8.2. Then $\widehat{\boldsymbol{\beta}}_{k+1}$ is obtained by weighted least squares, with weights $w_i = W(r_i/\widehat{\sigma}_k)$. It can be shown that if W is decreasing then $\widehat{\sigma}_k$ decreases at each step (see Salibian-Barrera and Yohai (2006) and Section 9.2). Since computing $\widehat{\sigma}$ consumes an major proportion of the computation time, it is important to do it economically, and in this regard one can, for example, start the iterations for $\widehat{\sigma}_k$ at the previous value $\widehat{\sigma}_{k-1}$. Other strategies are discussed in Salibian-Barrera and Yohai (2006).

5.7.1.2 Concentration steps for the LTS estimator

A local minimum of (5.25) can be attained iteratively using the "concentration step" (C-step) of Rousseeuw and van Driessen (2000). Given a candidate $\widehat{\boldsymbol{\beta}}_1$, let $\widehat{\boldsymbol{\beta}}_2$ be the LS estimator based on the data corresponding to the h smallest absolute residuals. It is proved in Section 9.3 that the trimmed L-scale $\widehat{\sigma}$ given by (5.25) is not larger for $\widehat{\boldsymbol{\beta}}_2$ than for $\widehat{\boldsymbol{\beta}}_1$. This procedure is exact, in the sense that after a finite number of steps it attains a value of $\widehat{\boldsymbol{\beta}}$ such that further steps do not decrease the values of $\widehat{\sigma}$. It can be shown that this $\widehat{\boldsymbol{\beta}}$ is a local minimum, but not necessarily a global one.

5.7.2 Starting values: the subsampling algorithm

The idea behind subsampling is to construct random candidate solutions to (5.17) using the sample points. More specifically, we construct candidates by fitting the model to randomly chosen subsets of the data. Since, intuitively, a good candidate should adjust well the clean portion of the sample, we need to find well-conditioned subsets of non-outlying points in order to obtain good starting points.

To compute the candidate solutions, we take subsamples J of size p from the data:

$$\{(\mathbf{x}_i, y_i) : i \in J\}, \quad J \subset \{1, ..., n\}, \quad \#(J) = p.$$

For each set J we find the vector $\boldsymbol{\beta}_J$ that satisfies the exact fit $\mathbf{x}_i'\boldsymbol{\beta}_J = y_i$, for $i \in J$. If a subsample is collinear, it is discarded and replaced by another. Since considering all $\binom{n}{p}$ subsamples would be prohibitive unless both n and p are rather small, we choose N of them at random: $\{J_k : k = 1, \ldots, N\}$. The initial value for the improvement steps is then taken to be the candidate that produces the best value of the object function: $\boldsymbol{\beta}_{J_{k_0}}$, where

$$k_0 = \arg \min_{1 \le k \le N} \hat{\sigma}(\mathbf{r}(\boldsymbol{\beta}_{J_k})). \tag{5.42}$$

We can now apply the algorithms in Section 5.7.1 to $\boldsymbol{\beta}_{J_{k_0}}$ to obtain a local minimum, which is taken as our estimator $\hat{\boldsymbol{\beta}}$.

Note that an alternative procedure would be to apply the local improvement iterations to *every* candidate $\boldsymbol{\beta}_{J_k}$, obtaining their corresponding local minimizers $\boldsymbol{\beta}^*_{J_k}$, $k = 1, \ldots, N$, and then select as our estimator $\hat{\boldsymbol{\beta}}$ the one with the best objective function. Although the resulting estimator will certainly not be worse (and will probably be better) than the previous one, its computational cost will be much higher. In Section 5.7.3 we discuss a better intermediate option.

Since the motivation for the subsampling method is to construct a good initial candidate by fitting a clean subsample of the data, one may need to take a large number N of random subsamples in order to find a clean one with high probability. Specifically, suppose the sample contains a proportion ε of outliers. The probability of an outlier-free subsample is $\alpha = (1 - \varepsilon)^p$, and the probability of at least one "good" subsample is $1 - (1 - \alpha)^N$. If we want this probability to be larger than $1 - \gamma$, we must have

$$\ln \gamma \ge N \ln(1 - \alpha) \approx -N\alpha$$

and hence

$$N \ge \frac{|\ln \gamma|}{|\ln(1 - (1 - \varepsilon)^p)|} \approx \frac{|\ln \gamma|}{(1 - \varepsilon)^p} \tag{5.43}$$

for p not too small. Therefore N must grow approximately exponentially with p. Table 5.3 gives the minimum N for $\gamma = 0.01$. Since the number of "good" (outlier-free) subsamples is binomial, the expected number of good samples is $N\alpha$, and so for $\gamma = 0.01$ the expected number of "good" subsamples is $|\ln 0.01| = 4.6$.

Table 5.3 Minimum N for $\delta = 0.01$

p	$\varepsilon = 0.1$	0.15	0.20	0.25	0.50
5	4	5	8	12	101
10	8	15	28	56	3295
15	14	35	90	239	105475
20	25	81	278	1013	3.38×10^6
30	74	420	2599	18023	3.46×10^9
40	216	2141	24214	320075	3.54×10^{12}
50	623	10882	225529	5.68×10^6	3.62×10^{15}

When this method is applied to an S-estimator, the following observation saves much computing time. Suppose we have examined $M-1$ subsamples and $\widehat{\sigma}_{M-1}$ is the current minimum. Now we draw the Mth subsample, which yields the candidate estimator $\boldsymbol{\beta}_M$. We may avoid the effort of computing the new scale estimator $\widehat{\sigma}_M$ in those cases where it will turn out to not be smaller than $\widehat{\sigma}_{M-1}$. The reason is as follows. If $\widehat{\sigma}_M < \widehat{\sigma}_{M-1}$, then since ρ is monotonic

$$n\delta = \sum_{i=1}^{n} \rho\left(\frac{r_i(\boldsymbol{\beta}_M)}{\widehat{\sigma}_M}\right) \geq \sum_{i=1}^{n} \rho\left(\frac{r_i(\boldsymbol{\beta}_M)}{\widehat{\sigma}_{M-1}}\right).$$

Thus if

$$n\delta \leq \sum_{i=1}^{n} \rho\left(\frac{r_i(\boldsymbol{\beta}_M)}{\widehat{\sigma}_{M-1}}\right), \tag{5.44}$$

we may discard $\boldsymbol{\beta}_M$ since $\widehat{\sigma}_M \geq \widehat{\sigma}_{M-1}$. Therefore $\widehat{\sigma}$ is computed only for those subsamples that do not verify condition (5.44).

Although the N given by (5.43) ensures that the approximate algorithm has the desired BP in a probabilistic sense, it does not imply that it is a good approximation to the exact estimator. Furthermore, because of the randomness of the subsampling procedure, the resulting estimator is *stochastic*; that is, repeating the computation may lead to another local minimum and hence to another $\widehat{\boldsymbol{\beta}}$, with the unpleasant consequence that repeating the computation may yield different results. In our experience, a carefully designed algorithm usually gives good results, and the above, infrequent but unpleasant, effects can be mitigated by increasing N as much as the available computing power will allow.

The subsampling procedure may be used to compute an approximate LMS estimator. Since total lack of smoothness precludes any kind of iterative improvement, the estimator is simply taken as the raw candidate $\boldsymbol{\beta}_{J_k}$ with smallest objective function. Usually this is followed by one-step reweighting (Section 5.9.1), which, besides improving the efficiency of the estimator, makes it more stable with respect to the randomness of subsampling. It must be recalled, however, that the resulting estimator is not asymptotically normal, and hence it is not possible to use it as a basis for approximate inference on the parameters. Since the resulting estimator is a weighted LS estimator, it would be intuitively attractive to apply classical LS inference as if these weights were constant, but this procedure is not valid. As was explained in Section 5.9.1, reweighting does not improve on the estimator's order of convergence.

5.7.3 A strategy for faster subsampling-based algorithms

With LTS or S-estimators, which allow iterative improvement steps as described in Section 5.7.1, it is possible to dramatically speed up the search for a global minimum. In the discussion below, "iteration" refers to one of the two iterative procedures described in that section (although the method to be described could be applied to any estimator that admits of iterative improvement steps). Consider the following two extreme strategies for combining the subsampling and the iterative parts of the minimization:

A Use the "best" result (5.42) of the subsampling as a starting point from which to iterate until convergence to a local minimum.
B Iterate to convergence from each of the N candidates $\widehat{\boldsymbol{\beta}}_{J_k}$ and keep the result with smallest $\widehat{\sigma}$.

Clearly strategy B would yield a better approximation of the absolute minimum than A, but is also much more expensive. An intermediate strategy, which depends on two parameters K_{iter} and K_{keep}, consists of the following steps:

1. For $k = 1, ..., N$, compute $\widehat{\boldsymbol{\beta}}_{J_k}$ and perform K_{iter} iterations, which yields the candidates $\widetilde{\boldsymbol{\beta}}_k$ with residual scale estimators $\widetilde{\sigma}_k$.
2. Only the K_{keep} candidates with smallest K_{keep} estimators $\widetilde{\sigma}_k$ are kept in storage, only needing to be updated when the current $\widetilde{\sigma}_k$ is lower than at least one of the current best K_{iter} values. Call these estimators $\widetilde{\boldsymbol{\beta}}_{(k)}$, $k = 1, .., K_{\text{keep}}$.
3. For $k = 1, .., K_{\text{keep}}$, iterate to convergence starting from $\widetilde{\boldsymbol{\beta}}_{(k)}$, obtaining the candidate $\widehat{\boldsymbol{\beta}}_k$ with residual scale estimator $\widehat{\sigma}_k$.
4. The final result is the candidate $\widehat{\boldsymbol{\beta}}_k$ with minimum $\widehat{\sigma}_k$.

Option A above corresponds to $K_{\text{iter}} = 0$ and $K_{\text{keep}} = 1$, while B corresponds to $K_{\text{iter}} = \infty$ and $K_{\text{keep}} = 1$. This strategy was first proposed by Rousseeuw and van Driessen (2000) for the LTS estimator (*Fast LTS*) with $K_{\text{iter}} = 2$ and $K_{\text{keep}} = 10$. As mentioned above, the general method can be used with any estimator that can be improved iteratively.

Salibian-Barrera and Yohai (2006) proposed a *Fast S-estimator*, based on this strategy. A theoretical study of the properties of this procedure seems impossible, but their simulations show that it is not worthwhile to increase K_{iter} and K_{keep} beyond values of 1 and 10, respectively. They also show that $N = 500$ gives reliable results at least for $p \leq 40$ and contamination fraction up to 10%. Their simulation also shows that Fast S is better than Fast LTS with respect both to mean squared errors and the probability of converging to a "wrong" solution. The simulation in Salibian-Barrera and Yohai (2006) also indicates that Fast LTS works better with $K_{\text{iter}} = 1$ than with $K_{\text{iter}} = 2$.

A further saving in time is obtained by replacing $\widehat{\sigma}$ in step 1 of the procedure by an approximation obtained by one step of the Newton–Raphson algorithm starting from the normalized median of absolute residuals.

Ruppert (1992) proposes a more complex random search method. However, the simulations by Salibian-Barrera and Yohai (2006) show that its behavior is worse than that of both the Fast S and the Fast LTS estimators.

5.7.4 Starting values: the Peña-Yohai estimator

The MM-estimator requires an initial estimator, for which we have chosen an S-estimator. Computing an S-estimator in turn requires initial values to start the iterations, which forces the user to use subsampling. While this approach is practical for small p, Table 5.3 shows that the number N_{sub} of required subsamples to obtain good candidates with high-probability increases rapidly with p, which makes the procedure impractical for large values of p. Of course, one can employ smaller values of N_{sub} than those given in the table, but then the estimator becomes unreliable. We shall now show a different approach that yields much better results than subsampling. It is based on a procedure initially proposed by Peña and Yohai (1999) for outlier detection, but which in fact yields a fast and reliable initial regression estimator.

Let $(\mathbf{X}, \mathbf{y})$ be a regression dataset, with $\mathbf{X} \in R^{n \times p}$. Call r_i the residuals from the LS estimator, and for $j = 1, .., n$ call $\widehat{\boldsymbol{\beta}}_{\text{LS}(j)}$ the LS estimator computed without observation j, and call $r_{i(j)} = y_i - \widehat{\boldsymbol{\beta}}'_{\text{LS}(j)}\mathbf{x}_i$ the ith residual using the LS estimator with observation j deleted. Then it can be shown that

$$r_{i(j)} = \frac{h_{ij} r_i}{1 - h_{ii}}, \tag{5.45}$$

where h_{ij} is the (i, j) element of the "hat matrix" $\mathbf{H} = \mathbf{X}(\mathbf{X}'\mathbf{X})^{-1}\mathbf{X}'$, defined in (4.29); see, for example, Belsley *et al.* (1980).

Define the sensitivity vectors $\mathbf{r}_i \in R^n$ with elements $r_{ij} = r_{i(j)}$, $i, j = 1, .., n$. Then $\mathbf{r}_i$ expresses the sensitivity of the prediction of y_i to the deletion of each observation. We define the sensitivityt matrix $\mathbf{R}$ as the $n \times n$ matrix with rows $\mathbf{r}'_i$. It follows from (5.45) that $\mathbf{R} = \mathbf{HW}$, where $\mathbf{W}$ is the diagonal matrix with elements $r_i/(1 - h_{ii})$. We can consider $\mathbf{R}$ as a data matrix with n "observations" of dimension n. Then we can try to find the most informative linear combinations of the columns of $\mathbf{R}$ using principal components. Let $\mathbf{v}_1, ...\mathbf{v}_n$ be the eigenvectors of

$$\mathbf{P} = \mathbf{R}'\mathbf{R} = \mathbf{W}\mathbf{H}^2\mathbf{W} = \mathbf{WHW}$$

corresponding to the eigenvalues $\lambda_1 \geq \lambda_2 \geq ... \geq \lambda_n$. Since **P** has rank p, only the first p eigenvalues are nonnull. Then the only informative principal components are

$$\mathbf{z}_i = \mathbf{R}\mathbf{v}_i, \ i = 1, ..., p.$$

Peña and Yohai (1999) show that a more convenient way to compute the $\mathbf{z}_i$s is as

$$\mathbf{z}_i = \mathbf{X}(\mathbf{X}'\mathbf{X})^{-1/2}\mathbf{X}'\mathbf{u}_i, \ i = 1, ..., p,$$

where $\mathbf{u}_1, ..., \mathbf{u}_p$ are the eigenvectors of the $p \times p$ matrix

$$\mathbf{Q} = (\mathbf{X}'\mathbf{X})^{-1/2}\mathbf{X}'\mathbf{W}\mathbf{X}(\mathbf{X}'\mathbf{X})^{-1/2}$$

corresponding to the eigenvalues $\lambda_1 \geq \lambda_2 \geq ... \geq \lambda_p$. Note that in this way we only need to compute the eigenvectors of the $p \times p$ matrix **Q** instead of those of the $n \times n$ matrix **P.** These "sensitivity principal components" $\mathbf{z}_1, ..., \mathbf{z}_p$ will be used to identify outliers. A set of natural candidates for the initial estimator will be found using the LS estimators obtained after deleting different sets of outliers found in this process. The initial estimator for the MM estimator will be the candidate whose residuals have a minimum M-scale. In what follows, we describe in detail the iterative procedure that will be used to obtain the initial estimator.

In each iteration k we obtain a set A_k of $3p + 1$ candidates for estimating $\boldsymbol{\beta}$. Then, in this iteration, we select the candidate $\boldsymbol{\beta}^{(k)}$ as

$$\boldsymbol{\beta}^{(k)} = \arg \min_{\boldsymbol{\beta} \in A_k} S(\mathbf{y} - \mathbf{X}\boldsymbol{\beta})$$

and let

$$s^{(k)} = \min_{\boldsymbol{\beta} \in A_k} S(\mathbf{y} - \mathbf{X}\boldsymbol{\beta}^{(k)})$$

where S is an M-scale with breakdown point equal to a given α. Now we will describe how to compute the sets A_i. Call z_{ji} $(i = 1, , .n)$ the coordinates of $\mathbf{z}_j$ and put $m = [\alpha n]$.

Iteration 1 The set A_1 includes the LS estimator, the L_1 estimator and, for each $j, 1 \leq j \leq p$, it also includes the LS estimator obtained after deleting the observations corresponding to the m largest z_{ji}, the m smallest z_{ji}, and the m largest $|z_{ji}|$,

Iteration $k + 1$ Suppose now that we have already completed iteration k. Then the residuals $\mathbf{r}^{(k)} = (r_1^{(k)}, ..., r_n^{(k)})' = \mathbf{y} - \mathbf{X}\boldsymbol{\beta}^{(k)}$ are computed and all observations j such that

$$|r_j^{(k)}| > Cs^{(k)}$$

are deleted, where C is a given constant. The remaining observations are used to compute new sensitivity principal components $\mathbf{z}_1, ...\mathbf{z}_p$. The set A_k is computed using the the same procedure that for A_1.

The iterations end when $\boldsymbol{\beta}^{(k+1)} = \boldsymbol{\beta}^{(k)}$. Ultimately, the initial estimator is defined by $\widetilde{\boldsymbol{\beta}} = \boldsymbol{\beta}^{(k_0)}$, where

$$k_0 = \arg\min_k s^{(k)}.$$

The recommended values are $\alpha = 0.5$ and $C = 2$. Peña and Yohai (1999) show that this estimator has breakdown point α for mass-point contamination, and simulations in the same paper show it to yield reliable results under different outlier configurations.

5.7.5 Starting values with numeric and categorical predictors

Consider a linear model of the form:

$$y_i = \mathbf{x}'_{1i}\boldsymbol{\beta}_1 + \mathbf{x}'_{2i}\boldsymbol{\beta}_2 + u_i, \quad i = 1, \ldots n, \tag{5.46}$$

where the $\mathbf{x}_{1i} \in R^{p_1}$ are 0–1 vectors, such as a model with some categorical variables as in the case of the example in Section 1.4.2, and the $\mathbf{x}_{2i} \in R^{p_2}$ are continuous random variables. The presence of the continuous variables would make it appropriate to use an MM estimator, since a monotone M-estimator would not be robust. However, the presence of the 0–1 variables is a source of possible difficulties with the initial estimator. If we employ subsampling, in an unbalanced structured design, there is a high probability that a subsampling algorithm yields collinear samples. For example, if there are five independent explanatory dummy variables that take the value 1 with probability 0.1, then the probability of selecting a noncollinear sample of size 5 is only 0.011! In any event the sub-sampling will be a waste if $p_2 \ll p_1$. If we instead employ the Peña–Yohai estimator, the result may be quite unreliable.

Our approach is based on the idea that if one knew $\boldsymbol{\beta}_2$ (respectively $\boldsymbol{\beta}_1$) in (5.46), it would be natural to use a monotone M-estimator (S-estimator) for the parameter $\boldsymbol{\beta}_1$ (the parameter $\boldsymbol{\beta}_2$). To carry out this idea we recall the well-known procedure to perform a bivariate regression through univariate regressions: first "orthogonalize" one of the predictors with respect to the other, and then compute the univariate regression of the response with respect to each of the predictors.

Let $M(\mathbf{X}, \mathbf{y})$ be a monotone regression M-estimator such as the L_1 estimator. The first step is to "remove the effect of $\mathbf{X}_1$ from $\mathbf{X}_2$ and $\mathbf{y}$". Let

$$\widetilde{\mathbf{X}}_2 = \mathbf{X}_2 - \mathbf{X}_1\mathbf{T}, \text{ and } \widetilde{\mathbf{y}} = \mathbf{y} - \mathbf{X}_1\mathbf{t}, \tag{5.47}$$

where $\mathbf{t} = M(\mathbf{X}_1, \mathbf{y})$ and the columns of $\mathbf{T} \in R^{p_1 \times p_2}$ are the regression vectors of the columns of $\mathbf{X}_2$ on $\mathbf{X}_1$.

Let $\widehat{\boldsymbol{\beta}}_2 = \arg\min_{\boldsymbol{\beta}_2} S(\widetilde{\mathbf{y}} - \widetilde{\mathbf{X}}_2\boldsymbol{\beta}_2 - \mathbf{X}_1\widetilde{\boldsymbol{\beta}}_1(\boldsymbol{\beta}_2))$, where $\widetilde{\boldsymbol{\beta}}_1(\boldsymbol{\beta}_2) = M(\mathbf{X}_1, \widetilde{\mathbf{y}} - \widetilde{\mathbf{X}}_2\boldsymbol{\beta}_2)$. In other words, $\widehat{\boldsymbol{\beta}}_2$ is the S-estimator obtained after adjusting for the 0–1 variables. The estimators $(\widehat{\boldsymbol{\beta}}_1, \widehat{\boldsymbol{\beta}}_2)$ are obtained by "back-transforming" from (5.47), so that $\mathbf{X}_1\widehat{\boldsymbol{\beta}}_1 + \mathbf{X}_2\widehat{\boldsymbol{\beta}}_2 = \mathbf{X}_1\widetilde{\boldsymbol{\beta}}_1 + \widetilde{\mathbf{X}}_2\widehat{\boldsymbol{\beta}}_2$, which yields $\widehat{\boldsymbol{\beta}}_1 = \widetilde{\boldsymbol{\beta}}_1 - \mathbf{T}\widehat{\boldsymbol{\beta}}_2$.

To compute $\widehat{\boldsymbol{\beta}}_2$ above we use Peña–Yohai candidates based on $\widetilde{\mathbf{X}}_2$ and $\widetilde{\mathbf{y}}$. For each of these candidates, we compute $\widetilde{\boldsymbol{\beta}}_1$ as above and select the candidate that yields the smallest M-scale of the residuals. In our numerical experiments we found that it is useful to further refine this best Peña–Yohai candidate employing the iterations described in Section 5.7.1.1. Finally, we use $(\widehat{\boldsymbol{\beta}}_1, \widehat{\boldsymbol{\beta}}_2)$ and its associated residual scale estimator as the initial point to compute an MM-estimator.

Maronna and Yohai (2000) proposed a similar idea, but employing subsampling for the S-estimator. Koller (2012) proposed a subsampling method for these models that avoids collinear subsamples. A naïve procedure would be to just apply the Peña–Yohai estimator to **X**. These alternatives were compared to the one proposed here by a thorough simulation study, which demonstrated that the proposed method outperforms its competitors.

Rousseeuw and Wagner (1994), and Hubert and Rousseeuw (1996, 1997) have also proposed other approaches for models with categorical predictors.

Example 5.4 *Each row of a dataset (from Hettich and Bay, 1999) is a set of 90 measurements at a river in Europe. There are 11 predictors. The first three are categorical: the season of the year, river size (small, medium and large) and fluid velocity (for low, medium and high). The other 8 are the concentrations of chemical substances. The response is the logarithm of the abundance of a certain class of algae. The tables and figures for this example are obtained with script* ***algae.R****.*

Figures 5.14 and 5.15 are the normal Q–Q plots of the residuals corresponding to the LS estimator and to the MS-estimator described above. The first gives the impression of short-tailed residuals, while the residuals from the robust fit indicate the existence of least two outliers.

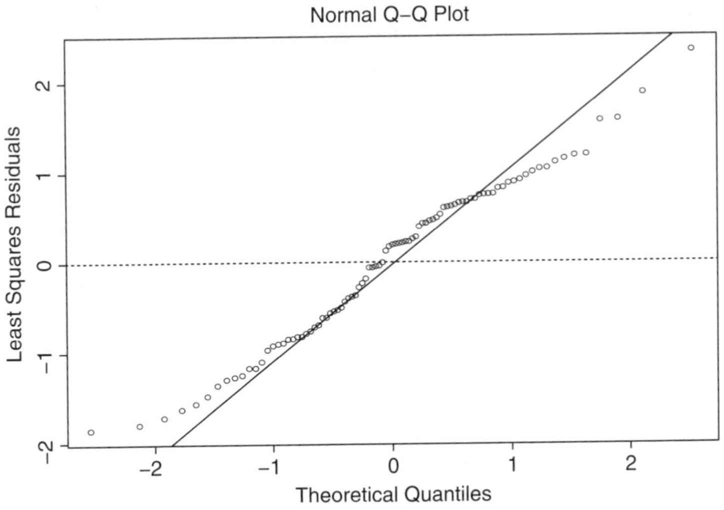

Figure 5.14 Algae data: normal Q–Q plot of LS residuals

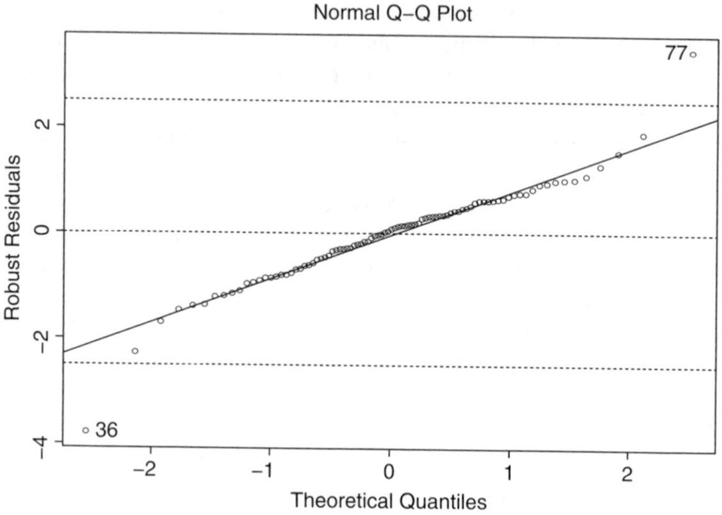

Figure 5.15 Algae data: normal Q–Q plot of robust residuals

Example 1.3 (continued) In the multiple linear regression in Section 1.4.2, the response variable was the rate of unemployment and the predictor variables were PA, GPA, HS, GHS, Region and Period. The last two are categorical variables with 22 and 2 parameters respectively, while the other predictors are continuous variables. The estimator used for that example was the MS-estimator. Figures 1.4 and 1.5 revealed that for these data the LS estimator found no outliers at all, while the MS-estimator found a number of large outliers. In this example three of the LS and MS-estimator t-statistics and p-values give opposite results using 0.05 as the level of the test, as shown in Table 5.4.

Table 5.4 Unemployment data: results from LS and MS estimators

Variable	Estimate	t value	p value
Region 20	MS	−1.0944	0.2811
	LS	−3.0033	0.0048
HS	MS	1.3855	0.1744
	LS	2.4157	0.0209
Period2	MS	2.1313	0.0400
	LS	0.9930	0.3273

For the “Region 20” level of the “Region” categorical variable and the HS variables, the LS fit declares these variables as significant while the robust fit declares them insignificant. The opposite is the case for the Period 2 level of the Period categorical variable. This shows that outliers can have a large influence on the classical test statistics of an LS fit.

5.7.6 Comparing initial estimators

We shall now compare two choices for the initial estimator of an MM-estimator with bisquare ρ. The first starts from an S-estimator computed with the strategy described in Section 5.7.2 and with N_{sub} subsamples (henceforth “S-Sub” for brevity). The second is to start from the Peña–Yohai estimator (henceforth “P-Y”).

For large p, the choice of the number of subsamples N_{sub} for the former has to be somewhat arbitrary, for if we wanted to ensure even a breakdown point of 0.15, the values of N_{sub} given by Table 5.3 for $n \geq 40$ would make the computation unfeasible.

Table 5.5 compares the computing times (in seconds) of both estimators for p between 10 and 100 and $n = 10p$. It is seen that the latter is much faster, and the difference between the estimators increases with p, despite the relatively low values of N_{sub}.

The left-hand half of Table 5.6 shows the finite-sample efficiencies of both estimators. It is seen that those of MM based on P-Y are always higher than those obtained from the S-estimator, and are reasonably close to the nominal one; the differences between both estimators are clearer when $n = 5p$. The last two columns show the respective maximum MSEs under 10% point-mass contamination, and it it seen that employing P-Y yields much lower values. In conclusion, using P-Y as initial estimator makes MM faster, more efficient and more robust. Another important feature of P-Y is that unlike subsampling it is *deterministic*, and therefore yields totally reproducible results. For all these reasons we recommend its routine use instead of subsampling.

Table 5.5 Computing times (in seconds) for the bisquare MM-estimator with S-Sub and P-Y starting points

p	n	N_{sub}	S-Sub	P-Y
10	100	1000	0.01	0.01
20	200	2000	0.06	0.05
30	300	3000	0.58	0.16
50	500	4000	7.68	0.71
80	800	4000	37.32	3.33
100	1000	4000	114.33	7.74

Table 5.6 Efficiencies and Max MSEs of MM-estimator with S-Sub and P-Y starting points

		Efficiencies		Max MSE	
p	n	S-Sub	P-Y	S-Sub	P-Y
10	50	0.727	0.827	1.63	0.98
	100	0.808	0.855	0.71	0.57
	200	0.838	0.857	0.47	0.44
20	100	0.693	0.834	1.51	0.96
	200	0.789	0.849	0.71	0.60
	400	0.825	0.851	0.49	0.45
50	250	0.682	0.862	2.10	1.06
	500	0.791	0.859	0.77	0.52
	1000	0.828	0.856	0.62	0.42
80	400	0.685	0.874	5.10	0.96
	800	0.791	0.863	3.22	0.64
	1600	0.827	0.857	2.41	0.49

5.8 Balancing asymptotic bias and efficiency

Defining the asymptotic bias of regression estimators requires a measure of the "size" of the difference between the value of an estimator, which for practical purposes we take to be the asymptotic value $\widehat{\boldsymbol{\beta}}_\infty$, and the true parameter value $\boldsymbol{\beta}$. We shall use an approach based on prediction. Consider an observation $(\mathbf{x}, y)$ from the model (5.1)–(5.2):

$$y = \mathbf{x}'\boldsymbol{\beta} + u, \quad \mathbf{x} \text{ and } u \text{ independent.}$$

The prediction error corresponding to $\widehat{\boldsymbol{\beta}}_\infty$ is

$$e = y - \mathbf{x}'\widehat{\boldsymbol{\beta}}_\infty = u - \mathbf{x}'(\widehat{\boldsymbol{\beta}}_\infty - \boldsymbol{\beta}).$$

Let $\mathrm{E}u^2 = \sigma^2 < \infty$, $\mathrm{E}u = 0$, and $\mathbf{V}_\mathbf{x} = \mathrm{E}\mathbf{x}\mathbf{x}'$. Then the mean squared prediction error is

$$\mathrm{E}e^2 = \sigma^2 + (\widehat{\boldsymbol{\beta}}_\infty - \boldsymbol{\beta})'\mathbf{V}_\mathbf{x}(\widehat{\boldsymbol{\beta}}_\infty - \boldsymbol{\beta}).$$

The second term is a measure of the increase in the prediction error due to the parameter estimation bias, and so we define the bias as

$$\mathrm{b}(\widehat{\boldsymbol{\beta}}_\infty) = \sqrt{(\widehat{\boldsymbol{\beta}}_\infty - \boldsymbol{\beta})'\mathbf{V}_\mathbf{x}(\widehat{\boldsymbol{\beta}}_\infty - \boldsymbol{\beta})}. \tag{5.48}$$

Note that if $\widehat{\boldsymbol{\beta}}$ is regression, scale and affine equivariant, this measure is *invariant* under the respective transformations; that is, $\mathrm{b}(\widehat{\boldsymbol{\beta}}_\infty)$ does not change when any of those transformations is applied to $(\mathbf{x}, y)$. If $\mathbf{V}_{\mathbf{x}}$ is a multiple of the identity – such as when the elements of $\mathbf{x}$ are i.i.d. zero mean normal – then (5.48) is a multiple of $\|\widehat{\boldsymbol{\beta}}_{\infty} - \boldsymbol{\beta}\|$, so in this special case the Euclidean norm is an adequate bias measure.

Now consider a model with intercept; that is,

$$\mathbf{x} = \begin{bmatrix} 1 \\ \underline{\mathbf{x}} \end{bmatrix}, \quad \boldsymbol{\beta} = \begin{bmatrix} \beta_0 \\ \boldsymbol{\beta}_1 \end{bmatrix}$$

and let $\boldsymbol{\mu} = \mathrm{E}\underline{\mathbf{x}}$ and $\mathbf{U} = \mathrm{E}\underline{\mathbf{x}}\underline{\mathbf{x}}'$, so that

$$\mathbf{V}_{\mathbf{x}} = \begin{bmatrix} 1 & \boldsymbol{\mu}' \\ \boldsymbol{\mu} & \mathbf{U} \end{bmatrix}.$$

For a regression and affine equivariant estimator, there is no loss of generality in assuming $\boldsymbol{\mu} = \mathbf{0}$ and $\beta_0 = 0$, and in this case

$$\mathrm{b}(\widehat{\boldsymbol{\beta}}_{\infty})^2 = \widehat{\beta}_{0,\infty}^2 + \widehat{\boldsymbol{\beta}}_{1,\infty}' \mathbf{U} \widehat{\boldsymbol{\beta}}_{1,\infty},$$

with the first term representing the contribution to bias of the intercept and the second that of the slopes.

A frequently used benchmark for comparing estimators is to assume that the joint distribution of $(\underline{\mathbf{x}}, y)$ belongs to a contamination neighborhood of a multivariate normal. By the affine and regression equivariance of the estimators, there is no loss of generality in assuming that this central normal distribution is $\mathrm{N}_{p+1}(\mathbf{0}, \mathbf{I})$. In this case it can be shown that the maximum biases of M-estimators do not depend on p. A proof is outlined in Section 5.13.5. The same is true of the other estimators treated in this chapter, except for GM-estimators, as set out in Section 5.11.1.

The maximum asymptotic bias of S-estimators can be derived from the results of Martin *et al.* (1989), and those of the LTS and LMS estimators from Berrendero and Zamar (2001). Table 5.7 compares the maximum asymptotic biases of LTS, LMS and the S-estimator with bisquare scale and three MM-estimators with bisquare ρ,

Table 5.7 Maximum bias of regression estimators for contamination ε

	ε				
	0.05	0.10	0.15	0.20	Eff.
LTS	0.63	1.02	1.46	2.02	0.07
S-E	0.56	0.88	1.23	1.65	0.29
LMS	0.53	0.83	1.13	1.52	0.0
MM (global)	0.78	1.24	1.77	2.42	0.95
MM (local)	0.56	0.88	1.23	1.65	0.95
MM (local)	0.56	0.88	1.23	1.65	0.85

in all cases with asymptotic BP equal to 0.5, when the joint distribution of $\mathbf{x}$ and y is in an ε-neighborhood of the multivariate normal $N_{p+1}(\mathbf{0}, \mathbf{I})$. One MM-estimator is given by the global minimum of (5.29) with normal distribution efficiency 0.95. The other two MM-estimators correspond to local minima of (5.29) obtained using the IRWLS algorithm starting from the S-estimator, with efficiencies 0.85 and 0.95. The LMS estimator has the smallest bias for all the values of ε considered, but also has zero asymptotic efficiency. It is remarkable that the maximum biases of both "local" MM-estimators are much lower than those of the "global" MM-estimator, and close to the maximum biases of the LMS estimator. This shows the importance of a good starting point. The fact that an estimator obtained as a local minimum starting from a very robust estimator may have a lower bias than one defined by the absolute minimum was pointed out by Hennig (1995), who also gave bounds for the bias of MM-estimators with general ρ-functions in contamination neighborhoods.

It is also curious that the two local MM-estimators with different efficiencies have the same maximum biases. To understand this phenomenon, we show in Figure 5.16 the asymptotic biases of the S- and MM-estimators for contamination fraction $\varepsilon = 0.2$ and point contamination located at (x_0, Kx_0) with $x_0 = 2.5$, as a function of the contamination slope K. It is seen that the bias of each estimator is worse than that of the LS estimator up to a certain value of K and then drops to zero. But the range of values where the MM-estimator with efficiency 0.95 has a larger bias than the LS estimator is greater than those for the 0.85 efficient MM-estimator and the S-estimator. This is the price paid for a higher normal efficiency. The MM-estimator with efficiency 0.85 is closer in behavior to the S-estimator than

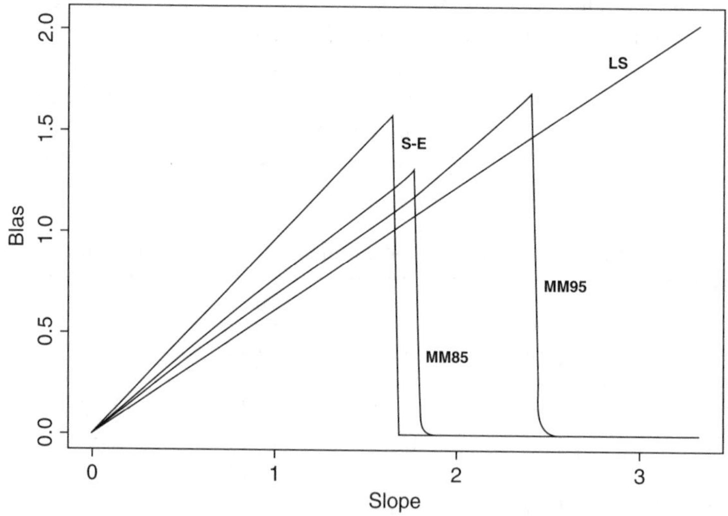

Figure 5.16 Biases of LS, S-estimator, and MM-estimators with efficiencies 0.85 and 0.95, as a function of the contamination slope, for $\varepsilon = 0.2$, when $x_0 = 2.5$

the one with efficiency 0.95. If one makes similar plots for other values of x_0, one finds that for $x_0 \geq 5$ the curves for the S-estimator and the two MM-estimators are very similar.

The former results, on MM-estimators, suggest a general approach for the choice of robust estimators. Consider in general an estimator $\widehat{\theta}$ with high efficiency, defined by the absolute minimum of a target function. We have an algorithm that, starting from any initial value θ_0, yields a local minimum of the target function, that we shall call $A(\theta_0)$. Assume also that we have an estimator θ^* with lower bias, although possibly with low efficiency. Define a new estimator $\widetilde{\theta}$ (call it the "approximate estimator") as the local minimum of the target function obtained by applying the algorithm starting from θ^*; that is, $\widetilde{\theta} = A(\theta^*)$. Then, in general, $\widetilde{\theta}$ has the same efficiency as $\widehat{\theta}$ under the model, while it has a lower bias than $\widehat{\theta}$ in contamination neighborhoods. If, in addition, θ^* is fast to compute, then $\widetilde{\theta}$ will also be faster than $\widehat{\theta}$. An instance of this approach in multivariate analysis will be seen in Section 6.8.5.

5.8.1 "Optimal" redescending M-estimators

In Section 3.5.4 we gave the solution to the Hampel dual problems for a one-dimensional parametric model, namely:

- finding the M-estimator minimizing the asymptotic variance subject to an upper bound on the gross-error sensitivity (GES), and
- minimizing the GES subject to an upper bound on the asymptotic variance.

This approach cannot be taken with regression M-estimators with random predictors since (5.13) implies that the GES is infinite. However, as we now show, it is possible to modify it in a suitable way.

Consider a regression estimator $\widehat{\boldsymbol{\beta}} = (\widehat{\beta}_0, \widehat{\boldsymbol{\beta}}_1)$, where $\widehat{\beta}_0$ corresponds to the intercept and $\widehat{\boldsymbol{\beta}}_1$ to the slopes. Yohai and Zamar (1997) showed that for an M-estimator $\widehat{\boldsymbol{\beta}}$ with bounded ρ, the maximum biases $\mathrm{MB}(\varepsilon, \widehat{\beta}_0)$ and $\mathrm{MB}(\varepsilon, \widehat{\boldsymbol{\beta}}_1)$ in an ε-contamination neighborhood are of order $\sqrt{\varepsilon}$. Therefore, the biases of these estimators are continuous at zero, which means that a small amount of contamination produces only a small change in the estimator. Because of this, the approach in Section 3.5.4 can be adapted to the present situation by replacing the GES with a different measure called the *contamination sensitivity* (CS), which is defined as

$$\mathrm{CS}(\widehat{\boldsymbol{\beta}}_j) = \lim_{\varepsilon \to 0} \frac{\mathrm{MB}(\varepsilon, \widehat{\boldsymbol{\beta}}_j)}{\sqrt{\varepsilon}} \quad (j = 0, 1).$$

Recall that the asymptotic covariance matrix of a regression M-estimator depends on ρ only through

$$v(\psi, F) = \frac{\mathrm{E}_F(\psi(u)^2)}{(\mathrm{E}_F \psi'(u))^2}$$

where $\psi = \rho'$ and F is the error distribution. We consider only the slopes $\boldsymbol{\beta}_1$, which are usually more important than the intercept. The analogues of the direct and dual Hampel problems can now be stated as finding the function ψ that

- minimizes $v(\psi, F)$ subject to the constraint $\mathrm{CS}(\widehat{\boldsymbol{\beta}}_1) \leq k_1$

or

- minimizes $\mathrm{CS}(\widehat{\boldsymbol{\beta}}_1)$ subject to $v(\psi, F) \leq k_2$,

where k_1 and k_2 are given constants.

Yohai and Zamar (1997) found that the optimal ψ for both problems has the form:

$$\psi_c(u) = \mathrm{sgn}(u)\left(-\frac{\varphi'(|u|) + c}{\varphi(|u|)}\right)^{+} \tag{5.49}$$

where φ is the standard normal density, c is a constant and $t^{+} = \max(t, 0)$ denotes the positive part of t. For $c = 0$ we have the LS estimator: $\psi(u) = u$.

Table 5.8 gives the values of c corresponding to different efficiencies, and Figure 5.17 shows the bisquare and optimal ψ-functions with efficiency 0.95. We observe that the optimal ψ increases almost linearly and then redescends much faster than the bisquare ψ. This optimal ψ-function is a smoothed, differentiable version of the hard-rejection function $\psi(u) = u\mathrm{I}(|u| \leq a)$ for some constant a. As such, it is not only good from the numerical optimization perspective, but also has the intuitive feature of making a rather rapid transition from its maximum absolute values to zero in the "flanks" of the nominal normal distribution. The latter is a region in which it is most difficult to tell whether a data point is an outlier or not: outside that transition region, outliers are clearly identified and rejected, and inside it. data values are left essentially unaltered. As a minor point, the reader should note that (5.49) implies that the optimal ψ has the curious feature of vanishing completely in a small interval around zero. For example, if $c = 0.013$ the interval is $(-0.032, 0.032)$, which is so small it is not visible in the figure.

Svarc *et al.* (2002) considered the two optimization problems stated above, but used the actual maximum bias $\mathrm{MB}(\varepsilon, \widehat{\boldsymbol{\beta}}_1)$ for a range of positive values of ε instead of the approximation given by the contamination sensitivity $\mathrm{CS}(\widehat{\boldsymbol{\beta}}_1)$. They calculated the optimal ψ and showed numerically that for $\varepsilon \leq 0.20$ it is almost identical to the one based on the contamination sensitivity. Therefore the optimal solution corresponding

Table 5.8 Constants for optimal estimator

Efficiency	0.80	0.85	0.90	0.95
c	0.060	0.044	0.028	0.013

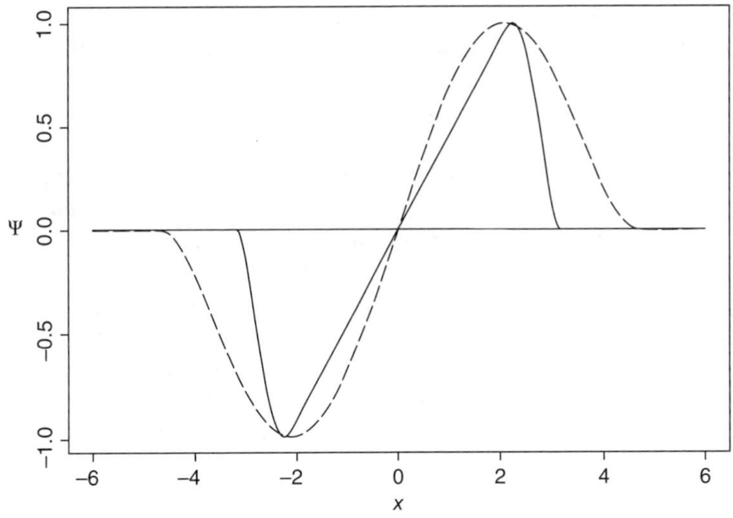

Figure 5.17 Optimal (–) and bisquare (....) psi-functions with efficiency 0.95

to an infinitesimal contamination is a good approximation to the one corresponding to $\varepsilon > 0$, at least for $\varepsilon \leq 0.20$.

The results of simulations in the next section show an unexpected property of this estimator, namely that under a very high nominal efficiency, such as 0.99, its behavior under contamination can be approximately as good as that of the bisquare estimator with a much lower efficiency. This fact can be explained by the form of its ψ-function, as shown in Figure 5.17, which coincides with that of the LS estimator up to a certain point, and then drops to zero.

5.9 Improving the efficiency of robust regression estimators

5.9.1 Improving efficiency with one-step reweighting

We have seen that estimators based on a robust scale cannot have both a high BP and high normal efficiency. As we have already discussed, one can obtain a desired normal efficiency by using an S-estimator as the starting point for an iterative procedure leading to an MM-estimator. In this section we consider a simpler alternative procedure, proposed by Rousseeuw and Leroy (1987), to increase the efficiency of an estimator $\widehat{\boldsymbol{\beta}}_0$ without decreasing its BP.

Let $\tilde{\sigma}$ be a robust scale of $\mathbf{r}(\widehat{\boldsymbol{\beta}}_0)$, say, the normalized median of absolute values (4.47). Then compute a new estimator $\widehat{\boldsymbol{\beta}}$, defined as the weighted LS estimator of the dataset with weights $w_i = W(r_i(\widehat{\boldsymbol{\beta}}_0)/\tilde{\sigma})$, where $W(t)$ is a decreasing function of

$|t|$. Rousseeuw and Leroy proposed the "weight function" W be chosen as the "hard rejection" function $W(t) = \mathrm{I}(|t| \leq k)$, with k equal to a γ-quantile of the distribution of $|x|$, where x has a standard normal distribution; for example, $\gamma = 0.975$. Under normality, this amounts to discarding a proportion of about $1 - \gamma$ of the points with largest absolute residuals.

He and Portnoy (1992) show that, in general, such reweighting methods preserve the order of consistency of $\widehat{\boldsymbol{\beta}}_0$, so in the standard situation where $\widehat{\boldsymbol{\beta}}_0$ is $\sqrt{n}$-consistent, then so is $\widehat{\boldsymbol{\beta}}$. Unfortunately, this means that because the LMS estimator is $n^{1/3}$-consistent, so is the reweighted LMS estimator.

In general, the reweighted estimator $\widehat{\boldsymbol{\beta}}$ is more efficient than $\widehat{\boldsymbol{\beta}}_0$, but its asymptotic distribution is complicated (and more so when W is discontinuous), and this makes it difficult to tune it for a given efficiency; in particular, it has to be noted than choosing $\gamma = 0.95$ for hard rejection does not make the asymptotic efficiency of $\widehat{\boldsymbol{\beta}}$ equal to 0.95: it continues to be zero. A better approach for increasing the efficiency is described in Section 5.9.2.

5.9.2 A fully asymptotically efficient one-step procedure

None of the estimators discussed so far can achieve full efficiency at the normal distribution and at the same time have a high BP and small maximum bias. We now discuss an adaptive one-step estimation method due to Gervini and Yohai (2002), which attains full asymptotic efficiency at the normal error distribution and at the same time has a high BP and small maximum bias. It is a weighted LS estimator computed from an initial estimator $\widehat{\boldsymbol{\beta}}_0$ with high BP, but rather than deleting the values larger than a fixed k, the procedure will keep a number N of observations $(\mathbf{x}_i, y_i)$ corresponding to the smallest values of $t_i = |r_i(\widehat{\boldsymbol{\beta}}_0)|/\widehat{\sigma}, i = 1, \ldots, n$, where N depends on the data, as will be described below. This N has the property that in large samples under normality it will have $N/n \to 1$, so that a vanishing fraction of data values will be deleted and full efficiency will be obtained.

Call G the distribution function of the absolute errors $|u_i|/\sigma$ under the normal model; that is,

$$G(t) = 2\Phi(t) - 1 = \mathrm{P}(|x| \leq t),$$

with $x \sim \Phi$, which is the standard normal distribution function. Let $t_{(1)} \leq \ldots \leq t_{(n)}$ denote the order statistics of the t_i. Let $\eta = G^{-1}(\gamma)$, where γ is a large value such as $\gamma = 0.95$. Define

$$i_0 = \min\{i : t_{(i)} \geq \eta\}, \quad q = \min_{i \geq i_0} \left(\frac{i-1}{G(t_{(i)})} \right), \tag{5.50}$$

and

$$N = [q] \tag{5.51}$$

where [.] denotes the integer part. The one-step estimator is the LS estimator of the observations corresponding to $t_{(i)}$ for $i \leq N$.

We now justify this procedure. The intuitive idea is to consider as potential outliers only those observations whose $t_{(i)}$ are not only greater than a given value, but also sufficiently larger than the corresponding order statistic of a sample from G. Note that if the data contain one or more outliers, then in a normal Q–Q plot of the $t_{(i)}$ against the respective quantiles of G, some large $t_{(i)}$ will appear well above the identity line, and we would delete it and all larger ones. The idea of the proposed procedure is to delete observations with large $t_{(i)}$ until the Q–Q plot of the remaining ones remains below the identity line, at least for large values of $|t_{(i)}|$. Since we are interested only in the tails of the distribution, we consider only values larger than some given η.

More precisely, for $N \leq n$, call G_N the empirical distribution function of $t_{(1)} \leq \ldots \leq t_{(N)}$:

$$G_N(t) = \frac{1}{N}\#\{t_{(i)} \leq t\}.$$

It follows that

$$G_N(t) = \frac{i-1}{N} \quad \text{for} \quad t_{(i-1)} \leq t < t_{(i)}$$

and hence each t in the half-open interval $[t_{(i-1)}, t_{(i)})$ is an α_i-quantile of G_N with

$$\alpha_i = \frac{i-1}{N}.$$

The α_i-quantile of G is $G^{-1}(\alpha_i)$. Then we look for N such that for $i_0 \leq i \leq N$ the α_i-quantile of G_N is not larger than that of G; that is

$$\text{for } i \in [i_0, N] : t_{(i-1)} \leq t < t_{(i)} \implies t \leq G^{-1}\left(\frac{i-1}{N}\right) \iff G(t) \leq \frac{i-1}{N} \tag{5.52}$$

Since G is continuous, (5.52) implies that

$$G(t_{(i)}) \leq \frac{i-1}{N} \quad \text{for } i_0 \leq i \leq N. \tag{5.53}$$

Also, since

$$i > N \implies \frac{i-1}{N} \geq 1 > G(t) \ \forall t,$$

the restriction $i \leq N$ may be dropped in (5.53), which can be seen to be equivalent to

$$N \leq \frac{i-1}{G(t_{(i)})} \quad \text{for } i \geq i_0 \iff N \leq q \tag{5.54}$$

with q defined in (5.50). We want the largest $N \leq q$, and since N is an integer, we ultimately get (5.51).

Gervini and Yohai show that under very general assumptions on $\widehat{\boldsymbol{\beta}}$ and $\widehat{\sigma}$, and regardless of the consistency rate of $\widehat{\boldsymbol{\beta}}$, these estimators attain the maximum BP and full asymptotic efficiency for normally distributed errors.

5.9.3 Improving finite-sample efficiency *and* robustness

Most theoretical results on the efficiency of robust estimators deal with their asymptotic efficiency – that is, the ratio between the asymptotic variances of the maximum likelihood estimator (MLE) and of the robust estimator – because it is much easier to deal with than the finite-sample efficiency, defined as the ratio between the mean squared errors (MSE) of the MLE and of the robust estimator, which is the one that really matters. However, unless the sample size is large enough, the finite-sample efficiency may be smaller than the asymptotic one. The last section dealt with a procedure for improving the asymptotic efficiency. Here we present a general approach proposed by Maronna and Yohai (2014), called *distance-constrained maximum likelihood* (DCML), which yields both a high finite-sample size efficiency and high robustness even for small n, and which may be applied to any parametric model.

Maronna and Yohai (2010) and Koller and Stahel (2011) have proposed partial solutions for this problem. A proposal by Bondell and Stefanski (2013) yields very high efficiencies, but at the cost of a serious loss of robustness.

5.9.3.1 Measuring finite-sample efficiency and robustness

Before presenting the DCML estimator, some definitions are necessary. The MSE of an univariate estimator was defined in (2.3). We now have to extend this concept to a regression estimator $\widehat{\boldsymbol{\beta}} \in R^p$. Consider a sample $\mathbf{Z} = \{(\mathbf{x}_i, y_i), i = 1, ..., n\}$ from the linear model (5.1) $y_i = \mathbf{x}_i'\boldsymbol{\beta}_0 + u_i$, with u_i independent of $\mathbf{x}_i$ and $\mathrm{E}u_i = 0$. The MSE of an estimator $\widehat{\boldsymbol{\beta}}$ of $\boldsymbol{\beta}_0$, based on $\mathbf{Z}$, is defined as

$$\mathrm{MSE}(\widehat{\boldsymbol{\beta}}) = \mathrm{E}(\widehat{\boldsymbol{\beta}} - \boldsymbol{\beta}_0)\mathbf{V}_{\mathbf{x}}(\widehat{\boldsymbol{\beta}} - \boldsymbol{\beta}_0)$$

where $\mathbf{V}_{\mathbf{x}} = \mathrm{E}\mathbf{x}\mathbf{x}'$.

This definition is based on a prediction approach. Consider an observation $\mathbf{z}_0 = (\mathbf{x}_0, y_0)$, independent of $\mathbf{Z}$, following the same model; that is, $y_0 = \mathbf{x}_0'\boldsymbol{\beta}_0 + u_0$. The prediction error of y_0 using $\widehat{\boldsymbol{\beta}}$ is

$$e = y_0 - \mathbf{x}_0'\widehat{\boldsymbol{\beta}} = u_0 - \mathbf{x}_0'(\widehat{\boldsymbol{\beta}} - \boldsymbol{\beta}_0).$$

Put $\sigma^2 = \mathrm{E}u_i^2$ and $\mathbf{X} = \{\mathbf{x}_1, ..., \mathbf{x}_n\}$.Then the mean squared prediction error is

$$\mathrm{E}e^2 = \sigma^2 + \mathrm{E}(\widehat{\boldsymbol{\beta}} - \boldsymbol{\beta}_0)'\mathbf{x}_0\mathbf{x}_0'(\widehat{\boldsymbol{\beta}} - \boldsymbol{\beta}_0) + 2\mathrm{E}u_0\mathbf{x}_0'(\widehat{\boldsymbol{\beta}} - \boldsymbol{\beta}_0).$$

The last term vanishes because of the independence of u_0 from the other elements. As to the second term, the independence between $\mathbf{X}$ and $\mathbf{x}_0$ yields

$$\begin{aligned}\mathrm{E}(\widehat{\boldsymbol{\beta}} - \boldsymbol{\beta}_0)'\mathbf{x}_0\mathbf{x}_0'(\widehat{\boldsymbol{\beta}} - \boldsymbol{\beta}_0) &= \mathrm{E}\{\mathrm{E}[(\widehat{\boldsymbol{\beta}} - \boldsymbol{\beta}_0)'\mathbf{x}_0\mathbf{x}_0'(\widehat{\boldsymbol{\beta}} - \boldsymbol{\beta}_0)|\mathbf{X}]\} \\ &= \mathrm{E}(\widehat{\boldsymbol{\beta}} - \boldsymbol{\beta}_0)'\mathbf{V}_{\mathbf{x}}(\widehat{\boldsymbol{\beta}} - \boldsymbol{\beta}_0),\end{aligned}$$

and therefore

$$\mathrm{E}e^2 = \sigma^2 + \mathrm{E}(\widehat{\boldsymbol{\beta}} - \boldsymbol{\beta}_0)'\mathbf{V}_{\mathbf{x}}(\widehat{\boldsymbol{\beta}} - \boldsymbol{\beta}_0).$$

The first term is the same for all estimators, while the second can be considered as a squared norm of the difference between the estimator and the true value, with respect to the quadratic form $\mathbf{V}_\mathbf{x}$. This norm is a more suitable measure of that difference than the Euclidean norm. In particular it is *invariant*, in the sense that if we replace $\mathbf{x}_i$ by $\mathbf{T}\mathbf{x}_i$, where $\mathbf{T}$ is any nonsingular matrix, then the MSE remains the same.

Then the finite-sample efficiency of $\widehat{\boldsymbol{\beta}}$ is defined as

$$\text{eff}(\widehat{\boldsymbol{\beta}}) = \frac{\text{MSE}(\widehat{\boldsymbol{\beta}}_{\text{MLE}})}{\text{MSE}(\widehat{\boldsymbol{\beta}})}. \tag{5.55}$$

While the asymptotic bias and variances can be in some cases calculated analytically, the finite-sample efficiency has to be computed through simulation.

We now deal with measuring robustness. Call $\text{MSE}_F(\widehat{\boldsymbol{\beta}})$ the MSE when the joint distribution of $(\mathbf{x}_i, y_i)$ is F. Given a "central model" F_0 (say, a model with normal errors), we want to see how the MSE varies when F is in some sense "near" F_0. To this end, consider a neighborhood of size ε of F_0, $\mathcal{F}(F_0, \varepsilon)$, as in (3.3). As a measure of the "worst" that can happen we take the maximum MSE in the neighborhood, and so define

$$\text{MaxMSE}(\widehat{\boldsymbol{\beta}}, F_0, \varepsilon) = \sup\{\text{MSE}_F(\widehat{\boldsymbol{\beta}}) \,:\, F \in \mathcal{F}(F_0, \varepsilon)\}. \tag{5.56}$$

Again, there is no analytical way to calculate MaxMSE, and therefore one must resort to simulation. The simplest choice for the family of contaminating distributions $\mathcal{G}$ is the one of point-mass distributions $\delta_{(\mathbf{x}_0, y_0)}$. Since this is an infinite set, one lets $(\mathbf{x}_0, y_0)$ range over a grid of values in order to approximate (5.56).

5.9.3.2 The DCML estimator

To define the DCML estimator proposed by Maronna and Yohai (2014), we need an initial robust estimator, not necessarily with high finite-sample efficiency. Then the estimators are defined by maximizing the likelihood function subject to the estimator being sufficiently close to the initial one. Doing so, we can expect that the resulting estimator will have the maximum possible finite-sample efficiency under the assumed model compatible with proximity to the initial robust estimator. This proximity guarantees the robustness of the new estimator.

The formulation of this proposal is as follows. Let D be a distance or discrepancy measure between densities. As a general notation, given a family of distributions with observation vector $\mathbf{z}$, parameter vector $\boldsymbol{\theta}$ and density $f(\mathbf{z}, \boldsymbol{\theta})$, put

$$d(\boldsymbol{\theta}_1, \boldsymbol{\theta}_2) = D(f(\mathbf{z}, \boldsymbol{\theta}_1), f(\mathbf{z}, \boldsymbol{\theta}_2)) \; .$$

Let $\mathbf{z}_i$, $i = 1, .., n$ be i.i.d. observations with distribution $f(\mathbf{z}, \boldsymbol{\theta})$, and let $\widehat{\boldsymbol{\theta}}_0$ be an initial robust estimator. Call $L(\mathbf{z}_1, ..., \mathbf{z}_n; \boldsymbol{\theta})$ the likelihood function. Then the proposal is to define an estimator $\widehat{\boldsymbol{\theta}}$ as

$$\widehat{\boldsymbol{\theta}} = \arg\max_{\boldsymbol{\theta}} L(\mathbf{z}_1, ..., \mathbf{z}_n; \boldsymbol{\theta}) \quad \text{with} \quad d(\widehat{\boldsymbol{\theta}}_0, \boldsymbol{\theta}) \leq \delta, \tag{5.57}$$

where δ is an adequately chosen constant that may depend on n. We shall call this proposal the "distance-constrained maximum likelihood' (DCML).

Several dissimilarity measures, such as the Hellinger distance, may be employed for this purpose. We shall employ as D the Kullback–Leibler (KL) divergence, because, as will be seen, this yields easily manageable results. The KL divergence between densities f_1 and f_2 is defined as

$$D(f_1,f_2) = \int_{-\infty}^{\infty} \log\left(\frac{f_1(\mathbf{z})}{f_2(\mathbf{z})}\right) f_1(\mathbf{z})\, d\mathbf{z}, \tag{5.58}$$

and therefore the d in (5.57) will be

$$d_{\mathrm{KL}}(\boldsymbol{\theta}_1,\boldsymbol{\theta}_2) = \int_{-\infty}^{\infty} \log\left(\frac{f(\mathbf{z},\boldsymbol{\theta}_1)}{f(\mathbf{z},\boldsymbol{\theta}_2)}\right) f(\mathbf{z},\boldsymbol{\theta}_1) d\mathbf{z}. \tag{5.59}$$

We now tailor our analysis to a linear model with random predictors. Consider the family of distributions of $\mathbf{z}=(\mathbf{x},y)$, where $\mathbf{x}\in R^p$ and $y\in R$, satisfying the model

$$y = \mathbf{x}'\boldsymbol{\beta} + \sigma u, \tag{5.60}$$

where $u \sim \mathrm{N}(0,1)$ is independent of $\mathbf{x}\in R^p$. Here $\boldsymbol{\theta}=(\boldsymbol{\beta},\sigma)$. Let $\widehat{\boldsymbol{\theta}}_0 = (\widehat{\boldsymbol{\beta}}_0,\widehat{\sigma}_0)$ be an initial robust estimator of regression and scale. We will actually consider σ as a nuisance parameter, and therefore we have

$$d_{\mathrm{KL}}(\boldsymbol{\beta}_0,\boldsymbol{\beta}) = \frac{1}{\sigma^2}(\boldsymbol{\beta}-\boldsymbol{\beta}_0)'\mathbf{V}(\boldsymbol{\beta}-\boldsymbol{\beta}_0) \tag{5.61}$$

with $\mathbf{V}=\mathrm{E}\mathbf{x}\mathbf{x}'$.

For $\boldsymbol{\beta}\in R^p$, the residuals from $\boldsymbol{\beta}$ are denoted $r_i(\boldsymbol{\beta}) = y_i - \mathbf{x}'\boldsymbol{\beta}$. Since σ is unknown, we replace it with its estimator $\widehat{\sigma}_0$. The natural estimator of $\mathbf{V}$ would be $\widehat{\mathbf{V}} = n^{-1}\mathbf{X}'\mathbf{X}$, where $\mathbf{X}$ is the $n\times p$ matrix with rows $\mathbf{x}_i'$.

Put, for any positive semidefinite matrix $\mathbf{U}$,

$$\widehat{d}_{\mathrm{KL,U}}(\boldsymbol{\beta}_0,\boldsymbol{\beta}) = \frac{1}{\widehat{\sigma}_0^2}(\boldsymbol{\beta}-\boldsymbol{\beta}_0)'\mathbf{U}(\boldsymbol{\beta}-\boldsymbol{\beta}_0) \tag{5.62}$$

It is immediately clear that (5.57) with $d=\widehat{d}_{\mathrm{KL,V}}$ is equivalent to minimizing $\sum_{i=1}^n r_i(\boldsymbol{\beta})^2$ subject to $\widehat{d}_{\mathrm{KL,C}}(\widehat{\boldsymbol{\beta}}_0,\boldsymbol{\beta}) \le \delta$. Call $\widehat{\boldsymbol{\beta}}_{\mathrm{LS}}$ the LSE. Put, for a general matrix $\mathbf{U}$:

$$\Delta_{\mathbf{U}} = \widehat{d}_{\mathrm{KL,U}}(\widehat{\boldsymbol{\beta}}_0,\widehat{\boldsymbol{\beta}}_{\mathrm{LS}}).$$

Define $\widehat{\boldsymbol{\beta}}$ as the minimizer of $\sum_{i=1}^n r_i(\boldsymbol{\beta})^2$ subject to $\widehat{d}_{\mathrm{KL},\widehat{\mathbf{V}}}(\boldsymbol{\beta}_0,\boldsymbol{\beta}) \le \delta$. In this case, the solution is explicit:

$$\widehat{\boldsymbol{\beta}} = t\widehat{\boldsymbol{\beta}}_{\mathrm{LS}} + (1-t)\widehat{\boldsymbol{\beta}}_0, \tag{5.63}$$

where $t=\min(1,\sqrt{\delta/\Delta_{\widehat{\mathbf{V}}}})$. We thus see that $\widehat{\boldsymbol{\beta}}$ is a convex combination of $\widehat{\boldsymbol{\beta}}_0$ and $\widehat{\boldsymbol{\beta}}_{\mathrm{LS}}$.

Table 5.9 Finite-sample efficiencies of MM, Gervini–Yohai and DCML estimators

p	n	MM-Bi85	G-Y	MM-Op99	DCML-Bi85	DCML-Op99
10	50	0.80	0.85	0.95	0.99	1.00
	100	0.82	0.89	0.97	0.99	1.00
	200	0.84	0.93	0.99	1.00	1.00
20	100	0.79	0.86	0.96	0.99	1.00
	200	0.82	0.92	0.97	0.99	1.00
	400	0.84	0.95	0.99	1.00	1.00
50	250	0.78	0.88	0.95	1.00	1.00
	500	0.82	0.94	0.98	1.00	1.00
	1000	0.84	0.97	0.99	1.00	1.00

Since $\widehat{\mathbf{V}}$ is not robust, we replace it with the matrix $\widetilde{\mathbf{V}}_{\mathbf{x}}$ defined in (5.3), and therefore we choose

$$t = \min\left(1, \sqrt{\frac{\delta}{\Delta_{\widetilde{\mathbf{V}}_{\mathbf{x}}}}}\right). \tag{5.64}$$

It is easy to show that if $\widehat{\boldsymbol{\beta}}_0$ is regression- and affine-equivariant, so is $\widehat{\boldsymbol{\beta}}$.

We have to choose δ in (5.57). We do so for the case when $\widehat{\boldsymbol{\beta}}_0$ is the MM-estimator with efficiency 0.85 and σ_0 is the residual M-scale. Then δ is chosen as

$$\delta_{p,n} = 0.3\frac{p}{n}. \tag{5.65}$$

To justify (5.65) note that under the model, the distribution of $nd_{\mathrm{KL}}(\widehat{\boldsymbol{\beta}}_0, \widehat{\boldsymbol{\beta}}_{\mathrm{LS}})$ is approximately that of vz, where $z \sim \chi^2_p$ and v is some constant, which implies that $\mathrm{E}d_{\mathrm{KL}}(\widehat{\boldsymbol{\beta}}_0, \widehat{\boldsymbol{\beta}}_{\mathrm{LS}}) \approx vp/n$. Therefore, in order to control the efficiency of $\widehat{\boldsymbol{\beta}}$, it seems reasonable to take δ of the form Kp/n for some K. The value $K = 0.3$ was arrived at after exploratory simulations aimed at striking a balance between efficiency and robustness. The behavior of the estimator is not very sensitive to the choice of the constant K; in fact, one may choose K between, say, 0.25 and 0.35 without serious effects.

It can be shown that for the estimators studied here, the finite-sample replacement BP of the DCML estimator $\widehat{\boldsymbol{\beta}}$ is at least that of the initial estimator $\widehat{\boldsymbol{\beta}}_0$.

Table 5.9 shows the results of a simulation comparing the efficiencies of

- the MM-estimator with bisquare ρ and asymptotic efficiency 0.85 (MM-Bi85);
- the MM-estimator with optimal ρ (Section 5.8.1) and asymptotic efficiency 0.99 (MM-Op99);
- the Gervini–Yohai (G-Y) estimator defined in Section 5.9.2 (with asymptotic efficiency 1);

- the DCML estimator with bisquare ρ starting from MM-Bi85;
- the DCML estimator with optimal ρ starting from MM-Op99;

for a linear model with normal errors, p normal predictors, and an intercept. The initial estimator employed for the MM and G-Y is described in Section 5.7.4.

It is seen that DCML-Op99 has the highest efficiency, followed in order by DCML-Bi85, MM-Op99, G-Y and MM-Bi85.

It should be noted that while the asymptotic efficiency of DCML does not depend on the predictors' distribution, the finite-sample efficiency does. For this reason Maronna and Yohai (2014) consider several other distributions for $\mathbf{X}$ with different degrees of heavy-tailedness, the results being similar to the ones shown here.

We now deal with the estimators' robustness. Table 5.10 gives the maximum mean squared errors (MSE) of the estimators, for 10% contaminated data with normal predictors. Again, DCML-Op99 is the best in all cases, closely followed by DCML-Bi85 and MM-Op99, which have similar performances, and then G-Y and MM-Bi85. It seems strange that the estimators with highest efficiency have the lowest MSEs. The reason is that the MSE is composed of bias and variance; and while, for example, MM-Bi85 may have a smaller bias than MM-Op99, it has a larger variance.

In addition, MM-Ops with efficiencies lower than 0.99 have larger MSEs than MM-Op99, and for this reason their results were not shown.

The DCML estimator uses as initial estimator an MM-estimator, which in turn is based on an S-estimator. It would be natural to ask why we do not use the S-estimator directly for DCML. The explanation is that since the S-estimator has a low efficiency, in order for DCML to have high efficiency, the value of δ must be much higher than (5.65), which greatly reduces the estimator's robustness. This fact has been confirmed by simulations.

Table 5.10 Maximum MSEs of MM, Gervini–Yohai and DCML estimators for 10% contamination

p	n	MM-Bi85	G-Y	MM-Op99	DCML-Bi85	DCML-Op99
10	50	1.10	1.03	0.89	0.92	0.86
	100	0.55	0.48	0.46	0.45	0.44
	200	0.44	0.37	0.37	0.36	0.36
20	100	1.28	1.16	0.98	1.02	0.94
	200	0.58	0.49	0.47	0.47	0.45
	400	0.46	0.38	0.39	0.39	0.38
50	250	1.38	1.18	1.03	1.04	0.96
	500	0.62	0.51	0.49	0.49	0.47
	1000	0.52	0.42	0.41	0.43	0.40

5.9.3.3 Confidence intervals based on the DCML estimator

The asymptotic distribution of the DCML estimator, derived in Maronna and Yohai (2014), is a very complicated non-normal distribution. For this reason, we propose a simple heuristic method to approximate the estimator's distribution with a normal one. We take the value t in (5.63) as if it were *fixed*. The asymptotic normal distribution of the resulting estimator can be calculated. Although there is no theoretical justification, simulations shows that the coverage probabilities of the resulting confidence intervals are a good approximation to the nominal one.

The joint asymptotic distribution of $\widehat{\boldsymbol{\beta}}_0$ and $\widehat{\boldsymbol{\beta}}_{\rm LS}$ can be derived from (3.49), and from this a straightforward calculation shows that the asymptotic covariance matrix of the estimator (5.63) under model (5.60) is

$$\mathbf{U} = \left(\frac{\sigma_0^2 a_1}{b^2} t^2 + a_2 (1-t)^2 + 2\frac{\sigma_0 c}{b} t(1-t) \right) \mathbf{C}^{-1}$$

where

$$a_1 = \mathrm{E}\psi(u)^2, \quad b = \mathrm{E}\psi(u), \quad c = \mathrm{E}\psi(u)u, \quad a_2 = \mathrm{E}u^2, \quad \mathbf{C} = \mathrm{E}\mathbf{x}\mathbf{x}'.$$

Then $\mathbf{U}$ can be estimated by

$$\widehat{\mathbf{U}} = \left(\frac{\widehat{\sigma}_0^2 \widehat{a}_1}{\widehat{b}^2} t^2 + \widehat{a}_2 (1-t)^2 + 2\frac{\widehat{\sigma}_0 \widehat{c}}{\widehat{b}} t(1-t) \right) \widehat{\mathbf{C}}_w^{-1}.$$

where

$$\widehat{a}_1 = \frac{1}{n} \sum_{i=1}^n \psi \left(\frac{r_i(\widehat{\boldsymbol{\beta}}_0)}{\widehat{\sigma}_o} \right)^2, \quad \widehat{b} = \frac{1}{n} \sum_{i=1}^n \psi' \left(\frac{r_i(\widehat{\boldsymbol{\beta}}_0)}{\widehat{\sigma}_o} \right),$$

$$\widehat{c} = \frac{1}{n} \sum_{i=1}^n \psi \left(\frac{r_i(\widehat{\boldsymbol{\beta}}_0)}{\widehat{\sigma}_o} \right) r_i(\widehat{\boldsymbol{\beta}}_0), \quad \widehat{a}_2 = S(r_i(\widehat{\boldsymbol{\beta}}_0))^2,$$

where S is an M-scale, normalized as shown at the end of Section 2.5. Actually, the natural estimators of a_2 would be $n^{-1} \sum_{i=1}^n r_i(\widehat{\boldsymbol{\beta}}_0)^2$, but then a large residual would yield overly large intervals for heavy-tailed errors.

Given $\widehat{\mathbf{U}}$, the asymptotic variance of a linear combination $\gamma = \mathbf{a}'\widehat{\boldsymbol{\beta}}$ can be estimated by $n^{-1}\mathbf{a}'\widehat{\mathbf{U}}\mathbf{a}$.

Table 5.11 shows the results of a simulation for a model with ten normal predictors plus intercept considering two error distributions: normal and Student with three degrees of freedom. The values are the average covering probabilities (for all 11 parameters) of the approximate confidence interval for elements of $\boldsymbol{\beta}$ computed as in (4.26), with nominal levels 0.90, 0.95 and 0.99. We see that the actual coverages are reasonably close to the nominal ones.

Smucler and Yohai (2015) use a different approach to obtain regression estimators that have simultaneously high efficiency and robustness for finite samples. The performance of this procedure is comparable to that of the DCML estimator.

Table 5.11 Actual coverage probabilities of DCML intervals for $p = 10$

		Levels		
Errors	n	0.90	0.95	0.99
Normal	50	0.92	0.96	0.99
	100	0.92	0.96	0.99
Student 3-DF	50	0.88	0.93	0.98
	100	0.86	0.92	0.98

5.9.4 Choosing a regression estimator

In Sections 5.4, 5.5 and 5.9 we considered several families of estimators. Estimators based on a residual scale are robust, but have low efficiency. MM-estimators can give (at least approximately) a given finite-sample efficiency without losing robustness. Both versions of DCML have simultaneously a higher efficiency and a lower MSE than the MMs, as seen in Tables 5.9 and 5.10. Also, the optimal ρ outperforms the bisquare ρ. Overall, DCML-Op99 appears the best in both efficiency and robustness, closely followed by DCML-Op85.

As to the initial estimators, Tables 5.5 and 5.6 show that the Peña–Yohai estimator outperforms subsampling in speed, efficiency and robustness. For these reasons we recommend the following procedure:

1. Compute the Peña-Yohai estimator.
2. Use it as starting estimator for an MM-estimator with optimal ρ and 99% efficiency.
3. Use this MM as initial estimator for the DCML estimator with optimal ρ.

5.10 Robust regularized regression

In this section we deal with robust versions of regression estimators that have better predictive accuracy than ordinary least squares (OLS) in situations where the predictors are highly correlated and/or the number of regressors p is large compared to the number of cases n, and even when $p > n$. The situation of prediction with $p >> n$ has become quite common, especially in chemometrics. One instance of such a situation appears when attempting to replace the laboratory determination of the amount of a given chemical compound in a sample of material, with methods based on cheaper and faster spectrographic measurements. These estimators work by putting a "penalty" on the coefficients, which produces a bias but at the same time decreases the variances of the estimators. These estimators are known as *regularized regression estimators*.

Consider a regression model with intercept (4.5) with data $(\mathbf{x}_i, y_i)$, $i = 1, ..., n$, where we put $\mathbf{x}_i' = (1, \underline{\mathbf{x}}_i')$ with $\underline{\mathbf{x}}_i \in R^{p-1}$ and $\boldsymbol{\beta} = (\beta_0, \widehat{\boldsymbol{\beta}}_1')'$ and call $\mathbf{X}$ the $n \times (p-1)$ matrix with ith row equal to $\underline{\mathbf{x}}_i$ Then then we can define a general class of regularized regression estimators by

$$\widehat{\boldsymbol{\beta}} = (\widehat{\beta}_0, \widehat{\boldsymbol{\beta}}_1')' = \arg \min_{\beta_0 \in R, \boldsymbol{\beta}_1 \in R^{p-1}} L(\boldsymbol{\beta}), \tag{5.66}$$

where L is of the form

$$L(\boldsymbol{\beta}) = s^2 \sum_{i=1}^{n} \rho\left(\frac{r_i(\boldsymbol{\beta})}{s}\right) + \lambda \sum_{j=1}^{p-1} h(\beta_j), \tag{5.67}$$

where $r_i(\boldsymbol{\beta}) = \mathbf{y}_i - \boldsymbol{\beta}'\mathbf{x}_i$, ρ is a loss function, s is a residual scale and h is a monotone function penalizing large coefficients. The parameter λ determines the severity of the penalization and is usually determined by $K-$ *fold cross validation,*described as follows. A set Λ of candidate λ's is chosen; the data are split into K subsets of approximately equal size; for each $\lambda \in \Lambda$, $K-1$ of the subsets are used in turn to compute the estimator, and the remaining one to compute the prediction errors. The set of n prediction errors for this λ is summarized by the MSE or a robust version thereof, and the λ yielding the smallest MSE is chosen.

When robustness is not an issue, ρ is usually chosen as a quadratic function: $\rho(r) = r^2$, and in this case $L(\boldsymbol{\beta})$ does not depend on s.

For an overview of this topic, see the book by Bühlmann and van de Geer (2011).

5.10.1 Ridge regression

The simplest and oldest of regularized estimators is the so-called *ridge regression* (henceforth RR), first proposed by Hoerl and Kennard (1970). The RR estimator is given by (5.66) and (5.67), with $\rho(r) = r^2$, $h(\beta) = \beta^2$ and $s = 1$; that is,

$$\widehat{\boldsymbol{\beta}} = (\widehat{\beta}_0, \widehat{\boldsymbol{\beta}}_1')' = \arg \min_{\beta_0 \in R,\ \boldsymbol{\beta}_1 \in R^{p-1}} L(\boldsymbol{\beta}) \tag{5.68}$$

with

$$L(\boldsymbol{\beta}) = \sum_{i=1}^{n} r_i^2(\boldsymbol{\beta}) + \lambda \sum_{j=1}^{p-1} \beta_j^2. \tag{5.69}$$

Ridge regression is particularly useful when it can be assumed that all coefficients have approximately the same order of magnitude (unlike "sparse" situations. when most coefficients are assumed to be null); see the comments on RR by Zou and Hastie (2005) and Frank and Friedman (1993).

Since RR is based on least squares, it is sensitive to atypical observations. Several approaches have been proposed to make RR robust towards outliers, the most interesting ones being the ones by Silvapulle (1991) and Simpson and Montgomery (1996). These estimators, however, present two drawbacks. The first is that they are not sufficiently robust towards "bad leverage points", especially for large p. The second is that they require an initial (standard) robust regression estimator. When $p > n$ this becomes impossible, and when $p < n$ but the ratio p/n is "large", the initial estimator has a low robustness, as measured by the BP.

In the next subsection we present a robust version of the RR estimator proposed by Maronna (2011). This can also be used when $p > n$, is robust when p/n is large, and is resistant to bad leverage points.

5.10.1.1 MM-estimation for ridge regression

To ensure both robustness and efficiency, under the normal model we employ the approach of MM-estimation (Yohai 1987). Start with an initial robust but possibly inefficient estimator $\widehat{\boldsymbol{\beta}}_{\text{ini}}$; from the respective residuals compute a robust scale estimator $\widehat{\sigma}_{\text{ini}}$. Then compute an M-estimator with *fixed* scale $\widehat{\sigma}_{\text{ini}}$, starting the iterations from $\widehat{\boldsymbol{\beta}}_{\text{ini}}$, and using a loss function that ensures the desired efficiency. Here "efficiency" will be loosely defined as "similarity with the classical RR estimator for the normal model".

Let $\widehat{\boldsymbol{\beta}}_{\text{ini}}$ be an initial estimator and let $\widehat{\sigma}_{\text{ini}}$ be an M–scale of $\mathbf{r}$:

$$\frac{1}{n}\sum_{i=1}^{n}\rho_0\left(\frac{r_i(\widehat{\boldsymbol{\beta}}_{\text{ini}})}{\widehat{\sigma}_{\text{ini}}}\right)=\delta, \tag{5.70}$$

where ρ_0 is a bounded ρ-function and δ is to be chosen. Then the MM-estimator for RR (henceforth RR-MM) is defined by (5.68) with

$$L(\boldsymbol{\beta})=\widehat{\sigma}^2_{\text{ini}}\sum_{i=1}^{n}\rho\left(\frac{r_i(\boldsymbol{\beta})}{\widehat{\sigma}_{\text{ini}}}\right)+\lambda\|\boldsymbol{\beta}_1\|^2, \tag{5.71}$$

where ρ is another bounded ρ-function such that $\rho \le \rho_0$. The factor $\widehat{\sigma}^2_{\text{ini}}$ before the summation is employed to make the estimator coincide with classical RR when $\rho(t) = t^2$.

We shall henceforth use

$$\rho_0(t)=\rho_{\text{bis}}\left(\frac{t}{c_0}\right),\quad \rho(t)=\rho_{\text{bis}}\left(\frac{t}{c}\right) \tag{5.72}$$

where ρ_{bis} denotes the bisquare ρ-function (2.38) and the constants $c_0 < c$ are chosen to control both robustness and efficiency.

Call ε^* the breakdown point (BDP) of $\widehat{\sigma}_{\text{ini}}$. It is not difficult to show that if $\lambda > 0$, the BDP of RR–MM is $\ge \varepsilon^*$. However, since the estimator is not regression-equivariant, this result is deceptive. As an extreme case, take $p = n - 1$ and $\varepsilon^* = 0.5$. Then the BDP of $\widehat{\boldsymbol{\beta}}$ is null for $\lambda = 0$, but is 0.5 for $\lambda = 10^{-6}$! Actually, for a given contamination rate less than 0.5, the maximum bias of the estimator remains bounded, but the bound tends to infinity when λ tends to zero.

The classical ridge regression estimator given by (5.68) and (5.69) will henceforth be denoted as RR–LS for brevity.

5.10.1.2 The iterative algorithm

Note that the loss function (5.69) is the same as the OLS loss for the augmented dataset $(\widetilde{\mathbf{X}}, \widetilde{\mathbf{y}})$ with

$$\widetilde{\mathbf{X}} = \begin{bmatrix} \mathbf{1}_n & \mathbf{X} \\ \mathbf{0}_{p-1} & \lambda \mathbf{I}_{p-1} \end{bmatrix}, \quad \widetilde{\mathbf{y}} = \begin{bmatrix} \mathbf{y} \\ \mathbf{0}_{p-1} \end{bmatrix}, \tag{5.73}$$

where, as a general notation, $\mathbf{I}_p$, $\mathbf{1}_p$ and $\mathbf{0}_p$ are respectively the p-dimensional identity matrix and the p-dimensional vectors of ones and of zeroes. Then the classical estimator RR–LS satisfies the "normal equations" for the augmented sample. This may be written

$$\widehat{\beta}_0 = \overline{y} - \overline{\mathbf{x}}'\widehat{\boldsymbol{\beta}}_1, \quad (\mathbf{X}'\mathbf{X} + \lambda \mathbf{I}_p)\widehat{\boldsymbol{\beta}}_1 = \mathbf{X}'(\mathbf{y} - \widehat{\beta}_0 \mathbf{1}_n), \tag{5.74}$$

where

$$\overline{\mathbf{x}} = \frac{1}{n}\sum_{i=1}^{n} \mathbf{x}_i, \quad \overline{\mathbf{y}} = \frac{1}{n}\sum_{i=1}^{n} y_i.$$

A similar system of equations is satisfied by RR–MM. Define, as in (2.31), $\psi(t) = \rho'(t)$ and $W(t) = \psi(t)/t$. Let

$$t_i = \frac{r_i}{\widehat{\sigma}_{\mathrm{ini}}}, \quad w_i = \frac{W(t_i)}{2}, \quad \mathbf{w} = (w_1, ..., w_n)', \quad \mathbf{W} = \mathrm{diag}(\mathbf{w}). \tag{5.75}$$

Setting the derivatives of (5.71) with respect to $\boldsymbol{\beta}$ to zero yields for RR-MM

$$\mathbf{w}'(\mathbf{y} - \widehat{\beta}_0 \mathbf{1}_n - \mathbf{X}\widehat{\boldsymbol{\beta}}_1) = 0 \tag{5.76}$$

and

$$(\mathbf{X}'\mathbf{W}\mathbf{X} + \lambda \mathbf{I}_p)\widehat{\boldsymbol{\beta}}_1 = \mathbf{X}'\mathbf{W}(\mathbf{y} - \widehat{\beta}_0 \mathbf{1}_n). \tag{5.77}$$

Therefore, RR–MM satisfies a weighted version of (5.74). Since for the chosen ρ, $W(t)$ is a decreasing function of $|t|$, observations with larger residuals will receive lower weights w_i.

As is usual in robust statistics, these "weighted normal equations" suggest an iterative procedure. Starting with an initial $\widehat{\boldsymbol{\beta}}$:

- Compute the residual vector $\mathbf{r}$ and the weights $\mathbf{w}$.
- Leaving $\widehat{\beta}_0$ and $\mathbf{w}$ fixed, compute $\widehat{\boldsymbol{\beta}}_1$ from (5.77).
- Recompute $\mathbf{w}$, and compute $\widehat{\beta}_0$ from (5.76).
- Repeat until the change in the residuals is small enough.

Recall that $\widehat{\sigma}_{\mathrm{ini}}$ remains fixed throughout. This procedure may be called "iterative reweighted RR". It can be shown that the objective function (6.46) descends at each iteration.

5.10.1.3 The initial estimator for RR–MM

As in the standard case, the initial estimator for RR–MM will be the one based on the minimization of a (penalized) robust scale (an S-estimator). We define an S-estimator for RR (henceforth RR–SE) by replacing in (5.67) the sum of squared residuals by a squared robust M–scale $\widehat{\sigma}$ (2.49), namely

$$L(\boldsymbol{\beta}) = n\widehat{\sigma}(\mathbf{r}(\boldsymbol{\beta}))^2 + \lambda\|\boldsymbol{\beta}_1\|^2, \tag{5.78}$$

where $\mathbf{r}(\boldsymbol{\beta}) = (r_1(\boldsymbol{\beta}),\ldots,r_n(\boldsymbol{\beta}))$. Here the factor n is used to make (5.78) coincide with (5.67) when $\widehat{\sigma}^2(\mathbf{r}) = \text{ave}_i(r_i^2)$, corresponding to $\rho(t) = t^2$ and $\delta = 1$ in (2.49).

A straightforward calculation shows that the estimator satisfies the "normal equations" (5.76)–(5.77), with $w_i = W(t_i)$ and λ replaced by $\lambda' = \lambda n^{-1} \sum_{i=1}^n w_i t_i^2$. These equations suggest an iterative algorithm similar to the one at the end of Section 5.10.1.2, with the difference that now the scale changes at each iteration; that is, now in the definition of t_i in (5.75), $\widehat{\sigma}_{\text{ini}}$ is replaced by $\widehat{\sigma}(\mathbf{r}(\widehat{\boldsymbol{\beta}}))$. Although it has not been possible to prove that the objective function descends at each iteration, this fact has been verified in all cases up to now. The initial estimator for SE–RR employs the procedure proposed by Peña and Yohai (1999) and described in Section 5.7.4, but applied to the augmented sample (5.73).

5.10.1.4 Choosing λ through cross–validation

In order to choose λ we must estimate the prediction error of the regularized estimator for different values of λ. Call CV(K) the K-fold cross validation process, which requires recomputing the estimator K times. For $K = n$ ("leave one out") we can use an approximation to avoid recomputing. Call $\widehat{y}_{-i}$ the fit of y_i computed without using the ith observation; that is, $y_{-i} = \widehat{\beta}_0^{(-i)} + \mathbf{x}_i'\widehat{\boldsymbol{\beta}}_1^{(-i)}$, where $(\widehat{\beta}_0^{(-i)}, \widehat{\boldsymbol{\beta}}_1^{(-i)\prime})$ is the estimator computed without observation i. Then a first-order Taylor approximation of the estimator yields the approximate prediction errors.

For this method to work properly, several technical details must be taken into account. They are omitted here for brevity, and are explained in the paper by Maronna (2011). An application of robust RR to functional regression is given by Maronna and Yohai (2013).

5.10.2 Lasso regression

One of the shortcomings of RR is that it cannot be employed in "sparse" situations, where it is assumed that a large part of the coefficients are null, since, although RR shrinks the coefficients toward zero, it does not produce exact zeros, and is thus not useful to select predictors.

To remedy this inconvenience, Tibshirani (1996) proposed a new choice for the penalization function h in (5.67). He introduced the *lasso* (least absolute shrinkage and selection operator) regression estimator, which is defined as:

$$L(\boldsymbol{\beta}) = \sum_{i=1}^{n} r_i^2(\boldsymbol{\beta}) + \lambda \sum_{j=1}^{p-1} |\beta_j| \tag{5.79}$$

In other words, $L(\boldsymbol{\beta})$ is of the form (5.67) with $\rho(r) = r^2$ and $h(\beta) = |\beta|$. It can be shown that using lasso regression, the number of null coefficients increases when λ increases.

Zou (2006) proposed a two-step modification of the lasso procedure called the *adaptive lasso*. In the first step, an estimator $\boldsymbol{\beta}^{(0)} = (\beta_0^{(0)}, \beta_1^{(0)}, ..., \beta_p^{(0)})$ (for example, a lasso estimator) is computed, and in the second step the adaptive lasso estimator is obtained as

$$\widehat{\boldsymbol{\beta}}_{\mathrm{A}} = \arg\min_{\boldsymbol{\beta}} \sum_{i=1}^{n} r_i^2(\boldsymbol{\beta}) + \lambda \sum_{j=1}^{p-1} \frac{|\beta_j|}{|\beta_j^{(0)}|} \tag{5.80}$$

(with the convention that $0/0 = 0$). Thus the coefficients that were small in the first step receive a higher penalty in the second. It also follows from (5.80) that the coefficients that are null in the first step are also null in the second.

There are two main algorithms to compute the lasso: least angle regression (Efron *et al.*, 2004) and the coordinate descent algorithm (Friedman et al., 2007). The first one is computationally similar to stepwise variable selection, while the second takes one coefficient β_i at a time to minimize the loss function.

5.10.2.1 Robust lasso

Since lasso regression uses a quadratic loss function, this estimator is very sensitive to outliers. There are several proposals in which the quadratic loss function is replaced by a convex ρ function, such as Huber's ρ function (2.28) (Li *et al.*, 2011) or $\rho(r) = |r|$. (Wang *et al.*, 2007). However, since these ρ functions are not bounded, these estimators are not robust. Alfons *et al.* (2013) present a lasso version of the trimmed least squares (LTS) estimator that was defined in Section 5.4.2. This estimator is robust, but very inefficient under Gaussian errors. Khan *et al.* (2007) propose a robust version of least angle regression, but they employ it as a tool for model selection rather than as an estimator.

Smucler and Yohai (2017) propose the MM lasso estimator, which is obtained by minimizing

$$L(\boldsymbol{\beta}) = \widehat{\sigma}_{\mathrm{ini}}^2 \sum_{i=1}^{n} \rho\left(\frac{r_i(\boldsymbol{\beta})}{\widehat{\sigma}_{\mathrm{ini}}}\right) + \lambda \sum_{i=1}^{p-1} |\beta_j|,$$

where ρ is a bounded ρ-function and $\widehat{\sigma}_{\text{ini}}$ is an initial scale, as in the case of MM-RR. Differentiating with respect to $\boldsymbol{\beta}$ yields the estimating equations

$$\sum_{i=1}^{n} \psi\left(\frac{r_i(\boldsymbol{\beta})}{\widehat{\sigma}_{\text{ini}}}\right) \mathbf{x}_{ij} + \lambda \, \text{sign}(\beta_j) \overset{\circ}{=}_j 0, \ 1 \leq j \leq p-1$$

$$\sum_{i=1}^{n} \psi\left(\frac{r_i(\boldsymbol{\beta})}{\widehat{\sigma}_{\text{ini}}}\right) = 0$$

where "$\overset{\circ}{=}_j$" means that the left-hand side changes sign when β_j crosses zero. This is equivalent to the system of equations

$$\sum_{i=1}^{n} r_i^*(\boldsymbol{\beta})\mathbf{x}_{ij}^* + \lambda\widehat{\sigma}_{\text{ini}}^2 \text{sign}(\beta_j) \overset{\circ}{=}_j 0, \ 1 \leq j \leq p-1$$

$$\sum_{i=1}^{n} r_i^*(\boldsymbol{\beta}) = 0 \tag{5.81}$$

where

$$r_i^*(\boldsymbol{\beta}) = y_i^* - \boldsymbol{\beta}'\mathbf{x}_i^*, \quad y_i^* = w_i^{1/2} y_i^*, \quad \mathbf{x}_i^* = w_i^{1/2}\mathbf{x}_i,$$

and w_i is as given in (5.75). This system of equations is very similar to that of lasso regression for the sample $(\mathbf{x}_1^*, y_1^*), \ldots.(\mathbf{x}_n^*, y_n^*)$ and penalization parameter $\lambda\widehat{\sigma}_{\text{ini}}^2$. Then, to compute the MM lasso, one can use the following iterative algorithm:

- Compute the initial estimator $\boldsymbol{\beta}^{(0)}$ and $\widehat{\sigma}_{\text{ini}}$ as in the case of MM–RR by the Peña–Yohai procedure.
- Once $\boldsymbol{\beta}^{(k)}$ is computed, compute the weights w_i, as in (5.75), using $\boldsymbol{\beta} = \boldsymbol{\beta}^{(k)}$; compute the transformed sample $(\mathbf{x}_i^*, y_i^*)$, $i = 1, ..n$, and obtain $\boldsymbol{\beta}^{(k+1)}$ by solving the lasso problem for this sample with penalization parameter $\lambda\widehat{\sigma}_{\text{ini}}^2$.
- Stop when $||\boldsymbol{\beta}^{(k+1)} - \boldsymbol{\beta}^{(k)}||/||\boldsymbol{\beta}^{(k)}|| \leq \varepsilon$.

More details of the algorithm can be found in Smucler and Yohai (2017).

An adaptive MM lasso estimator is defined by minimizing the loss function

$$L(\boldsymbol{\beta}) = \arg\min_{\boldsymbol{\beta}} \sum_{i=1}^{n} \rho\left(\frac{r_i(\boldsymbol{\beta})}{\widehat{\sigma}_{\text{ini}}}\right) + \lambda \sum_{j=1}^{p-1} \frac{|\beta_j|}{|\beta_j^{(0)}|}.$$

This estimator can be computed using an approach similar to the one used for the MM-lasso, using a simple transformation of the x_{ij}s (see Problem 5.11).

The penalization parameter λ may be chosen by cross validation. For this purpose, a grid of equally spaced candidates is chosen in the interval $[0, \lambda_{\max}]$, where $\lambda_{\max}$ is a value large enough so that the lasso estimator has all the components equal to 0. For each candidate, a robust scale of the residuals is computed, and the candidate with minimum scale is chosen. We can, for example, use an M-scale or a τ-scale. More details on the algorithm and the cross validation procedure can be found in the paper by Smucler and Yohai (2017).

5.10.2.2 Oracle properties

Consider a sparse linear regression model with $\boldsymbol{\beta}_1 = (\boldsymbol{\beta}_{11}.\boldsymbol{\beta}_{12})$, where $\boldsymbol{\beta}_{11} \in R^q$ and $\boldsymbol{\beta}_{12} = \mathbf{0}_m$ with $m = p - q - 1$. It is said that an estimator $\widehat{\boldsymbol{\beta}}$ of $\boldsymbol{\beta}$ has an oracle-type property if has behavior similar to that of the estimator computed assuming that $\boldsymbol{\beta}_{12} = \mathbf{0}$; that is, when the true submodel is known. This property can be formalized in different ways: some approaches look at the behavior of the estimator itself, and others at the prediction accuracy that can be achieved with the estimator. We will use here the characterization of Fan and Li (2001). According to their definition, a regularized estimator of $\widehat{\boldsymbol{\beta}} = (\widehat{\beta}_0, \widehat{\boldsymbol{\beta}}_{11}, \widehat{\boldsymbol{\beta}}_{12})$ of $\boldsymbol{\beta}$ has the oracle property if it satisfies:

- $\lim_{n\to\infty} \mathrm{P}(\widehat{\boldsymbol{\beta}}_{12} = \mathbf{0}) = \mathbf{0}$
- $n^{1/2}(\widehat{\beta}_{0,}\widehat{\boldsymbol{\beta}}_{11}) \to^D N_{q+1}(0, \Sigma^*)$,

where Σ^* is the asymptotic covariance of the non-penalized estimator assuming that $\boldsymbol{\beta}_{12} = \mathbf{0}$.

It was proved by Smucler and Yohai (2017) that if λ is chosen depending of n so that $n^{1/2}\lambda \to 0$, then, under very general conditions, these properties are satisfied by the adaptive MM lasso. In this case, the asymptotic distribution of $n^{1/2}(\widehat{\beta}_{0,}\widehat{\boldsymbol{\beta}}_{11})$ is the same as if the MM-estimator without penalization were applied only to the first $p - 1$ predictors.

5.10.3 Other regularized estimators

It can be proved that the lasso estimator has at most $\min(n, p)$ nonnull components. When p is much larger than n this may be a limitation for selecting all the relevant variables. To overcome this limitation, Zou and Hastie (2005) proposed the so-called *elastic net estimator,* which is defined by minimizing

$$L(\boldsymbol{\beta}) = \sum_{i=1}^{n} r_i^2(\boldsymbol{\beta}) + \lambda_1 \sum_{j=1}^{p-1} |\beta_j| + \lambda_2 \sum_{j=1}^{p-1} \beta_j^2. \tag{5.82}$$

A robust version of this estimator can also be obtained through the MM approach.

In some cases there are groups of covariables that have to be selected or rejected in a block. This happens, for example, when there is a categorical variable that is coded as $k - 1$ binary covariables, where k is the number of categories. In that case, one would like either to select all the variables or to reject all the variables in the group; that is, to make either all of the coefficients or none of the coefficients in the group equal to zero. Yuan and Lin (2006) proposed an approach called *group lasso* to deal with this case.

5.11 *Other estimators

5.11.1 Generalized M-estimators

In this section we treat a family of estimators which is of historical importance. The simplest way to robustify a monotone M-estimator is to downweight the influential $\mathbf{x}_i$ to prevent them from dominating the estimating equations. Hence we may define an estimator by

$$\sum_{i=1}^{n} \psi\left(\frac{r_i(\boldsymbol{\beta})}{\widehat{\sigma}}\right) \mathbf{x}_i W(d(\mathbf{x}_i)) = \mathbf{0} \tag{5.83}$$

where W is a weight function and $d(\mathbf{x})$ is some measure of the "largeness" of $\mathbf{x}$. Here ψ is monotone and $\widehat{\sigma}$ is simultaneously estimated by an M-estimating equation of the form (5.18). For instance, to fit a straight line $y_i = \beta_0 + \beta_1 x_i + \varepsilon_i$, we may choose

$$d(x_i) = \frac{|x_i - \widehat{\mu}_x|}{\widehat{\sigma}_x} \tag{5.84}$$

where $\widehat{\mu}_x$ and $\widehat{\sigma}_x$ are respectively robust location and dispersion statistics of the x_i, such as the median and MAD. In order to bound the effect of influential points, W must be such that $W(t)t$ is bounded.

More generally, we may let the weights depend on the residuals as well as the predictor variables and use a *generalized* M-estimator (*GM-estimator*) $\widehat{\boldsymbol{\beta}}$ defined by

$$\sum_{i=1}^{n} \eta\left(d(\mathbf{x}_i), \frac{r_i(\widehat{\boldsymbol{\beta}})}{\widehat{\sigma}}\right) \mathbf{x}_i = \mathbf{0}, \tag{5.85}$$

where for each s, $\eta(s, r)$ is a nondecreasing and bounded ψ-function of r, and $\widehat{\sigma}$ is obtained by a simultaneous M-scale estimator equation of the form

$$\frac{1}{n}\sum_{i=1}^{n} \rho_{\text{scale}}\left(\frac{r_i(\boldsymbol{\beta})}{\widehat{\sigma}}\right) = \delta.$$

Two particular forms of GM-estimator have been of primary interest in the literature. The first is the estimator (5.83), which corresponds to the choice $\eta(s, r) = W(s)\psi(r)$ and is called a "Mallows estimator" (Mallows, 1975). The second form is the choice

$$\eta(s, r) = \frac{\psi(sr)}{s}, \tag{5.86}$$

which was first proposed by Schweppe *et al.* (1970) in the context of electric power systems.

The GM-estimator with the Schweppe function (5.86) is also called the "Hampel–Krasker–Welsch" estimator (Krasker and Welsch, 1982). When ψ is Huber's ψ_k, it is a solution to Hampel's problem (Section 3.5.4). See Section 5.13.6 for details.

Note that the function $d(\mathbf{x})$ in (5.84) depends on the data, and for this reason it will be better to denote it by $d_n(\mathbf{x})$. The most usual way to measure largeness is as a generalization of (5.84). Let $\widehat{\boldsymbol{\mu}}_n$ and $\widehat{\boldsymbol{\Sigma}}_n$ be a robust location vector and robust scatter matrix; these will be looked at in more detail in Chapter 6. Then d_n is defined as

$$d_n(\mathbf{x}) = (\mathbf{x} - \widehat{\boldsymbol{\mu}}_n)'\widehat{\boldsymbol{\Sigma}}_n^{-1}(\mathbf{x} - \widehat{\boldsymbol{\mu}}_n). \tag{5.87}$$

In the case where $\widehat{\boldsymbol{\mu}}_n$ and $\widehat{\boldsymbol{\Sigma}}_n$ are the sample mean and covariance matrix, $\sqrt{d_n(\mathbf{x})}$ is known as the Mahalanobis distance.

Assume that $\widehat{\boldsymbol{\mu}}_n$ and $\widehat{\boldsymbol{\Sigma}}_n$ converge in probability to $\boldsymbol{\mu}$ and $\boldsymbol{\Sigma}$ respectively. With this assumption, it can be shown that if the errors are symmetric, then the influence function of a GM-estimator for the model (5.1)–(5.2) is

$$\mathrm{IF}((\mathbf{x}_0, y_0), F) = \eta\left(d(\mathbf{x}_0), \frac{y_0 - \mathbf{x}_0'\boldsymbol{\beta}}{\sigma}\right)\mathbf{B}^{-1}\mathbf{x}_0 \tag{5.88}$$

with

$$\mathbf{B} = -\mathrm{E}\dot{\eta}\left(d(\mathbf{x}), \frac{u}{\sigma}\right)\mathbf{x}\mathbf{x}', \quad \dot{\eta}(s,r) = \frac{\partial\eta(s,r)}{\partial r} \tag{5.89}$$

and

$$d(\mathbf{x}) = (\mathbf{x} - \boldsymbol{\mu})'\boldsymbol{\Sigma}^{-1}(\mathbf{x} - \boldsymbol{\mu}).$$

Hence the IF is the same as would be obtained from (3.48) using d instead of d_n. It can be shown that $\widehat{\boldsymbol{\beta}}$ is asymptotically normal, and as a consequence of 5.88, the asymptotic covariance matrix of $\widehat{\boldsymbol{\beta}}$ is

$$\mathbf{B}^{-1}\mathbf{C}\mathbf{B}^{-1'} \tag{5.90}$$

with

$$\mathbf{C} = \mathrm{E}\eta\left(d(\mathbf{x}), \frac{y - \mathbf{x}'\boldsymbol{\beta}}{\sigma}\right)^2\mathbf{x}\mathbf{x}'.$$

It follows from (5.88) that GM-estimators have several attractive properties:

- If $\eta(s,r)s$ is bounded, then their IF is bounded.
- The same condition ensures a positive BP (Maronna *et al.*, 1979).
- They are defined by estimating equations, and hence easy to compute like ordinary monotone M-estimators.

However, GM-estimators also have several drawbacks:

- Their efficiency depends on the distribution of $\mathbf{x}$: if $\mathbf{x}$ is heavy-tailed they cannot be simultaneously very efficient and very robust.
- Their BP is less than 0.5 and is quite low for large p. For example, if $\mathbf{x}$ is multivariate normal, the BP is $O(p^{-1/2})$ (Maronna *et al.*, 1979).
- A further drawback is that the simultaneous estimation of σ reduces the BP, especially for large p (Maronna and Yohai, 1991).

For these reasons, GM-estimators, although much examined in the literature, are not a good choice, except perhaps for small p. See also Mili *et al.* (1996) and Pires *et al.* (1999).

5.11.2 Projection estimators

Note that the residuals of the LS estimator are uncorrelated with any linear combination of the predictors. In fact, the normal equations (4.13) imply that for any $\boldsymbol{\lambda} \in R^p$, the LS regression of the residuals r_i on the projections $\boldsymbol{\lambda}'\mathbf{x}_i$ is zero, since $\sum_{i=1}^n r_i \boldsymbol{\lambda}'\mathbf{x}_i = 0$. The LS estimator of regression through the origin is defined for $\mathbf{z} = (z_1, \ldots, z_n)'$ and $\mathbf{y} = (y_1, \ldots, y_n)'$ as

$$b(\mathbf{z}, \mathbf{y}) = \frac{\sum_{i=1}^n z_i y_i}{\sum_{i=1}^n z_i^2},$$

and it follows that the LS estimator $\widehat{\boldsymbol{\beta}}$ satisfies

$$b(\mathbf{X}\boldsymbol{\lambda}, \mathbf{r}(\widehat{\boldsymbol{\beta}})) = 0 \ \forall \ \boldsymbol{\lambda} \in R^p, \ \boldsymbol{\lambda} \neq \mathbf{0}. \tag{5.91}$$

A robust regression estimator could be obtained by (5.91) using for b a *robust* estimator of regression through the origin. But in general it is not possible to obtain equality in (5.91). As a result, we must content ourselves with making b "as small as possible". Let $\widehat{\sigma}$ be a robust scale estimator, such as $\widehat{\sigma}(\mathbf{z}) = \mathrm{Med}(|\mathbf{z}|)$. Then the projection estimators for regression ("P-estimators") proposed by Maronna and Yohai (1993) are defined as

$$\widehat{\boldsymbol{\beta}} = \arg\min_{\beta} \left(\max_{\boldsymbol{\lambda} \neq \mathbf{0}} |b(\mathbf{X}\boldsymbol{\lambda}, \mathbf{r}(\widehat{\boldsymbol{\beta}}))| \widehat{\sigma}(\mathbf{X}\boldsymbol{\lambda}) \right), \tag{5.92}$$

which means that the residuals are "as uncorrelated as possible" with all projections. Note that the condition $\boldsymbol{\lambda} \neq \mathbf{0}$ can be replaced by $||\boldsymbol{\lambda}|| = 1$. The factor $\widehat{\sigma}(\mathbf{X}\boldsymbol{\lambda})$ is needed to make the regression estimator scale equivariant.

The "median of slopes" estimator for regression through the origin is defined as the conditional median

$$b(\mathbf{x}, \mathbf{y}) = \mathrm{Med}\left(\frac{y_i}{x_i} \middle| x_i \neq 0 \right). \tag{5.93}$$

Martin *et al.* (1989) extended Huber's minimax result for the median (Section 3.8.5) showing that (5.93) minimizes asymptotic bias among regression invariant estimators. Maronna and Yohai (1993) studied P-estimators with b given by (5.93), which they called MP estimators, and found that their maximum asymptotic bias is lower than that of MM- and S-estimators. They have an $n^{-1/2}$ consistency rate, but are not asymptotically normal, which makes their use difficult for inference.

Maronna and Yohai (1993) show that if the $\mathbf{x}_i$ are multivariate normal, then the maximum asymptotic bias of P-estimators does not depend on p, and is not larger than twice the minimax asymptotic bias for all regression equivariant estimators.

Numerical computation of P-estimators is difficult because of the nested optimization in (5.92). An approximate solution can be found by reducing the searches over $\boldsymbol{\beta}$ and $\boldsymbol{\lambda}$ to finite sets. A set of N candidate estimators $\boldsymbol{\beta}_k, k = 1, \ldots, N$ is obtained by subsampling, as in Section 5.7.2, and from these the candidate directions are computed as

$$\boldsymbol{\lambda}_{jk} = \frac{\boldsymbol{\beta}_j - \boldsymbol{\beta}_k}{\|\boldsymbol{\beta}_j - \boldsymbol{\beta}_k\|}, \quad j \neq k.$$

Then (5.92) is replaced by

$$\widehat{\boldsymbol{\beta}} = \arg\min_k \left(\max_{j \neq k} |b(\mathbf{X}\boldsymbol{\lambda}_j, \mathbf{r}(\widehat{\boldsymbol{\beta}}_k))| \widehat{\sigma}(\mathbf{X}\boldsymbol{\lambda}_j) \right).$$

The resulting approximate estimator is regression and affine equivariant. In principle the procedure requires $N(N-1)$ evaluations, but this can be reduced to $O(N \log N)$ by the trick described in Maronna and Yohai (1993).

5.11.3 Constrained M-estimators

Mendes and Tyler (1996) define *constrained M-estimators* (CM-estimators for short), as in (4.36)

$$(\widehat{\boldsymbol{\beta}}, \widehat{\sigma}) = \arg\min_{\boldsymbol{\beta}, \sigma} \left\{ \frac{1}{n} \sum_{i=1}^{n} \rho\left(\frac{r_i(\boldsymbol{\beta})}{\sigma} \right) + \ln \sigma \right\} \tag{5.94}$$

with the restriction

$$\frac{1}{n} \sum_{i=1}^{n} \rho\left(\frac{r_i(\boldsymbol{\beta})}{\sigma} \right) \leq \varepsilon, \tag{5.95}$$

where ρ is a bounded ρ-function and $\varepsilon \in (0, 1)$. Note that if ρ is bounded, (5.94) cannot be handled without restrictions, for then $\sigma \to 0$ would yield a trivial solution.

Mendes and Tyler show that CM-estimators are M-estimators with the same ρ. Thus they are asymptotically normal, with a normal distribution efficiency that depends only on ρ (but not on ε), and hence the efficiency can be made arbitrarily high. Mendes and Tyler also show that for a continuous distribution the solution asymptotically attains the bound (5.95), so that $\widehat{\sigma}$ is an M-scale of the residuals. It follows that the estimator has an asymptotic BP equal to $\min(\varepsilon, 1 - \varepsilon)$, and taking $\varepsilon = 0.5$ yields the maximum BP.

5.11.4 Maximum depth estimators

Maximum regression depth estimators were introduced by Rousseeuw and Hubert (1999). Define the *regression depth* of $\boldsymbol{\beta} \in R^p$ with respect to a sample $(\mathbf{x}_i, y_i)$ as

$$d(\boldsymbol{\beta}) = \frac{1}{n} \min_{\boldsymbol{\lambda} \neq \mathbf{0}} \# \left\{ \frac{r_i(\boldsymbol{\beta})}{\boldsymbol{\lambda}' \mathbf{x}_i} < 0, \boldsymbol{\lambda}' \mathbf{x}_i \neq 0 \right\}, \tag{5.96}$$

where $\boldsymbol{\lambda} \in R^p$. Then the *maximum depth regression estimator* is defined as

$$\widehat{\boldsymbol{\beta}} = \arg\max_{\boldsymbol{\beta}} d(\boldsymbol{\beta}). \tag{5.97}$$

The solution need not be unique. Since only the direction matters, the infimum in (5.96) may be restricted to $\{\|\boldsymbol{\lambda}\| = 1\}$. Like the P-estimators of Section 5.11.2, maximum depth estimators are based on the univariate projections $\boldsymbol{\lambda}'\mathbf{x}_i$ of the predictors. In the case of regression through the origin, $\widehat{\boldsymbol{\beta}}$ coincides with the median of slopes given by (5.93). But when $p > 1$, the maximum asymptotic BP of these estimators at the linear model (5.1) is 1/3, and for an arbitrary joint distribution of $(\mathbf{x}, y)$ it can only be asserted to be $\geq 1/(p+1)$.

Adrover *et al.* (2002) discuss the relationships between maximum depth and P-estimators. They derive the asymptotic bias of the former and compare it to that of the MP-estimators defined in Section 5.11.2. Both biases turn out to be similar for moderate contamination (in particular, the GESs are equal), while the MP-estimator is better for large contamination. They define an approximate algorithm for computing the maximum depth estimator, based on an analogous idea already studied for the MP-estimator.

5.12 Other topics

5.12.1 The exact fit property

The so-called *exact fit property* states essentially that if a proportion α of observations lies exactly on a subspace, and $1 - \alpha$ is less than the BP of a regression and scale equivariant estimator, then the fit given by the estimator coincides with the subspace. More precisely, let the FBP of $\widehat{\boldsymbol{\beta}}$ be $\varepsilon^* = m^*/n$, and let the dataset contain q points such that $y_i = \mathbf{x}_i'\boldsymbol{\gamma}$ for some $\boldsymbol{\gamma}$. We prove in Section 5.13.3 that if $q \geq n - m^*$ then $\widehat{\boldsymbol{\beta}} = \boldsymbol{\gamma}$. For example, in the location case, if more than half the sample points are concentrated at x_0, then the median coincides with x_0. In practice if a sufficiently large number q of observations satisfy an approximate linear fit $y_i \approx \mathbf{x}_i'\boldsymbol{\gamma}$ for some $\boldsymbol{\gamma}$, then the estimator coincides approximately with that fit: $\widehat{\boldsymbol{\beta}} \approx \boldsymbol{\gamma}$.

The exact fit property implies that if a dataset comprises two linear substructures, an estimator with a high BP will choose to fit one of them, and this will allow the other to be discovered through the analysis of the residuals. A nonrobust estimator such as LS will instead try to make a compromise fit, with the undesirable result that the existence of two structures passes unnoticed.

Example 5.5 *To illustrate this point, we generate 100 points lying approximately on a straight line with slope = 1, and another 50 points with slope = −1. The tables and figures for this example are obtained with script* ***ExactFit.R****. Figure 5.18 shows the fits corresponding to the LS and MM-estimators. It is seen that the latter fits the bulk of the data, while the former does not fit any of the two structures.*

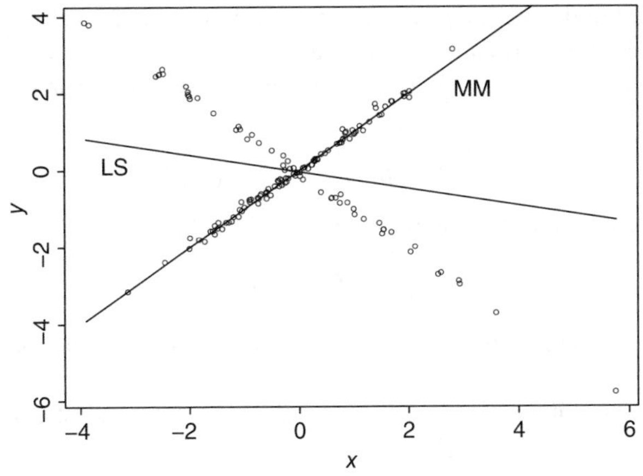

Figure 5.18 Artificial data: fits by the LS and MM-estimators

5.12.2 Heteroskedastic errors

The asymptotic theory for M-estimators, which includes S- and MM-estimators, has been derived under the assumption that the errors are i.i.d. and hence homoskedastic.

These assumptions do not always hold in practice. When the y_i are time series or spatial variables, the errors may be correlated. Moreover, in many cases the variability of the error may depend on the explanatory variables; in particular, the conditional variance of y given $\mathbf{x}$ may depend on $\boldsymbol{\beta}'\mathbf{x} = \mathrm{E}(y|\mathbf{x})$.

In fact, the assumptions of independent and homoskesdastic errors are not necessary for the consistency and asymptotic normality of M-estimators; it can be shown that these properties hold under much weaker conditions. Nevertheless we can mention two problems:

- The estimators may have lower efficiency than others that take into account the correlation or heteroskedasticy of the errors.
- The asymptotic covariance matrix of M-estimators may be different from $v\mathbf{V}_{\mathbf{x}}^{-1}$ given in (5.14), which was derived assuming i.i.d. errors. Therefore, the estimator $\widetilde{\mathbf{V}}_{\mathbf{x}}$ given in Section 5.6 would not converge to the true asymptotic covariance matrix of $\widehat{\boldsymbol{\beta}}$.

We deal with these problems in the next two subsections.

5.12.2.1 Modelling heteroskedasticity for M-estimators

To improve the efficiency of M-estimators under heteroskedasticity, the dependence of the error scale on $\mathbf{x}$ should be included in the model. For instance, we can replace model (5.1) by

$$y_i = \boldsymbol{\beta}'\mathbf{x}_i + h(\boldsymbol{\lambda}, \boldsymbol{\beta}'\mathbf{x})u_i,$$

where the u_i are i.i.d. and independent of $\mathbf{x}_i$, and $\boldsymbol{\lambda}$ is an additional vector parameter. In this case the error scales are proportional to $h^2(\lambda, \boldsymbol{\beta}'\mathbf{x})$.

Observe that if we knew $h(\boldsymbol{\lambda}, \boldsymbol{\beta}'\mathbf{x})$, then the transformed variables

$$y_i^* = \frac{y_i}{h(\boldsymbol{\lambda}, \boldsymbol{\beta}'\mathbf{x})}, \quad \mathbf{x}_i^* = \frac{\mathbf{x}_i}{h(\boldsymbol{\lambda}, \boldsymbol{\beta}'\mathbf{x})}$$

would follow the homoskedastic regression model

$$y_i^* = \boldsymbol{\beta}'\mathbf{x}_i^* + u_i.$$

This suggests the following procedure to obtain robust estimators of $\boldsymbol{\beta}$ and $\boldsymbol{\lambda}$.

- **(i)** Compute an initial robust estimator $\widehat{\boldsymbol{\beta}}_0$ for homoskedastic regression; for example, $\widehat{\boldsymbol{\beta}}_0$ may be an MM-estimator.
- **(ii)** Compute the residuals $r_i(\widehat{\boldsymbol{\beta}}_0)$.
- **(iii)** Use these residuals to obtain an estimator $\widehat{\boldsymbol{\lambda}}$ of $\boldsymbol{\lambda}$. For example, if

  $$h(\boldsymbol{\lambda}, t) = \exp(\lambda_1 + \lambda_2|t|),$$

  then λ can be estimated by a robust linear fit of $\log(|r_i(\widehat{\boldsymbol{\beta}}_0)|)$ on $|\widehat{\boldsymbol{\beta}}_0'\mathbf{x}_i|$.
- **(iv)** Compute a robust estimator for homoskedastic regression based on the transformed variables

  $$y_i^* = \frac{y_i}{h(\widehat{\boldsymbol{\lambda}}, \widehat{\boldsymbol{\beta}}_0'\mathbf{x})}, \quad \mathbf{x}_i^* = \frac{\mathbf{x}_i}{h(\widehat{\boldsymbol{\lambda}}, \widehat{\boldsymbol{\beta}}_0'\mathbf{x})}$$

Steps (i)–(iv) may be iterated.

Robust methods for heteroskedastic regression have been proposed by Carroll and Ruppert (1982) who used monotone M-estimators; by Giltinan *et al.* (1986) who employed GM-estimators; and by Bianco *et al.* (2000) and Bianco and Boente (2002) who defined estimators with high BP and bounded influence starting with an initial MM-estimator followed by one Newton–Raphson step of a GM-estimator.

5.12.2.2 Estimating the asymptotic covariance matrix under heteroskedastic errors

Simpson *et al.* (1992) proposed an estimator for the asymptotic covariance matrix of regression GM-estimators that does not require homoskedasticity but requires symmetry of the error distribution. Salibian-Barrera (2000) and Croux *et al.* (2003) proposed a method to estimate the asymptotic covariance matrix of a regression M-estimator requiring neither homoskedasticity nor symmetry. This method can also be applied to simultaneous M-estimators of regression of scale (which includes S-estimators) and to MM-estimators.

We shall give some details of the method for the case of MM-estimators. Let $\widehat{\boldsymbol{\gamma}}$ and $\widehat{\sigma}$ be the initial S-estimator used to compute the MM-estimator and the corresponding scale estimator, respectively. Since $\widehat{\gamma}$ and $\widehat{\sigma}$ are M-estimators of regression and of scale, they satisfy equations of the form

$$\sum_{i=1}^{n} \psi_0\left(\frac{r_i(\widehat{\gamma})}{\widehat{\sigma}}\right)\mathbf{x}_i = \mathbf{0}, \tag{5.98}$$

$$\frac{1}{n}\sum_{i=1}^{n}\left[\rho_0\left(\frac{r_i(\widehat{\gamma})}{\widehat{\sigma}}\right) - \delta\right] = 0, \tag{5.99}$$

with $\psi_0 = \rho_0'$. The final MM-estimator $\widehat{\boldsymbol{\beta}}$ is a solution of

$$\sum_{i=1}^{n} \psi\left(\frac{r_i(\widehat{\boldsymbol{\beta}})}{\widehat{\sigma}}\right)\mathbf{x}_i = \mathbf{0}. \tag{5.100}$$

To explain the proposed method we need to express the system (5.98)–(5.100) as a unique set of M-estimating equations. To this end, let the vector $\boldsymbol{\gamma}$ represent the values taken on by $\widehat{\boldsymbol{\gamma}}$. Let $\mathbf{z} = (\mathbf{x}, y)$. For $\boldsymbol{\gamma}, \boldsymbol{\beta} \in R^p$ and $\sigma \in R$, put $\boldsymbol{\alpha} = (\boldsymbol{\gamma}', \sigma, \boldsymbol{\beta}')' \in R^{2p+1}$, and define the function

$$\boldsymbol{\Psi}(z, \alpha) = (\boldsymbol{\Psi}_1(\mathbf{z}, \boldsymbol{\alpha}), \Psi_2(\mathbf{z}, \boldsymbol{\alpha}), \boldsymbol{\Psi}_3(\mathbf{z}, \boldsymbol{\alpha}))$$

where

$$\boldsymbol{\Psi}_1(\mathbf{z}, \boldsymbol{\alpha}) = \psi_1\left(\frac{y - \boldsymbol{\gamma}'\mathbf{x}}{\sigma}\right)\mathbf{x}$$

$$\Psi_2(\mathbf{z}, \boldsymbol{\alpha}) = \rho_1\left(\frac{y - \boldsymbol{\gamma}'\mathbf{x}}{\sigma}\right) - \delta$$

$$\boldsymbol{\Psi}_3(\mathbf{z}, \boldsymbol{\alpha}) = \psi_2\left(\frac{y - \boldsymbol{\beta}'\mathbf{x}}{\sigma}\right)\mathbf{x}.$$

Then $\widehat{\boldsymbol{\alpha}} = (\widehat{\boldsymbol{\gamma}}, \widehat{\sigma}, \widehat{\boldsymbol{\beta}})$is an M-estimator satisfying

$$\sum_{i=1}^{n} \boldsymbol{\Psi}(\mathbf{z}_i, \widehat{\alpha}) = 0,$$

and therefore according to (3.48), its asymptotic covariance matrix is

$$\mathbf{V} = \mathbf{A}^{-1}\mathbf{B}\mathbf{A}'^{-1}$$

with

$$\mathbf{A} = \mathrm{E}[\boldsymbol{\Psi}(\mathbf{z}, \alpha)\boldsymbol{\Psi}(\mathbf{z}, \alpha)']$$

and

$$\mathbf{B} = \mathrm{E}\left(\frac{\partial\boldsymbol{\Psi}(\mathbf{z}, \alpha)}{\partial\boldsymbol{\alpha}}\right),$$

where the expectation is calculated under the model $y = \mathbf{x}'\boldsymbol{\beta} + u$ and taking $\boldsymbol{\gamma} = \boldsymbol{\beta}$.

Then $\mathbf{V}$ can be estimated by

$$\widehat{\mathbf{V}} = \widehat{\mathbf{V}} = \widehat{\mathbf{B}}^{-1}\widehat{\mathbf{A}}\widehat{\mathbf{B}}^{-1\prime}$$

where $\widehat{\mathbf{A}}$ and $\widehat{\mathbf{B}}$ are the the empirical versions of $\mathbf{A}$ and $\mathbf{B}$ obtained by replacing $\boldsymbol{\alpha}$ with $\widehat{\alpha}$.

Observe that the only requirement for $\widehat{\mathbf{V}}$ to be a consistent estimator of $\mathbf{V}$ is that the observations $(\mathbf{x}_1, y_1), \ldots, (\mathbf{x}_n, y_n)$ be i.i.d., but this does not require any condition on the conditional distribution of y given $\mathbf{x}$, for example homoskedasticity. For the justification of the above procedure see Remark 3 and point (viii) of Theorem 6 of Fasano *et al.* (2012).

Croux *et al.* (2003) also consider estimators of $\mathbf{V}$ when the errors are not independent.

5.12.3 A robust multiple correlation coefficient

In a multiple linear regression model, the R^2 statistic measures the proportion of the variation in the dependent variable accounted for by the explanatory variables. It is defined for a model with intercept (4.5) as

$$R^2 = \frac{\mathrm{Var}(y) - \mathrm{E}u^2}{\mathrm{Var}(y)}, \tag{5.101}$$

and it can be estimated by

$$\widehat{R}^2 = \frac{S_0^2 - S^2}{S_0^2} \tag{5.102}$$

with

$$S^2 = \sum_{i=1}^{n} r_i^2, \quad S_0^2 = \sum_{i=1}^{n} (y_i - \overline{y})^2, \tag{5.103}$$

where r_i are the LS residuals. Note that $\overline{y}$ is the LS estimator of the regression coefficients under model (4.5) with the restriction $\boldsymbol{\beta}_1 = 0$.

Recall that $S^2/(n - p^*)$ and $S_0^2/(n - 1)$ are unbiased estimators of the error variance for the complete model, and for the model with $\boldsymbol{\beta}_1 = 0$, respectively. To take the degrees of freedom into account, an adjusted coefficient R_a^2 is defined by

$$\widehat{R}_a^2 = \frac{S_0^2/(n-1) - S^2/(n-p^*)}{S_0^2/(n-1)}, \tag{5.104}$$

which is equivalent to

$$\widehat{R}_a^2 = \frac{n-1}{n-p^*}R^2 + \frac{1-p^*}{n-p^*}. \tag{5.105}$$

If, instead of the LS estimator, we use an M-estimator based on a bounded ρ-function ρ and a scale σ defined as in (5.7), a robust version of R^2 can be defined as

$$R^2_\rho = \frac{A_0 - A_1}{(1 - A_1)A_0} \tag{5.106}$$

with

$$A_0 = \min_{\beta_0 \in R} \mathrm{E}\rho\left(\frac{y - \beta_0}{\sigma}\right), \quad A_1 = \mathrm{E}\rho\left(\frac{u}{\sigma}\right),$$

where the factor $(1 - A_1)$ in the denominator ensures that $0 \le \widehat{R}^2_\rho \le 1$. Note that we are assumming that $\rho \in [0, 1]$; if ρ is the classical quadratic function, then (5.106) does not coincide with (5.101).

We can estimate R^2_ρ by

$$\widehat{R}^2_\rho = \frac{S^2_{\rho 0} - S^2_\rho}{(1 - S^2_\rho/n)S^2_{\rho 0}},$$

where

$$S^2_\rho = \min_{\beta \in R^p} \sum \rho\left(\frac{r_i(\beta)}{\widehat{\sigma}}\right), \quad S^2_{\rho 0} = \min_{\beta_0 \in R} \sum_{i=1}^{p} \rho\left(\frac{y_i - \beta_0}{\widehat{\sigma}}\right),$$

and where $\widehat{\sigma}$ is a robust estimate of the error escale.

However $\widehat{R}^2_\rho$ is asymptotically biased as an estimator of R^2. In fact it is easy to see that in the case that y and u are normal, we have $R^2_\rho = g(R^2)$, where g is a continuous and strictly monotone function different from the identity. The function g is plotted in Figure 5.19. Then we can obtain an estimator of R^2 that is unbiased in the case of normal data, as $\widehat{R}^2_{\rho,\mathrm{un}} = g^{-1}(\widehat{R}^2_\rho)$. An adjusted estimator of R^2 for finite samples can be obtained by replacing, $\widehat{R}^2$ in the right-hand side of (5.105) by $\widehat{R}^2_{\rho,\mathrm{un}}$.

To compute the function g, we can assume, without loss of generality, that $\sigma = 1$. Since in this case $u \sim \mathrm{N}(0, 1)$ and $y \sim \mathrm{N}(E(y), 1/\sqrt{1 - R^2})$, we have $A_1 = E(\rho(u))$ and $A_0 = E(\rho(u/\sqrt{1 - R^2})$.

Croux and Dehon (2003) have considered alternative definitions of a robust R^2. They proposed an estimator of R^2 of the form

$$\widehat{R}^2 = 1 - \frac{s^2(r_1(\widehat{\boldsymbol{\beta}}), \ldots, r_n(\widehat{\boldsymbol{\beta}}))}{s^2(y_1 - \widehat{\mu}, \ldots, y_n - \widehat{\mu})}, \tag{5.107}$$

where s is a robust scale estimator, $\widehat{\boldsymbol{\beta}}$ a robust regression estimator of $\boldsymbol{\beta}$ and $\widehat{\mu}$ a robust location estimator applied to the sample $y_1, \ldots, y_n$. If s and $\widehat{\mu}$ are consistent for the standard deviation and mean at the normal distribution, then $\widehat{R^2}$ is consistent for the classical R^2 for normal models. If $\widehat{\boldsymbol{\beta}}$ and $\widehat{\mu}$ are S-estimators of regression and location that minimize the scale s, then this R^2 is nonnegative. This need not happen in all cases if other estimators are employed.

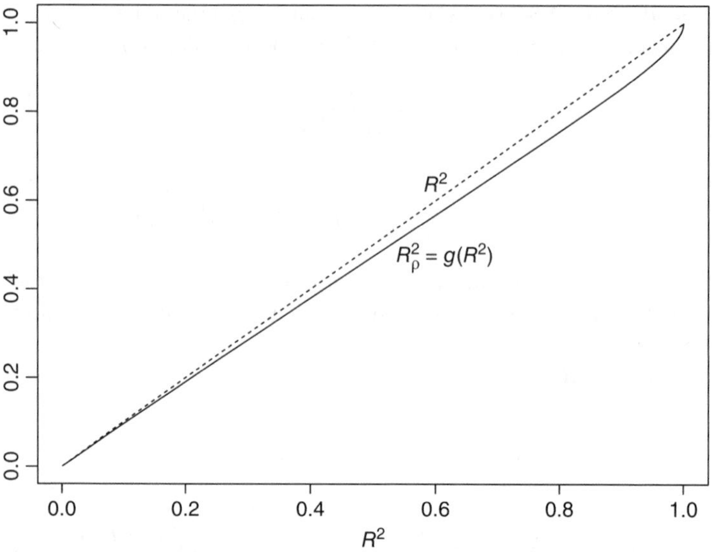

Figure 5.19 Relationship between R^2 and R^2_ρ

The estimator R_ρ is especially adapted to express the proportion of explained variability of regession M-estimators. Since the regression estimators proposed in this book all have the form of M-estimators, we recommend this correlation coefficient rather than (5.107).

5.13 *Appendix: proofs and complements

5.13.1 The BP of monotone M-estimators with random X

We assume σ known and equal to one. The estimator verifies

$$\psi(y_1 - \mathbf{x}_1'\hat{\boldsymbol{\beta}})\mathbf{x}_1 + \sum_{i=2}^{n} \psi(y_i - \mathbf{x}_i'\hat{\boldsymbol{\beta}})\mathbf{x}_i = \mathbf{0}. \tag{5.108}$$

Let y_1 and $\mathbf{x}_1$ tend to infinity in such a way that $y_1/\|\mathbf{x}_1\| \to \infty$. If $\hat{\boldsymbol{\beta}}$ remained bounded, we would have

$$y_1 - \mathbf{x}_1'\hat{\boldsymbol{\beta}} \geq y_1 - \|\mathbf{x}_1\|\|\hat{\boldsymbol{\beta}}\| = \|\mathbf{x}_1\| \left(\frac{y_1}{\|\mathbf{x}_1\|} - \|\hat{\boldsymbol{\beta}}\| \right) \to \infty.$$

Since ψ is nondecreasing, $\psi(y_1 - \mathbf{x}_1'\hat{\boldsymbol{\beta}})$ would tend to $\sup \psi > 0$, and hence the first term in (5.108) would tend to infinity, while the sum would remain bounded.

5.13.2 Heavy-tailed x

The behavior of the estimators under heavy-tailed **x** is most easily understood when the estimator is the LS estimator and $p = 1$; that is,

$$y_i = \beta x_i + u_i,$$

where $\{x_i\}$ and $\{u_i\}$ are independent i.i.d. sequences. Then

$$\widehat{\beta}_n = \frac{\sum_{i=1}^n x_i y_i}{T_n} \quad \text{with} \quad T_n = \sum_{i=1}^n x_i^2.$$

Assume $\mathrm{E}u_i = 0$ and $\mathrm{Var}(u_i) = 1$. Then

$$\mathrm{Var}(\widehat{\beta}_n|\mathbf{X}) = \frac{1}{T_n} \quad \text{and} \quad \mathrm{E}(\widehat{\beta}_n|\mathbf{X}) = \beta,$$

and hence, by a well-known property of the variance; see, for example, Feller (1971):

$$\mathrm{Var}(\sqrt{n}\widehat{\beta}_n) = n\{\mathrm{E}[\mathrm{Var}(\widehat{\beta}_n|\mathbf{X})] + \mathrm{Var}[\mathrm{E}(\widehat{\beta}_n|\mathbf{X})]\} = \mathrm{E}\frac{1}{T_n/n}.$$

If $a = \mathrm{E}x_i^2 < \infty$, the law of large numbers implies that $T_n/n \rightarrow_p a$, and, under suitable conditions on the x_i, this implies that

$$\mathrm{E}\frac{1}{T_n/n} \rightarrow \frac{1}{a}, \tag{5.109}$$

hence

$$\mathrm{Var}(\sqrt{n}\widehat{\beta}_n) \rightarrow \frac{1}{a}.$$

In other words, $\widehat{\beta}$ is $\sqrt{n}$-consistent.

If, instead, $\mathrm{E}x_i^2 = \infty$, then $T_n/n \rightarrow_p \infty$, which implies that

$$\mathrm{Var}(\sqrt{n}\widehat{\beta}_n) \rightarrow 0,$$

and hence $\widehat{\beta}$ tends to β at a higher rate than $\sqrt{n}$.

A simple sufficient condition for (5.109) is that $x_i \geq \alpha$ for some $\alpha > 0$, for then $n/T_n \leq 1/\alpha^2$ and (5.109) holds by the bounded convergence theorem (Theorem 10.6). But the result can be shown to hold under more general assumptions.

5.13.3 Proof of the exact fit property

Define for $t \in R$:

$$\mathbf{y}^* = \mathbf{y} + t(\mathbf{y} - \mathbf{X}\boldsymbol{\gamma}).$$

Then the regression and scale equivariance of $\widehat{\boldsymbol{\beta}}$ implies

$$\widehat{\boldsymbol{\beta}}(\mathbf{X}, \mathbf{y}^*) = \widehat{\boldsymbol{\beta}}(\mathbf{X}, \mathbf{y}) + t(\widehat{\boldsymbol{\beta}}(\mathbf{X}, \mathbf{y}) - \boldsymbol{\gamma}).$$

Since for all t, $\mathbf{y}^*$ has at least $q \geq n - m^*$ values in common with $\mathbf{y}$, the above expression must remain bounded, and this requires $\widehat{\boldsymbol{\beta}}(\mathbf{X}, \mathbf{y}) - \boldsymbol{\gamma} = \mathbf{0}$.

5.13.4 The BP of S-estimators

It will be shown that the finite BP of an S-estimator defined in Section 5.4.1 does not depend on **y**, and that its maximum is given by (5.19)–(5.20).

This result has been proved by Rousseeuw and Leroy (1987) and Mili and Coakley (1996) under slightly more restricted conditions. The main result of this section is the following.

Theorem 5.1 *Let m^* be as in (5.9) and $m^*_{\max}$ as in (5.20). Call $m(\delta)$ the largest integer $< n\delta$. Then:*

(a) $m^* \leq n\delta$
(b) *if* $[n\delta] \leq m^*_{\max}$, *then* $m^* \geq m(\delta)$.

It follows from this result that if $n\delta$ is not an integer and $\delta \leq m^*_{\max}/n$, then $m^* = [n\delta]$, and hence the δ given by (5.21) yields $m^* = m^*_{\max}$.

To prove the theorem we first need an auxiliary result.

Lemma 5.1 *Consider any sequence* $\mathbf{r}_N = (r_{N,1}, ..., r_{N,n})$ *with* $\sigma_N = \hat{\sigma}(\mathbf{r}_N)$. *Then*

(i) *Let* $C = \{i : |r_{N,i}| \to \infty\}$. *If* $\#(C) > n\delta$, *then* $\sigma_N \to \infty$,
(ii) *Let* $D = \{i : |r_{N,i}| \text{ is bounded}\}$. *If* $\#(D) > n - n\delta$, *then* σ_N *is bounded.*

Proof of the Lemma

(i) Assume σ_N bounded. Then the definition of σ_N implies

$$n\delta \geq \lim_{N\to\infty} \sum_{i\in C} \rho\left(\frac{r_{N,i}}{\sigma_N}\right) = \#(C) > n\delta,$$

which is a contradiction.

(ii) To show that σ_N remains bounded, assume that $\sigma_N \to \infty$. Then $r_{N,i}/\sigma_N \to 0$ for $i \in D$, which implies

$$n\delta = \lim_{N\to\infty} \sum_{i=1}^{n} \rho\left(\frac{r_{N,i}}{\sigma_N}\right) = \lim_{N\to\infty} \sum_{i\notin D} \rho\left(\frac{r_{N,i}}{\sigma_N}\right) \leq n - \#(D) < n\delta.$$

which is a contradiction.

Proof of (a): It will be shown that $m^* \leq n\delta$. Let $m > n\delta$. Take $C \subset \{1, ..., n\}$ with $\#(C) = m$. Let $\mathbf{x}_0 \in R^p$ with $\|\mathbf{x}_0\| = 1$. Given a sequence $(\mathbf{X}_N, \mathbf{y}_N)$, define for $\boldsymbol{\beta} \in R^p$

$$\mathbf{r}_N(\boldsymbol{\beta}) = \mathbf{y}_N - \mathbf{X}_N\boldsymbol{\beta}.$$

Take $(\mathbf{X}_N, \mathbf{y}_N)$ such that

$$(\mathbf{x}_{N,i}, y_{N,i}) = \begin{cases} (N\mathbf{x}_0, N^2) & \text{if} \quad i \in C \\ (\mathbf{x}_i, y_i) & \text{otherwise} \end{cases} \tag{5.110}$$

It will be shown that the estimator $\widehat{\boldsymbol{\beta}}_N$ based on $(\mathbf{X}_N, \mathbf{Y}_N)$ cannot be bounded.

Assume first that $\widehat{\boldsymbol{\beta}}_N$ is bounded, which implies that $|r_{N,i}| \to \infty$ for $i \in C$. Then part (i) of the lemma implies that $\widehat{\sigma}(\mathbf{r}_N(\widehat{\boldsymbol{\beta}}_N)) \to \infty$. Since $n\delta/m < 1 = \rho(\infty)$, condition R3 of Definition 2.1 implies that there is a single value γ such that

$$\rho\left(\frac{1}{\gamma}\right) = \frac{n\delta}{m}. \tag{5.111}$$

It will be shown that

$$\frac{1}{N^2}\widehat{\sigma}(\mathbf{r}_N(\widehat{\boldsymbol{\beta}}_N)) \to \gamma. \tag{5.112}$$

In fact,

$$n\delta = \sum_{i \notin C} \rho\left(\frac{y_i - \mathbf{x}_i'\widehat{\boldsymbol{\beta}}_N}{\widehat{\sigma}_N}\right) + \sum_{i \in C} \rho\left(\frac{N^2 - N\mathbf{x}_0'\widehat{\boldsymbol{\beta}}_N}{\widehat{\sigma}_N}\right).$$

The first sum tends to zero. The second one is

$$m\rho\left(\frac{1 - N^{-1}\mathbf{x}_0'\widehat{\boldsymbol{\beta}}_N}{N^{-2}\widehat{\sigma}_N}\right).$$

The numerator of the fraction tends to one. If a subsequence $\{N_j^{-2}\widehat{\sigma}_{N_j}\}$ has a (possibly infinite) limit t, then it must fulfill $n\delta = m\rho(1/t)$, which proves (5.112).

Now define $\widetilde{\boldsymbol{\beta}}_N = \mathbf{x}_0\, N/2$, so that $\mathbf{r}_N(\widetilde{\boldsymbol{\beta}}_N)$ has elements

$$\widetilde{r}_{N,i} = \frac{N^2}{2} \text{ for } i \in C, \quad \widetilde{r}_{N,i} = y_i - \mathbf{x}_0'\mathbf{x}_i\, \frac{N}{2} \text{ otherwise.}$$

Since $\#\{i : |\widetilde{r}_{N,i}| \to \infty\} = n$, part (i) of the lemma implies that $\widehat{\sigma}(\mathbf{r}_N(\widetilde{\boldsymbol{\beta}}_N)) \to \infty$, and proceeding as in the proof of (5.112) yields

$$\frac{1}{N^2}\widehat{\sigma}(\mathbf{r}_N(\widetilde{\boldsymbol{\beta}}_N)) \to \frac{\gamma}{2},$$

and hence

$$\widehat{\sigma}(\mathbf{r}_N(\widetilde{\boldsymbol{\beta}}_N)) < \widehat{\sigma}(\mathbf{r}_N(\widehat{\boldsymbol{\beta}}_N))$$

for large N, so that $\widehat{\boldsymbol{\beta}}_N$ cannot minimize σ.

Proof of (b): Let $m \le m(\delta) < n\delta$, and consider a contamination sequence in a set C of size m. It will be shown that the corresponding estimator $\widehat{\boldsymbol{\beta}}_N$ is bounded. Assume first that $\widehat{\boldsymbol{\beta}}_N \to \infty$. Then

$$i \notin C, \quad |r_{N,i}(\widehat{\boldsymbol{\beta}}_N)| \to \infty \quad \Longrightarrow \quad \widehat{\boldsymbol{\beta}}_N'\, \mathbf{x}_{N,i} \neq 0,$$

and hence

$$\#\{i : |r_{N,i}(\widehat{\boldsymbol{\beta}}_N)| \to \infty\} \geq \#\{\widehat{\boldsymbol{\beta}}_N' \mathbf{x}_{N,i} \neq 0, i \notin C\}$$
$$= n - \#(\{i : \widehat{\boldsymbol{\beta}}_N' \mathbf{x}_{N,i} = 0\} \cup C).$$

The Bonferroni inequality implies that

$$\#(\{i : \widehat{\boldsymbol{\beta}}_N' \mathbf{x}_{N,i} = 0\} \cup C) \leq \#\{i : \widehat{\boldsymbol{\beta}}_N' \mathbf{x}_{N,i} = 0\} + \#(C),$$

and $\#\{i : \widehat{\boldsymbol{\beta}}_N' \mathbf{x}_{N,i} = 0\} \leq k^*(\mathbf{X})$ by (4.56). Hence

$$\#\{i : |r_{N,i}(\widehat{\boldsymbol{\beta}}_N)| \to \infty\} \geq n - k^*(\mathbf{X}) - m.$$

Now (4.58) implies that $n - k^*(\mathbf{X}) \geq 2m^*_{\max} + 1$, and since

$$m \leq m(\delta) \leq [n\delta] \leq m^*_{\max},$$

we have

$$\#\{i : |r_{N,i}(\widehat{\boldsymbol{\beta}}_N)| \to \infty\} \geq 1 + m^*_{\max} \geq 1 + [n\delta] > n\delta,$$

which by part (i) of the Lemma implies $\widehat{\sigma}(\mathbf{r}_N(\widehat{\boldsymbol{\beta}}_N)) \to \infty$.

Assume now that $\widehat{\boldsymbol{\beta}}_N$ is bounded. Then

$$\#\{i : |r_{N,i}(\widehat{\boldsymbol{\beta}}_N)| \to \infty\} \leq m < n\delta,$$

which by part (ii) of the lemma implies $\widehat{\sigma}(\mathbf{r}_N(\widehat{\boldsymbol{\beta}}_N))$ is bounded. Hence $\widehat{\boldsymbol{\beta}}_N$ cannot be unbounded. This completes the proof of the finite BP.

The least quantile estimator corresponds to the scale given by $\rho(t) = \mathrm{I}(|t| > 1)$, and according to Problem 2.14 it has $\widehat{\sigma} = |r|_{(h)}$, where $|r|_{(i)}$ are the ordered absolute residuals and $h = n - [n\delta]$. The optimal choice of δ in (5.21) yields $h = n - m^*_{\max}$, and formal application of the theorem would imply that this h yields the maximum FBP. In fact, the proof of the theorem does not hold because ρ is discontinuous and hence does not fulfill (5.111), but the proof can be reworked for $\widehat{\sigma} = |r|_{(h)}$ to show that the result also holds in this case.

As for the asymptotic BP, a proof similar to but much simpler than that of Theorem 5.1, with averages replaced by expectations, shows that in the asymptotic case $\varepsilon^* \leq \delta$ and if $\delta \leq (1 - \alpha)/2$ then $\varepsilon^* \geq \delta$. It follows that $\varepsilon^* = \delta$ for $\delta \leq (1 - \alpha)/2$, and this proves that the maximum asymptotic BP is (5.22).

5.13.5 Asymptotic bias of M-estimators

Let $F = \mathcal{D}(\mathbf{x}, y)$ be $\mathrm{N}_{p+1}(\mathbf{0}, \mathbf{I})$. We shall first show that the asymptotic bias under point-mass contamination of M-estimators and of estimators that minimize a robust scale does not depend on the dimension p. To simplify the exposition, we consider only the case of an M-estimator with known scale $\sigma = 1$. Call $(\mathbf{x}_0, y_0)$ the contamination location. The asymptotic value of the estimator is given by

$$\widehat{\boldsymbol{\beta}}_\infty = \arg\min_{\boldsymbol{\beta}} L(\boldsymbol{\beta}),$$

where

$$L(\boldsymbol{\beta}) = (1-\varepsilon)\mathrm{E}_F\rho(y - \mathbf{x}'\boldsymbol{\beta}) + \varepsilon\rho(y_0 - \mathbf{x}_0'\boldsymbol{\beta}). \tag{5.113}$$

Since $\mathcal{D}(y - \mathbf{x}'\boldsymbol{\beta}) = \mathrm{N}(0, 1 + \|\boldsymbol{\beta}\|^2)$ under F, we have

$$\mathrm{E}_F\rho(y - \mathbf{x}'\boldsymbol{\beta}) = g(\|\boldsymbol{\beta}\|),$$

where

$$g(t) = \mathrm{E}\rho\left(z\sqrt{1+t^2}\right), \quad z \sim \mathrm{N}(0,1).$$

It is easy to show that g is an increasing function.

By the affine equivariance of the estimator, $\|\widehat{\boldsymbol{\beta}}_\infty\|$ does not change if we take $\mathbf{x}_0$ along the first coordinate axis; in other words, of the form $\mathbf{x}_0 = (x_0, 0, .., 0)$. Therefore,

$$L(\boldsymbol{\beta}) = (1-\varepsilon)g(\|\boldsymbol{\beta}\|) + \varepsilon\rho(y_0 - x_0\beta_1),$$

where β_1 is the first coordinate of $\boldsymbol{\beta}$.

Given $\boldsymbol{\beta} = (\beta_1, \beta_2, \ldots, \beta_p)$, with $\beta_j \neq 0$ for some $j \geq 2$, the vector $\widetilde{\boldsymbol{\beta}} = (\beta_1, 0, \ldots, 0)$ has $\|\widetilde{\boldsymbol{\beta}}\| < \|\boldsymbol{\beta}\|$, which implies $g(\|\widetilde{\boldsymbol{\beta}}\|) < g(\|\boldsymbol{\beta}\|)$ and $L(\widetilde{\boldsymbol{\beta}}) < L(\boldsymbol{\beta})$. Then, we may restrict the search to the vectors of the form $(\beta_1, 0, \ldots, 0)$, for which

$$L(\boldsymbol{\beta}) = L_1(\beta_1) = (1-\varepsilon)g(\beta_1) + \varepsilon\rho(y_0 - x_0\beta_1),$$

and therefore the value minimizing $L_1(\beta_1)$ depends only on x_0 and y_0, and not on p, which proves the initial assertion.

It follows that the maximum asymptotic bias for point-mass contamination does not depend on p. Actually, it can be shown that the maximum asymptotic bias for unrestricted contamination coincides with the former, and hence does not depend on p either.

The same results hold for M-estimators with the previous scale, and for S-estimators, but the details are more involved.

5.13.6 Hampel optimality for GM-estimators

We now deal with general M-estimators for regression through the origin ($y = \beta x + u$), defined by

$$\sum_{i=1}^{n} \Psi(x_i, y_i; \beta) = 0,$$

with Ψ of the form

$$\Psi(x, y; \beta) = \eta(x, y - x\beta)x.$$

Assume σ is known and equal to one. It follows that the influence function is

$$\mathrm{IF}((x_0, y_0), F) = \frac{1}{b}\eta(x_0, y_0 - x_0\beta)x_0,$$

where

$$b = -\mathrm{E}\dot{\eta}(x, y - \beta x)x^2,$$

with $\dot{\eta}$ defined in (5.89), and hence the GES is

$$\gamma^* = \sup_{x_0, y_0} |\mathrm{IF}((x_0, y_0), F)| = \frac{1}{b} \sup_{s>0} K(s), \quad \text{with} \quad K(s) = \sup_r |\eta(s, r)|.$$

The asymptotic variance is

$$v = \frac{1}{b^2}\mathrm{E}\eta(x, y - x\beta)^2 x^2$$

The direct and dual Hampel problems can now be stated as minimizing v subject to a bound on γ^*, and minimizing γ^* subject to a bound on v, respectively.

Let F correspond to the model (5.1)–(5.2), with normal us. The MLE corresponds to

$$\Psi_0(x, y; \beta) = (y - x\beta)x.$$

Since the estimators are equivariant, it suffices to treat the case of $\beta = 0$. Proceeding as in Section 3.8.7, it follows that the solutions to both problems,

$$\tilde{\Psi}(x, y; \beta) = \tilde{\boldsymbol{\eta}}(x, y - x\beta)x,$$

have the form $\tilde{\Psi}(x, y; \beta) = \psi_k(\Psi_0(x, y; \beta))$ for some $k > 0$, where ψ_k is Huber's ψ, which implies that $\tilde{\eta}$ has the form (5.86).

The case $p > 1$ is more difficult to deal with, since $\boldsymbol{\beta}$ – and hence the IF – are multidimensional. But the present reasoning gives some justification for the use of (5.86).

5.13.7 Justification of RFPE*

We are going to give a heuristic justification of (5.39). Let $(\mathbf{x}_i, y_i)$, $i = 0, 1, ..., n$ be i.i.d. and satisfy the model

$$y_i = \mathbf{x}_i' \boldsymbol{\beta} + u_i \quad (i = 0, ..., n), \tag{5.114}$$

where u_i and $\mathbf{x}_i$ are independent, and

$$\mathrm{E}\psi\left(\frac{u_i}{\sigma}\right) = 0. \tag{5.115}$$

Call $C_0 = \{j : \beta_j \neq 0\}$ the set of variables that actually have some predictive power. Given $C \subseteq \{1, \ldots, p\}$ let

$$\boldsymbol{\beta}_C = (\beta_j, j \in C), \quad \mathbf{x}_{iC} = (x_{ij} : j \in C), \quad i = 0, .., n.$$

Put $q = \#(C)$ and call $\widehat{\boldsymbol{\beta}}_C \in R^q$ the estimator based on $\{(\mathbf{x}_{iC}, y_i), i = 1, \ldots, n\}$. Then the residuals are $r_i = y_i - \widehat{\boldsymbol{\beta}}'_C \mathbf{x}_{iC}$ for $i = 1, \ldots, n$.

Assume that $C \supseteq C_0$. Then $\mathbf{x}'_i \boldsymbol{\beta} = \mathbf{x}'_{iC} \boldsymbol{\beta}_C$ and hence the model (5.114) can be rewritten as

$$y_i = \mathbf{x}'_{iC} \boldsymbol{\beta}_C + u_i, \quad i = 0, ..., n. \tag{5.116}$$

Put $\boldsymbol{\Delta} = \widehat{\boldsymbol{\beta}}_C - \boldsymbol{\beta}_C$. A second-order Taylor expansion yields

$$\rho\left(\frac{y_0 - \widehat{\boldsymbol{\beta}}'_C \mathbf{x}_{0C}}{\sigma}\right) = \rho\left(\frac{u_0 \quad - \mathbf{x}'_{0C}\boldsymbol{\Delta}}{\sigma}\right) \tag{5.117}$$

$$\approx \rho\left(\frac{u_0}{\sigma}\right) - \psi(\frac{u_0}{\sigma})\frac{\mathbf{x}'_{0C}\boldsymbol{\Delta}}{\sigma} + \frac{1}{2}\psi'(\frac{u_0}{\sigma})\left(\frac{\mathbf{x}'_{0C}\boldsymbol{\Delta}}{\sigma}\right)^2.$$

The independence of u_0 and $\widehat{\boldsymbol{\beta}}_C$, and (5.115) yield

$$\mathrm{E}\psi\left(\frac{u_0}{\sigma}\right)\mathbf{x}'_{0C}\boldsymbol{\Delta} = \mathrm{E}\psi\left(\frac{u_0}{\sigma}\right)\mathrm{E}(\mathbf{x}'_{0C}\boldsymbol{\Delta}) = 0. \tag{5.118}$$

According to (5.14), we have, for large n,

$$\mathcal{D}(\sqrt{n}\boldsymbol{\Delta}) \approx \mathrm{N}\left(0, \frac{\sigma^2 A}{B^2}\mathbf{V}^{-1}\right),$$

where

$$A = \mathrm{E}\psi^2\left(\frac{u}{\sigma}\right), \quad B = \mathrm{E}\psi'\left(\frac{u}{\sigma}\right), \quad \mathbf{V} = \mathrm{E}(\mathbf{x}_{0C}\mathbf{x}'_{0C}).$$

Since $u_0, \boldsymbol{\Delta}$ and $\mathbf{x}_0$ are independent we have

$$\mathrm{E}\psi'\left(\frac{u_0}{\sigma}\right)\left(\frac{\boldsymbol{\Delta}'\mathbf{x}_{0C}}{\sigma}\right)^2 = \mathrm{E}\psi'\left(\frac{u_0}{\sigma}\right)\mathrm{E}\left(\frac{\boldsymbol{\Delta}'\mathbf{x}_{0C}}{\sigma}\right)^2$$

$$\approx B\frac{A}{nB^2}\mathrm{E}\mathbf{x}'_{0C}\mathbf{V}^{-1}\mathbf{x}_{0C}. \tag{5.119}$$

Let $\mathbf{U}$ be any matrix such that $\mathbf{V} = \mathbf{U}\mathbf{U}'$, and hence such that

$$\mathrm{E}(\mathbf{U}^{-1}\mathbf{x}_{0C})(\mathbf{U}^{-1}\mathbf{x}_{0C})' = \mathbf{I}_q,$$

where $\mathbf{I}_q$ is the $q \times q$ identity matrix. Then

$$\mathrm{E}\mathbf{x}'_{0C}\mathbf{V}^{-1}\mathbf{x}_{0C} = \mathrm{E}\|\mathbf{U}^{-1}\mathbf{x}_{0C}\|^2 = \mathrm{trace}(\mathbf{I}_q) = q, \tag{5.120}$$

and hence (5.117), (5.118) and (5.119) yield

$$\mathrm{RFPE}(C) \approx \mathrm{E}\rho\left(\frac{u_0}{\sigma}\right) + \frac{q}{2n}\frac{A}{B} \tag{5.121}$$

To estimate RFPE(C) using (5.121) we need to estimate $\mathrm{E}\rho(u_0/\sigma)$. A second-order Taylor expansion yields

$$\frac{1}{n}\sum_{i=1}^{n}\rho\left(\frac{r_i}{\widehat{\sigma}}\right) = \frac{1}{n}\sum_{i=1}^{n}\rho\left(\frac{u_i - \mathbf{x}_{iC}'\boldsymbol{\Delta}}{\widehat{\sigma}}\right) \tag{5.122}$$

$$\approx \frac{1}{n}\sum_{i=1}^{n}\rho\left(\frac{u_i}{\widehat{\sigma}}\right) - \frac{1}{n\widehat{\sigma}}\sum_{i=1}^{n}\psi\left(\frac{u_i}{\widehat{\sigma}}\right)\mathbf{x}_{iC}'\boldsymbol{\Delta}$$

$$+ \frac{1}{2n\widehat{\sigma}^2}\sum_{i=1}^{n}\psi'\left(\frac{u_i}{\widehat{\sigma}}\right)(\mathbf{x}_{iC}'\boldsymbol{\Delta})^2.$$

The estimator $\widehat{\boldsymbol{\beta}}_C$ satisfies the equation

$$\sum_{i}^{n}\psi\left(\frac{r_{iC}}{\widehat{\sigma}}\right)\mathbf{x}_{iC} = \mathbf{0}, \tag{5.123}$$

and a first-order Taylor expansion of (5.123) yields

$$\mathbf{0} = \sum_{i=1}^{n}\psi\left(\frac{r_{iC}}{\widehat{\sigma}}\right)\mathbf{x}_{iC} = \sum_{i=1}^{n}\psi\left(\frac{u_i - \mathbf{x}_{iC}'\boldsymbol{\Delta}}{\widehat{\sigma}}\right)\mathbf{x}_{iC}$$

$$\approx \sum_{i=1}^{n}\psi\left(\frac{u_i}{\widehat{\sigma}}\right)\mathbf{x}_{iC} - \frac{1}{\widehat{\sigma}}\sum_{i=1}^{n}\psi'\left(\frac{u_i}{\widehat{\sigma}}\right)(\mathbf{x}_{iC}'\boldsymbol{\Delta})x_{iC},$$

and hence

$$\sum_{i}^{n}\psi\left(\frac{u_i}{\widehat{\sigma}}\right)\mathbf{x}_{iC} \approx \frac{1}{\widehat{\sigma}}\sum_{i=1}^{n}\psi'\left(\frac{u_i}{\widehat{\sigma}}\right)(\mathbf{x}_{iC}'\boldsymbol{\Delta})x_{iC}, \tag{5.124}$$

Replacing (5.124) in (5.122) yields

$$\frac{1}{n}\sum_{i=1}^{n}\rho\left(\frac{r_i}{\widehat{\sigma}}\right) \approx \frac{1}{n}\sum_{i=1}^{n}\rho\left(\frac{u_i}{\widehat{\sigma}}\right) - \frac{1}{2n\widehat{\sigma}^2}\sum_{i=1}^{n}\psi'\left(\frac{u_i}{\widehat{\sigma}}\right)(\mathbf{x}_{iC}'\boldsymbol{\Delta})^2.$$

Since

$$\frac{1}{n}\sum_{i=1}^{n}\psi'\left(\frac{u_i}{\widehat{\sigma}}\right)\mathbf{x}_{iC}\mathbf{x}_{iC}' \to_p \mathrm{E}\psi'\left(\frac{u_0}{\sigma}\right)\mathrm{E}(\mathbf{x}_{0C}\mathbf{x}_{0C}') = B\mathbf{V},$$

we obtain using (5.120)

$$\frac{1}{n}\sum_{i=1}^{n}\rho\left(\frac{r_i}{\widehat{\sigma}}\right) \approx \frac{1}{n}\sum_{i=1}^{n}\rho\left(\frac{u_i}{\widehat{\sigma}}\right) - \frac{B}{2\widehat{\sigma}^2}\boldsymbol{\Delta}'\mathbf{V}\boldsymbol{\Delta}$$

$$= \frac{1}{n}\sum_{i=1}^{n}\rho\left(\frac{u_i}{\widehat{\sigma}}\right) - \frac{A}{2Bn}q,$$

Hence by the law of large numbers and the consistency of $\widehat{\sigma}$

$$\mathrm{E}\rho\left(\frac{u_0}{\sigma}\right) \approx \frac{1}{n}\sum_{i=1}^{n}\rho\left(\frac{u_i}{\widehat{\sigma}}\right) \approx \frac{1}{n}\sum_{i=1}^{n}\rho\left(\frac{r_i}{\widehat{\sigma}}\right) + \frac{Aq}{2Bn} \tag{5.125}$$

and finally, inserting (5.125) in (5.121) yields

$$\mathrm{RFPE}(C) \approx \frac{1}{n}\sum_{i=1}^{n}\rho\left(\frac{r_i}{\widehat{\sigma}}\right) + \frac{Aq}{Bn} \approx \frac{1}{n}\sum_{i=1}^{n}\rho\left(\frac{r_i}{\widehat{\sigma}}\right) + \frac{\widehat{A}q}{\widehat{B}n} = \mathrm{RFPE}^*(C).$$

5.14 Recommendations and software

For linear regression with random predictors or fixed predictors with high leverage points, we recommend the MM-estimator with optimal ρ using the Peña–Yohai estimator as initial value (Section 5.9.4), implemented in **lmrobdetMM** (`RobStatTM`); and the DCML estimator (Section 5.9.3.2) with optimal ρ, implemented in **lmrobdetDCML**(`RobStatTM`), which uses the output from **lmrobdetMM** to boost its efficiency.

To compute regularized estimators for the linear model we recommend the function **pense** (`pense`), which computes an MM version of the elastic net estimator. As particular cases, it can compute the MM ridge and MM lasso.

For model selection we recommend **step.lmrobdetMM** (`RobStatTM`), which uses the RFPE criterion (Section 5.6.2), and is based on the output from **lmrobdetMM.**

Other possible options to compute robust estimators for linear regression are:

- **initPY** (`pyinit`) computes the Peña–Yohai estimator defined in Section 5.7.4, which is used as initial estimator of **lmrobdetMM**.
- **lmrob** (`robustbase`) and **lmRob** (`robust`) are general programs that compute MM- and other robust estimators. Both use as initial value an S-estimator (Section 5.4.1). The first employs subsampling (Section 5.7.2) while the second allows the user to choose between subsampling and Peña–Yohai.

5.15 Problems

5.1. Show that S-estimators are regression, affine and scale equivariant.

5.2. The **stack loss** dataset (Brownlee, 1965, p. 454) given in Table 5.12 are observations from 21 days' operation of a plant for the oxidation of ammonia, a stage in the production of nitric acid. The predictors X_1, X_2, X_3 are respectively the air flow, the cooling water inlet temperature, and the acid concentration, and the response Y is the stack loss. Fit a linear model to these data using the LS estimator, the MM-estimators with efficiencies 0.95 and 0.80, and the DCML estimator, and compare the results. Fit the residuals against the day. Is there a pattern?

Table 5.12 Stack loss data

Day	X_1	X_2	X_3	Y
1	80	27	58.9	4.2
2	80	27	58.8	3.7
3	75	25	59.0	3.7
4	62	24	58.7	2.8
5	62	22	58.7	1.8
6	62	23	58.7	1.8
7	62	24	59.3	1.9
8	62	24	59.3	2.0
9	58	23	58.7	1.5
10	58	18	58.0	1.4
11	58	18	58.9	1.4
12	58	17	58.8	1.3
13	58	18	58.2	1.1
14	58	19	59.3	1.2
15	50	18	58.9	0.8
16	50	18	58.6	0.7
17	50	19	57.2	0.8
18	50	19	57.9	0.8
19	50	20	58.0	0.9
20	56	20	58.2	1.5
21	70	20	59.1	1.5

5.3. The dataset **alcohol** (Romanelli *et al.*, 2001) gives, for 44 aliphatic alcohols, the logarithm of their solubility together with six physicochemical characteristics. The interest is in predicting the solubility. Compare the results of using the LS, MM and DCML estimators to fit the log-solubility as a function of the characteristics.

5.4. The dataset **waste** (Chatterjee and Hadi, 1988) contains, for 40 regions, the solid waste and five variables on land use. Fit a linear model to these data using the LS, L_1, MM and DCML estimators. Draw the respective Q–Q plots of residuals, and the plots of residuals against fitted values, and compare the estimators and the plots.

5.5. Show that the "median of slopes" estimator (5.93) is a GM-estimator (5.85).

5.6. For the "median of slopes" estimator and the model $y_i = \beta x_i + u_i$, calculate the following, assuming that $\mathrm{P}(x = 0) = 0$:

(a) the asymptotic BP
(b) the influence function and the gross-error sensitivity
(c) the maximum asymptotic bias (hint: use (3.68)).

5.7. Show that when using the shortcut (5.44), the number of times that the M-scale is computed has expectation $\sum_{i=1}^{N}(1/i) \leq \ln N$, where N is the number of subsamples.

5.8. The minimum α-quantile regression estimator is defined for $\alpha \in (0, 1)$ as the value of β minimizing the $\alpha-$quantile of $|y - \mathbf{x}'\beta|$. Show that this estimator is an S-estimator for the scale given by $\rho(u) = \mathrm{I}(|u| > 1)$ and $\delta = (1 - \alpha)$. Find its asymptotic breakdown point.

5.9. For each $\boldsymbol{\beta}$ let $c(\boldsymbol{\beta})$ the minimum c such that

$$\#\{i : \boldsymbol{\beta}'\mathbf{x}_i - c \leq y_i \leq \boldsymbol{\beta}'\mathbf{x}_i + c\} \geq n/2.$$

Show that the LMS estimator minimizes $c(\boldsymbol{\beta})$.

5.10. Let $\{(\mathbf{x}_1, y_1), \ldots, (\mathbf{x}_n, y_n)\}$ be a regression dataset, and $\widehat{\boldsymbol{\beta}}$ an S-estimator with finite BP equal to ε^*. Let $D \subset (1, .., n)$ with $\#(D) < n\varepsilon^*$. Show that there exists K such that:

(a) $\widehat{\boldsymbol{\beta}}$ as a function of the y_i is constant if the y_i with $i \notin D$ remain fixed and those with $i \in D$ are changed in any way such that $|y_i| \geq K$.

(b) there exists $\widehat{\sigma}$ depending only on D such that $\widehat{\boldsymbol{\beta}}$ verifies

$$\sum_{i \notin D} \rho\left(\frac{r_i(\widehat{\boldsymbol{\beta}})}{\widehat{\sigma}}\right) = \min.$$

Then:

(c) Discuss why property (a) does not mean that the that the value of the estimator is the same as if we omit the points $(\mathbf{x}_i, y_i)$ with $i \in D$.

(d) Show that properties (a)–(c) also hold for MM-estimators.

5.11. Show that the adaptive MM-lasso can be computed by transforming the regressors as $x_{ij}^* = |\beta_j|^0 x_{ij}$.

6

Multivariate Analysis

6.1 Introduction

Multivariate analysis deals with situations in which several variables are measured on each experimental unit. In most cases of interest it is known or assumed that some form of relationship exists among the variables, and hence that considering each of them separately would entail a loss of information. Some possible goals of the analysis are:

- reduction of dimensionality (principal components, factor analysis, canonical correlation);
- identification (discriminant analysis);
- explanatory models (multivariate linear model).

The reader is referred to Seber (1984) and Johnson and Wichern (1998) for further details.

A p-variate observation is now a vector $\mathbf{x} = (x_1, \ldots, x_p)' \in R^p$, and a distribution F now means a distribution on R^p. In the classical approach, location of a p-variate random variable $\mathbf{x}$ is described by the expectation $\boldsymbol{\mu} = \mathrm{E}\mathbf{x} = (\mathrm{E}x_1, \ldots, \mathrm{E}x_n)'$, and scatter is described by the covariance matrix

$$\mathbf{Var}(\mathbf{x}) = \mathrm{E}((\mathbf{x} - \boldsymbol{\mu})(\mathbf{x} - \boldsymbol{\mu})').$$

It is well known that $\mathbf{Var}(\mathbf{x})$ is symmetric and positive semidefinite, and that for each constant vector $\mathbf{a}$ and matrix $\mathbf{A}$

$$\mathrm{E}(\mathbf{Ax} + \mathbf{a}) = \mathbf{A}\ \mathrm{E}\mathbf{x} + \mathbf{a}, \quad \mathbf{Var}(\mathbf{Ax} + \mathbf{a}) = \mathbf{A}\mathbf{Var}(\mathbf{x})\mathbf{A}'. \tag{6.1}$$

Robust Statistics: Theory and Methods (with R), Second Edition.
Ricardo A. Maronna, R. Douglas Martin, Victor J. Yohai and Matías Salibián-Barrera.

Companion website: www.wiley.com/go/maronna/robust

Classical multivariate methods of estimation are based on the assumption of an i.i.d. sample of observations $X = \{\mathbf{x}_1, \ldots, \mathbf{x}_n\}$ with each $\mathbf{x}_i$ having a p-variate normal $\mathrm{N}_p(\boldsymbol{\mu}, \boldsymbol{\Sigma})$ distribution with density

$$f(\mathbf{x}) = \frac{1}{(2\pi)^{p/2}\sqrt{|\boldsymbol{\Sigma}|}} \exp\left(-\frac{1}{2}(\mathbf{x}-\boldsymbol{\mu})'\boldsymbol{\Sigma}^{-1}(\mathbf{x}-\boldsymbol{\mu})\right), \tag{6.2}$$

where $\boldsymbol{\Sigma} = \mathbf{Var}(\mathbf{x})$ and $|\boldsymbol{\Sigma}|$ stands for the determinant of $\boldsymbol{\Sigma}$. The contours of constant density are the elliptical surfaces

$$\{\mathbf{z} : (\mathbf{z}-\boldsymbol{\mu})'\boldsymbol{\Sigma}^{-1}(\mathbf{z}-\boldsymbol{\mu}) = c\}.$$

Assuming $\mathbf{x}$ is multivariate normal implies that for any constant vector $\mathbf{a}$, all linear combinations $\mathbf{a}'\mathbf{x}$ are normally distributed. It also implies that since the conditional expectation of one coordinate with respect to any group of coordinates is a linear function of the latter, the type of dependence among variables is linear. Thus methods based on multivariate normality will yield information only about linear relationships among coordinates. As in the univariate case, the main reason for assuming normality is simplicity.

It is known that under the normal distribution (6.2), the MLEs of $\boldsymbol{\mu}$ and $\boldsymbol{\Sigma}$ for a sample X are respectively the sample mean and sample covariance matrix

$$\overline{\mathbf{x}} = \mathrm{ave}(X) = \frac{1}{n}\sum_{i=1}^{n}\mathbf{x}_i, \quad \mathbf{Var}(X) = \mathrm{ave}\{(X-\overline{\mathbf{x}})(X-\overline{\mathbf{x}})'\}.$$

The sample mean and sample covariance matrix share the behavior of the distribution mean and covariance matrix under affine transformations, namely (6.1), for each vector $\mathbf{a}$ and matrix $\mathbf{A}$

$$\mathrm{ave}(\mathbf{A}X + \mathbf{a}) = \mathbf{A}\mathrm{ave}(X) + \mathbf{a}, \quad \mathbf{Var}(\mathbf{A}X + \mathbf{a}) = \mathbf{A}\mathbf{Var}(X)\mathbf{A}',$$

where $\mathbf{A}X + \mathbf{a}$ is the dataset $\{\mathbf{A}\mathbf{x}_i + \mathbf{a},\ i = 1, \ldots, n\}$. This property is known as the *affine equivariance* of the sample mean and covariances.

Just as in the univariate case, a few atypical observations may completely alter the sample means and/or covariances. Worse still, a multivariate outlier need not be an outlier in any of the coordinates considered separately.

Example 6.1 *The dataset in Seber (1984; Table 9.12) contains biochemical measurements on 12 men with similar weights. The data are plotted in Figure 6.1. The tables and figures for this example are obtained with script* ***biochem.R****.*

We see in Figure 6.1 that observation 3, which has the lowest phosphate value, stands out clearly from the rest. However, Figure 6.2, which shows the normal QQ plot of phosphate, does not reveal any atypical value, and the same occurs in the QQ plot of chloride (not shown). Thus the atypical character of observation 3 is visible only when considering both variables simultaneously.

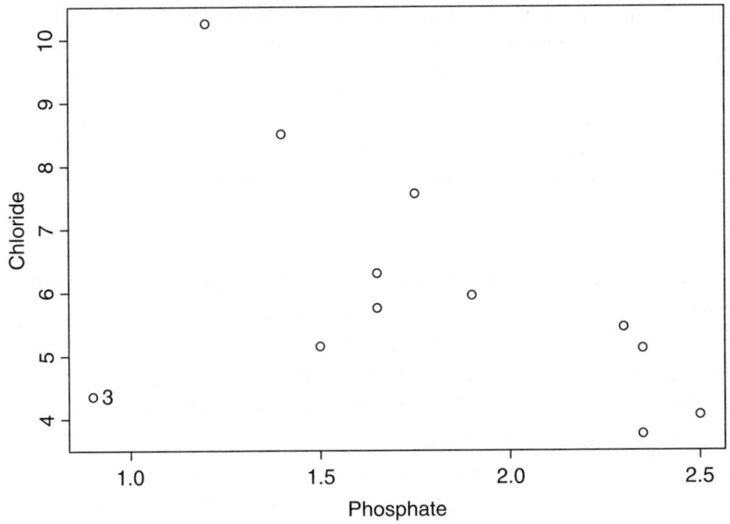

Figure 6.1 Biochemical data

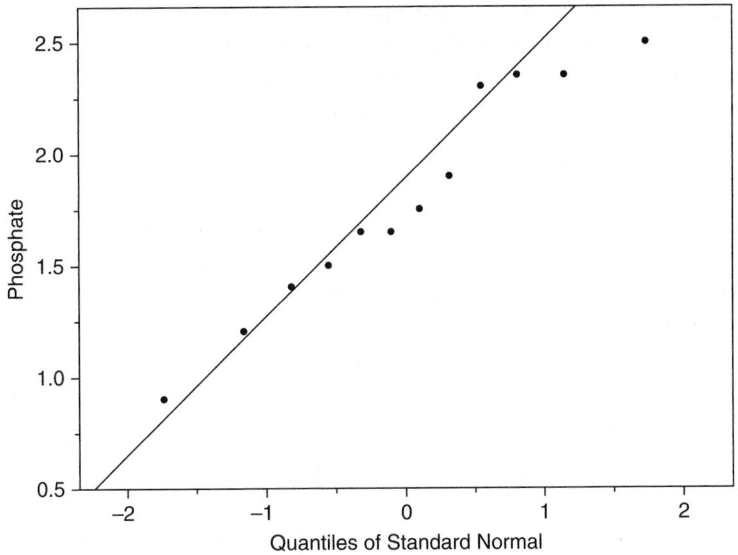

Figure 6.2 Normal QQ plot for phosphate

Table 6.1 below shows that omitting this observation has no major effect on means or variances, but the correlation almost doubles in magnitude; that is, the influence of the outlier has been to decrease the correlation by a factor of two relative to the situation without the outlier.

Table 6.1 The effect of omitting a bivariate outlier

	Means		Vars.		Correl.
Complete data	1.79	6.01	0.26	3.66	−0.49
Without obs. 3	1.87	6.16	0.20	3.73	−0.80

Here we have an example of an observation that is not a one-dimensional outlier in either coordinate but strongly affects the results of the analysis. This example shows the need for robust substitutes for the mean vector and covariance matrix, and this will be the main theme of this chapter. The substitutes are generally referred to as multivariate *location vectors* and *scatter matrices*. The latter are also called *robust covariance matrices* in the literature.

Some methods in multivariate analysis make no use of means or covariances; example include Breiman *et al.*'s (1984) nonparametric classification and regression trees (CART) methods. To some extent, such (nonequivariant) methods have a certain built-in robustness. But if we want to retain the simplicity of the normal distribution as the "nominal" model, with corresponding linear relationships, elliptical distributional shapes and affine equivariance for the bulk of the data, then the appropriate approach is to consider slight or moderate departures from normality.

Let $(\widehat{\boldsymbol{\mu}}(X), \widehat{\boldsymbol{\Sigma}}(X))$ be location and scatter estimators corresponding to a sample $X = \{\mathbf{x}_1, ..., \mathbf{x}_n\}$. Then the estimators are affine equivariant if

$$\widehat{\boldsymbol{\mu}}(\mathbf{A}X + \mathbf{b}) = \mathbf{A}\widehat{\boldsymbol{\mu}}(X) + \mathbf{b}, \quad \widehat{\boldsymbol{\Sigma}}(\mathbf{A}X + \mathbf{a}) = \mathbf{A}\widehat{\boldsymbol{\Sigma}}\mathbf{A}'. \tag{6.3}$$

Affine equivariance is a desirable property of an estimator. The reasons are given in Section 6.17.1. This is, however, not a mandatory property, and may in some cases be sacrificed for other properties such as computational speed; an instance of this trade-off is considered in Section 6.9.1.

As in the univariate case, one may consider outlier detection methods. The squared *Mahalanobis distance* between the vectors $\mathbf{x}$ and $\boldsymbol{\mu}$ with respect to the matrix $\boldsymbol{\Sigma}$ is defined as

$$d(\mathbf{x}, \boldsymbol{\mu}, \boldsymbol{\Sigma}) = (\mathbf{x} - \boldsymbol{\mu})'\boldsymbol{\Sigma}^{-1}(\mathbf{x} - \boldsymbol{\mu}). \tag{6.4}$$

For simplicity, d will be sometimes referred to as "distance", although it should be kept in mind that it is actually a *squared* distance. Then the multivariate analogue of t_i^2, where $t_i = (x_i - \overline{x})/s$ is the univariate outlyingness measure in (1.3), is $D_i = d(\mathbf{x}_i, \overline{\mathbf{x}}, \mathbf{C})$, with $\mathbf{C} = \mathbf{Var}(X)$. When $p = 1$ we have $D_i = t_i^2 n/(n-1)$. It is known (Seber, 1984) that if $\mathbf{x} \sim \mathrm{N}_p(\boldsymbol{\mu}, \boldsymbol{\Sigma})$, then $d(\mathbf{x}, \boldsymbol{\mu}, \boldsymbol{\Sigma}) \sim \chi_p^2$. Thus, assuming the estimators $\overline{\mathbf{x}}$ and $\mathbf{C}$ are close to their true values, we may examine the QQ plot of D_i against the quantiles of a χ_p^2 distribution and delete observations for which D_i is "too high". This approach may be effective when there is a single outlier but, as in the case of location, it can be ineffective when n is small (recall Section 1.3) and, as in regression, several outliers may mask one another.

Example 6.2 *The following dataset is a part of one given by Hettich and Bay (1999). It contains, for each of 59 wines grown in the same region in Italy, the quantities of 13 constituents. The original purpose of the analysis (de Vel* et al. *1993) was to classify wines from different cultivars by means of these measurements. In this example we treat cultivar one. The tables and figures for this example are obtained with script* ***wine.R****.*

The upper row of Figure 6.3 shows the plots of the classical squared distances as a function of observation number, and their QQ plot with respect to the χ_p^2 distribution. No clear outliers stand out.

The lower row shows the results of using a robust estimator (called the "MM"), to be defined in Section 6.4.4. At least seven points stand out clearly. The failure of

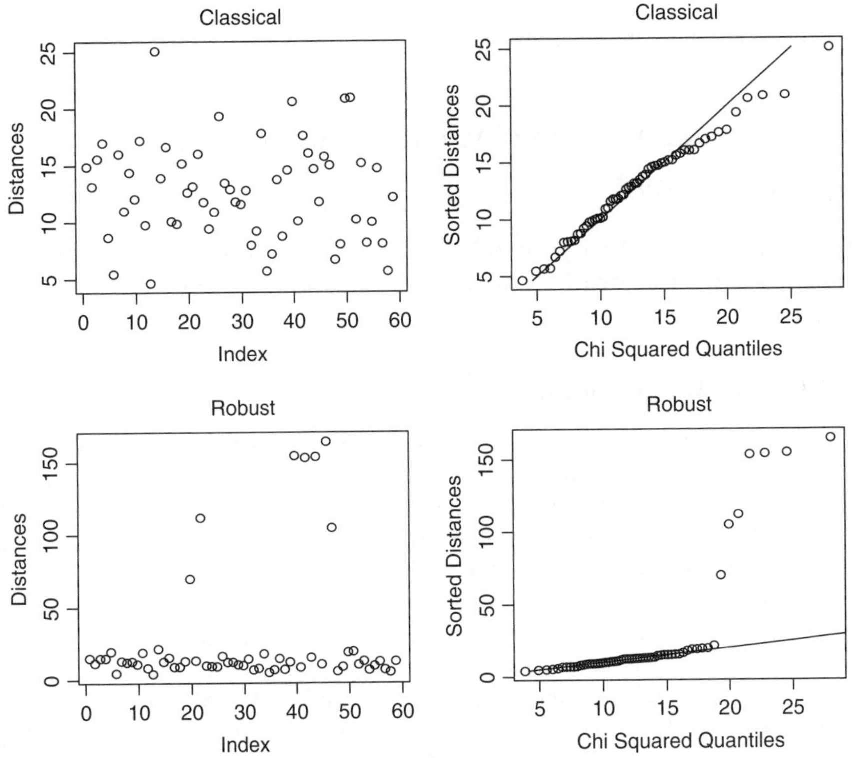

Figure 6.3 Wine example: Mahalanobis distances versus index number for classical and robust estimators (left), and QQ plots of distances (right)

the classical analysis in the upper row of Figure 6.3 shows that several outliers may "mask" one another. These seven outliers have a strong influence on the results of the analysis.

Simple robust estimators of multivariate location can be obtained by applying a robust univariate location estimator to each coordinate, but this lacks affine equivariance. For scatter, there are simple robust estimators of the covariance between two variables (*pairwise covariances*) that could be used to construct a *robust covariance matrix* (see Devlin *et al.*, 1981; Huber and Ronchetti, 2009).

Apart from not being equivariant, the resulting matrix may not be positive semidefinite. See, however, Section 6.9 for an approach that ensures positive definiteness and "approximate" equivariance. Nonequivariant procedures may also lack robustness when the data are very collinear (Section 6.7). In subsequent sections we shall discuss a number of equivariant location and scatter estimators.

Note that if the matrix $\widehat{\boldsymbol{\Sigma}}$ with elements $\sigma_{jk}, j, k = 1, \ldots, p$, is a "robust covariance matrix", then the matrix $\mathbf{R}$ with elements

$$r_{jk} = \frac{\sigma_{jk}}{\sqrt{\sigma_{jj}\sigma_{kk}}} \tag{6.5}$$

is a robust analog of the correlation matrix.

6.2 Breakdown and efficiency of multivariate estimators

The concepts of breakdown point (BP) and efficiency will be necessary to understand the advantages and drawbacks of the different families of estimators discussed in this chapter.

6.2.1 Breakdown point

To define the BP of $(\widehat{\boldsymbol{\mu}}, \widehat{\boldsymbol{\Sigma}})$ based on the ideas in Section 3.2, we must establish the meaning of "bounded, and also bounded away from the boundary of the parameter space". For the location vector, the parameter space is a finite-dimensional Euclidean space, and so the statement means simply that $\widehat{\boldsymbol{\mu}}$ remains in a bounded set. However, the scatter matrix has a more complex parameter space consisting of the set of symmetric nonnegative definite matrices. Each such matrix is characterized by the matrix of its eigenvalues. Thus "$\widehat{\boldsymbol{\Sigma}}$ bounded, and also bounded away from the boundary" is equivalent to the eigenvalues being bounded away from zero and infinity.

From a more intuitive point of view, recall that if $\boldsymbol{\Sigma} = \mathbf{Var}(\mathbf{x})$ and $\mathbf{a}$ is a constant vector, then $\mathrm{Var}(\mathbf{a}'\mathbf{x}) = \mathbf{a}'\boldsymbol{\Sigma}\mathbf{a}$. Hence if $\boldsymbol{\Sigma}$ is any robust scatter matrix, then $\sqrt{\mathbf{a}'\boldsymbol{\Sigma}\mathbf{a}}$ can be considered as a robust measure of scatter of the linear combination $\mathbf{a}'\mathbf{x}$. Let $\lambda_1(\boldsymbol{\Sigma}) \geq \ldots \geq \lambda_p(\boldsymbol{\Sigma})$ be the eigenvalues of $\boldsymbol{\Sigma}$ in descending order, and $\mathbf{e}_1, \ldots, \mathbf{e}_p$ the corresponding eigenvectors. It is a fact of linear algebra that for any symmetric matrix

$\mathbf{\Sigma}$, the minimum (resp. maximum) of $\mathbf{a}'\mathbf{\Sigma}a$ over $\|\mathbf{a}\| = 1$ is equal to λ_p (λ_1) and this minimum is attained for $\mathbf{a} = \mathbf{e}_p$ $(\mathbf{e}_1)$. If we are interested in linear relationships among the variables, then it is dangerous not only that the largest eigenvalue becomes too large ("explosion") but also that the smallest one becomes too small ("implosion"). The first case is caused by outliers (observations far away from the bulk of the data), the second by "inliers" (observations concentrated at some point or in general on a region of lower dimensionality).

For $0 \leq m \leq n$, call $\mathcal{Z}_m$ the set of "samples" $Z = \{\mathbf{z}_1, \ldots, \mathbf{z}_n\}$, such that $\#\{\mathbf{z}_i = \mathbf{x}_i\} = n\text{-}m$, and call $\widehat{\boldsymbol{\mu}}(Z)$ and $\widehat{\mathbf{\Sigma}}(Z)$ the location vector and scatter matrix estimators based on the sample Z. The finite breakdown point (FBP) of $(\widehat{\boldsymbol{\mu}}, \widehat{\mathbf{\Sigma}})$ is defined as $\varepsilon^* = m^*/n$, where m^* is the largest m such that there are finite positive a, b, c such that

$$\|\widehat{\boldsymbol{\mu}}(Z)\| \leq a \text{ and } b \leq \lambda_p(\widehat{\mathbf{\Sigma}}(Z)) \leq \lambda_1(\widehat{\mathbf{\Sigma}}(Z)) \leq c$$

for all $Z \in \mathcal{Z}_m$.

For theoretical purposes it may be simpler to work with the asymptotic BP. An ε-contamination neighborhood $\mathcal{F}(F, \varepsilon)$ of a multivariate distribution F is defined as in (3.3). Applying Definition 3.2 and (3.20) to the present context we have that the asymptotic BP of $(\widehat{\boldsymbol{\mu}}, \widehat{\mathbf{\Sigma}})$ is the largest $\varepsilon^* \in (0, 1)$ for which there exist finite positive a, b, c such that the following holds for all G:

$$\|\widehat{\boldsymbol{\mu}}_\infty((1-\varepsilon)F + \varepsilon G)\| \leq a,$$

$$b \leq \lambda_p(\widehat{\mathbf{\Sigma}}((1-\varepsilon)F + \varepsilon G)) \leq \lambda_1(\widehat{\mathbf{\Sigma}}((1-\varepsilon)F + \varepsilon G)) \leq c.$$

In some cases we may restrict G to range over point-mass distributions, and in that case we use the terms "point-mass contamination neighborhoods" and "point-mass breakdown point".

6.2.2 The multivariate exact fit property

A result analogous to that of Section 5.12.1 holds for multivariate location and scatter estimation. Let the FBP of the affine equivariant estimator $(\widehat{\boldsymbol{\mu}}, \widehat{\mathbf{\Sigma}})$ be $\varepsilon^* = m^*/n$. Let the dataset contain q points on a hyperplane $H = \{\mathbf{x} : \boldsymbol{\beta}'\mathbf{x} = \gamma\}$ for some $\boldsymbol{\beta} \in R^p$ and $\gamma \in R$. If $q \geq n - m^*$ then $\widehat{\boldsymbol{\mu}} \in H$, and $\widehat{\mathbf{\Sigma}}\boldsymbol{\beta} = \mathbf{0}$. The proof is given in Section 6.17.8.

6.2.3 Efficiency

The asymptotic efficiency of $(\widehat{\boldsymbol{\mu}}, \widehat{\mathbf{\Sigma}})$ is defined as in (3.46). Call $(\widehat{\boldsymbol{\mu}}_n, \widehat{\mathbf{\Sigma}}_n)$ the estimators for a sample of size n, and let $(\widehat{\boldsymbol{\mu}}_\infty, \widehat{\mathbf{\Sigma}}_\infty)$ be their asymptotic values. All estimators considered in this chapter are consistent for the normal distribution in the following sense: if $\mathbf{x}_i \sim \mathrm{N}_p(\boldsymbol{\mu}, \mathbf{\Sigma})$ then $\widehat{\boldsymbol{\mu}}_\infty = \boldsymbol{\mu}$ and $\widehat{\mathbf{\Sigma}}_\infty = c\mathbf{\Sigma}$ where c is a constant (if $c = 1$ we have the usual definition of consistency). This result will be seen to hold for the larger family of *elliptical* distributions, to be defined later. Most estimators defined in this chapter are also asymptotically normal:

$$\sqrt{n}(\hat{\boldsymbol{\mu}}_n - \hat{\boldsymbol{\mu}}_\infty)\rightarrow_d \mathrm{N}_p(\mathbf{0}, \mathbf{V}_\mu), \qquad \sqrt{n}\mathrm{vec}(\hat{\boldsymbol{\Sigma}}_n - \hat{\boldsymbol{\Sigma}}_\infty)\rightarrow_d \mathrm{N}_q(\mathbf{0}, \mathbf{V}_\Sigma),$$

where $q = p(p+1)/2$ and, for a symmetric matrix $\boldsymbol{\Sigma}$, $\mathrm{vec}(\boldsymbol{\Sigma})$ is the vector containing the q elements of the upper triangle of $\boldsymbol{\Sigma}$. The matrices $\mathbf{V}_\mu$ and $\mathbf{V}_\Sigma$ are the asymptotic covariance matrices of $\hat{\boldsymbol{\mu}}$ and $\hat{\boldsymbol{\Sigma}}$. In general, the estimator can be defined in such a way that $c = 1$ for a given model, say the multivariate normal.

We consider the efficiency of $\hat{\boldsymbol{\mu}}$ when the data have a $\mathrm{N}_p(\boldsymbol{\mu}, \boldsymbol{\Sigma})$ distribution. In Section 6.17.2 it is shown that an affine equivariant location estimator $\hat{\boldsymbol{\mu}}$ has an asymptotic covariance matrix of the form

$$\mathbf{V}_\mu = \upsilon\boldsymbol{\Sigma}, \tag{6.6}$$

where υ is a constant depending on the estimator. In the case of the normal distribution MLE $\overline{\mathbf{x}}$ we have $\upsilon = 1$ and the matrix $\mathbf{V}_0$ in (3.46) is simply $\boldsymbol{\Sigma}$, which results in $\mathbf{V}_\mu^{-1}\mathbf{V}_0 = \upsilon^{-1}\mathbf{I}$ and $\mathrm{eff}(\hat{\boldsymbol{\mu}}) = 1/\upsilon$. Thus the normal distribution efficiency of an affine equivariant location estimator is independent of $\boldsymbol{\mu}$ and $\boldsymbol{\Sigma}$. With three exceptions, considered in Sections 6.9.1, 6.12 and 6.13,the location estimators treated in this chapter are affine equivariant.

The efficiency of $\hat{\boldsymbol{\Sigma}}$ is much more complicated and will not be discussed here. It has been dealt with by Tyler (1983) in the case of the class of M-estimators, which are defined in the next section.

6.3 M-estimators

Multivariate M-estimators will now be defined, as in Section 2.3, by generalizing MLEs. Recall that in the univariate case it was possible to define separate robust equivariant estimators of location and of scatter. This is more complicated to do in the multivariate case and, if we want equivariant estimators, it is better to estimate location and scatter simultaneously. We shall develop the multivariate analog of simultaneous M-estimators (2.71)–(2.72). Recall that a multivariate normal density has the form

$$f(\mathbf{x}, \boldsymbol{\mu}, \boldsymbol{\Sigma}) = \frac{1}{\sqrt{|\boldsymbol{\Sigma}|}} h(d(\mathbf{x}, \boldsymbol{\mu}, \boldsymbol{\Sigma})) \tag{6.7}$$

where $h(s) = c\exp(-s/2)$, with $c = (2\pi)^{-p/2}$ and $d(\mathbf{x}, \boldsymbol{\mu}, \boldsymbol{\Sigma}) = (\mathbf{x} - \boldsymbol{\mu})'\boldsymbol{\Sigma}^{-1}(\mathbf{x} - \boldsymbol{\mu})$. We note that the level sets of f are ellipsoidal surfaces. In fact, for any choice of positive h such that f integrates to one, the level sets of f are ellipsoids, and so any density of this form is called *elliptically symmetric* (henceforth "elliptical" for short). In the special case where $\boldsymbol{\mu} = 0$ and $\boldsymbol{\Sigma} = c\mathbf{I}$, a density of the form (6.7) is called *spherically symmetric* or *radial* (henceforth "spherical" for short). It is easy to verify that the distribution $\mathcal{D}(\mathbf{x})$ is elliptical if and only if for some constant vector

a and matrix **A**, $\mathcal{D}(\mathbf{A}(\mathbf{x}-\mathbf{a}))$ is spherical. An important example of a nonnormal elliptical distribution is the p-variate *Student distribution* with ν degrees of freedom $(0<\nu<\infty)$, which will be denoted as $T_{p,\nu}$, and is obtained by the choice

$$h(s)=\frac{c}{(s+\nu)^{(p+\nu)/2}} \tag{6.8}$$

where c is a constant. The case $\nu=1$ is called the multivariate Cauchy density, and the limiting case $\nu\to\infty$ yields the normal distribution. If the mean (resp. the covariance matrix) of an elliptical distribution exists, then it is equal to $\boldsymbol{\mu}$ (resp. a multiple of $\boldsymbol{\Sigma}$) (Problem 6.1). More details on elliptical distributions are given in Section 6.17.9.

Let $\mathbf{x}_1,\ldots,\mathbf{x}_n$ be an i.i.d. sample from an f of the form (6.7) in which h is assumed everywhere positive. To calculate the MLE of $\boldsymbol{\mu}$ and $\boldsymbol{\Sigma}$, note that the likelihood function is

$$L(\boldsymbol{\mu},\boldsymbol{\Sigma})=\frac{1}{|\boldsymbol{\Sigma}|^{n/2}}\prod_{i=1}^{n}h(d(\mathbf{x}_i,\boldsymbol{\mu},\boldsymbol{\Sigma})),$$

and maximizing $L(\boldsymbol{\mu},\boldsymbol{\Sigma})$ is equivalent to

$$-2\ln\ L(\boldsymbol{\mu},\boldsymbol{\Sigma})=n\ln|\widehat{\boldsymbol{\Sigma}}|+\sum_{i=1}^{n}\rho(d_i)=\min, \tag{6.9}$$

where

$$\rho(s)=-2\ln\ h(s)\text{ and }d_i=d(\mathbf{x}_i,\widehat{\boldsymbol{\mu}},\widehat{\boldsymbol{\Sigma}}). \tag{6.10}$$

Differentiating with respect to $\boldsymbol{\mu}$ and $\boldsymbol{\Sigma}$ yields the system of estimating equations (see Section 6.17.3 for details):

$$\sum_{i=1}^{n}W(d_i)(\mathbf{x}_i-\widehat{\boldsymbol{\mu}})=\mathbf{0} \tag{6.11}$$

$$\frac{1}{n}\sum_{i=1}^{n}W(d_i)(\mathbf{x}_i-\widehat{\boldsymbol{\mu}})(\mathbf{x}_i-\widehat{\boldsymbol{\mu}})'=\widehat{\boldsymbol{\Sigma}}, \tag{6.12}$$

with $W=\rho'$. For the normal distribution we have $W\equiv 1$, which yields the sample mean and sample covariance matrix for $\widehat{\boldsymbol{\mu}}$ and $\widehat{\boldsymbol{\Sigma}}$. For the multivariate Student distribution (6.8) we have

$$W(d)=\frac{p+\nu}{d+\nu}. \tag{6.13}$$

In general, we define M-estimators as solutions of

$$\sum_{i=1}^{n}W_1(d_i)(\mathbf{x}_i-\widehat{\boldsymbol{\mu}})=\mathbf{0} \tag{6.14}$$

$$\frac{1}{n}\sum_{i=1}^{n}W_2(d_i)(\mathbf{x}_i-\widehat{\boldsymbol{\mu}})(\mathbf{x}_i-\widehat{\boldsymbol{\mu}})'=\widehat{\boldsymbol{\Sigma}}, \tag{6.15}$$

where the functions W_1 and W_2 need not be equal. Note that by (6.15) we may interpret $\widehat{\boldsymbol{\Sigma}}$ as a weighted covariance matrix, and by (6.14) we can express $\widehat{\boldsymbol{\mu}}$ as the weighted mean

$$\widehat{\boldsymbol{\mu}} = \frac{\sum_{i=1}^{n} W_1(d_i)\mathbf{x}_i}{\sum_{i=1}^{n} W_1(d_i)}, \tag{6.16}$$

with weights depending on an outlyingness measure d_i. This is similar to (2.32) in that with $w_i = W_1(d_i)$ we can express $\widehat{\boldsymbol{\mu}}$ as a weighted mean with data-dependent weights.

Existence and uniqueness of solutions were treated by Maronna (1976) and more generally by Tatsuoka and Tyler (2000). Uniqueness of the solutions of (6.14)–(6.15) requires that $dW_2(d)$ be a nondecreasing function of d. To understand the reason for this condition, note that an M-scale estimator of a univariate sample $\mathbf{z}$ may be written as the solution of

$$\delta = \text{ave}\left(\rho\left(\frac{\mathbf{z}}{\widehat{\sigma}}\right)\right) = \text{ave}\left(\left(\frac{\mathbf{z}}{\widehat{\sigma}}\right) W\left(\frac{\mathbf{z}}{\widehat{\sigma}}\right)\right),$$

where $W(t) = \rho(t)/t$. Thus the condition on the monotonicity of $dW_2(d)$ is the multivariate version of the requirement that the ρ-function of a univariate M-scale be monotone.

We shall call an M-estimator of location and scatter *monotone* if $dW_2(d)$ is nondecreasing, and *redescending* otherwise. Monotone M-estimators are defined as solutions to the estimating equations (6.14)–(6.15), while redescending ones must be defined by the minimization of some objective function, as happens with S-estimators or CM-estimators, to be defined in Sections 6.4 and 6.16.2 respectively. Huber and Ronchetti (2009) consider a slightly more general definition of monotone M-estimators. For practical purposes, monotone estimators are essentially unique, in the sense that *all* solutions to the M-estimating equations are consistent estimators.

It is proved by Huber and Ronchetti (2009; Ch. 8) that if the $\mathbf{x}_i$ are i.i.d. with distribution F, then under general assumptions when $n \to \infty$, monotone M-estimators, defined as any solution $\widehat{\boldsymbol{\mu}}$ and $\widehat{\boldsymbol{\Sigma}}$ of (6.14) and (6.15), converge in probability to the solution $(\widehat{\boldsymbol{\mu}}_\infty, \widehat{\boldsymbol{\Sigma}}_\infty)$ of

$$\text{E}W_1(d)(\mathbf{x} - \widehat{\boldsymbol{\mu}}_\infty) = \mathbf{0}, \tag{6.17}$$

$$\text{E}W_2(d)(\mathbf{x} - \widehat{\boldsymbol{\mu}}_\infty)(\mathbf{x} - \widehat{\boldsymbol{\mu}}_\infty)' = \widehat{\boldsymbol{\Sigma}}_\infty \tag{6.18}$$

where $d = d(\mathbf{x}, \widehat{\boldsymbol{\mu}}_\infty, \widehat{\boldsymbol{\Sigma}}_\infty)$. They also prove that $\sqrt{n}(\widehat{\boldsymbol{\mu}} - \widehat{\boldsymbol{\mu}}_\infty, \widehat{\boldsymbol{\Sigma}} - \widehat{\boldsymbol{\Sigma}}_\infty)$ tends to a multivariate normal distribution, It is easy to show that M-estimators are affine equivariant (Problem 6.2) and so if $\mathbf{x}$ has an elliptical distribution (6.7), the asymptotic covariance matrix of $\widehat{\boldsymbol{\mu}}$ has the form (6.6) (see Sections 6.17.1 and 6.17.7).

6.3.1 Collinearity

If the data are collinear – that is, all points lie on a hyperplane H – the sample covariance matrix is singular and $\overline{\mathbf{x}} \in H$. It follows from (6.16) that since $\widehat{\boldsymbol{\mu}}$ is a linear combination of elements of H, it lies in H. Furthermore, (6.15) shows that $\widehat{\boldsymbol{\Sigma}}$ must be singular. In fact, if a sufficiently large proportion of the observations lie on a hyperplane, $\widehat{\boldsymbol{\Sigma}}$ must be singular (Section 6.2.2). But in this case $\widehat{\boldsymbol{\Sigma}}^{-1}$, and hence the d_i, do not exist and the M-estimator is not defined.

To make the estimator well defined in all cases, it suffices to extend the definition in (6.4) as follows. Let $\lambda_1 \geq \lambda_2 \geq \ldots \geq \lambda_p$ and $\mathbf{b}_j$ ($j = 1, \ldots, p$) be the eigenvalues and eigenvectors of $\widehat{\boldsymbol{\Sigma}}$. For a given $\mathbf{x}$, let $z_j = \mathbf{b}_j'(\mathbf{x} - \widehat{\boldsymbol{\mu}})$. Since $\mathbf{b}_1, \ldots, \mathbf{b}_p$ are an orthonormal basis, we have

$$\mathbf{x} - \widehat{\boldsymbol{\mu}} = \sum_{j=1}^{p} z_j \mathbf{b}_j.$$

Then, if $\widehat{\boldsymbol{\Sigma}}$ is not singular, we have (Problem 6.12):

$$d(\mathbf{x}, \widehat{\boldsymbol{\mu}}, \hat{\boldsymbol{\Sigma}}) = \sum_{j=1}^{p} \frac{z_j^2}{\lambda_j}. \tag{6.19}$$

On the other hand, if $\widehat{\boldsymbol{\Sigma}}$ is singular, its smallest q eigenvalues are zero and in this case we define

$$d(\mathbf{x}, \widehat{\boldsymbol{\mu}}, \hat{\boldsymbol{\Sigma}}) = \begin{cases} \sum_{j=1}^{p-q} z_j^2 / \lambda_j & \text{if} \quad z_{p-q+1} = \ldots = z_p = 0 \\ \infty & \text{else} \end{cases} \tag{6.20}$$

which may be seen as the limit case of (6.19) when $\lambda_j \downarrow 0$ for $j > p - q$.

Note that d_i enters (6.14)–(6.15) only through the functions W_1 and W_2, which tend to zero at infinity, so this extended definition simply excludes those points that do not belong to the hyperplane spanned by the eigenvectors corresponding to the positive eigenvalues of $\widehat{\boldsymbol{\Sigma}}$.

6.3.2 Size and shape

If one scatter matrix is a scalar multiple of another – that is, $\boldsymbol{\Sigma}_2 = k\boldsymbol{\Sigma}_1$ – we say that they have the same *shape*, but different *sizes*. Several important features of the distribution, such as correlations, principal components and linear discriminant functions, depend only on shape.

Let $\widehat{\boldsymbol{\mu}}_\infty$ and $\widehat{\boldsymbol{\Sigma}}_\infty$ be the asymptotic values of location and scatter estimators at an elliptical distribution F defined in (6.7). It is shown in Section 6.17.2 that in this case $\widehat{\boldsymbol{\mu}}_\infty$ is equal to the center of symmetry $\boldsymbol{\mu}$, and $\widehat{\boldsymbol{\Sigma}}_\infty$ is a constant multiple of $\boldsymbol{\Sigma}$, with the proportionality constant depending on F and on the estimator. This situation is similar to the scaling problem in (2.63) and at the end of Section 2.5. Consider in particular an M-estimator at the distribution $F = \mathrm{N}_p(\boldsymbol{\mu}, \boldsymbol{\Sigma})$. By the equivariance of the estimator, we may assume that $\mu = \mathbf{0}$ and $\boldsymbol{\Sigma} = \mathbf{I}$. Then $\widehat{\boldsymbol{\mu}}_\infty = \mathbf{0}$ and $\widehat{\boldsymbol{\Sigma}}_\infty = c\mathbf{I}$, and hence $d(\mathbf{x}, \widehat{\boldsymbol{\mu}}_\infty, \widehat{\boldsymbol{\Sigma}}_\infty) = \|\mathbf{x}\|^2/c$. Taking the trace in (6.18) yields

$$\mathrm{E}W_2\left(\frac{\|\mathbf{x}\|^2}{c}\right)\|\mathbf{x}\|^2 = pc.$$

Since $\|\mathbf{x}\|^2$ has a χ^2_p distribution, we obtain a consistent estimator of the covariance matrix $\boldsymbol{\Sigma}$ in the normal case by replacing $\widehat{\boldsymbol{\Sigma}}$ by $\widehat{\boldsymbol{\Sigma}}/c$, with c defined as the solution of

$$\int_0^\infty W_2\left(\frac{z}{c}\right)\frac{z}{c}g(z)dz = p, \tag{6.21}$$

where g is the density of the χ^2_p distribution.

Another approach to estimating the size of $\boldsymbol{\Sigma}$ is based on noting that if $\mathbf{x} \sim \mathrm{N}(\boldsymbol{\mu}, \boldsymbol{\Sigma})$, then $d(\mathbf{x}, \boldsymbol{\mu}, \boldsymbol{\Sigma}) \sim \chi^2_p$, and the fact that $\boldsymbol{\Sigma} = c\widehat{\boldsymbol{\Sigma}}_\infty$ implies

$$cd(\mathbf{x}, \boldsymbol{\mu}, \boldsymbol{\Sigma}) = d(\mathbf{x}, \boldsymbol{\mu}, \widehat{\boldsymbol{\Sigma}}_\infty).$$

Hence the empirical distribution of

$$\{d(\mathbf{x}_1, \widehat{\boldsymbol{\mu}}, \widehat{\boldsymbol{\Sigma}}), \ldots, d(\mathbf{x}_n, \widehat{\boldsymbol{\mu}}, \widehat{\boldsymbol{\Sigma}})\}$$

will resemble that of $d(\mathbf{x}, \widehat{\boldsymbol{\mu}}_\infty, \widehat{\boldsymbol{\Sigma}}_\infty)$, which is $c\chi^2_p$, and so we may estimate c robustly with

$$\widehat{c} = \frac{\mathrm{Med}\{d(\mathbf{x}_1, \widehat{\boldsymbol{\mu}}, \widehat{\boldsymbol{\Sigma}}), \ldots, d(\mathbf{x}_n, \widehat{\boldsymbol{\mu}}, \widehat{\boldsymbol{\Sigma}})\}}{\chi^2_p(0.5)}, \tag{6.22}$$

where $\chi^2_p(\alpha)$ denotes the α-quantile of the χ^2_p distribution.

6.3.3 Breakdown point

It is intuitively clear that robustness of the estimators requires that no term dominates the sums in (6.14)–(6.15), and to achieve this we assume

$$W_1(d)\sqrt{d}, \quad W_2(d) \quad \text{and } W_2(d)d \text{ are bounded for } d \geq 0. \tag{6.23}$$

Let

$$K = \sup_d W_2(d)d. \tag{6.24}$$

We first consider the asymptotic BP, which is easier to deal with. The "weak part" of joint M-estimators of $\boldsymbol{\mu}$ and $\boldsymbol{\Sigma}$ is the estimator $\widehat{\boldsymbol{\Sigma}}$, for if we take $\boldsymbol{\Sigma}$ as known, then it is not difficult to prove that the asymptotic BP of $\widehat{\boldsymbol{\mu}}$ is 1/2 (see Section 6.17.4.1). On the other hand, in the case where $\boldsymbol{\mu}$ is known, the following result was obtained by Maronna (1976). If the underlying distribution F_0 attributes zero mass to any hyperplane, then the asymptotic BP of a *monotone* M-estimator of $\mathbf{\Sigma}$ with W_2 satisfying (6.23) is

$$\varepsilon^* = \min\left(\frac{1}{K}, 1 - \frac{p}{K}\right). \tag{6.25}$$

See Section 6.17.4.2 for a simplified proof. The above expression has a maximum value of $1/(p+1)$, attained at $K = p + 1$, and hence

$$\varepsilon^* \leq \frac{1}{p+1}. \tag{6.26}$$

See also Tyler, 1990.

Tyler (1987) proposed a monotone M-estimator with $W_2(d) = p/d$, which corresponds to the multivariate t-distribution MLE weights (6.13) with degrees of freedom $\nu \downarrow 0$. Tyler showed that the BP of this estimator is $\varepsilon^* = 1/p$, which is slightly larger than the bound (6.26). This result is not a contradiction of (6.26), since W_2 is not defined at zero and hence does not satisfy (6.23). Unfortunately this unboundedness may make the estimator unstable.

It is useful to understand the form of the breakdown under the assumptions (6.23). Take $F = (1-\varepsilon)F_0 + \varepsilon G$, where G is any contaminating distribution. First let G be concentrated at $\mathbf{x}_0$. Then the term $1/K$ in (6.25) is obtained by letting $\mathbf{x}_0 \to \infty$, and the term $1 - p/K$ is obtained by letting $\mathbf{x}_0 \to \boldsymbol{\mu}$. Now consider a general G. For the joint estimation of $\boldsymbol{\mu}$ and $\mathbf{\Sigma}$, Tyler shows that if $\varepsilon > \varepsilon^*$ and one lets G tend to $\delta_{\mathbf{x}_0}$ then $\boldsymbol{\mu} \to \mathbf{x}_0$ and $\lambda_p(\mathbf{\Sigma}) \to 0$; that is, inliers can make $\mathbf{\Sigma}$ nearly singular.

The FBP is similar, but the details are more involved (Tyler, 1990). Define a sample to be in *general position* if no hyperplane contains more than p points. Davies (1987) showed that the maximum FBP of any equivariant estimator for a sample in general position is $m^*_{\max}/n$, with

$$m^*_{\max} = \left[\frac{n-p}{2}\right]. \tag{6.27}$$

It is therefore natural to search for estimators whose BP is nearer to this maximum BP than that of monotone M-estimators.

6.4 Estimators based on a robust scale

Just as with the regression estimators of Section 5.4, where we aimed at making the residuals "small", we shall define multivariate estimators of location and scatter that make the distances d_i "small". To this end, we look for $\widehat{\boldsymbol{\mu}}$ and $\widehat{\mathbf{\Sigma}}$ minimizing some measure of "largeness" of $d(\mathbf{x}, \widehat{\boldsymbol{\mu}}, \widehat{\mathbf{\Sigma}})$. If follows from (6.4) that this can be trivially done by letting the smallest eigenvalue of $\widehat{\mathbf{\Sigma}}$ tend to infinity. To prevent this, we impose the constraint $|\widehat{\mathbf{\Sigma}}| = 1$. Call $\mathcal{S}_p$ the set of symmetric positive definite $p \times p$ matrices. For a dataset X, call $\mathbf{d}\,(X, \widehat{\boldsymbol{\mu}}, \widehat{\mathbf{\Sigma}})$ the vector with elements $d(\mathbf{x}_i, \widehat{\boldsymbol{\mu}}, \widehat{\mathbf{\Sigma}})$,

$i = 1, ..., n$, and let $\widehat{\sigma}$ be a robust scale estimator. Then we define the estimators $\widehat{\boldsymbol{\mu}}$ and $\widehat{\boldsymbol{\Sigma}}$ by

$$\widehat{\sigma}(\mathbf{d}(X, \widehat{\boldsymbol{\mu}}, \widehat{\boldsymbol{\Sigma}})) = \min \text{ with } \widehat{\boldsymbol{\mu}} \in R^p, \ \widehat{\boldsymbol{\Sigma}} \in \mathcal{S}_p, \ |\widehat{\boldsymbol{\Sigma}}| = 1. \tag{6.28}$$

It is easy to show that the estimators defined by (6.28) are equivariant. An equivalent formulation of the above goal is to minimize $|\widehat{\boldsymbol{\Sigma}}|$ subject to a bound on $\widehat{\sigma}$ (Problems 6.7–6.9).

6.4.1 The minimum volume ellipsoid estimator

The simplest case of (6.28) is to mimic the approach that results in the LMS in Section 5.4, and let $\widehat{\sigma}$ be the sample median. The resulting location and scatter matrix estimator is called the *minimum volume ellipsoid* (MVE) estimator (Rousseeuw 1985). The name stems from the fact that among all ellipsoids $\{\mathbf{x} : d(\mathbf{x}, \boldsymbol{\mu}, \boldsymbol{\Sigma}) \leq 1\}$ containing at least half of the data points, the one given by the MVE estimator has minimum volume; that is, the minimum $|\boldsymbol{\Sigma}|$. The consistency rate of the MVE is the same slow rate as the LMS, namely only $n^{-1/3}$, and hence is very inefficient (Davies, 1992).

6.4.2 S-estimators

To overcome the inefficiency of the MVE we consider a more general class of estimators called S-estimators (Davies, 1987), defined by (6.28), taking for $\widehat{\sigma}$ an M-scale estimator that satisfies

$$\frac{1}{n}\sum_{i=1}^{n} \rho\left(\frac{d_i}{\widehat{\sigma}}\right) = \delta, \tag{6.29}$$

where ρ is a smooth bounded ρ-function. The same reasoning as in (5.24) shows that an S-estimator $(\widehat{\boldsymbol{\mu}}, \widehat{\boldsymbol{\Sigma}})$ is an M-estimator, in the sense that for any $\tilde{\boldsymbol{\mu}}, \tilde{\boldsymbol{\Sigma}}$ with $|\tilde{\boldsymbol{\Sigma}}| = 1$ and $\widehat{\sigma} = \widehat{\sigma}(\mathbf{d}(X, \widehat{\boldsymbol{\mu}}, \widehat{\boldsymbol{\Sigma}}))$,

$$\sum_{i=1}^{n} \rho\left(\frac{d(\mathbf{x}_i, \widehat{\boldsymbol{\mu}}, \widehat{\boldsymbol{\Sigma}})}{\widehat{\sigma}}\right) \leq \sum_{i=1}^{n} \rho\left(\frac{d(\mathbf{x}_i, \tilde{\boldsymbol{\mu}}, \tilde{\boldsymbol{\Sigma}})}{\widehat{\sigma}}\right). \tag{6.30}$$

If ρ is differentiable, it can be shown (Section 6.17.5) that the solution to (6.28) must satisfy estimating equations of the form (6.14)–(6.15), namely

$$\sum_{i=1}^{n} W\left(\frac{d_i}{\widehat{\sigma}}\right)(\mathbf{x}_i - \widehat{\boldsymbol{\mu}}) = \mathbf{0}, \tag{6.31}$$

$$\frac{1}{n}\sum_{i=1}^{n} W\left(\frac{d_i}{\widehat{\sigma}}\right)(\mathbf{x}_i - \widehat{\boldsymbol{\mu}})(\mathbf{x}_i - \widehat{\boldsymbol{\mu}})' = c\widehat{\boldsymbol{\Sigma}}, \tag{6.32}$$

where

$$W = \rho' \text{ and } \widehat{\sigma} = \widehat{\sigma}(d_1, ..., d_n), \tag{6.33}$$

and c is a scalar such that $|\widehat{\boldsymbol{\Sigma}}| = 1$. Note, however, that if ρ is bounded (as is the usual case), $dW(d)$ cannot be monotone (Problem 6.5); in fact, for the estimators usually employed, $W(d)$ vanishes for large d. Therefore, the estimator is not a monotone M-estimator, and so the estimating equations yield only *local* minima of $\widehat{\sigma}$.

The choice $\rho(d) = d$ yields the average of the d_i as scale estimator. In this case, $W \equiv 1$ and hence

$$\widehat{\boldsymbol{\mu}} = \overline{\mathbf{x}}, \quad \widehat{\boldsymbol{\Sigma}} = \frac{\mathbf{C}}{|\mathbf{C}|^{1/p}} \tag{6.34}$$

where $\mathbf{C}$ is the sample covariance matrix. For this choice of scale estimator it follows that

$$\sum_{i=1}^{n} (\mathbf{x}_i - \overline{\mathbf{x}})'\boldsymbol{\Sigma}^{-1}(\mathbf{x}_i - \overline{\mathbf{x}}) \leq \sum_{i=1}^{n} (\mathbf{x}_i - \boldsymbol{\nu})'\mathbf{V}^{-1}(\mathbf{x}_i - \boldsymbol{\nu}) \tag{6.35}$$

for all $\boldsymbol{\nu}$ and $\mathbf{V}$ with $|\mathbf{V}| = 1$.

It can be shown (Davies, 1987) that if ρ is differentiable, then for S-estimators the distribution of $\sqrt{n}(\widehat{\boldsymbol{\mu}} - \widehat{\boldsymbol{\mu}}_\infty, \widehat{\boldsymbol{\Sigma}} - \widehat{\boldsymbol{\Sigma}}_\infty)$ tends to a multivariate normal.

Similarly to Section 5.4, it can be shown that the maximum FBP (6.27) is attained for S-estimators by taking in (6.29):

$$n\delta = m^*_{\max} = \left[\frac{n-p}{2}\right].$$

We define the bisquare multivariate S-estimator as the one with scale given by (6.29), with

$$\rho(t) = \min\{1, 1 - (1-t)^3\}, \tag{6.36}$$

which has weight function

$$W(t) = 3(1-t)^2 \mathrm{I}(t \leq 1). \tag{6.37}$$

The reason for this definition is that in the univariate case the bisquare scale estimator – call it $\widehat{\eta}$ for notational convenience – based on centered data x_i with location $\widehat{\mu}$, is the solution of

$$\frac{1}{n}\sum_{i=1}^{n} \rho_{bisq}\left(\frac{x_i - \widehat{\mu}}{\widehat{\eta}}\right) = \delta, \tag{6.38}$$

where $\rho_{bisq}(t) = \min\{1, 1 - (1-t^2)^3\}$. Since $\rho_{bisq}(t) = \rho(t^2)$ for the ρ defined in (6.36), it follows that (6.38) is equivalent to

$$\frac{1}{n}\sum_{i=1}^{n} \rho\left(\frac{(x_i - \widehat{\mu})^2}{\widehat{\sigma}}\right) = \delta$$

with $\widehat{\sigma} = \widehat{\eta}^2$. Now $d(\mathbf{x}, \boldsymbol{\mu}, \boldsymbol{\Sigma})$ is the normalized *squared* distance between $\mathbf{x}$ and $\boldsymbol{\mu}$, which explains the use of ρ.

6.4.3 The MCD estimator

Another possibility is to use a trimmed scale for $\widehat{\sigma}$ instead of an M-scale, as was done to define the LTS estimator in Section 5.4.2. Let $d_{(1)} \leq \ldots \leq d_{(n)}$ be the ordered values of the squared distances $d_i = d(\mathbf{x}_i, \boldsymbol{\mu}, \boldsymbol{\Sigma})$, and for $1 \leq h < n$ define the trimmed scale of the squared distances as

$$\widehat{\sigma} = \sum_{i=1}^{h} d_{(i)}. \tag{6.39}$$

An estimator $(\widehat{\boldsymbol{\mu}}, \widehat{\boldsymbol{\Sigma}})$ defined by (6.28) with this trimmed scale is called a *minimum covariance determinant* (MCD) estimator (Rousseeuw 1985). The reason for the name is the following: for each ellipsoid $\{\mathbf{x} : d(\mathbf{x}, \mathbf{t}, \mathbf{V}) \leq 1\}$ containing at least h data points, compute the covariance matrix $\mathbf{C}$ of the data points in the ellipsoid. If $(\widehat{\boldsymbol{\mu}}, \widehat{\boldsymbol{\Sigma}})$ is an MCD estimator, then the ellipsoid with $\mathbf{t} = \widehat{\boldsymbol{\mu}}$ and $\mathbf{V}$ equal to a scalar multiple of $\widehat{\boldsymbol{\Sigma}}$ minimizes $|\mathbf{C}|$.

As in the case of the LTS estimator in Section 5.4, the maximum BP of the MCD estimator is attained by taking $h = n - m^*_{\max}$, with $m^*_{\max}$ as defined in (6.27).

Note that increasing h increases the efficiency, but at the cost of decreasing the BP. Paindaveine and Van Bever (2014) show that the asymptotic efficiency of MCD is very low.

6.4.4 S-estimators for high dimension

Consider the multivariate S-estimator with a bisquare ρ-function. Table 6.2 gives the asymptotic efficiencies of the scatter matrix vector under normality for different dimensions p and $n = 10p$ (efficiency will be defined in (6.80)).

It is seen that the efficiency is low for small p, but it approaches one for large p. The same thing happens with the location vector. It is shown in Section 6.17.6 that this behavior holds for any S-estimator with a continuous weight function $W = \rho'$. This may seem like good news. However, the proof shows that for large p all observations, except those that are extremely far away from the bulk of the data, have approximately the same weight, and hence the estimator is approximately equal to the sample mean and sample covariance matrix. As a result, observations outlying enough to be dangerous may also have nearly maximum weight, and as a result, the bias can be very large (bias is defined in Section 6.7). It will be seen later

Table 6.2 Efficiencies of the S-estimator with bisquare weights for dimension p

p	2	5	10	20	30	40	50
Efficiency	0.427	0.793	0.930	0.976	0.984	0.990	0.992

that this increase in efficiency and decrease in robustness with large p does not occur with the MVE.

Rocke (1996) pointed out the problem just described, and proposed that the ρ-function change with dimension to prevent both the efficiency from increasing to values arbitrarily close to one and, correspondingly, the bias becoming arbitrarily large. He proposed a family of ρ-functions with the property that when $p \to \infty$ the function ρ approaches the step function $\rho(d) = \mathrm{I}(d > 1)$. The latter corresponds to the scale estimator $\hat{\sigma} = \mathrm{Med}\,(\mathbf{d})$ and so the limiting form of the estimator for large dimensions is the MVE estimator.

Put, for brevity, $d = d(\mathbf{x}, \hat{\boldsymbol{\mu}}_\infty, \hat{\boldsymbol{\Sigma}}_\infty)$. It is shown in Section 6.17.6 that if $\mathbf{x}$ is normal, then for large p

$$\mathcal{D}\left(\frac{d}{\sigma}\right) \approx \mathcal{D}\left(\frac{z}{p}\right) \quad \text{with } z \sim \chi^2_p$$

and hence d/σ is increasingly concentrated around one. For large p, the χ^2_p distribution is approximately symmetric, with $\chi^2_p(0.5) \approx \mathrm{E}z = p$ and $\chi^2_p(1-\alpha) - p \approx p - \chi^2_p(\alpha)$. Figure 6.4 shows the densities of z/p for $p = 10$ and 100, scaled to facilitate the comparison. Note that when p increases, the density becomes more concentrated around one.

To have a sufficiently high (but not too high) efficiency, we should give a high weight to the values of d/σ near one and downweight the extreme ones. A simple way to do this is to have $W(t) = 0$ for t between the α- and the $(1-\alpha)$-quantiles of d/σ for some $\alpha \in (0, 1)$. We now define a smooth ρ-function such that the resulting

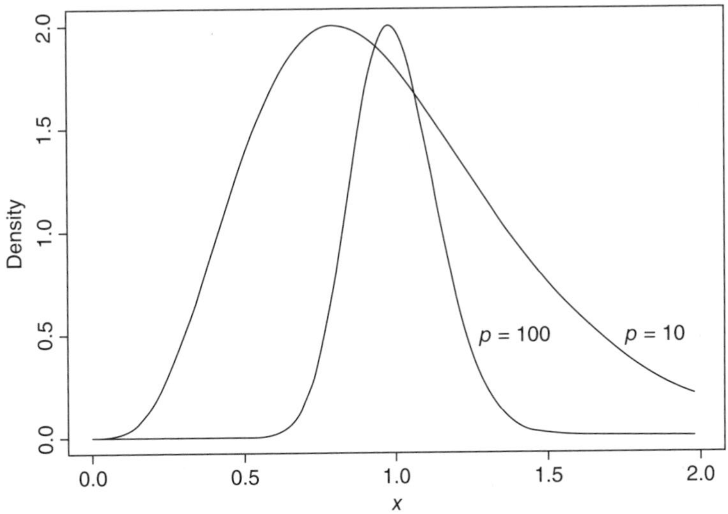

Figure 6.4 Densities of z/p for $p = 10$ and 100

weight function $W(t)$ vanishes for $t \notin [1-\gamma, 1+\gamma]$, where γ depends on α. Let

$$\gamma = \min\left(\frac{\chi_p^2(1-\alpha)}{p} - 1, 1\right), \tag{6.40}$$

where $\chi_p^2(\alpha)$ denotes the α-quantile of χ_p^2. Define

$$\rho(t) = \begin{cases} 0 & \text{for} \quad 0 \le t \le 1-\gamma \\ \left(\frac{t-1}{4\gamma}\right)\left[3 - \left(\frac{t-1}{\gamma}\right)^2\right] + \frac{1}{2} & \text{for } 1-\gamma < t < 1+\gamma \\ 1 & \text{for} \quad t \ge 1+\gamma \end{cases} \tag{6.41}$$

which has as derivative the weight function

$$W(t) = \frac{3}{4\gamma}\left[1 - \left(\frac{t-1}{\gamma}\right)^2\right] \mathrm{I}(1-\gamma \le t \le 1+\gamma). \tag{6.42}$$

Figure 6.5 shows the ρ- and W-functions corresponding to $\alpha = 0.005$ for $p = 10$ and 100. To simplify viewing, the weight functions are scaled so that $\max_t W(t) = 1$. When p increases, the interval on which W is positive shrinks, and ρ tends to the step function corresponding to the MVE.

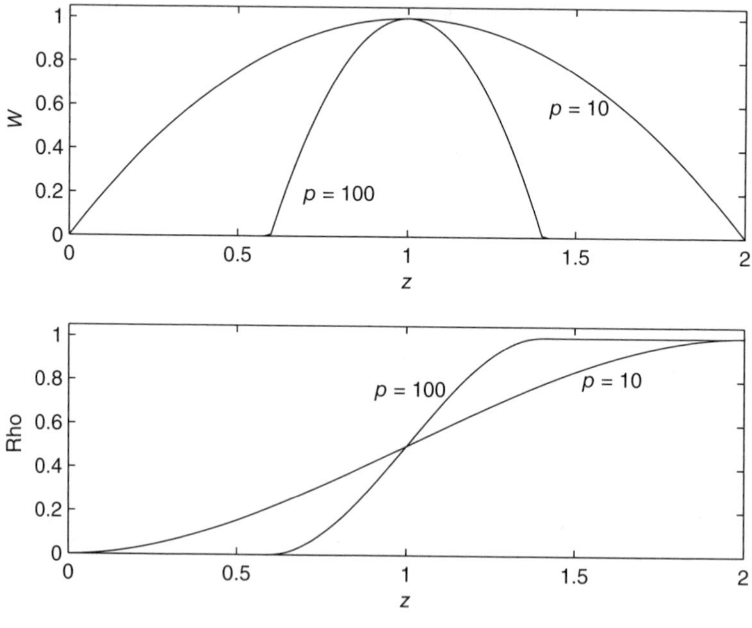

Figure 6.5 ρ- and W-functions of Rocke estimator with $\alpha = 0.005$ for $p = 10$ and 100

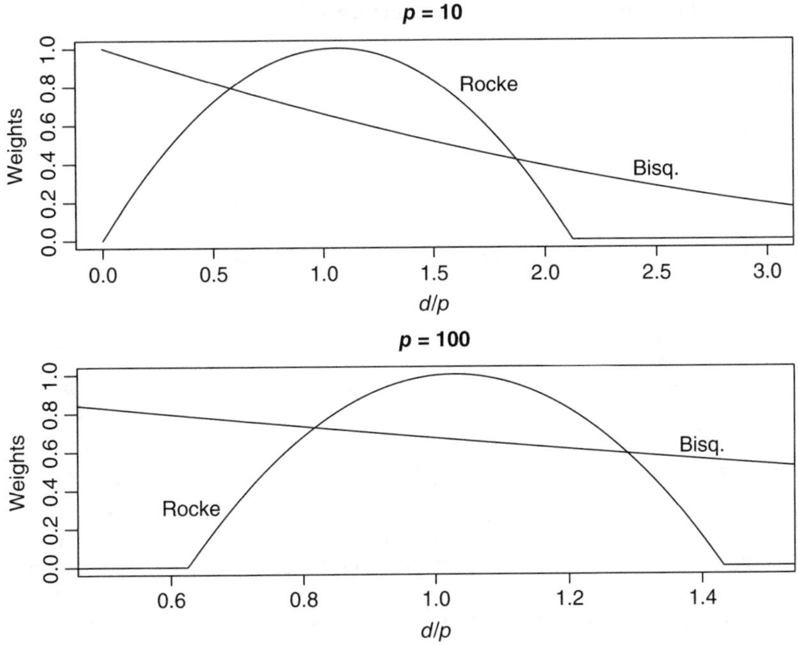

Figure 6.6 Weight functions of Rocke (with 0.9 efficiency) and bisquare estimators

The parameter α allows the user to control the efficiency (see more in Section 6.10). Figure 6.6 shows for $p = 10$ and 100 and $n = 10p$ the weights of the Rocke estimator with finite-sample efficiency equal to 0.9, and those of the bisquare estimator, as a function of dp where d is the squared Mahalanobis distance. It is seen that the bisquare weight function descends slowly when $p = 10$, and is almost constant when $p = 100$, while the Rocke function assigns high weights to the bulk of the data and rapidly downweights data far from it, except possibly for very small values of p.

Rocke's *biflat* family of weight functions is the squared values of the W in (6.42). It is smoother at the endpoints but gives less weight to inner points.

Example 6.3 *The following data are part of a study on shape recognition. An ensemble of shape-feature extractors for the 2D silhouettes of different vehicles. The purpose is to classify a given silhouette as one of four types of vehicle, using a set of 18 features extracted from the silhouette. The tables and figures for this example are obtained with script **vehicle.R**.*

The vehicle may be viewed from one of many different angles. The features are extracted from the silhouettes by an image processing system, which extracts a combination of scale-independent features utilizing both measures based on classical

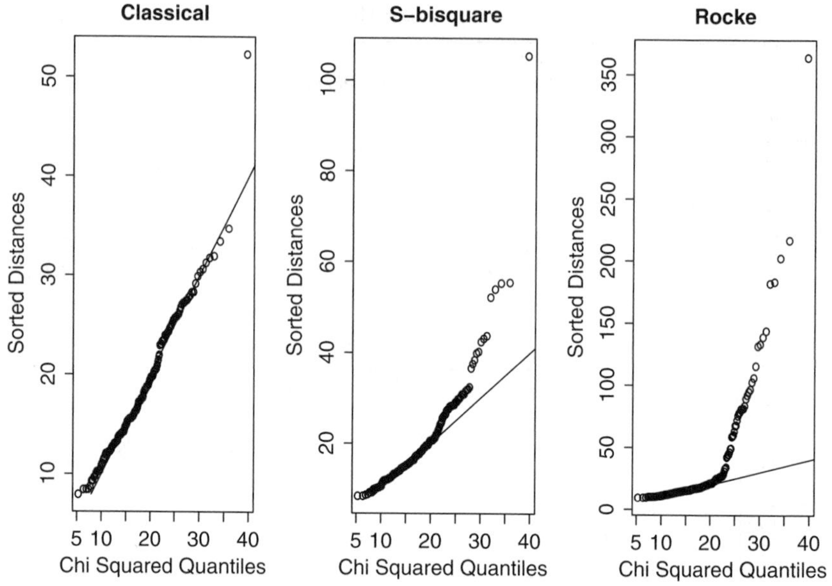

Figure 6.7 Vehicle data. QQ-plots of squared Mahalanobis distances

moments, such as scaled variance, skewness and kurtosis about the major/minor axes, and heuristic measures, such as hollows, circularity, rectangularity and compactness. The data were collected at the Turing Institute, Glasgow, and are available at https://archive.ics.uci.edu/ml/datasets/Statlog+(Vehicle+Silhouettes). Here we deal with the "van" type, which has $n = 217$ cases.

Figure 6.7 shows the chi-squared QQ-plots of the squared Mahalanobis distances from the classical estimator and the bisquare and Rocke S estimators. It is seen that the classical estimator finds no outliers, the bisquare S about 12, and the Rocke S about 30. The largest 10 distances for bisquare and Rocke correspond to the same cases.

A more detailed analysis of the data shows that for 10 of the 18 variables, the values corresponding to the 30 cases pinpointed by the Rocke S are either the 12% lower or upper extreme. We can hence consider those cases as lying on a "corner" of the dataset, and we may therefore conclude that the Rocke S has revealed more structure than the bisquare S.

6.4.5 τ-estimators

The same approach used in Section 5.4.3 to obtain robust regression estimators with controllable efficiency can be employed for multivariate estimation. Multivariate τ-estimators were proposed and studied by Lopuhaä (1991).

This approach requires two functions, ρ_1 and ρ_2. For given $(\boldsymbol{\mu}, \boldsymbol{\Sigma})$ call $\sigma_0(\boldsymbol{\mu}, \boldsymbol{\Sigma})$ the solution of

$$\frac{1}{n}\sum_{i=1}^{n}\rho_1\left(\frac{d(\mathbf{x}_i, \boldsymbol{\mu}, \boldsymbol{\Sigma})}{\sigma_0}\right) = \delta.$$

Then the estimator $(\widehat{\boldsymbol{\mu}}, \widehat{\boldsymbol{\Sigma}})$ is defined as the minimizer of the "τ-scale"

$$\sigma(\boldsymbol{\mu}, \boldsymbol{\Sigma}) = \sigma_0(\boldsymbol{\mu}, \boldsymbol{\Sigma})\frac{1}{n}\sum_{i=1}^{n}\rho_2\left(\frac{d(\mathbf{x}_i, \boldsymbol{\mu}, \boldsymbol{\Sigma})}{\sigma_0(\boldsymbol{\mu}, \boldsymbol{\Sigma})}\right),$$

with $|\widehat{\boldsymbol{\Sigma}}| = 1$. Here

$$\rho_2(t) = \rho_1\left(\frac{t}{c}\right), \tag{6.43}$$

where c is chosen to regulate the efficiency.

Originally, τ-estimators were proposed to obtain estimators with higher efficiency than S-estimators for small p, which required $c > 1$; but for large p we need $c < 1$ in order to decrease the efficiency.

It can be shown that $(\widehat{\boldsymbol{\mu}}, \widehat{\boldsymbol{\Sigma}})$ satisfies estimating equations of the form (6.31)–(6.32) with $\widehat{\sigma} = \sigma(\widehat{\boldsymbol{\mu}}, \widehat{\boldsymbol{\Sigma}})$, where W depends on ρ_1' and ρ_2'.

6.4.6 One-step reweighting

In order to increase the efficiency of a given estimator $(\widehat{\boldsymbol{\mu}}, \widehat{\boldsymbol{\Sigma}})$, a one-step reweighting procedure can be used in a similar way as shown in Section 5.9.1. Let W be a weight function. Given the estimator $(\widehat{\boldsymbol{\mu}}, \widehat{\boldsymbol{\Sigma}})$, define new estimators $\tilde{\boldsymbol{\mu}}, \tilde{\boldsymbol{\Sigma}}$ as a weighted mean vector and weighted covariance matrix with weights $W(d_i)$, where the d_i are the squared distances corresponding to $\widehat{\boldsymbol{\mu}}$ and $\widehat{\boldsymbol{\Sigma}}$. The most popular function is hard rejection, corresponding to $W(t) = \mathrm{I}(t \leq k)$, where k is chosen with the same criterion as in Section 5.9.1. For $\widehat{c}$ defined in (6.22), the distribution of $d_i/\widehat{c}$ is approximately χ^2_p under normality, and hence choosing $k = \widehat{c}\chi^2_{p,\beta}$ will reject approximately a fraction $1 - \beta$ of the "good" data if there are no outliers. It is customary to take $\beta = 0.95$ or 0.975. If the scatter matrix estimator is singular, we proceed as in Section 6.3.1.

Although no theoretical results are known, simulations have showed this procedure improves the bias and efficiency of the MVE and MCD estimators. But it cannot be asserted that such improvement happens with *any* estimator.

Croux and Haesbroeck (1999, Tables VII and VIII) computed the finite-sample efficiencies of the reweighted MCD; although they are much higher than for the "raw" estimator, they are still low if one wants a high BP.

6.5 MM-estimators

We now present a family of estimators with controllable efficiency based on the same principle as the regression MM-estimators described in Section 5.5, namely: start with a very robust but possibly inefficient estimator, and use it to compute a scale of the Mahalanobis distances and as a starting point to compute an M-estimator whose ρ-function has a tuning parameter. This was proposed by Lopuhaä (1992) and Tatsuoka and Tyler (2000) as a means of *increasing* the low efficiency of S estimators for small p. Here we give a simplified version of the latter approach.

Let $(\widehat{\boldsymbol{\mu}}_0, \widehat{\boldsymbol{\Sigma}}_0)$ be an initial estimator. Put $d_i^0 = d(\mathbf{x}_i, \widehat{\boldsymbol{\mu}}_0, \widehat{\boldsymbol{\Sigma}}_0)$ and call S the respective M-scale

$$\frac{1}{n}\sum_{i=1}^{n}\rho\left(\frac{d_i^0}{S}\right)=\delta. \tag{6.44}$$

The estimator is defined by $(\widehat{\boldsymbol{\mu}}, \widehat{\boldsymbol{\Sigma}})$ with $|\widehat{\boldsymbol{\Sigma}}| = 1$ such that

$$\sum_{i=1}^{n}\rho\left(\frac{d_i}{cS}\right)=\min, \tag{6.45}$$

where $d_i = d(\mathbf{x}_i, \widehat{\boldsymbol{\mu}}, \widehat{\boldsymbol{\Sigma}})$ and the constant c is chosen to control efficiency.

It can be shown that the solution satisfies the equations

$$\frac{1}{n}\sum_{i=1}^{n}W\left(\frac{d_i}{cS}\right)(\mathbf{x}-\boldsymbol{\mu})(\mathbf{x}-\boldsymbol{\mu})'=\boldsymbol{\Sigma} \tag{6.46}$$

$$\frac{1}{n}\sum_{i=1}^{n}W\left(\frac{d_i}{cS}\right)(\mathbf{x}-\boldsymbol{\mu})=\mathbf{0}$$

with $W = \rho'$,

In fact, it is not necessary to obtain the absolute minimum in (6.45). As with the regression MM-estimators in Section 5.5, it is possible to show that any solution of (6.46) with the objective function (6.45) is lower that for the initial estimator, has the same asymptotic behavior as the absolute minimum and has BP at least as high as the initial estimator.

As to the ρ-function, there are better choices than the traditional bisquare. Muler and Yohai (2002) employ a different ρ for time-series estimation, with a weight function that is constant up to a certain value, and then descends rapidly but smoothly to zero. We shall call this *smoothed hard rejection* (SHR). Its version for multivariate estimation has a weight function

$$W_{\mathrm{SHR}}(d)=\begin{cases}1 & \text{if} \quad d\le 4\\ q(d) & \text{if} \quad 4<d\le 9\\ 0 & \text{if} \quad d>9\end{cases}, \tag{6.47}$$

where

$$q(d) = -1.944 + 1.728d - 0.312d^2 + 0.016d^3$$

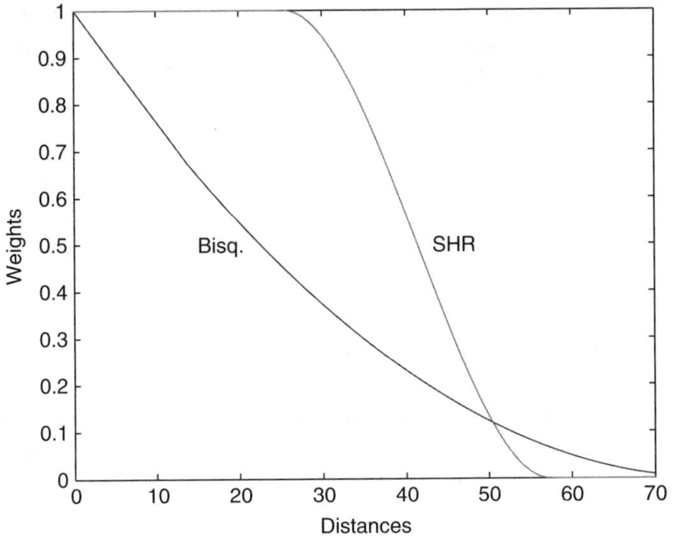

Figure 6.8 Bisquare and SHR weight functions

This is such that W is continuous and differentiable at $d = 4$ and $d = 9$. The respective ρ function is

$$\rho_{\text{SHR}}(d) = \frac{1}{6.494} \begin{cases} d & \text{if} \quad d \leq 4 \\ s(d) & \text{if} \quad 4 < d \leq 9 \\ 6.494 & \text{if} \quad d > 9 \end{cases}, \tag{6.48}$$

where

$$s(d) = 3.534 - 1.944d + 0.864d^2 - 0.104d^3 + 0.004d^4.$$

Figure 6.8 shows the bisquare and SHR weight functions, the former scaled for consistency and the latter for 90% efficiency for the normal, with $p = 30$. It is seen that SHR yields a smaller cutoff point, while giving more weight to small distances. The advantages of SHR will be demonstrated in Section 6.10.

6.6 The Stahel–Donoho estimator

Recall that the simplest approach to the detection of outliers in a univariate sample is the one given in Section 1.3: for each data point compute an "outlyingness measure" (1.4) and identify those points having a "large" value of this measure. The key idea for the extension of this approach to the multivariate case is that a multivariate outlier should be an outlier in *some* univariate projection. More precisely, given a direction $\mathbf{a} \in R^p$ with $\|\mathbf{a}\| = 1$, denote by $\mathbf{a}'X = \{\mathbf{a}'\mathbf{x}_1, \ldots, \mathbf{a}'\mathbf{x}_n\}$ the projection of the dataset

X along $\mathbf{a}$. Let $\widehat{\mu}$ and $\widehat{\sigma}$ be robust univariate location and scatter statistics, say the median and MAD respectively. The outlyingness with respect to X of a point $\mathbf{x} \in R^p$ along $\mathbf{a}$ is defined, as in (1.4), by

$$t(\mathbf{x}, \mathbf{a}) = \frac{\mathbf{x}'\mathbf{a} - \widehat{\mu}(\mathbf{a}'X)}{\widehat{\sigma}(\mathbf{a}'X)}.$$

The outlyingness of $\mathbf{x}$ is then defined by

$$t(\mathbf{x}) = \max_{\mathbf{a}} |t(\mathbf{x}, \mathbf{a})|. \tag{6.49}$$

In the above maximum, $\mathbf{a}$ ranges over the set $\{\|\mathbf{a}\| = 1\}$, but in view of the equivariance of $\widehat{\mu}$ and $\widehat{\sigma}$, it is equivalent to take the set $\{\mathbf{a} \neq \mathbf{0}\}$.

The Stahel–Donoho estimator, proposed by Stahel (1981) and Donoho (1982), is a weighted mean and covariance matrix, where the weight of $\mathbf{x}_i$ is a nonincreasing function of $t(\mathbf{x}_i)$. More precisely, let W_1 and W_2 be two weight functions, and define

$$\widehat{\boldsymbol{\mu}} = \frac{1}{\sum_{i=1}^n w_{i1}} \sum_{i=1}^n w_{i1}\mathbf{x}_i, \tag{6.50}$$

$$\widehat{\boldsymbol{\Sigma}} = \frac{1}{\sum_{i=1}^n w_{i2}} \sum_{i=1}^n w_{i2}(\mathbf{x}_i - \widehat{\boldsymbol{\mu}})(\mathbf{x}_i - \widehat{\boldsymbol{\mu}})' \tag{6.51}$$

with

$$w_{ij} = W_j(t(\mathbf{x}_i)),\ j = 1, 2. \tag{6.52}$$

If $\mathbf{y}_i = \mathbf{A}\mathbf{x}_i + \mathbf{b}$, then it is easy to show that $t(\mathbf{y}_i) = t(\mathbf{x}_i)$ (t is invariant) and hence the estimators are equivariant.

In order that no term dominates in (6.50)–(6.51) it is clear that the weight functions must satisfy the conditions

$$tW_1(t) \text{ and } t^2W_2(t) \text{ are bounded for } t \geq 0. \tag{6.53}$$

It can be shown (see Maronna and Yohai, 1995) that under (6.53) the asymptotic BP is 1/2. For the FBP, Tyler (1994) and Gather and Hilker (1997) show that the estimator attains the maximum BP given by (6.27) if $\widehat{\boldsymbol{\mu}}$ is the sample median and the scale is

$$\widehat{\sigma}(z) = \frac{1}{2}(\tilde{z}_k + \tilde{z}_{k+1})$$

where $\tilde{z}_i$ denotes the ordered values of $|z_i - \mathrm{Med}(\mathbf{z})|$ and $k = [(n + p)/2]$.

The asymptotic normality of the estimator was shown by Zuo *et al.* (2004a), and its influence function and maximum asymptotic bias were derived by Zuo *et al.* (2004b).

Note that since the W_j in (6.52) may include tuning constants, these estimators have a controllable efficiency. There are several proposals for the form of these functions. A family of weight functions used in the literature is the "Huber weights"

$$W^H_{c,k}(t) = \min\left(1, \left(\frac{c}{t}\right)^k\right) \tag{6.54}$$

where $k \geq 2$ in order to satisfy (6.53). Maronna and Yohai (1995) used

$$W_1 = W_2 = W^H_{c,2} \text{ with } c = \sqrt{\chi^2_{p,\beta}} \tag{6.55}$$

in (6.52) with $\beta = 0.95$. Zuo *et al.* (2004a) proposed the family of weights

$$W^Z_{c,k}(t) = \min\left\{1, 1 - \frac{1}{b}\exp\left[-k\left(1 - \frac{1}{c(1+t)}\right)^2\right]\right\} \tag{6.56}$$

where $c = \text{Med}(1/(1 + t(\mathbf{x})))$, k is a tuning parameter and $b = 1 - e^{-k}$.

Another choice is the SHR family, defined in (6.47), the advantages of which are demonstrated in Section 6.10.

Simulations suggest that one-step reweighting does not improve the Stahel–Donoho estimator.

6.7 Asymptotic bias

This section deals with the definition of asymptotic bias for multivariate location and scatter estimators. We now deal with data from a contaminated distribution $F = (1 - \varepsilon)F_0 + \varepsilon G$, where F_0 describes the "typical" data. In order to define bias, we have to define which are the "true" parameters to be estimated. For concreteness, assume $F_0 = \text{N}_p(\boldsymbol{\mu}_0, \boldsymbol{\Sigma}_0)$, but note that the following discussion applies to any other elliptical distribution. Let $\widehat{\boldsymbol{\mu}}_\infty$ and $\widehat{\boldsymbol{\Sigma}}_\infty$ be the asymptotic values of location and scatter estimators.

Defining a single measure of bias for a multidimensional estimator is more complicated than in Section 3.3. Assume first that $\boldsymbol{\Sigma}_0 = \mathbf{I}$. In this case, the symmetry of the situation makes it natural to choose the Euclidean norm $\|\widehat{\boldsymbol{\mu}}_\infty - \boldsymbol{\mu}_0\|$ as a reasonable bias measure for location. For the scatter matrix, size is relatively easy to adjust, by means of (6.21) or (6.22), and it will be most useful to focus on shape. Thus we want to measure the discrepancy between $\widehat{\boldsymbol{\Sigma}}_\infty$ and scalar multiples of $\mathbf{I}$. The simplest way to do so is with the *condition number*, which is defined as the ratio of the largest to the smallest eigenvalue,

$$\text{cond}(\widehat{\boldsymbol{\Sigma}}_\infty) = \frac{\lambda_1(\widehat{\boldsymbol{\Sigma}}_\infty)}{\lambda_p(\widehat{\boldsymbol{\Sigma}}_\infty)}.$$

The condition number equals 1 if and only if $\widehat{\boldsymbol{\Sigma}}_\infty = c\mathbf{I}$ for some $c \in R$. Other functions of the eigenvalues may be used for measuring shape discrepancies, such as the likelihood ratio test statistic for testing sphericity (Seber, 1984), which is the ratio of the arithmetic to the geometric mean of the eigenvalues:

$$\frac{\mathrm{trace}(\widehat{\boldsymbol{\Sigma}}_\infty)}{|\widehat{\boldsymbol{\Sigma}}_\infty|^{1/p}}.$$

It is easy to show that in the special case of a spherical distribution, the asymptotic value of an equivariant $\widehat{\boldsymbol{\Sigma}}$ is a scalar multiple of $\mathbf{I}$ (Problem 6.3), and so in this case there is no shape discrepancy.

For the case of an equivariant estimator and a general $\boldsymbol{\Sigma}_0$, we want to define bias so that it is invariant under affine transformations; that is, the bias does not change if $\mathbf{x}$ is replaced by $\mathbf{Ax} + \mathbf{b}$. To this end we "normalize" the data so that it has identity scatter matrix. Let $\mathbf{A}$ be any matrix such that $\mathbf{A}'\mathbf{A} = \boldsymbol{\Sigma}_0^{-1}$, and define $\mathbf{y} = \mathbf{Ax}$. Then $\mathbf{y}$ has mean $\mathbf{A}\boldsymbol{\mu}_0$ and identity scatter matrix, and if the estimators $\widehat{\boldsymbol{\mu}}$ and $\widehat{\boldsymbol{\Sigma}}$ are equivariant then then their asymptotic values based on data $\mathbf{y}_i$ are $\mathbf{A}\widehat{\boldsymbol{\mu}}_\infty$ and $\mathbf{A}\widehat{\boldsymbol{\Sigma}}_\infty\mathbf{A}'$ respectively, where $\widehat{\boldsymbol{\mu}}_\infty$ and $\widehat{\boldsymbol{\Sigma}}_\infty$ are their values based on data $\mathbf{x}_i$. Since their respective discrepancies are given by

$$\|\mathbf{A}\widehat{\boldsymbol{\mu}}_\infty - \mathbf{A}\boldsymbol{\mu}_0\|^2 = (\widehat{\boldsymbol{\mu}}_\infty - \boldsymbol{\mu}_0)'\boldsymbol{\Sigma}_0^{-1}(\widehat{\boldsymbol{\mu}}_\infty - \boldsymbol{\mu}_0) \text{ and } \mathrm{cond}(\mathbf{A}\widehat{\boldsymbol{\Sigma}}_\infty\mathbf{A}')$$

and noting that $\mathbf{A}\widehat{\boldsymbol{\Sigma}}_\infty\mathbf{A}'$ has the same eigenvalues as $\boldsymbol{\Sigma}_0^{-1}\widehat{\boldsymbol{\Sigma}}_\infty$, it is natural to define

$$\mathrm{bias}(\widehat{\boldsymbol{\mu}}) = \sqrt{(\widehat{\boldsymbol{\mu}}_\infty - \boldsymbol{\mu}_0)'\boldsymbol{\Sigma}_0^{-1}(\widehat{\boldsymbol{\mu}}_\infty - \boldsymbol{\mu}_0)} \text{ and } \mathrm{bias}(\widehat{\boldsymbol{\Sigma}}) = \mathrm{cond}(\boldsymbol{\Sigma}_0^{-1}\widehat{\boldsymbol{\Sigma}}_\infty). \tag{6.57}$$

It is easy to show that if the estimators are equivariant, then (6.57) does not depend upon either $\boldsymbol{\mu}_0$ or $\boldsymbol{\Sigma}_0$. Therefore, to evaluate equivariant estimators, we may, without loss of generality, take $\boldsymbol{\mu}_0 = \mathbf{0}$ and $\boldsymbol{\Sigma}_0 = \mathbf{I}$.

Adrover and Yohai (2002) computed the maximum asymptotic biases of several robust estimators. However, in order to compare the performances of different estimators we shall, in Section 6.10, employ a measure that takes both bias and variability into account.

6.8 Numerical computing of multivariate estimators

6.8.1 Monotone M-estimators

Equations (6.15) and (6.16) yield an iterative algorithm similar to the one for regression in Section 4.5. Start with initial estimators $\widehat{\boldsymbol{\mu}}_0$ and $\widehat{\boldsymbol{\Sigma}}_0$, for example the vector of coordinate-wise medians and the diagonal matrix with the squared normalized MADs of the variables in the diagonal. At iteration k, let $d_{ki} = d(\mathbf{x}_i, \widehat{\boldsymbol{\mu}}_k, \widehat{\boldsymbol{\Sigma}}_k)$ and compute

$$\widehat{\boldsymbol{\mu}}_{k+1} = \frac{\sum_{i=1}^n W_1(d_{ki})\mathbf{x}_i}{\sum_{i=1}^n W_1(d_{ki})}, \quad \widehat{\boldsymbol{\Sigma}}_{k+1} = \frac{1}{n}\sum_{i=1}^n W_2(d_{ki})(\mathbf{x}_i - \widehat{\boldsymbol{\mu}}_{k+1})(\mathbf{x}_i - \widehat{\boldsymbol{\mu}}_{k+1})'. \quad (6.58)$$

If, at some iteration, $\widehat{\boldsymbol{\Sigma}}_k$ becomes singular, it suffices to compute the d_i through (6.20). The convergence of the procedure is established in Section 9.5. Since the solution is unique for monotone M-estimators, the starting values influence only the number of iterations but not the end result.

6.8.2 Local solutions for S-estimators

Since local minima of $\widehat{\sigma}$ are solutions of the M-estimating equations (6.31)–(6.32), a natural procedure to minimize $\widehat{\sigma}$ is to use the iterative procedure (6.58) to solve the equations, with $W_1 = W_2$ equal to $W = \rho'$, as stated in (6.33). It must be recalled that since $tW(t)$ is redescending, this pair of equations yields only a *local* minimum of σ, and hence the starting values chosen are essential. Assume for the moment that we have the initial $\widehat{\boldsymbol{\mu}}_0$ and $\widehat{\boldsymbol{\Sigma}}_0$ (their computation is considered in Section 6.8.5).

At iteration k, call $\widehat{\boldsymbol{\mu}}_k$ and $\widehat{\boldsymbol{\Sigma}}_k$ the current values and compute

$$d_{ki} = d(\mathbf{x}_i, \widehat{\boldsymbol{\mu}}_k, \widehat{\boldsymbol{\Sigma}}_k), \; \widehat{\sigma}_k = \widehat{\sigma}(d_{k1}, \ldots, d_{kn}), \; w_{ki} = W\left(\frac{d_{ki}}{\widehat{\sigma}_k}\right). \quad (6.59)$$

Then compute

$$\widehat{\boldsymbol{\mu}}_{k+1} = \frac{\sum_{i=1}^n w_{ki}\mathbf{x}_i}{\sum_{i=1}^n w_{ki}}, \; \widehat{\mathbf{C}}_k = \sum_{i=1}^n w_{ki}(\mathbf{x}_i - \widehat{\boldsymbol{\mu}}_{k+1})(\mathbf{x}_i - \widehat{\boldsymbol{\mu}}_{k+1})', \; \widehat{\boldsymbol{\Sigma}}_{k+1} = \frac{\widehat{\mathbf{C}}_k}{|\widehat{\mathbf{C}}_k|^{1/p}}. \quad (6.60)$$

It is shown in Section 9.6 that if the weight function W is nonincreasing, then $\widehat{\sigma}_k$ decreases at each step. One can then stop the iteration when the relative change $(\widehat{\sigma}_k - \widehat{\sigma}_{k+1})/\widehat{\sigma}_k$ is below a given tolerance. Experience shows that since the decrease of $\widehat{\sigma}_k$ is generally slow, it is not necessary to recompute it at each step, but at, say, every tenth iteration.

If W is not monotonic, the iteration steps (6.59)–(6.60) are not guaranteed to cause a decrease in $\widehat{\sigma}_k$ at each step. However, the algorithm can be modified to ensure a decrease at each iteration. Since the details are involved, they are deferred to Section 9.6.1.

6.8.3 Subsampling for estimators based on a robust scale

The obvious procedure to generate an initial approximation for an estimator defined by (6.28) is follow the general approach for regression described in Section 5.7.2, in which the minimization problem is replaced by a finite one, in which the candidate estimators are sample means and covariance matrices of subsamples. To obtain a finite set of candidate solutions, take a subsample of size $p + 1$, $\{\mathbf{x}_i : i \in J\}$, where the set $J \subset \{1, \ldots, n\}$ has $p + 1$ elements, and compute

$$\widehat{\boldsymbol{\mu}}_J = \text{ave}_{i \in J}(\mathbf{x}_i) \text{ and } \widehat{\boldsymbol{\Sigma}}_J = \frac{\widehat{\mathbf{C}}_J}{|\widehat{\mathbf{C}}_J|^{1/p}}, \tag{6.61}$$

where $\widehat{\mathbf{C}}_J$ is the covariance matrix of the subsample; and let

$$\mathbf{d}_J = \{d_{Ji} : i = 1, \ldots, n\}, \quad d_{Ji} = d(\mathbf{x}_i, \widehat{\boldsymbol{\mu}}_J, \widehat{\boldsymbol{\Sigma}}_J). \tag{6.62}$$

The problem of minimizing $\widehat{\sigma}$ is thus replaced by the finite problem of minimizing $\widehat{\sigma}(\mathbf{d}_J)$ over J. Since choosing all $\binom{n}{p+1}$ subsamples is prohibitive unless both n and p are rather small, we choose N of them at random, $\{J_k : k = 1, \ldots, N\}$, and the estimators are $\widehat{\boldsymbol{\mu}}_{J_k}, \widehat{\boldsymbol{\Sigma}}_{J_k}$ with

$$k^* = \arg \min_{k=1,\ldots,N} \widehat{\sigma}(\mathbf{d}_{J_k}). \tag{6.63}$$

If the sample contains a proportion ε of outliers, the probability of at least one "good" subsample is $1 - (1 - \alpha)^N$, where $\alpha = (1 - \varepsilon)^{p+1}$. If we want this probability to be larger than $1 - \delta$ we must have

$$N \geq \frac{|\ln \delta|}{|\ln(1 - \alpha)|} \approx \frac{|\ln \delta|}{(1 - \varepsilon)^{p+1}}. \tag{6.64}$$

See Table 5.3 for the values of N required as a function of p and ε.

A simple but effective improvement of the subsampling procedure is as follows. For subsample J with distances $\mathbf{d}_J$ defined in (6.62), let $\tau = \text{Med}(\mathbf{d}_J)$ and compute

$$\boldsymbol{\mu}_J^* = \text{ave}\{\mathbf{x}_i : d_{Ji} \leq \tau\}, \quad \mathbf{C}_J^* = \mathbf{Var}\{\mathbf{x}_i : d_{Ji} \leq \tau\}, \quad \boldsymbol{\Sigma}_J^* = \frac{\mathbf{C}_J^*}{|\mathbf{C}_J^*|^{1/p}}. \tag{6.65}$$

Then use $\boldsymbol{\mu}_J^*$ and $\mathbf{C}_J^*$ instead of $\widehat{\boldsymbol{\mu}}_J$ and $\widehat{\mathbf{C}}_J$. The motivation for this idea is that a subsample of $p + 1$ points is too small to yield reliable means and covariances, and so it is desirable to enlarge the subsample in a suitable way. This is done by selecting the half-sample with the smallest distances. Although no theoretical results are known for this method, our simulations show that this extra effort yields a remarkable improvement in the behavior of the estimators, with only a small increase in computational time. In particular, for the MVE estimator, the minimum scale obtained this way is always much smaller than that obtained with the original subsamples. For example, the ratio is about 0.3 for $p = 40$ and 500 subsamples.

6.8.4 The MVE

We have seen that the objective function of S-estimators can be decreased by iteration steps, and the same thing happens with the MCD (Section 6.8.6). However, no such improvements are known for the MVE, which makes the outcome of the subsampling procedure the only available approximation to the estimator.

The simplest approach to this problem is to use directly the "best" subsample given by (6.63). However, in view of the success of the improved subsampling method given by (6.65), we make it our method of choice for computing the MVE.

An exact method for the MVE was proposed by Agulló (1996) but, since it is not feasible except for small n and p, we do not describe it here.

6.8.5 Computation of S-estimators

Once we have initial values $\widehat{\boldsymbol{\mu}}_0$ and $\widehat{\boldsymbol{\Sigma}}_0$, an S-estimator is computed by means of the iterative procedures described in Section 6.8.2. We present two approaches to compute $\widehat{\boldsymbol{\mu}}_0$ and $\widehat{\boldsymbol{\Sigma}}_0$.

The simplest approach is to obtain initial values of $\boldsymbol{\mu}_0, \boldsymbol{\Sigma}_0$ through subsampling, and then apply the iterative algorithm. Much better results are obtained by following the same principles as the strategy described for regression in Section 5.7.3. This approach was first employed for multivariate estimation by Rousseeuw and van Driessen (1999) (see Section 6.8.6).

Our preferred approach, however, proceeds as was done in Section 5.5 for computing the MM-estimators of regression; that is, start the iterative algorithm from a bias-robust but possibly inefficient estimator. The MVE appears to be a candidate for this task. Adrover and Yohai (2002) computed the maximum asymptotic biases of the MVE, MCD, Stahel–Donoho and Rocke estimators, and concluded that the MVE has the smallest maximum bias for $p \geq 10$.

It is important to note that although the MVE estimator has the unattractive feature of a slow $n^{-1/3}$ rate of consistency, this feature does not affect the efficiency of the local minimum that is the outcome of the iterative algorithm, since it satisfies the M-estimating equations (6.31)–(6.32); if equations (6.17)–(6.18) for the $\widehat{\boldsymbol{\mu}}_\infty, \widehat{\boldsymbol{\Sigma}}_\infty$ have a unique solution, then *all* solutions of (6.31)–(6.32) converge to $(\widehat{\boldsymbol{\mu}}_\infty, \widehat{\boldsymbol{\Sigma}}_\infty)$ with a rate of order $n^{-1/2}$.

The MVE computed using the improved method (6.65) is an option for the initial estimator. However, it will be seen in Section 6.10 that a more complex procedure, to be described in Section 6.9.2, yields better results.

Other numerical algorithms have been proposed: by Ruppert (1992) and Woodruff and Rocke (1994).

6.8.6 The MCD

Rousseeuw and van Driessen (1999) found an iterative algorithm for the MCD, based on the following fact. Given any $\widehat{\boldsymbol{\mu}}_1$ and $\widehat{\boldsymbol{\Sigma}}_1$, let d_i be the corresponding squared distances. Then compute $\widehat{\boldsymbol{\mu}}_2$ and $\widehat{\mathbf{C}}$ as the sample mean and covariance matrix of the data with the h smallest of the d_is, and set $\widehat{\boldsymbol{\Sigma}}_2 = \widehat{\mathbf{C}}/|\widehat{\mathbf{C}}|^{1/p}$. Then $\widehat{\boldsymbol{\mu}}_2$ and $\widehat{\boldsymbol{\Sigma}}_2$ yield a lower value of $\widehat{\sigma}$ in (6.39) than $\widehat{\boldsymbol{\mu}}_1$ and $\widehat{\boldsymbol{\Sigma}}_1$. This is called the *concentration step* ("C-step" in the above paper), and a proof of the above reduction in $\widehat{\sigma}$ is given in Section 9.6.2. In this case, the modification (6.65) is not necessary, since the concentration steps already perform this sort of modification. The overall strategy then is as follows: for each of N candidate solutions obtained by subsampling, perform, say, two of the above steps, keep the 10 out of N that yield the smallest values of the criterion, and starting from each of them iterate the C-steps to convergence.

6.8.7 The Stahel–Donoho estimator

No exact algorithm for the Stahel–Donoho estimator is known. To approximate the estimator we need a large number of directions, and these can be obtained by subsampling. For each subsample $J = \{\mathbf{x}_{i_1}, \ldots, \mathbf{x}_{i_p}\}$ of size p, let $\mathbf{a}_J$ be a vector of norm 1 orthogonal to the hyperplane spanned by the subsample. The unit length vector $\mathbf{a}_J$ can be obtained by applying the QR orthogonalization procedure (see, for example, Chambers (1977) to $\{\mathbf{x}_{i_1} - \overline{\mathbf{x}}_J, \ldots, \mathbf{x}_{i_{p-1}} - \overline{\mathbf{x}}_J, \mathbf{b}\}$, where $\overline{\mathbf{x}}_J$ is the average of the subsample and $\mathbf{b}$ is any vector not collinear with $\overline{\mathbf{x}}_J$. Then we generate N subsamples $J_1, \ldots, J_N$ and replace (6.49) by

$$\widehat{t}(\mathbf{x}) = \max_k t(\mathbf{x}, \mathbf{a}_{J_k}).$$

It is easy to show that $\widehat{t}$ is invariant under affine transformations, and hence the approximate estimator is equivariant.

6.9 Faster robust scatter matrix estimators

Estimators based on a subsampling approach will be too slow when p is large; for example, of the order of a few hundred. We now present two faster methods for high-dimensional data. These are based on projections. The first is deterministic and is based on pairwise robust covariances. The second combines deterministic directions that yield extreme values of the kurtosis with random ones obtained by an elaborate procedure.

6.9.1 Using pairwise robust covariances

Much faster estimators can be obtained if equivariance is given up. The simplest approaches for location and dispersion are respectively to apply a robust location estimator to each coordinate and a robust estimator of covariance to each pair of variables. Such pairwise robust covariance estimators are easy to compute, but unfortunately the resulting scatter matrix lacks affine equivariance and positive definiteness. Besides, such estimators for location and scatter may lack both bias robustness and high normal efficiency if the data are very correlated. This is because the coordinate-wise location estimators need to incorporate the correlation structure for full efficiency for the normal distribution, and because the pairwise covariance estimators may fail to downweight higher-dimensional outliers.

A simple way to define a robust covariance between two random variables x, y is by truncation or rejection. Let ψ be a bounded monotone or redescending ψ-function, and $\mu(.)$ and $\sigma(.)$ robust location and scatter statistics. Then robust correlations and covariances can be defined as

$$\mathrm{RCov}(x,y) = \sigma(x)\sigma(y)\mathrm{E}\left[\psi\left(\frac{x-\mu(x)}{\sigma(x)}\right)\psi\left(\frac{y-\mu(y)}{\sigma(y)}\right)\right], \tag{6.66}$$

$$\mathrm{RCorr}(x,y) = \frac{\mathrm{RCov}(x,y)}{[\mathrm{RCov}(x,x)\mathrm{RCov}(y,y)]^{1/2}}. \tag{6.67}$$

See Huber and Ronchetti (2009; Sec 8.2–8.3). This definition satisfies $\mathrm{RCorr}(x,x) = 1$. When $\psi(x) = \mathrm{sgn}(x)$ and μ is the median, (6.67) and (6.70) are called the *quadrant* correlation and covariance estimators. The sample versions of (6.66) and (6.67) are obtained by replacing the expectation by the average, and μ and σ by their estimators $\hat{\mu}$ and $\hat{\sigma}$.

These estimators are not consistent under a given model. In particular, if $\mathcal{D}(x,y)$ is bivariate normal with correlation ρ and ψ is monotone, then the value ρ_R of $\mathrm{RCorr}(x,y)$ is an increasing function $\rho_R = g(\rho)$ of ρ, which can be computed (Problem 6.11). Then, the estimator $\hat{\rho}_R$ of ρ_R can be corrected to ensure consistency for the normal model by using the inverse transformation $\hat{\rho} = g^{-1}(\hat{\rho}_R)$.

Another robust pairwise covariance, initially proposed by Gnanadesikan and Kettenring (1972) and later studied by Devlin *et al.* (1981), is based on the identity

$$\mathrm{Cov}(x,y) = \frac{1}{4}(\mathrm{SD}\ (x+y)^2 - \mathrm{SD}\ (x-y)^2). \tag{6.68}$$

The proposal defined a robust correlation by replacing the standard deviation by a robust scatter σ (they chose a trimmed standard deviation):

$$\mathrm{RCorr}(x,y) = \frac{1}{4}\left(\sigma\left(\frac{x}{\sigma(x)} + \frac{y}{\sigma(y)}\right)^2 - \sigma\left(\frac{x}{\sigma(x)} - \frac{y}{\sigma(y)}\right)^2\right). \tag{6.69}$$

A robust covariance is defined by

$$\mathrm{RCov}(x,y) = \sigma(x)\sigma(y)\mathrm{RCorr}(x,y). \tag{6.70}$$

The latter satisfies

$$\mathrm{RCov}(t_1x, t_2y) = t_1t_2\mathrm{RCov}(x,y) \qquad \text{for all } t_1, t_2 \in R \tag{6.71}$$

and

$$\mathrm{RCov}(x,x) = \sigma(x)^2.$$

Note that dividing x and y by their σs in (6.69) is required for (6.71) to hold.

The above pairwise robust covariances can be used in the obvious way to define a "robust correlation (or covariance) matrix" of a random vector $\mathbf{x} = (x_1, \ldots, x_p)'$. The resulting scatter matrix is symmetric but not necessarily positive semidefinite, and is not affine equivariant. Genton and Ma (1999) calculated the influence function and asymptotic efficiency of the estimators of such matrices. It can be shown that the above correlation matrix estimator is consistent if $\mathcal{D}(\mathbf{x})$ is an elliptical distribution, and a proof is given in Section 6.17.10.

Maronna and Zamar (2002) show that a simple modification of Gnanadesikan and Kettenring's approach yields a positive definite matrix and "approximately equivariant" estimators of location and scatter. Recall that if $\mathbf{\Sigma}$ is the covariance matrix of the p-dimensional random vector $\mathbf{x}$, and σ denotes the standard deviation, then

$$\sigma(\mathbf{a}'\mathbf{x})^2 = \mathbf{a}'\mathbf{\Sigma}a \tag{6.72}$$

for all $\mathbf{a} \in R^p$. The lack of positive semidefiniteness of the Gnanadesikan–Kettenring matrix is overcome by a modification that forces (6.72) for a robust σ and a set of "principal directions", and is based on the observation that the eigenvalues of the covariance matrix are the variances along the directions given by the respective eigenvectors.

Let $\mathbf{X} = [x_{ij}]$ be an $n \times p$ data matrix with rows $\mathbf{x}_i'$, $i = 1, \ldots, n$, and columns $\mathbf{x}^j$, $j = 1, \ldots, p$. Let $\widehat{\sigma}(.)$ and $\widehat{\mu}(.)$ be robust univariate location and scatter statistics. For a data matrix $\mathbf{X}$, we shall define a robust scatter matrix estimator $\widehat{\mathbf{\Sigma}}(\mathbf{X})$ and a robust location vector estimator $\widehat{\boldsymbol{\mu}}(\mathbf{X})$ by the following computational steps:

1. First compute a normalized data matrix $\mathbf{Y}$ with columns $\mathbf{y}^j = \mathbf{x}^j / \widehat{\sigma}(\mathbf{x}^j)$, and hence with rows

$$\mathbf{y}_i = \mathbf{D}^{-1}\mathbf{x}_i \quad (i = 1, \ldots, n) \text{ where } \mathbf{D} = \mathrm{diag}(\widehat{\sigma}(\mathbf{x}^1), \ldots, \widehat{\sigma}(\mathbf{x}^p)). \tag{6.73}$$

2. Compute a robust "correlation matrix" $\mathbf{U} = [U_{jk}]$ of $\mathbf{X}$ as the "covariance matrix" of $\mathbf{Y}$ by applying (6.69) to the columns of $\mathbf{Y}$:

$$U_{jj} = 1, \; U_{jk} = \frac{1}{4}\left[\widehat{\sigma}(\mathbf{y}^j + \mathbf{y}^k)^2 - \widehat{\sigma}(\mathbf{y}^j - \mathbf{y}^k)^2\right] \quad (j \neq k).$$

3. Compute the eigenvalues λ_j and eigenvectors $\mathbf{e}_j$ of $\mathbf{U}$ $(j = 1, \ldots, p)$, and let $\mathbf{E}$ be the matrix whose columns are the $\mathbf{e}_j$. It follows that $\mathbf{U} = \mathbf{E}\mathbf{\Lambda}\mathbf{E}'$, where $\mathbf{\Lambda} = \mathrm{diag}(\lambda_1, \ldots, \lambda_p)$. Here the λ_i need not be nonnegative. This is the "principal component decomposition" of $\mathbf{Y}$.
4. Compute the matrix $\mathbf{Z}$ with

$$\mathbf{z}_i = \mathbf{E}'\mathbf{y}_i = \mathbf{E}'\mathbf{D}^{-1}\mathbf{x}_i, \; (i = 1, \ldots, n) \tag{6.74}$$

so that $(\mathbf{z}^1, \ldots, \mathbf{z}^p)$ are the "principal components" of $\mathbf{Y}$.

5. Compute $\widehat{\sigma}(\mathbf{z}^j)$ and $\widehat{\mu}(\mathbf{z}^j)$ for $j = 1, ..., p$, and set

$$\mathbf{\Gamma} = \text{diag}\left(\widehat{\sigma}(\mathbf{z}^1)^2, ..., \widehat{\sigma}(\mathbf{z}^p)^2\right), \quad \boldsymbol{\nu} = (\widehat{\mu}(\mathbf{z}^1), ..., \widehat{\mu}(\mathbf{z}^p))'.$$

Here the elements of $\mathbf{\Gamma}$ are nonnegative. Being "principal components" of $\mathbf{Y}$, the $\mathbf{z}^j$ should be approximately uncorrelated with covariance matrix $\mathbf{\Gamma}$.

6. Now transform back to $\mathbf{X}$ with

$$\mathbf{x}_i = \mathbf{A}\mathbf{z}_i, \quad \text{with} \quad \mathbf{A} = \mathbf{D}\mathbf{E}. \tag{6.75}$$

and finally define

$$\widehat{\mathbf{\Sigma}}(\mathbf{X}) = \mathbf{A}\mathbf{\Gamma}\mathbf{A}', \ \widehat{\boldsymbol{\mu}}(\mathbf{X}) = \mathbf{A}\boldsymbol{\nu}. \tag{6.76}$$

The justification for the last equation is that, if $\boldsymbol{\nu}$ and $\mathbf{\Gamma}$ were the mean and covariance matrix of $\mathbf{Z}$, since $\mathbf{x}_i = \mathbf{A}\mathbf{z}_i$ the mean and covariance matrix of $\mathbf{X}$ would be given by (6.76).

Note that (6.73) makes the estimator scale equivariant, and that (6.76) replaces the λ_i, which may be negative, with the "robust variances" $\sigma(\mathbf{z}^j)^2$ of the corresponding directions. The reason for defining $\widehat{\boldsymbol{\mu}}$ as in (6.76) is that it is better to apply a coordinate-wise location estimator to the approximately uncorrelated $\mathbf{z}^j$ and then transform back to the $\mathbf{X}$-coordinates than to apply a coordinate-wise location estimator directly to the $\mathbf{x}^j$s.

The procedure can be iterated in the following way. Put $\widehat{\boldsymbol{\mu}}_{(0)} = \widehat{\boldsymbol{\mu}}(\mathbf{X})$ and $\widehat{\mathbf{\Sigma}}_{(0)} = \widehat{\mathbf{\Sigma}}(\mathbf{X})$. At iteration k, we have $\widehat{\boldsymbol{\mu}}_{(k)}$ and $\widehat{\mathbf{\Sigma}}_{(k)}$, whose computation has required computing a matrix $\mathbf{A}$, as in (6.75). Call $\mathbf{Z}_{(k)}$ the matrix with rows $\mathbf{z}_i = \mathbf{A}^{-1}\mathbf{x}_i$. Then $\widehat{\boldsymbol{\mu}}_{(k+1)}$ and $\widehat{\mathbf{\Sigma}}_{(k+1)}$ are obtained by computing $\widehat{\mathbf{\Sigma}}$ and $\widehat{\boldsymbol{\mu}}$ for $\mathbf{Z}_{(k)}$ and then expressing them back in the original coordinate system. More precisely, we define

$$\widehat{\mathbf{\Sigma}}_{(k+1)}(\mathbf{X}) = \mathbf{A}\widehat{\mathbf{\Sigma}}_{(k)}(\mathbf{Z})\mathbf{A}', \ \widehat{\boldsymbol{\mu}}_{(k+1)}(\mathbf{X}) = \mathbf{A}\widehat{\boldsymbol{\mu}}_{(k)}(\mathbf{Z}). \tag{6.77}$$

The reason for iterating is that the first step works very well when the data have low correlations; and the $\mathbf{z}^j$s are (hopefully) less correlated than the original variables. The resulting estimator will be called the "orthogonalized Gnanadesikan–Kettenring estimator" (OGK).

A final step is convenient both to increase the estimator's efficiency and to make it "more equivariant". The simplest and fastest option is the reweighting procedure in Section 6.4.6. But it is much better to use this estimator as the starting point for the iterations of an S-estimator.

Since a large part of the computing effort is consumed by the univariate estimators $\widehat{\mu}$ and $\widehat{\sigma}$, they must be fast. The experiments by Maronna and Zamar (2002) showed that it is desirable that $\widehat{\mu}$ and $\widehat{\sigma}$ be both bias robust and efficient for the normal distribution in order for $\widehat{\mathbf{\Sigma}}$ and $\widehat{\boldsymbol{\mu}}$ to perform satisfactorily. To this end, the scatter estimator $\widehat{\sigma}$ is defined in a way similar to the τ-scale estimator (5.27), which is a truncated standard deviation, and the location estimator $\widehat{\mu}$ is a weighted mean.

More precisely, let $\widehat{\mu}_0$ and $\widehat{\sigma}_0$ be the median and MAD. Let W be a weight function and ρ a ρ-function. Let $w_i = W((x_i - \widehat{\mu}_0)/\widehat{\sigma}_0)$ and

$$\widehat{\mu} = \frac{\sum_i x_i w_i}{\sum_i w_i}, \quad \widehat{\sigma}^2 = \frac{\sigma_0^2}{n} \sum_i \rho\left(\frac{x_i - \widehat{\mu}}{\widehat{\sigma}_0}\right).$$

An adequate balance of robustness and efficiency is obtained with W the bisquare weight function (2.57) with $k = 4.5$, and ρ as the bisquare ρ (2.38) with $k = 3$.

It is shown by Maronna and Zamar (2002) that if the BPs of $\widehat{\mu}$ and $\widehat{\sigma}$ are not less than ε then so is the BP of $(\widehat{\boldsymbol{\mu}}, \widehat{\boldsymbol{\Sigma}})$, as long as the data are not collinear. Simulations in their paper show that two is an adequate number of iterations (6.77), and that further iterations do not seem to converge and yield no improvement.

An implementation of the OGK estimator for applications to data mining was discussed by Alqallaf *et al.* (2002), using the quadrant correlation estimator. A reason for focusing on the quadrant correlation was the desire to operate on huge datasets that are too large to fit in computer memory. A fast bucketing algorithm can be used to compute this estimator on “streaming” input data (data read into the computer sequentially from a database). The median and MAD estimators were used for robust location and scatter because there are algorithms for the approximate computation of order statistics from a single pass on large streaming datasets (Manku *et al.* 1999).

6.9.2 The Peña–Prieto procedure

The *kurtosis* of a random variable x is defined as

$$\text{Kurt}\,(x) = \frac{\text{E}(x - \text{E}x)^4}{\text{SD}\,(x)^4}.$$

Peña and Prieto (2007) propose an equivariant procedure based on the following observation. A distribution is called *unimodal* if its density has a maximum at some point x_0, and is increasing for $x < x_0$ and decreasing for $x > x_0$. Then it can be shown that the kurtosis is a measure of both heavy-tailedness and unimodality. It follows that, roughly speaking, for univariate data a small proportion of outliers increases the kurtosis, since it makes the data tails heavier, and a large proportion decreases the kurtosis, since it makes the data more bimodal.

Hence Peña and Prieto look for projections that either maximize or minimize the kurtosis, and use them in a way similar to the Stahel–Donoho estimator.

At the same time, they point out that ensuring a high probability of getting “good” directions (those that detect outliers) by ordinary subsampling, as in Section 6.8.7, can be very low unless the number of directions is extremely large. For this reason they also employ a number of random “specific directions”, which are obtained by a kind of stratified sampling. The intuitive idea is that if one selects two observations at random, project the data in the direction defined by them and then selects the sub-sample from the extremes of the projected data, one can increase the probability of generating "good" directions, because the proportion of good or bad observations in the extremes is expected to be greater than in the whole sample.

The procedure is complex, but it may be summarized as follows for p-dimensional data:

1. In Stage I of the procedure, two sets of p directions $\mathbf{a}$ are found; one corresponding to local maxima and the other to local minima of the kurtosis.
2. In Stage II, a number L of directions are generated, each defined by two random points; the space is cut into a number of "slices" orthogonal to each direction, and a subsample is chosen at random from each slice, which defines a direction.
3. The outlyingness of each data point is measured through (6.49), with the vector $\mathbf{a}$ ranging over the directions computed in Stages I and II.
4. Points with outlyingness above a given threshold are transitorily deleted, and (1)–(2) are iterated on the remaining points until no more deletions take place.
5. The sample mean and covariance matrix of the remaining points is computed.
6. Deleted points whose Mahalanobis distances are below a threshold are again included, and steps (4)–(5) are repeated until no more inclusions take place.

The procedure is very fast for high dimensions. Maronna (2017) found out that in certain extreme situations when the contamination rate is "high" (≥ 0.2) and the ratio n/p is "low" (< 10), KSD may be unstable and yield useless values. For this reason he proposed two simple modifications that largely correct this drawback with only a small increase in computing time.

Although there are no full theoretical results for the breakdown point of KSD, Maronna and Yohai (2017) show that its asymptotic BP is 0.5 for point-mass contamination under elliptical distributions, and the simulations by Peña and Prieto (2007) also suggest that it has a high FBP.

In Section 6.10 we shall employ this procedure as a starting point for the computation of MM-estimators and of estimators based on robust scales.

6.10 Choosing a location/scatter estimator

So far, we have described several types of robust location/scale estimators, and the purpose of this section is to give guidelines for choosing among them, taking into account their efficiency, robustness and computing times. Our results are based on an extensive simulation study by Maronna and Yohai (2017).

The first issue is how we measure the performance of an estimator $(\hat{\boldsymbol{\mu}}, \hat{\boldsymbol{\Sigma}})$ given a central model $N_p(\boldsymbol{\mu}_0, \boldsymbol{\Sigma}_0)$. As in Section 5.9.3.2, we shall employ the Kullback–Leibler divergence, defined in (5.59) (henceforth D for short). It is straightforward to show that in the normal family, for $\boldsymbol{\mu}$ with known Σ, we have

$$D(\hat{\boldsymbol{\mu}}) = (\hat{\boldsymbol{\mu}} - \boldsymbol{\mu}_0)'\boldsymbol{\Sigma}_0^{-1}(\hat{\boldsymbol{\mu}} - \boldsymbol{\mu}_0), \tag{6.78}$$

and for Σ with known μ we have

$$D(\widehat{\mathbf{\Sigma}}) = \mathrm{trace}\,(\mathbf{\Sigma}_0^{-1}\widehat{\mathbf{\Sigma}}) - \log|\mathbf{\Sigma}_0^{-1}\widehat{\mathbf{\Sigma}}| - p. \tag{6.79}$$

The normal finite-sample efficiency is then defined as

$$\mathrm{eff}\,(\widehat{\boldsymbol{\mu}}) = \frac{\mathrm{E}D\,(\overline{\mathbf{x}})}{\mathrm{E}D\,(\widehat{\boldsymbol{\mu}})}, \qquad \mathrm{eff}\,(\widehat{\mathbf{\Sigma}}) = \frac{\mathrm{E}D\,(\mathbf{C})}{\mathrm{E}D\,(\widehat{\mathbf{\Sigma}})}, \tag{6.80}$$

where $\overline{\mathbf{x}}$ and $\mathbf{C}$ are the sample mean and covariance matrix, computed from samples of size n.

The elements to be compared are:

- *The estimators*: We consider four types of estimators for which the efficiency can be controlled without affecting the BP: Rocke, MM, τ and Stahel-Donoho. We add for completeness four other estimators with uncontrollable efficiency: the S-estimator (S-E), the MVE, the MCD with one-step reweighting, and Peña and Prieto's "KSD" estimator described in Section 6.9.2. In all cases except for the KSD the scales were tuned to attain the maximum FBP (6.27). As explained at the end of Section 6.9.2, there is strong evidence to believe that the FBP of KSD is also near the maximum one given by (6.27). All scatter estimators were corrected for "size" by means of (6.22),
- *The starting values*: For Rocke, MM, τ and S-E we compare the MVE and KSD estimators as starting values. For Stahel–Donoho we compare the directions supplied by KSD with those obtained by subsampling.
- *The $\rho-$ (or $W-$) functions*: For MM, τ and S-E we compare the bisquare and SHR ρs; for Stahel–Donoho we compare SHR and Huber weights (6.55).

6.10.1 Efficiency

For all estimators with controllable efficiency, the tuning constants for each estimator are chosen to make the finite-sample efficiency of $\widehat{\mathbf{\Sigma}}$ equal to 0.90. This priority of $\widehat{\mathbf{\Sigma}}$ over $\widehat{\boldsymbol{\mu}}$ is due to the fact that, as explained in Section 6.17.4, robustly estimating scatter is "more difficult" than location. The simulations showed that in all cases $\mathrm{eff}\,(\widehat{\boldsymbol{\mu}}) \geq \mathrm{eff}\,(\widehat{\mathbf{\Sigma}})$. A preliminary simulation was performed for each estimator. Its tuning constants were computed for $n = Kp$ with $K = 5$, 10 and 20 and p between 5 and 50, and were then fitted as simple functions of n and p. Relevant values are given in Section 6.10.4.

A simulation was run to estimate (6.80), replacing the expected values $\mathrm{E}D$ by their Monte Carlo averages $\overline{D}$ computed over 1000 samples of size n from $\mathrm{N}_p(\boldsymbol{\mu}_0, \mathbf{\Sigma}_0)$ with $(\boldsymbol{\mu}_0, \mathbf{\Sigma}_0) = (\mathbf{0}, \mathbf{I})$; since all estimators are equivariant, the results do not depend on $(\boldsymbol{\mu}_0, \mathbf{\Sigma}_0)$. The simulation showed the efficiency cannot be controlled in all cases:

- For $p = 10$ the efficiency of the Rocke scatter estimator is 0.73, and is still lower for smaller p. The explanation is that when α tends to zero, the estimator does not tend to the covariance matrix unless p is large enough.

- The minimum efficiency of the τ-estimators over all constants c tends to 1 with increasing p, for both ρ-functions. In particular, it is > 0.95 for $p \geq 50$. The reason is that when c is small, the τ-scale approaches the M-scale, and therefore the τ-estimator approaches the S-estimator.
- The simulations under contamination, described in the next section, showed that when $p \leq 10$, the MM- estimator with efficiency 0.90 is too sensitive to outliers, and therefore lower efficiencies were chosen for those cases.

As explained in Sections 6.4.1, 6.4.3 and 6.4.6, the efficiency of the MVE is very low, and that of MCD is low unless one accepts a very low BP.

As to KSD, its efficiency depends on n and p: it is less than 0.5 for $n = 5p$ and less than 0.8 for $n = 10p$.

6.10.2 Behavior under contamination

To assess the estimators' robustness, a simulation was run with contaminated data from $N_p(\mathbf{0}, \mathbf{I})$. For each estimator and scenario, the average $\overline{D}$ of (6.78) and (6.79) was computed. Given the contamination rate $\varepsilon \in (0, 1)$ let $m = [n\varepsilon]$, where n is the sample size. The first coordinate x_{i1} of $\mathbf{x}_i$ $(i = 1, ..., m)$ is replaced by $\gamma x_{i1} + K$, where K is the outliers' size and the constant γ determines the scatter of the outliers. Here K is varied between 1 and 12 in order to find the maximum $\overline{D}$ for all estimators. The values chosen for ε were 0.1 and 0.2, and those for γ were 0 and 0.5.

The simulations were run for $p = 5$, 10, 15, 20 and 30, and $n = mp$ with $m = 5$, 10 and 20. Each estimator was evaluated by its maximum $\overline{D}$. The most important conclusions are the following:

- The price paid for the high efficiency of S-E is a large loss of robustness.
- KSD is always better than MVE as a starting estimator for MM and τ.
- KSD is generally better than subsampling for Stahel–Donoho.
- The SHR ρ is always better than the bisquare ρ for both MM and τ.
- The SHR weights are better than Huber weights for Stahel–Donoho.
- In all situations, the best estimators are MM and τ with SHR ρ, Rocke, and Stahel–Donoho, all starting from KSD.
- Although the results for $\gamma = 0$ and 0.5 are different, the comparisons among estimators are almost the same.
- The relative performances of the estimators for location and scatter are similar.
- The relative performances of the estimators for $n = 5p$, $10p$ and $20p$ are similar.

Table 6.3 shows a reduced version of the results, for $n = 10p$ and $\gamma = 0$, of the maximum $\overline{D}$s of the scatter estimators corresponding to MM and τ (both with SHR ρ), Rocke and Stahel–Donoho (S-D), all starting from KSD. For completeness, we add S-E with KSD start, and MCD. The results for estimators with efficiency less than 0.9 are shown in italics.

Table 6.3 Maximum mean Kullback–Leibler divergences of scatter matrices, for $n = 10p$ and $\gamma = 0$. Italics denote estimators with less than 90% efficiency

p	ε	MM	τ	Rocke	S-D	S-E	MCD
5	0.1	*0.85*	0.89	*0.95*	0.99	*1.09*	*1.99*
	0.2	*2.27*	2.46	*2.61*	4.53	*4.38*	*17.58*
10	0.1	*1.67*	*1.77*	*1.43*	1.61	*3.54*	*6.66*
	0.2	*3.88*	*4.53*	*3.48*	7.94	*11.26*	*21.89*
15	0.1	2.38	2.98	1.95	2.26	*6.68*	*12.53*
	0.2	5.68	7.85	4.47	12.31	*19.82*	*28.33*
20	0.1	3.32	4.59	2.49	3.00	*10.03*	*16.46*
	0.2	7.90	12.62	3.17	17.09	*25.41*	*32.04*
30	0.1	5.34	8.56	3.03	4.64	*18.39*	*17.66*
	0.2	14.21	20.71	5.61	29.66	*49.14*	*34.02*

It is seen that:

- The performance of S-D is competitive for $\varepsilon = 0.1$, but is poor for $\varepsilon = 0.2$.
- For $p < 10$, MM has the best overall performance.
- For $p \geq 10$, Rocke has the best overall performance.
- The MCD has poor performance.

Figure 6.9 shows the values of $\overline{D}$ as a function of the outlier size K for some of the estimators in the case $p = 20$, $n = 200$ and $\gamma = 0$. Here "MM-SHR" stands for "MM with SHR ρ". All estimators in the lower panel start from KSD. The plot confirms the superiority of Rocke with KSD start.

6.10.3 Computing times

We compare the computing times of the Rocke estimator with MVE and KSD starts. The results are the average of 20 runs with normal samples, on a PC with Intel TM12 Duo CPU and 3.01 GHz clock speed. The values of n were $5p$, $10p$ and $20p$, with p between 20 and 100.

The rational way to choose the number of subsamples N_{sub} for the MVE would be to ensure a given (probabilistic) breakdown point. But according to (6.64), the values of N_{sub} that ensure a breakdown point of just 0.15 is 18298 for $p = 50$ and 6.18×10^7 for $p = 100$l; these are impractically large. For this reason, N_{sub} was chosen to increase more slowly so as to yield feasible computing times, namely as $N_{\text{sub}} = 50p$.

Table 6.4 displays the results, where for brevity we only show the values for $p = 20$, 50, 80 and 100. It is seen that Rocke+KSD is faster than Rocke+MVE.

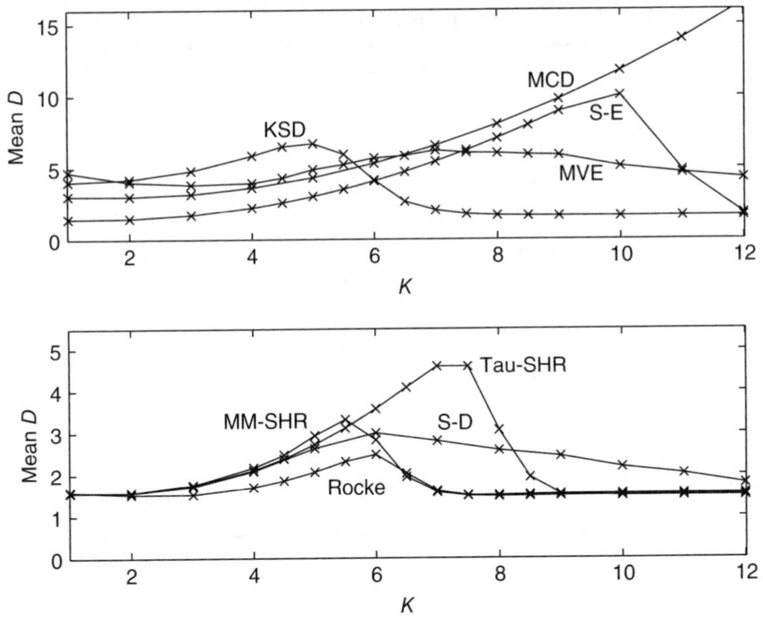

Figure 6.9 Mean D for $p = 20$, $n = 200$, $\varepsilon = 0.1$ and $\gamma = 0$ with outlier size K

6.10.4 Tuning constants

For the MM-estimator with KSD start and SHR, c is approximated by

$$c = a + \frac{b}{p} + c\frac{p}{n},$$

with $a = 0.612$, $b = 4.504$ and $c = -1.112$.

For the Rocke estimator with KSD, the value of α which yields 90% efficiency is approximated for $p \geq 15$ by

$$\alpha = ap^b n^c$$

with $a = 0.00216$, $b = -1.0078$, $c = 0.8156$.

6.10.5 Conclusions

The Rocke estimator has a controllable efficiency for $p \geq 15$. With equal efficiencies, the Rocke estimator with KSD start outperforms all its competitors. Its computing time is competitive for $p \leq 100$. Bearing in mind the trade-off between robustness and efficiency, we recommend the Rocke estimator for $p \geq 10$ and the MM-estimator with SHR for $p < 10$ (in both cases with KSD start).

Table 6.4 Mean computing times of estimators

p	n	Rocke+MVE (s)	Rocke+KSD (s)
20	100	0.62	0.09
	200	0.98	0.10
	400	1.31	0.15
50	250	5.03	0.46
	500	6.54	1.02
	1000	12.72	2.59
80	400	14.55	4.20
	800	22.46	10.01
	1600	65.45	16.73
100	500	28.86	27.56
	1000	74.01	55.69
	2000	152.06	65.76

6.11 Robust principal components

Principal components analysis (PCA) is a widely used method for dimensionality reduction. Let $\mathbf{x}$ be a p-dimensional random vector with mean $\boldsymbol{\mu}$ and covariance matrix $\boldsymbol{\Sigma}$. The first principal component is the univariate projection of maximum variance; more precisely, it is the linear combination $\mathbf{x}'\mathbf{b}_1$, where $\mathbf{b}_1$ (called the first *principal direction*) is the vector $\mathbf{b}$ such that

$$\mathrm{Var}(\mathbf{b}'\mathbf{x}) = \max \text{ subject to } \|\mathbf{b}\| = 1. \tag{6.81}$$

The second principal component is $\mathbf{x}'\mathbf{b}_2$, where $\mathbf{b}_2$ (the second principal direction) satisfies (6.81) with $\mathbf{b}_2'\mathbf{b}_1 = 0$, and so on. Call $\lambda_1 \geq \lambda_2 \geq \ldots \geq \lambda_p$ the eigenvalues of $\boldsymbol{\Sigma}$. Then $\mathbf{b}_1, \ldots, \mathbf{b}_p$ are the respective eigenvectors and $\mathrm{Var}(\mathbf{b}_j'\mathbf{x}) = \lambda_j$. The number q of components can be chosen on the basis of the "proportion of unexplained variance"

$$\frac{\sum_{j=q+1}^{p} \lambda_j}{\sum_{j=1}^{p} \lambda_j}. \tag{6.82}$$

PCA can be viewed in an alternative geometric form in the spirit of regression modeling. Consider finding a q-dimensional hyperplane H such the orthogonal distance of $\mathbf{x}$ to H is the "smallest", in the following sense. Call $\widehat{\mathbf{x}}_H$ the point of H closest in Euclidean distance to $\mathbf{x}$; that is, such that

$$\widehat{\mathbf{x}}_H = \arg\min_{\mathbf{z}\in H} \|\mathbf{x} - \mathbf{z}\|.$$

Then we look for H^* such that

$$\mathrm{E}\,\|\mathbf{x}-\widehat{\mathbf{x}}_{H^*}\|^2=\min. \tag{6.83}$$

It can be shown (Seber, 1984) that H^* contains the mean $\boldsymbol{\mu}$ and has the directions of the first q eigenvectors $\mathbf{b}_1,\ldots,\mathbf{b}_q$, and so H^* is the set of translated linear combinations of $\mathbf{b}_1,\ldots,\mathbf{b}_q$:

$$H^*=\left\{\boldsymbol{\mu}+\sum_{k=1}^{q}\alpha_k\mathbf{b}_k\ :\ \alpha_1,\ldots,\alpha_q\in R\right\}. \tag{6.84}$$

Then

$$z_j=(\mathbf{x}-\boldsymbol{\mu})'\mathbf{b}_j,\quad (j=1,\ldots,q) \tag{6.85}$$

are the coordinates of the centered $\mathbf{x}$ in the coordinate system of the $\mathbf{b}_j$, and

$$d_H=\sum_{j=q+1}^{p}z_j^2=\|\mathbf{x}-\widehat{\mathbf{x}}_H\|^2,\quad d_C=\sum_{j=1}^{q}z_j^2 \tag{6.86}$$

are the squared distances from $\mathbf{x}$ to H and from $\widehat{\mathbf{x}}_H$ to $\boldsymbol{\mu}$, respectively.

Note that the results of PCA are not invariant under general affine transformations, in particular under changes in the units of the variables. Doing PCA implies that we consider the Euclidean norm to be a sensible measure of distance, and this may require a previous rescaling of the variables. PCA is, however, invariant under *orthogonal* transformations; that is, transformations that do not change Euclidean distances.

Given a dataset $X=\{\mathbf{x}_1,\ldots,\mathbf{x}_n\}$, the sample principal components are computed by replacing $\boldsymbol{\mu}$ and $\boldsymbol{\Sigma}$ by the sample mean and covariance matrix. For each observation $\mathbf{x}_i$, we compute the scores $z_{ij}=(\mathbf{x}_i-\overline{\mathbf{x}})'\mathbf{b}_j$ and the distances

$$d_{\widehat{H},i}=\sum_{j=q+1}^{p}z_{ij}^2=\left\|\mathbf{x}_i-\widehat{\mathbf{x}}_{\widehat{H},i}\right\|^2,\quad d_{C.i}=\sum_{j=1}^{q}z_{ij}^2, \tag{6.87}$$

where $\widehat{H}$ is the estimated hyperplane. A simple data analytic tool, similar to the plot of residuals versus fitted values in regression, is to plot $d_{\widehat{H},i}$ against $d_{C,i}$.

As can be expected, outliers may have a distorting effect on the results. For instance, in Example 6.1, the first principal component of the correlation matrix of the data explains 75% of the variability, while after deleting the atypical point it explains 90%. The simplest way to deal with this problem is to replace $\overline{\mathbf{x}}$ and $\mathbf{Var}(X)$ with robust estimators $\widehat{\boldsymbol{\mu}}$ and $\widehat{\boldsymbol{\Sigma}}$ of multivariate location and scatter. Campbell (1980) uses M-estimators. Croux and Haesbroeck (2000) discuss several properties of this approach. Note that the results depend only on the shape of $\widehat{\boldsymbol{\Sigma}}$ (Section 6.3.2).

However, better results can be obtained by taking advantage of the particular features of PCA. For affine equivariant estimation, the "natural" metric is that given by squared Mahalanobis distances $d_i = (\mathbf{x}_i - \hat{\boldsymbol{\mu}})'\hat{\boldsymbol{\Sigma}}^{-1}(\mathbf{x}_i - \hat{\boldsymbol{\mu}})$, which depend on the data through $\hat{\boldsymbol{\Sigma}}^{-1}$, while for PCA we have a fixed metric given by Euclidean distances. This implies that the concept of outliers changes. In the first case, an outlier that should be downweighted is a point with a large squared Mahalanobis distance to the center of the data, while in the second it is a point with a large Euclidian distance to the hyperplane ("large" as compared to the majority of points). For instance, consider two independent variables with zero means and standard deviations 10 and 1, and $q = 1$. The first principal component corresponds to the first coordinate axis. Two data values, one at (100, 1) and one at (10, 10), have identical large squared Mahalanobis distances of 101, but their Euclidean distances to the first axis are 1 and 10 respectively. The second one would be harmful to the estimation of the principal components, but the first one is a "good" point for that purpose.

Boente (1983, 1987) studied M-estimators for PCA. An alternative approach to robust PCA is to replace the variance in (6.81) by a robust scale. This approach was first proposed by Li and Chen (1985), who found serious computational problems. Croux and Ruiz-Gazen (1996) proposed an approximation based on a finite number of directions. Hubert *et al.* (2012) propose a method that combines projection pursuit ideas with robust scatter matrix estimation.

The next sections describe two of our preferred approaches to robust PCA. One is based on robust fitting of a hyperplane H by minimization of a robust scale, while the other is a simple and fast "spherical" principal components method that works well for large datasets.

6.11.1 Spherical principal components

In this section we describe a simple but effective approach proposed by Locantore *et al.* (1999). Let $\mathbf{x}$ have an elliptical distribution (6.7), in which case if $\mathbf{Var}(\mathbf{x})$ exists it is a constant multiple of $\boldsymbol{\Sigma}$. Let $\mathbf{y} = (\mathbf{x} - \boldsymbol{\mu})/\|\mathbf{x} - \boldsymbol{\mu}\|$; that is, $\mathbf{y}$ is the normalization of $\mathbf{x}$ to the surface of the unit sphere centered at $\boldsymbol{\mu}$. Boente and Fraiman (1999) showed that the eigenvectors $\mathbf{t}_1, \ldots, \mathbf{t}_p$ (but not the eigenvalues!) of the covariance matrix of $\mathbf{y}$ (that is, its principal axes) coincide with those of $\boldsymbol{\Sigma}$. They showed furthermore that if $\sigma(.)$ is any scatter statistic, then the values $\sigma(\mathbf{x}'\mathbf{t}_j)^2$ are proportional to the eigenvalues of $\boldsymbol{\Sigma}$. Proofs are given in Section 6.17.11.

This result is the basis for a simple robust approach to PCA, called *spherical principal components* (SPC). Let $\hat{\boldsymbol{\mu}}$ be a robust multivariate location estimator, and compute

$$\mathbf{y}_i = \begin{cases} (\mathbf{x}_i - \hat{\boldsymbol{\mu}})/\|\mathbf{x}_i - \hat{\boldsymbol{\mu}}\| & \text{if} \quad \mathbf{x}_i \neq \hat{\boldsymbol{\mu}} \\ \mathbf{0} & \text{otherwise} \end{cases}.$$

Let $\hat{\mathbf{V}}$ be the sample covariance matrix of the $\mathbf{y}_i$s with corresponding eigenvectors $\mathbf{b}_j$ $(j = 1, \ldots, p)$. Now compute $\hat{\lambda}_j = \hat{\sigma}(\mathbf{x}'\mathbf{b}_j)^2$, where $\hat{\sigma}$ is a robust scatter estimator (such as the MAD). Call $\hat{\lambda}_{(j)}$ the sorted λs, $\hat{\lambda}_{(1)} \geq \ldots \geq \hat{\lambda}_{(p)}$, and $\mathbf{b}_{(j)}$ the corresponding eigenvectors. Then the first q principal directions are given by the $\mathbf{b}_{(j)}$s, $j = 1, \ldots, q$, and the respective "proportion of unexplained variance" is given by (6.82), where λ_j is replaced by $\widehat{\lambda}_{(j)}$.

In order for the resulting robust PCA to be invariant under orthogonal transformations of the data, it is not necessary that $\widehat{\mu}$ be affine equivariant, but only orthogonal equivariant; that is, such that $\widehat{\mu}(\mathbf{T}X) = \mathbf{T}\widehat{\mu}(X)$ for all *orthogonal* $\mathbf{T}$. The simplest choice for $\widehat{\boldsymbol{\mu}}$ is the "space median":

$$\widehat{\boldsymbol{\mu}} = \arg\min_{\boldsymbol{\mu}} \sum_{i=1}^{n} ||\mathbf{x}_i - \boldsymbol{\mu}||.$$

Note that this is an M-estimator since it corresponds to (6.9) with $\rho(t) = \sqrt{t}$ and $\boldsymbol{\Sigma} = \mathbf{I}$. Thus the estimator can be easily computed through the first equation in (6.58), with $W_1(t) = 1/\sqrt{t}$, and starting with the coordinate-wise medians. It follows from Section 6.17.4.1 that this estimator has BP = 0.5.

This procedure is deterministic and very fast, and it can be computed with collinear data without any special adjustments. Despite its simplicity, simulations by Maronna (2005) show that this SPC method performs reasonably well.

6.11.2 Robust PCA based on a robust scale

The proposed approach is based on replacing the expectation in (6.83) by a robust M-scale (Maronna, 2005). For given q-dimensional hyperplane H, call $\widehat{\sigma}(H)$ an M-scale estimator of the $d_{H,i}$ in (6.87); that is, $\widehat{\sigma}(H)$ satisfies the M-scale equation

$$\frac{1}{n}\sum_{i=1}^{n} \rho\left(\frac{\|\mathbf{x}_i - \widehat{\mathbf{x}}_{H,i}\|^2}{\widehat{\sigma}(H)}\right) = \delta. \tag{6.88}$$

Then we search for $\widehat{H}$ having the form of the right-hand-side of (6.84), such that $\widehat{\sigma}(\widehat{H})$ is minimum. For a given H let

$$\widehat{\boldsymbol{\mu}} = \frac{1}{\sum_{i=1}^{n} w_i} \sum_{i=1}^{n} w_i \mathbf{x}_i, \quad \widehat{\mathbf{V}} = \sum_{i=1}^{n} w_i (\mathbf{x}_i - \widehat{\boldsymbol{\mu}})(\mathbf{x}_i - \widehat{\boldsymbol{\mu}})', \tag{6.89}$$

with

$$w_i = W\left(\frac{\|\mathbf{x}_i - \widehat{\mathbf{x}}_{H,i}\|^2}{\widehat{\sigma}}\right) \tag{6.90}$$

where $W = \rho'$. It can be shown by differentiating (6.88) with respect to $\boldsymbol{\mu}$ and $\mathbf{b}_j$ that the optimal $\widehat{H}$ has the form (6.84), where $\boldsymbol{\mu} = \widehat{\boldsymbol{\mu}}$ and $\mathbf{b}_1, \ldots, \mathbf{b}_q$ are the eigenvectors of $\widehat{\mathbf{V}}$ corresponding to its q largest eigenvalues. In other words, the hyperplane is defined by a weighted mean of the data and the principal directions of a weighted covariance matrix, where points distant from $\widehat{H}$ receive small weights.

This result suggests an iterative procedure, in the spirit of the iterative reweighting approach of Sections 6.8.1 and 6.8.2. Starting with some initial $\widehat{H}_0$, compute the weights with (6.90), then compute $\widehat{\boldsymbol{\mu}}$ and $\widehat{\mathbf{V}}$ with (6.89) and the corresponding principal components, which yield a new $\widehat{H}$. It follows from the results by Boente (1983) that if W is nondecreasing, then σ decreases at each step of this procedure and the method converges to a local minimum.

Since this estimator minimizes an M-scale, it will henceforth be denoted by M-S for brevity.

There remains the problem of starting values. Simulations in Maronna (2005) suggest that the SPCs described in Section 6.11.1 are a fast and reliable starting point, and for this reason this will be our chosen initial procedure.

A similar estimator can be based on an L-scale. For $h < n$, compute the L-scale estimator

$$\widehat{\sigma}(H) = \sum_{i=1}^{h} d_{H,(i)},$$

where the $d_{H,(i)}$s are the ordered values of $\|\mathbf{x}_i - \widehat{\mathbf{x}}_{H,i}\|^2$. Then the hyperplane $\widehat{H}$ minimizing $\widehat{\sigma}(H)$ corresponds to the principal components of (6.89) with $w_i = \mathrm{I}(d_{\widehat{H},i} \leq d_{\widehat{H},(h)})$. This amounts to " trimmed" principal components, in which the $\mathbf{x}_i$s with the h smallest values of $d_{\widehat{H},i}$ are trimmed. The analogous iterative procedure converges to a local minimum. The results obtained in the simulations in Maronna (2005) for the L-scale are not as good as those corresponding to the M-scale.

Example 6.4 *The following dataset from Hettich and Bay (1999) corresponds to a study in automatic vehicle recognition (Siebert, 1987). Each of the 218 rows corresponds to a view of a bus silhouette, and contains 18 attributes of the image. The SDs are in general much larger than the respective MADNs. The latter vary between 0 (for variable 9) to 34. Hence it was decided to exclude variable 9 and divide the remaining variables by their MADNs. The tables and figures for this example are obtained with script* ***bus.R****.*

Table 6.5 shows the proportions of unexplained variability (6.82) as a function of the number q of components, for the classical PCA and for M-S.

It would seem that since the classical method has smaller unexplained variability than the robust method, classical PCAs give a better representation. However, this is not the case. Table 6.6 gives the quantiles of the distances $d_{H,i}$ in (6.87) for $q = 3$, and Figure 6.10 compares the logs of the respective ordered values (the log scale was used because of the extremely large outliers).

It is seen in the figure that the hyperplane from the robust fit has in general smaller distances to the data points, except for some clear outliers. On the other hand, in Table 6.5 the classical estimator seems to perform better than the robust one. The reason is that the two estimators use *different* measures of variability. The classical

Table 6.5 Bus data: proportion of unexplained variability for q components

q	Classical	M-S
1	0.188	0.451
2	0.083	0.182
3	0.044	0.114
4	0.026	0.081
5	0.018	0.054
6	0.012	0.039

Table 6.6 Bus data: quantiles of distances to hyperplane

	0.1	0.2	0.3	0.4	0.5	0.6	0.7	0.8	0.9	Max
Classical	1.86	2.28	2.78	3.23	3.74	4.37	5.45	6.47	8.17	23
Robust	1.20	1.59	1.90	2.17	2.59	2.92	3.47	4.33	5.48	1055

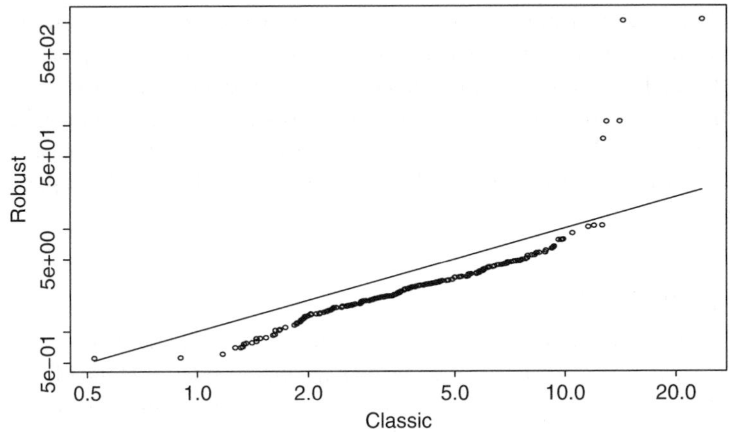

Figure 6.10 Distances to the hyperplane from classic and robust estimators, in log-scale. The straight line is the identity

procedure uses variances, which are influenced by the outliers, and so large outliers in the direction of the first principal axes will inflate the corresponding variances and hence increase their proportion of explained variability. On the other hand the robust estimator uses a robust measure, which is free of this drawback, and gives a more accurate measure of the unexplained variability for the bulk of the data.

6.12 Estimation of multivariate scatter and location with missing data

A frequent problem is that some of the cells of the $n \times p$ data matrix $\mathbf{X} = [\mathbf{x}_{ij}]$ are missing. The robust estimation of the scatter matrix and multivariate location in this case is a challenging problem. In the case of multivariate normal data, the maximum likelihood estimator can be computed by means of the expectation-maximization (EM) algorithm; see Dempster *et al.* (1977) and Little and Rubin (2002). However, as generally happens with the maximum likelihood estimators for normal data, these estimators are not robust.

Throughout this section it is assumed that the missing cells are completely at random; that is, the np events $\{x_{ij}$ is missing$\}$ are independent. It is desirable to have estimators that are both consistent for normal data and highly robust. Robust estimators for this case have been proposed by Little and Smith (1987), Little (1988) and Frahm and Jaekel (2010). However, all of these have low BPs. The one proposed by Cheng and Victoria-Feser (2002) is not consistent for normal data and is sensitive towards clustered outliers.

Danilov *et al.* (2012) proposed *generalized S* (GS) estimators for estimating scatter and multivariate location. These estimators are defined by the minimization of a weighted M-scale of Mahalanobis distances. The Mahalanobis distances for each observation depend only on the non-missing components and are based on a marginal covariance matrix standardized so that its determinant is one. To compute these estimators, the authors proposed a weighted EM- type algorithm (Dempster *et al.*, 1977) where the weights are based on a redescending score function. They prove that under general conditions these estimators are shape-consistent for elliptical data and shape- and size- consistent for normal data. We now give a more detailed description of this approach.

6.12.1 Notation

Let $\mathbf{x}_i = (x_{i1}, ..., x_{ip})'$, $1 \le i \le n$, be p-dimensional i.i.d. random vectors. Let $\mathbf{u}_i = (u_{i1}, ..., u_{ip})'$, $1 \le i \le n$, be independent p-dimensional vectors of zeros and ones, where $u_{ij} = 1$ if x_{ij} is observed and $u_{ij} = 0$ if it is missing. We also assume that $\mathbf{u}_i$ and $\mathbf{x}_i$ are independent (which corresponds to the "missing at random" assumption).

Given $\mathbf{x} = (x_1, ..., x_p)'$ and a vector of zeros and ones $\mathbf{u} = (u_1, ..., u_p)'$, let $\mathbf{x}^{(\mathbf{u})}$ be the vector whose components are all the x_is such that $u_i = 1$, and let

$$p(\mathbf{u}) = \sum_{j=1}^{p} u_j. \tag{6.91}$$

In other words, $\mathbf{x}^{(\mathbf{u})}$ is a vector of dimension $p(\mathbf{u})$ formed with the available entries of $\mathbf{x}$. We assume the following identifiability condition: given $1 \le j < k \le p$, there exists at least one $\mathbf{u}_i$, $1 \leq i \leq n$, with $u_{ij} = u_{ik} = 1$. Let $\mathcal{U}_p = \{\mathbf{u} : (u_1, ..., u_p)', u_i \in \{0, 1\}\}$. Then, given a $p \times p$ positive definite matrix $\boldsymbol{\Sigma}$ and $\mathbf{u} \in \mathcal{U}_p$, we denote by $\boldsymbol{\Sigma}^{(\mathbf{u})}$ the submatrix of $\boldsymbol{\Sigma}$ formed with the rows and columns corresponding to $u_i = 1$. Finally, we set $\boldsymbol{\Sigma}^{*(\mathbf{u})} = \boldsymbol{\Sigma}^{(\mathbf{u})}/|\boldsymbol{\Sigma}^{(\mathbf{u})}|^{1/p(\mathbf{u})}$, so that $|\boldsymbol{\Sigma}^{*(\mathbf{u})}| = 1$.

Given a data point $(\mathbf{x}, \mathbf{u})$, a center $\mathbf{m} \in R^p$ and a $p \times p$ positive definite scatter matrix $\boldsymbol{\Sigma}$, the *partial square Mahalanobis distance* is given by

$$d(\mathbf{x}, \mathbf{u}, \mathbf{m}, \boldsymbol{\Sigma}) = (\mathbf{x}^{(\mathbf{u})} - \mathbf{m}^{(\mathbf{u})})'(\boldsymbol{\Sigma}^{(\mathbf{u})})^{-1}(\mathbf{x}^{(\mathbf{u})} - \mathbf{m}^{(\mathbf{u})}). \tag{6.92}$$

6.12.2 GS estimators for missing data

It is easy to show that the Davies S-estimators for complete data introduced in Section 6.4.2 can also be defined as follows. Suppose that $n > 2p$ and let $\rho : R_+ \to R_+$ be a ρ-function. Given $\mathbf{m} \in R^p$ and a $p \times p$ positive definite matrix $\boldsymbol{\Sigma}$, let $S_n(\mathbf{m}, \boldsymbol{\Sigma})$ be the solution in s to the equation

$$\frac{1}{n}\sum_{i=1}^{n} \rho\left(\frac{d(\mathbf{x}_i, \mathbf{m}, \boldsymbol{\Sigma})}{c_p s}\right) = 0.5,$$

where c_p is such that

$$E\left(\rho\left(\frac{||\mathbf{X}||^2}{c_p}\right)\right) = 0.5, \tag{6.93}$$

where $\mathbf{X}$ has an elliptical density f_0 centered at 0 and scatter matrix equal to the identity. Usually f_0 is chosen such that $f_0(||\mathbf{x}||^2)$ is the standard multivariate normal density. Then the S-estimator $(\widehat{\mathbf{m}}_n, \widehat{\boldsymbol{\Sigma}}_n)$ is given by

$$\begin{aligned}(\widehat{\mathbf{m}}_n, \tilde{\boldsymbol{\Sigma}}_n) &= \arg\min_{\mathbf{m}, |\boldsymbol{\Sigma}|=1} S_n(\mathbf{m}, \boldsymbol{\Sigma}),\\ \hat{s}_n &= S_n(\widehat{\mathbf{m}}_n, \tilde{\boldsymbol{\Sigma}}_n),\\ \widehat{\boldsymbol{\Sigma}}_n &= \hat{s}_n \tilde{\boldsymbol{\Sigma}}_n.\end{aligned} \tag{6.94}$$

We now generalize the definition of S-estimators for the case of incomplete data. Let $\widehat{\boldsymbol{\Omega}}_n$ be a $p \times p$ positive definite initial estimator for $\boldsymbol{\Sigma}_0$. Given $\mathbf{m} \in R^p$ and a $p \times p$ positive definite matrix $\boldsymbol{\Sigma}$, let $S_n^*(\mathbf{m}, \boldsymbol{\Sigma})$ be defined by

$$\sum_{i=1}^{n} c_{p(\mathbf{u}_i)} \rho\left(\frac{d(\mathbf{x}_i^{(\mathbf{u}_i)}, \mathbf{m}^{(\mathbf{u}_i)}, \boldsymbol{\Sigma}^{*(\mathbf{u}_i)})}{s\left|\widehat{\boldsymbol{\Omega}}_n^{(\mathbf{u}_i)}\right|^{1/p(\mathbf{u}_i)} c_{p(\mathbf{u}_i)}}\right) = \frac{1}{2}\sum_{i=1}^{n} c_{p(\mathbf{u}_i)}, \tag{6.95}$$

where c_p is defined in (6.93) and $p(\mathbf{u})$ in (6.91). First, location and scatter shape component estimators $(\widehat{\mathbf{m}}_n, \tilde{\boldsymbol{\Sigma}}_n)$ are defined by

$$(\widehat{\mathbf{m}}_n, \tilde{\boldsymbol{\Sigma}}_n) = \arg\min_{\mathbf{m}, \boldsymbol{\Sigma}} S_n^*(\mathbf{m}, \boldsymbol{\Sigma}). \tag{6.96}$$

Note that $S_n^*(\mathbf{m}, t\boldsymbol{\Sigma}) = S_n^*(\mathbf{m}, \boldsymbol{\Sigma})$ for all $t > 0$, and therefore $\widetilde{\boldsymbol{\Sigma}}_n$ only estimates the shape component of $\boldsymbol{\Sigma}_0$. The GS-estimator of scatter for $\boldsymbol{\Sigma}_0$ (shape and size) is

$$\widehat{\boldsymbol{\Sigma}}_n = \widehat{s}_n \widetilde{\boldsymbol{\Sigma}}_n, \tag{6.97}$$

with $\widehat{s}_n$ defined by

$$\sum_{i=1}^{n} c_{p(\mathbf{u}_i)} \rho \left(\frac{d\left(\mathbf{x}_i^{(\mathbf{u}_i)}, \widehat{\mathbf{m}}_n^{(\mathbf{u}_i)}, \widetilde{\boldsymbol{\Sigma}}_n^{(\mathbf{u}_i)}\right)}{c_{p(\mathbf{u}_i)} \widehat{s}_n} \right) = \frac{1}{2} \sum_{i=1}^{n} c_{p(\mathbf{u}_i)}. \tag{6.98}$$

The estimating equations and the iterative algorithm for computing the GS estimator are given in Sections 6.17.12 and 6.17.12.1 respectively.

The GS-estimator is available in the package GSE for R. This package includes the function EM that computes the EM-algorithm for Gaussian data, and the functions GSE and GRE that compute the GS estimators with Tukey bisquare and Rocke-type loss functions respectively (see Leung *et al.*, 2017).

Example 6.5 *We consider again the wine dataset described in Example 6.2. We eliminate cells at random from this dataset with probability 0.2. The total number of eliminated cells is 125, which represents approximately 17% of the total. The tables and figures for this example are obtained with script* ***wine1.R***.

In Figure 6.11 we compare the adjusted squared Mahalanobis distances obtained with the EM estimator for Gaussian data and the GS-estimator based on a bisquare ρ-function. Since there are missing data, the squared Mahalanobis distances under normality follow χ^2 distributions with different degrees of freedom. Therefore, in order to make fair comparisons, they are adjusted as follows. Suppose that d is the partial squared Mahalanobis distance from a row where only q variables out of p are observed; that is, $p - q$ are missing. Then the adjusted squared Mahalanobis distance (referred as the "adjusted distance") is defined by $d^* = \chi_p^{2-1}(\chi_q^2(d))$, where χ_p^2 is the χ^2 distribution function with p degrees of freedom. For each estimator we show the plot of the adjusted distances against the observation index and a QQ-plot of the adjusted distances. We observe that while the Gaussian procedure does not detect any outlier, the robust procedure detects seven outliers. An observation is considered as an outlier if its adjusted distance is larger than $\chi_p^{2-1}(0.999)$.

6.13 Robust estimators under the cellwise contamination model

The model for contamination considered up to now consists of replacing a proportion of *rows* of the data matrix $\mathbf{X}$ with outliers ("casewise contamination").

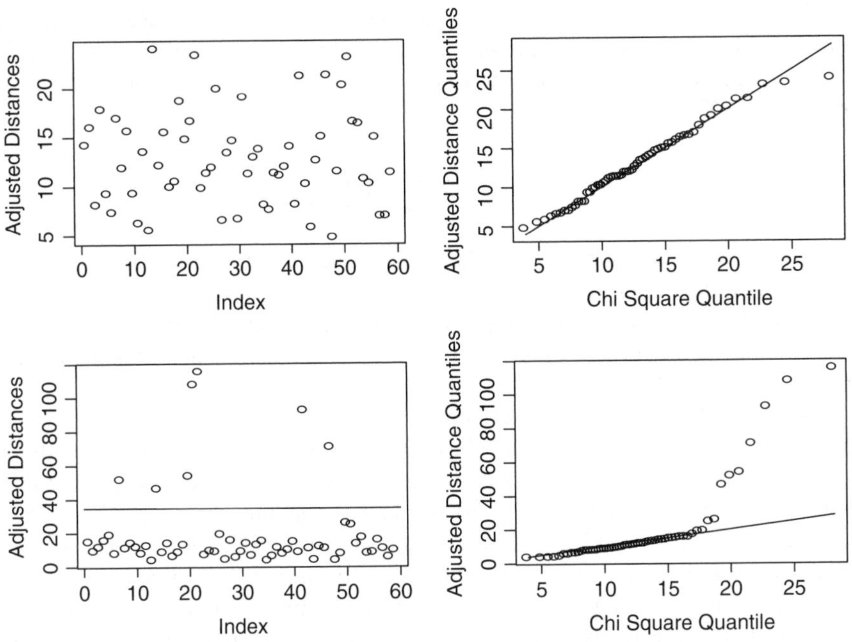

Figure 6.11 Wine example: comparison of adjusted squared Mahalanobis distances of the Gaussian EM-estimator (above) and the GS-estimator (below): In the left-hand column the adjusted distances are plotted against the index number and in the right-hand column their QQ plots are shown

Alqallaf *et al.* (2009) proposed a different model of outlier contamination. Instead of contaminating a proportion of rows (cases), each *cell* of $\mathbf{X}$ is contaminated independently with a given probability ε. This is called *cellwise* (or *independent*) contamination. Then, even if ε is small, as p increases, the probability that at least one cell of a row is an outlier tends to 1. In fact, this probability is equal to $\eta(\varepsilon.(,p) = 1 - (1-\varepsilon)^p$. Therefore, since the affine equivariant estimators already presented have BP of at most 0.5, they are not robust against this type of contamination. An upper bound ε^* for the cellwise contamination BP can be obtained by solving $\eta(\varepsilon^*, p) = 0.5$. The values of ε^* are shown in Table 6.7.

Table 6.7 Minimal fraction of cellwise contamination that causes breakdown of an affine equivariant estimator

	Dimension								
	1	2	3	4	5	10	15	20	100
ε^*	0.50	0.29	0.21	0.16	0.13	0.07	0.05	0.03	0.01

To deal with this type of outlier, Alqallaf *et al.* (2009) replace the original sample with pseudo observations obtained as follows. Let $\widehat{\mu}_j$ and $\widehat{\sigma}_j$, $1 \leq j \leq p$ be robust estimators of location and scale of the jth variable. Then the matrix of pseudo observations $\mathbf{X}^* = (x^*_{ij})$ is defined by

$$x^*_{ij} = \widehat{\sigma}_j \psi \left(\frac{x_{ij} - \widehat{\mu}_j}{\widehat{\sigma}_j} \right) + \widehat{\mu}_j,$$

where ψ is a ψ−function, such as the Huber function $\psi(x) = \max(\min(x, c), -c)$, where c is a conveniently chosen constant. They then estimate the scatter matrix and location vector using the sample covariance and sample mean of $\mathbf{X}^*$. These estimators exhibit good behavior under the cellwise contamination model, but they may be much affected by case outliers.

Danilov (2010) proposed treating outlying cells as if they were missing observations. They then applied a casewise robust procedure for missing data to the modified dataset. Farcomeni (2014) developed a related robust procedure for clustering, called *snipping*. When the number of clusters is taken as one, this procedure gives robust estimators of multivariate location and scatter. The idea is to treat a given fraction α of cell values as it they were missing. Once the αnp "missing" observations are fixed, the location and covariance estimators are obtained by maximizing the likelihood under normality. Finally the αnp "missing" cells are chosen in order to maximize the likelihood. Therefore, computing these estimators requires solving a combinatorial optimization problem of high computational complexity. For this reason, an exact solution is not feasible and therefore the authors give only approximate algorithms. Agostinelli *et al.* (2015) simulated these approximate snipping estimators and found that they work quite well for independent outlier contamination but they are not sufficiently robust under case contamination.

Agostinelli et al. (2015) proposed a two-step procedure for the case of independently contaminated data. In the first step, the filter to detect univariate outliers introduced in Gervini and Yohai (2002) and described in Section 5.9.2 is applied to each column of $\mathbf{X}$. In the second step, all detected outliers are considered as missing and the GS-estimator for missing data described in Section 6.12.2 is applied to the resulting incomplete dataset. This procedure is called the *two steps GS* (TSGS) estimator. It has been proved that the TSGS-estimator is consistent for normal data.

Leung *et al.* (2017) gave a new version of the TSGS-estimators that performs better for casewise outliers when $p \geq 10$ and for cellwise outliers when the data are highly correlated. This is achieved using a new filter to detect outliers in the first step. This new filter is a combination of a consistent bivariate filter proposed by Leung *et al.* (2017) and another filter called DDC proposed by Rousseeuw and van den Bossche (2016). It is available in the GSE package mentioned in Section 6.12.2 and can be based on the bisquare or Rocke ρ-functions.

Cellwise contamination also affects the behavior of robust PCA. Candés (2011) proposes a method that is robust towards independent contamination and is very fast for high dimensions, but its BP for row-wise contamination is zero. Maronna and Yohai (2008) propose an approach that is robust for both types of contamination, but

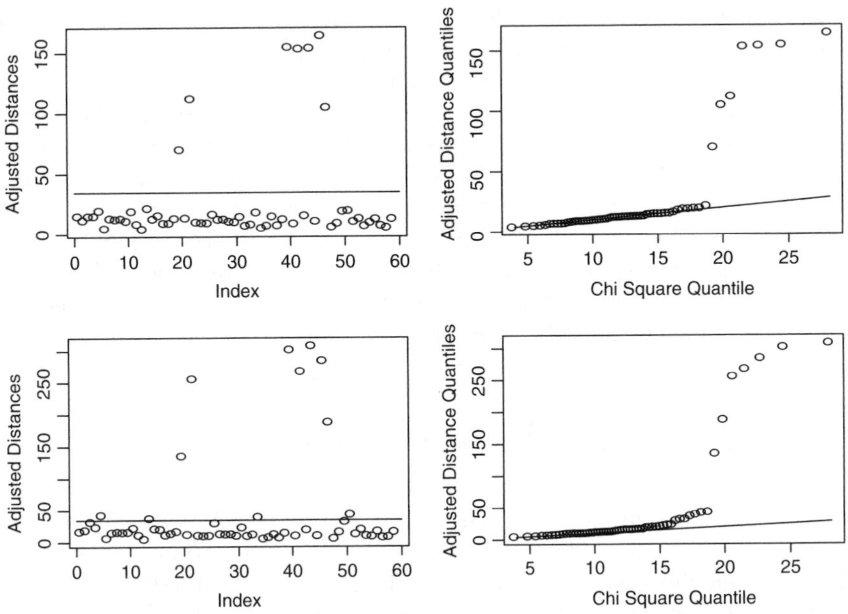

Figure 6.12 Wine example: comparison of the squared Mahalanobis distances of the MM- (above) and the TSGS-estimators (below); left column: the distances plotted against index number; right, QQ plots

not fast enough for large p. The problem of combining resistance to cellwise outliers, resistance to casewise outliers, and high computational speed remains a difficult one.

Example 6.6 *In this example we compare the results obtained with the MM-estimator (see Section 6.5) and TSGS-estimators for the wine dataset employed in Example 6.5. The tables and figures for this example are obtained with script **wine1.R**.*

The results are shown in Figure 6.12. For each estimator, we plot the squared Mahalanobis distances against the observation index and the corresponding QQ-plot. We observe that while the MM-estimator detects seven outliers, the TSGS-estimator detects ten. The cutoff point for a squared Mahalanobis distance to indicate an outlier is again $\chi_p^{2-1}(0.999)$.

6.14 Regularized robust estimators of the inverse of the covariance matrix

Many applications require estimating the precision matrix $\mathbf{\Sigma}^{-1}$, where $\mathbf{\Sigma}$ is the covariance matrix. This occurs, for example, when computing Mahalanobis distances, for discrimination procedures in graphical models (see, for example, Lauritzen, 1996) and in mean-variance portfolio optimization (Scherer, 2015). Suppose that the rows of the matrix $\mathbf{X}$ are i.i.d. observations of a multivariate normal distribution. In this case, the maximum likelihood estimator of the covariance matrix $\boldsymbol{\Sigma}$ is the sample covariance $\mathbf{S}$. When p/n is close to one, $\mathbf{S}$ is nearly singular, and if $p \geq n$ it becomes singular, and therefore in these situations it is not possible to obtain a stable estimator of $\boldsymbol{\Sigma}^{-1}$. Several approaches have been proposed to overcome this problem: the quadratic approximation method for sparse inverse covariance learning (Hsieh *et al.*, 2011), the constrained L_1-minimization for inverse matrix estimation (CLIME) (Cai *et al.*, 2011) and the constrained maximum likelihood estimator (CMLE) (Banerjee *et al.*, 2008; Yuan and Lin, 2007; Friedman *et al.*, 2008). The CMLE of the precision matrix $\boldsymbol{\Sigma}^{-1}$ is defined as the matrix $\mathbf{T} = (t_{ij})$, given by

$$\mathbf{T} = \arg \max_{\mathbf{T} \in R^{p \times p}, \mathbf{T} \succ 0} \left\{ \log(\det(\mathbf{T})) - \mathrm{tr}(\mathbf{ST}) - \lambda \sum_{i=1}^{p} \sum_{j=1}^{p} |t_{ij}| \right\},$$

where λ is a fixed given number, and $\mathbf{A} \succ 0$ denotes that $\mathbf{A}$ is positive definite. Note that maximizing $\log(\det(\mathbf{T})) - \mathrm{tr}(\mathbf{ST})$ is equivalent to maximizing the normal likelihood, and this occurs when $\mathbf{T} = \mathbf{S}^{-1}$. Therefore the CMLE estimator is, as in the case of the regression LASSO, a penalized maximum likelihood estimator. The penalization term $\lambda \sum_{i=1}^{p} \sum_{j=1}^{p} |t_{ij}|$ prevents $\mathbf{T}$ from having large cell values. The penalty parameter λ can be chosen by maximizing the likelihood function, computed by means of cross-validation. An efficient algorithm to compute the CMLE, called *graphical lasso* (GLASSO), was proposed by Friedman *et al.* (2008).

Ollerer and Croux (2015) and Tarr *et al.* (2015) proposed robustifying the GLASSO estimators by replacing $\mathbf{S}$ by robust estimators, thus obtaining a higher resistance towards contamination.

6.15 Mixed linear models

Mixed linear models (MLM) are linear models in which the expectations of the observations are linear functions of a parameter vector $\boldsymbol{\beta}$, but unlike the ordinary linear models, the covariance matrix has a special structure that depends on further parameters.

Consider, for example, an ANOVA model with one fixed and one random factor, in which the observations y_{ij} verify

$$y_{ij} = \beta_j + a_i + u_{ij}, \quad 1 \leq i \leq n, \quad 1 \leq j \leq q,$$

where the β_is are unknown parameters, the a_is are random variables and u_{ij} is an error term. The standard assumptions are that all a_is and u_{ij}s are independent, with $a_i \sim \mathrm{N}(0, \tau^2)$ and $u_{ij} \sim \mathrm{N}(0,\sigma^2)$. Then for $1 \leq i \leq n$ and $1 \leq j \leq q$

$$\mathrm{E}y_{ij} = \beta_{0k}, \ \mathrm{E}y_{ij}^2 = \tau^2 + \sigma^2, \ \mathrm{E}y_{ij}y_{ih} = \tau^2 \quad \text{for } j \neq h.$$

Putting $\mathbf{y}_i = (y_{i1}, ...y_{iq})'$, it follows that

$$\mathbf{y}_i \sim \mathrm{N}_q\left(\boldsymbol{\beta}, \sigma^2\left(\mathbf{I}_q + \frac{\tau^2}{\sigma^2}\mathbf{V}\right)\right),$$

where $\mathbf{V}$ is the $q \times q$ matrix with all its elements equal to one.

This situation is a particular case of a class of MLMs that will be dealt with in this section. The data are $(\mathbf{X}_i, \mathbf{y}_i)$, $i = 1, ..., n$, where the $q \times p$ matrices $\mathbf{X}_i$ can be fixed or random, and $\mathbf{y}_i \in R^q$ are random vectors. Call $\boldsymbol{\mu}$ and $\boldsymbol{\Sigma}$ the mean vector and covariance matrix of $\mathbf{y}_i$ in the case $\mathbf{X}_i$ is fixed; if $\mathbf{X}_i$ is random, $\boldsymbol{\mu}$ and $\boldsymbol{\Sigma}$ are the conditional means and covariances given $\mathbf{X}$. It is assumed that $\boldsymbol{\mu}$ and $\boldsymbol{\Sigma}$ depend on unknown parameters in the following way:

$$\mu(\mathbf{X}, \boldsymbol{\beta}) = \mathbf{X}\boldsymbol{\beta}, \quad \boldsymbol{\Sigma}(\eta, \boldsymbol{\gamma}) = \eta\left(\mathbf{V}_0 + \sum_{j=1}^{J} \gamma_j \mathbf{V}_j\right), \tag{6.99}$$

where $\boldsymbol{\beta} \in R^p$, $\boldsymbol{\gamma} \in R^J$ and η are unknown parameters, with $\gamma_i \geq 0$ and $\eta > 0$; and $\mathbf{V}_j$ $(j = 0, ..., J)$ are known positive-definite $q \times q$ matrices.

It is seen that the ANOVA model described above is an instance of (6.99) with $\mathbf{X}_i = \mathbf{I}_q, \eta = \sigma^2, \mathbf{V}_0 = \mathbf{I}_q, \mathbf{V}_1 = \mathbf{V}, J = 1$ and $\gamma_1 = \tau^2/\sigma^2$.

This class of models includes, among others, ANOVA models with repeated measurements, models with random nested design and models for longitudinal data. It contains models of the form

$$\mathbf{y}_i = \mathbf{X}_i\boldsymbol{\beta} + \sum_{j=1}^{J} \mathbf{Z}_j \boldsymbol{a}_{ij} + \mathbf{e}_i, \ 1 \leq i \leq n, \tag{6.100}$$

where the $\mathbf{X}_i$s are as above, $\mathbf{Z}_j$, $1 \leq j \leq J$, are $p \times q_j$ known design matrices for the random effects, $\boldsymbol{a}_{ij}$ are independent q_j-dimensional random vectors with distribution $\mathrm{N}_{q_j}(0, \sigma_j^2 \mathbf{I}_{q_j})$, and $\mathbf{e}_i$ $(1 \leq i \leq n)$ are p-dimensional error vectors with distribution $\mathrm{N}_p(0, \sigma^2 \mathbf{I}_p)$. We are then in the framework (6.99) with $\eta_0 = \sigma^2, \boldsymbol{\gamma} = (\gamma_1, \ldots, \gamma_J)'$ with $\gamma_j = \sigma_j^2/\sigma^2 > 0$ and $\mathbf{V}_j = \mathbf{Z}_j \mathbf{Z}_j'$, $1 \leq j \leq J$.

An instance of (6.100) is longitudinal analysis models, in which, for a random sample of n individuals, one response and q covariates are measured at fixed times $t_1, ..., t_p$. Let y_{ij} be the response for individual i at time t_j, and call x_{ijk} the value of covariate j at time t_k for the individual i. It is assumed that

$$y_{ij} = \sum_{k=1}^{p} \beta_k x_{ijk} + a_i + u_{ij}$$

where $\boldsymbol{\beta} = (\beta_1, ...\beta_p)'$ is an unknown vector, a_i, the effect of individual i, is a $\mathrm{N}(0, \tau^2)$ random variable, and the error u_{ijk} is $\mathrm{N}(0, \sigma^2)$. It is assumed that all the u_{ij}s and a_is are independent. Let $\mathbf{y}_i = (y_{i1}, ..., y_{iq})'$, $\mathbf{X}_i$ the $p \times q$ matrix whose (j, k) element is x_{ijk},

$\mathbf{u}_i = (u_{i1}, ..., u_{iq})'$ and $\mathbf{z}$ the q -dimensional vector with all elements equal to one. Then we can write

$$\mathbf{y}_i = \mathbf{x}_i\boldsymbol{\beta} + za_i + \mathbf{u}_i, 1 \leq i \leq n,$$

which has the form (6.100).

6.15.1 Robust estimation for MLM

Consider a sample $(\mathbf{X}_1, \mathbf{y}_1), ...(\mathbf{X}_n, \mathbf{y}_n)$ of the MLM. If the $\mathbf{X}_i$ are fixed, it is assumed that the distribution of $\mathbf{y}_i$ belongs to an elliptical family (6.7) with location vector $\boldsymbol{\mu}$ and scatter matrix $\boldsymbol{\Sigma}$ given by (6.99). If the $\mathbf{X}_i$ are random, the same applies to the conditional distribution of $\mathbf{y}_i$ given $\mathbf{X}_i$.

Most commonly, $\boldsymbol{\beta}, \boldsymbol{\gamma}$ and η are estimated by maximum likelihood (ML) or restricted maximum likelihood (REML), assuming that the elliptical family is the multivariate normal; see for example (Searle *et al.*, 1992). However, as can be expected, these estimators are not robust. Several robust estimators for MLM with different degrees of generality have been proposed (Fellner, 1986; Richardson and Welsh, 1995; Stahel and Welsh, 1997). Copt and Victoria-Feser (2006) proposed an S-estimator similar to the one described in Section 6.4.2 that can be applied to the model (6.99). The S-estimators are extended to a more general class of models by Chervoneva and Vishnyakov (2011). While all these proposals are robust for casewise contamination, they cannot cope with $\mathbf{y}_i$s affected by independent contamination (Section 6.13) when q is large. Koller (2013) proposed an estimator, denoted hereafter SMDM, applicable to a very general class of MLMs obtained by bounding the terms of the maximum likelihood equations of Henderson *et al.* (1959).

Agostinelli and Yohai (2016) introduced a new class of robust estimators called *composite* τ *-estimators* for MLM, which are robust under both casewise and cellwise (or independent) contamination. These estimators are defined similarly to the composite likelihood estimators proposed by Lindsay (1988). For p-dimensional observations $\mathbf{y}_i$, the composite likelihood estimators are based on the likelihood of all subvectors of dimension p^* for some $p^* < p$. The composite τ-estimators are defined by minimizing a τ-scale (see Section 6.4.5) of the Mahalanobis distances of the two-dimensional subvectors of the $\mathbf{y}_i$s. Their asymptotic breakdown point is 0.5 for the casewise contamination and is 0.29 for independent contamination. S-estimators and composite τ-estimators will be described below.

6.15.2 Breakdown point of MLM estimators

As in Section 6.13, we can consider two types of contaminations for MLM: casewise and cellwise. Let F be the common distribution of $(\mathbf{X}_i, \mathbf{y}_i)$, $1 \leq i \leq n$, $\mathbf{x}_{ij}$ the jth row of $\mathbf{X}_i$ and y_{ij} the jth component of $\mathbf{y}_i$. A casewise contamination of size ε occurs when $(\mathbf{X}_i, \mathbf{y}_i)$ is replaced by

$$(\mathbf{X}_i^*, \mathbf{y}_i^*) = \delta_i(\mathbf{X}_i, \mathbf{y}_i) + (1 - \delta_i)(\mathbf{Z}_i, \mathbf{w}_i),$$

where the δ_i are independent Bernoulli random variables with success probability $(1-\varepsilon)$, and $(\mathbf{Z}_i, \mathbf{w}_i)$, $1 \le i \le n$, are i.i.d. with an arbitrary joint distribution G. Moreover, it is required that the three sets of variables, $(\mathbf{X}_i, \mathbf{y}_i)_{1\le i\le n}$, $(\mathbf{Z}_i, \mathbf{w}_i)_{1\le i\le n}$ and $(\delta_i)_{1\le i\le n}$, are independent. We denote by $F_\varepsilon^{\text{case}}$ the distribution of $(\mathbf{X}_i^*, \mathbf{y}_i^*)$. Note that $F_\varepsilon^{\text{case}}(\mathbf{X}, \mathbf{y}) = (1-\varepsilon)F(\mathbf{X}, \mathbf{y}) + \varepsilon G(\mathbf{X}, \mathbf{y})$ is determined by G.

Instead, we have cellwise contamination of size ε when $(\mathbf{x}_{ij}, y_{ij})$ is replaced by

$$(\mathbf{x}_i^*, y_{ij}^*) = \delta_{ij}(\mathbf{x}_{ij}, y_{ij}) + (1-\delta_{ij})(\mathbf{z}_i, w_{ij}), 1 \le i \le n, 1 \le j \le q,$$

where the δ_{ij} are independent Bernoulli random variables with success probability $(1-\varepsilon)$, and $(\mathbf{z}_{ij}, w_{ij})$, $1 \le i \le n, 1 \le j \le q$, are independent with an arbitrary joint distribution G_j. Once again, the three sets of variables, $(\mathbf{x}_{ij}, \mathbf{y}_{ij})_{1\le i\le n, 1\le j\le q}$, $(\mathbf{z}_{ij}, w_{ij})_{1\le i\le n, 1\le j\le q}$ and $(\delta_{ij})_{1\le i\le n, 1\le j\le q}$, are required to be independent. Let $F_\varepsilon^{\text{cell}}$ be the distribution of $(\mathbf{X}_i^*, \mathbf{y}_i^*)$, where the jth row of $\mathbf{X}_i^*$ is $\mathbf{x}_{ij}^*$ and the jth coordinate of $\mathbf{y}_i^*$ is y_{ij}^*. Clearly $F_\varepsilon^{\text{cell}}$ depends on $G_1, ...G_q$.

Let $\theta = \eta\gamma$, and consider an estimator $(\widehat{\beta}, \widehat{\theta})$ of (β, θ) with asymptotic value $(\widehat{\beta}_\infty, \widehat{\theta}_\infty)$. The *case contamination asymptotic breakdown point* of $\widehat{\beta}$ is defined by

$$\varepsilon_{\text{case}}(\widehat{\beta}_\infty) = \inf\left\{\varepsilon : \sup_G \|\beta_\infty(F_\varepsilon^{\text{case}})\| = \infty\right\}. \tag{6.101}$$

Two *asymptotic casewise breakdown points* of $\widehat{\theta}_\infty$ can be defined: one to infinity, denoted by $\varepsilon_{\text{case}}^+(\widehat{\theta}_\infty)$, and the other to 0, denoted by $\varepsilon_{\text{case}}^-(\widehat{\theta}_\infty)$. They are given by

$$\varepsilon_{\text{case}}^+(\widehat{\theta}_\infty) = \inf\left\{\varepsilon : \sup_G \|\widehat{\theta}_\infty(F_\varepsilon^{\text{case}})\| = \infty\right\} \tag{6.102}$$

$$\varepsilon_{\text{case}}^-(\widehat{\theta}_\infty) = \inf\left\{\varepsilon : \inf_G \|\widehat{\theta}_\infty(F_\varepsilon^{\text{case}})\| = 0\right\}. \tag{6.103}$$

Three similar asymptotic breakdown points can be similarly defined for cellwise contamination

$$\varepsilon_{\text{cell}}(\widehat{\beta}_\infty) = \inf\left\{\varepsilon : \sup_{G_1,\dots,G_q} \|\widehat{\beta}_\infty(F_\varepsilon^{\text{cell}})\| = \infty\right\} \tag{6.104}$$

$$\varepsilon_{\text{cell}}^+(\widehat{\theta}_\infty) = \inf\left\{\varepsilon : \sup_{G_1,\dots,G_q} \|\widehat{\theta}_\infty(F_\varepsilon^{\text{cell}})\| = \infty\right\} \tag{6.105}$$

$$\varepsilon_{\text{cell}}^-(\widehat{\theta}_\infty) = \inf\left\{\varepsilon : \inf_{G_1,\dots,G_q} \|\widehat{\theta}_\infty(F_\varepsilon^{\text{cell}})\| = 0\right\}. \tag{6.106}$$

6.15.3 S-estimators for MLMs

The class of S-estimators proposed by Copt and Victoria-Feser (2006) for the model (6.99) are defined as follows. Let S be an M-scale of the form (6.29) with ρ -function ρ, and call s_0 the solution of

$$\mathrm{E}\rho\left(\frac{v}{s_0}\right) = \delta, \tag{6.107}$$

where $v \sim \chi^2_p$. Put

$$m_i(\boldsymbol{\beta}, \eta, \boldsymbol{\gamma}) = d(\mathbf{y}_i, \boldsymbol{\mu}(\mathbf{X}, \boldsymbol{\beta}), \quad \boldsymbol{\Sigma}(\eta, \boldsymbol{\gamma})),$$

where d denotes the squared Mahalanobis distance (6.4) and $\boldsymbol{\mu}$ and $\boldsymbol{\Sigma}$ are given by (6.99).

Then then the S-estimator for the MLM is defined by

$$(\hat{\boldsymbol{\beta}}, \hat{\eta}, \hat{\boldsymbol{\gamma}}) = \arg\ \min_{\eta,\gamma}\ \det \boldsymbol{\Sigma}(\eta, \boldsymbol{\gamma})$$

subject to

$$S(m_1(\boldsymbol{\beta}, \eta, \boldsymbol{\gamma}), ..., m_n(\boldsymbol{\beta}, \eta, \boldsymbol{\gamma})) = s_0.$$

Put

$$\boldsymbol{\Sigma}^*(\eta) = \frac{\boldsymbol{\Sigma}(\eta, \boldsymbol{\gamma})}{\det(\boldsymbol{\Sigma}(\eta, \boldsymbol{\gamma}))^{1/p}}, \tag{6.108}$$

which depends only on $\boldsymbol{\gamma}$. It is easy to show that the S-estimators for a MLM can be also defined by

$$(\hat{\boldsymbol{\beta}}, \hat{\boldsymbol{\gamma}}) = \arg \min_{\beta,\gamma} S\left\{d(\mathbf{y}_1, \boldsymbol{\mu}(\mathbf{X}, \boldsymbol{\beta}), \quad \boldsymbol{\Sigma}^*(\boldsymbol{\gamma})), ..., d(\mathbf{y}_n, \boldsymbol{\mu}(\mathbf{X}, \boldsymbol{\beta}), \quad \boldsymbol{\Sigma}^*(\boldsymbol{\gamma}))\right\},$$

$$\hat{\eta} = \frac{1}{s_0} S\left\{d(\mathbf{y}_1, \boldsymbol{\mu}(\mathbf{X}, \boldsymbol{\beta}), \quad \boldsymbol{\Sigma}(1, \boldsymbol{\gamma})), ..., d(\mathbf{y}_n, \boldsymbol{\mu}(\mathbf{X}, \boldsymbol{\beta}), \quad \boldsymbol{\Sigma}(1, \boldsymbol{\gamma}))\right\}.$$

Copt and Victoria-Feser (2006) showed that the asymptotic breakdown point of the S-estimator for casewise contamination in the three cases defined in (6.101)–(6.103) are equal to $\min(\delta, 1 - \delta)$ with δ in (6.107). However, as in the case of scatter estimation (6.13), the three cellwise breakdown points defined in (6.104)–(6.106) tend to 0 when $q \to \infty$.

6.15.4 Composite τ-estimators

Given a vector $\mathbf{a} = (a_1, \ldots, a_p)^\top$, a $p \times p$ matrix $\mathbf{A}$ and a couple (j, l) of indices $(1 \leq j < l \leq p)$, put $\mathbf{a}^{jl} = (a_j, a_l)^\top$ and denote by $\mathbf{A}_{jl}$ the submatrix

$$\mathbf{A}_{jl} = \begin{pmatrix} a_{jj} & a_{jl} \\ a_{lj} & a_{ll} \end{pmatrix}.$$

In a similar way, given a $p \times k$ matrix $\mathbf{X}$ we denote by $\mathbf{X}^{jl}$ the matrix of dimension $2 \times k$ built by using rows j and l of $\mathbf{X}$.

Pairwise squared Mahalanobis distances are defined by

$$m_i^{jl}(\boldsymbol{\beta}, \boldsymbol{\gamma}) = d(\mathbf{y}_i^{jl}, \boldsymbol{\mu}_i^{jl}(\boldsymbol{\beta}), \boldsymbol{\Sigma}_{jl}^*(\boldsymbol{\gamma})),$$

with $\boldsymbol{\Sigma}^*$ defined as in (6.108).

Let ρ_1 and ρ_2 be two bounded and continuously differentiable ρ-functions such that $2\rho_2(u) - \rho_2'(u)u \geq 0$. Call $s_{jl}(\boldsymbol{\beta}, \boldsymbol{\gamma})$ the M-scale defined as solution of

$$\frac{1}{n} \sum_{j=1}^{n} \rho_1 \left(\frac{m_j^{jl}(\boldsymbol{\beta}, \boldsymbol{\gamma})}{s_{jl}(\boldsymbol{\beta}, \boldsymbol{\gamma})} \right) = \delta \tag{6.109}$$

and $\tau_{jl}(\boldsymbol{\beta}, \boldsymbol{\gamma})$ the τ-scale (see Section 6.4.5) given by

$$\tau_{jl}(\boldsymbol{\beta}, \boldsymbol{\gamma}) = s_{jl}(\boldsymbol{\beta}, \boldsymbol{\gamma}) \frac{1}{n} \sum_{j=1}^{n} \rho_2 \left(\frac{m_j^{jl}(\boldsymbol{\beta}, \boldsymbol{\gamma})}{s_{jl}(\boldsymbol{\beta}, \boldsymbol{\gamma})} \right). \tag{6.110}$$

Put

$$L(\boldsymbol{\beta}, \boldsymbol{\gamma}) = \sum_{j=1}^{p-1} \sum_{l=j+1}^{p} \tau_{jl}(\boldsymbol{\beta}, \boldsymbol{\gamma}). \tag{6.111}$$

Then the composite τ-estimators of $\boldsymbol{\beta}$ and $\boldsymbol{\gamma}$ are defined by

$$(\widehat{\boldsymbol{\beta}}, \widehat{\boldsymbol{\gamma}}) = \arg\min_{\boldsymbol{\beta}, \boldsymbol{\gamma}} L(\boldsymbol{\beta}, \boldsymbol{\gamma}), \tag{6.112}$$

and the estimator $\widehat{\eta}$ of η by solving

$$\frac{2}{p(p-1)n} \sum_{i=1}^{n} \sum_{j=1}^{p-1} \sum_{l=j+1}^{p} \rho_1 \left(\frac{\mathbf{m}_i^{jl}(\widehat{\boldsymbol{\beta}}, \widehat{\boldsymbol{\gamma}})}{s_0} \right) = \delta, \tag{6.113}$$

where s_0 is defined by

$$\mathrm{E}\rho_1 \left(\frac{v}{s_0} \right) = \delta, \tag{6.114}$$

where $v \sim \chi_2^2$. As a particular case, when $\rho_2 = \rho_1$ we have the class of composite S-estimators. Agostinelli and Yohai (2016) considered ρ_k, $k = 1, 2$ in the SHR family, as defined in Section 6.5. These are of the form $\rho_k(d) = \rho_{\mathrm{SHR}}(d/c_k)$, where ρ_{SHR} is defined in (6.48) and c_k are given constants. They found that the choice $c_1 = 1$ and $c_2 = 1.64$ yields a good trade-off between robustness and efficiency.

It can be shown that the three asymptotic casewise breakdown points defined in (6.101)–(6.103) of the composite τ-estimator are equal to $\min(\delta, 1 - \delta)$. Moreover, the three cellwise breakdown points defined in (6.104)–(6.106) are given as the solution to $1 - (1 - \varepsilon)^2 = \min(\delta, 1 - \delta)$. If $\delta = 0.5$, the three cellwise breakdown points are equal to 0.29. Looking at Table 6.7, we observe that this value is the same as the maximum asymptotic breakdown point of equivariant scatter estimators for two-dimensional data. This result can be explained by the fact that the composite τ-estimator is based on subvectors of dimension two. These results hold for any value of q. Agostinelli and Yohai (2016) also showed that under very general conditions the composite τ-estimators are consistent and asymptotically normal.

Computational aspects and algorithms for the composite τ-estimator are discussed in the Supplementary Material of Agostinelli and Yohai (2016). These algorithms are implemented in the R package `robustvarComp`, which is available in the Comprehensive R Archive Network.

A simulation study in Agostinelli and Yohai (2016) confirms that composite τ-estimators exhibit good behavior under both types of outlier contamination, while Copt and Victoria-Feser's S-estimators as well as Koller's SMDM estimator can only cope with casewise outliers.

Example 6.7 *We study the behavior of the different estimators when applied to a real dataset collected by researchers at the University of Michigan (Anderson D.K.* et al., *2009). This is a longitudinal study of 214 children with neural development disorders. Outliers are present in the couples rather than in the units, and the composite τ-estimator yields quite different results to the maximum likelihood and S-estimators. The tables and figures for this example are obtained with script* ***autism.R***.

The children in the study were divided into three diagnostic groups (d) at the age of two: autism, pervasive developmental disorder and nonspectrum children. An index measuring the degree of socialization was obtained after collecting information at the ages (a) of 2, 3, 5, 9, and 13 years. Since not all children were measured at all ages, 41 children for whom complete data were available were selected. We analyze this data using a regression model with random coefficients where the socialization index y is explained by the age a, its square, the factor variable d and the interactions between a and d, plus an intercept. Let $\mathbf{a} = (a_1, a_2, a_3, a_4, a_5) = (2, 3, 5, 9, 13)$, $d_{(k)i}, k = 1, 2, 3,\ 1 \leq i \leq 41$ the indicator of diagnostic j for child i at age 2, and call y_{ij} the socialization index of child i at age a_j for $1 \leq i \leq 41$ and $1 \leq j \leq 6$. Then it is assumed that

$$y_{ij} = b_{i1} + b_{i2}a_j + b_{i3}a_j^2 + \beta_4 d_{(1)i} + \beta_5 d_{(2)i}$$
$$+ \beta_6 a_j \times d_{(1)i} + \beta_7 a_j \times d_{(2)i} + \beta_8 a_j^2 \times d_{(1)i} + \beta_9 a_j^2 \times d_{(2)i} + \varepsilon_{ij},$$

for $1 \leq i \leq 41, 1 \leq j \leq 5$, where b_{i1}, b_{i2}, b_{i3} are i.i.d. random coefficients with mean $(\beta_1, \beta_2, \beta_3)$ and covariance matrix

$$\mathbf{\Sigma}_b = \begin{pmatrix} \sigma_{11} & \sigma_{1a} & \sigma_{1a^2} \\ \sigma_{1a} & \sigma_{aa} & \sigma_{aa^2} \\ \sigma_{1a^2} & \sigma_{aa^2} & \sigma_{a^2a^2} \end{pmatrix}.$$

Here $\beta_4, \ldots, \beta_9$ are fixed coefficients and the ϵ_{ij} are i.i.d. random errors, independent of the random coefficients, with zero mean and variance $\sigma_{\epsilon e}$. Then the model may be rewritten in terms of (6.99) with $p = 5$, $n = 41$, $J = 5$ and $k = 9$, $\mathbf{y}_i = (y_{i1}, ..., y_{i5})'$,

$$\mathbf{X}_i = (\mathbf{e}, \mathbf{a}, \mathbf{c}, d_{(1)i}\mathbf{e}, d_{(2)i}\mathbf{e}, d_{(1)i}\mathbf{a}, d_{(2)i}\mathbf{a}, d_{(1)i}\mathbf{c}, d_{(2)i}\mathbf{c}),$$

where $\mathbf{e}$ is a vector of dimension 5 with all the elements equal to one and $\mathbf{c} = (a_1^2, ..., a_5^2)$. The variance and covariance structure $\boldsymbol{\Sigma}(\eta, \boldsymbol{\gamma}) = \eta(I + \sum_{j=1}^{J} \gamma_j \mathbf{V}_j)$ has the following components:

$$\mathbf{V}_1 = \mathbf{ee}', \mathbf{V}_2 = \mathbf{aa}', \mathbf{V}_3 = \mathbf{cc}', \mathbf{V}_4 = \mathbf{ea}' + \mathbf{ae}', \mathbf{V}_5 = \mathbf{ec}' + \mathbf{ce}', \quad \mathbf{V}_6 = \mathbf{ac}' + \mathbf{ca}',$$

$$\eta = \sigma_{\epsilon\epsilon}, \gamma_1 = \sigma_{11}/\sigma_{\epsilon\epsilon}, \gamma_2 = \sigma_{aa}/\sigma_{\epsilon\epsilon}, \gamma_3 = \sigma_{a^2a^2}/\sigma_{\epsilon\epsilon},$$

$$\gamma_4 = \sigma_{1a}/\sigma_{\epsilon\epsilon}, \gamma_5 = \sigma_{1a^2}/\sigma_{\epsilon\epsilon}, \gamma_6 = \sigma_{aa^2}/\sigma_{\epsilon\epsilon}.$$

The parameters were estimated using the following methods:

- restricted maximum likelihood (ML);
- composite τ;
- the Copt and Victoria-Feser (2006) S-estimator (CVFS), using a Rocke ρ-function with asymptotic rejection point $\alpha = 0.1$;
- the SMDM estimator defined in Koller (2013);
- composite τ-estimator with $c_1 = 1$ and $c_2 = 1.64$.

Table 6.8 shows the estimators and standard deviations for the fixed effects parameters using different methods, while Table 6.9 reports the estimators of the random effect terms. ML, S and SMDM behave similarly but differently from the composite τ method. The main differences are in the estimation of the random effects terms, both in size (error variance component) and shape (correlation components). Composite τ assigns a large part of the total variance to the random components while the other methods assign it to the error term. We observe that variances estimated by composite τ are in general larger than those estimated by the other methods. On the other

Table 6.8 Autism dataset. Estimators of the fixed effects parameters. The p-values are reported between brackets

Method	Int.	a	a^2	$s_{(1)}$	$s_{(2)}$	$a \times s_{(1)}$	$a \times s_{(2)}$	$a^2 \times s_{(1)}$	$a^2 \times s_{(2)}$
Max. lik.	12.847	6.851	−0.062	−5.245	−2.154	−6.345	−4.512	0.133	0.236
	[0.000]	[0.000]	[0.579]	[0.041]	[0.325]	[0.000]	[0.000]	[0.446]	[0.121]
Composite τ	12.143	6.308	−0.089	−5.214	−4.209	−5.361	−3.852	0.082	0.061
	[0.000]	[0.000]	[0.329]	[0.000]	[0.012]	[0.000]	[0.001]	[0.578]	[0.677]
S Rocke	10.934	7.162	−0.107	−4.457	−0.108	−5.769	−4.995	0.094	0.419
	[0.000]	[0.001]	[0.666]	[0.049]	[0.957]	[0.002]	[0.000]	[0.688]	[0.011]
SMDM	12.346	6.020	0.001	−5.192	−2.173	−5.190	−3.870	0.046	0.151
	[0.000]	[0.000]	[0.992]	[0.010]	[0.213]	[0.000]	[0.000]	[0.781]	[0.300]

Table 6.9 Autism dataset. Estimates of the random effects parameters

Method	σ_{11}	σ_{aa}	$\sigma_{a^2a^2}$	σ_{1a}	σ_{1a^2}	σ_{aa^2}	$\sigma_{\varepsilon\varepsilon}$
Maximum likelihood	2.643	2.328	0.102	0.775	0.429	−0.038	51.360
Compositeτ	9.362	9.670	0.052	−4.019	−0.002	−0.327	5.164
S Rocke	9.467	3.373	0.222	2.170	1.062	−0.349	22.209
SMDM	5.745	0.092	0.115	0.727	0.813	0.103	25.385

side, the estimators for the error variance obtained with the composite τ-estimator are smaller than those obtained with the other methods. As a consequence, the inference based on the composite τ-estimator concludes that the regression coefficients corresponding to the diagnostic variables $d_{(1)}$ and $d_{(2)}$ are significant, while for the other estimators they are not significant.

To better understand the reasons for these discrepancies, we investigate cell, couple and row outliers. For a given dimension $1 \leq q \leq p$, we define as q-dimensional outliers those q-dimensional observations such that the corresponding squared Mahalanobis distance is greater than a the α-quantile of χ^2_q for a given α. In particular we call cell, couple and row outliers respectively the 1-dimensional, 2-dimensional and p-dimensional outliers. The composite τ procedure identifies 33 couple outliers out of 410 couples (8%) at $\alpha = 0.999$. The number of rows with at least one couple of outliers is 12 out of 41; that is, 28% of the rows are contaminated, and this is a too high percentage of contamination for the S and SMDM procedures.

6.16 *Other estimators of location and scatter

6.16.1 Projection estimators

Note that, if $\widehat{\mathbf{\Sigma}}$ is the sample covariance matrix of $\mathbf{x}$, and $\widehat{\sigma}(.)$ denotes the sample SD, then

$$\widehat{\sigma}(\mathbf{a}'\mathbf{x})^2 = \mathbf{a}'\widehat{\mathbf{\Sigma}}\mathbf{a} \quad \forall \ \mathbf{a} \in R^p. \tag{6.115}$$

It would be desirable to have a robust $\widehat{\mathbf{\Sigma}}$ fulfilling (6.115) when $\widehat{\sigma}$ is a robust scatter like the MAD. It can be shown that the SD is the only scatter measure satisfying (6.115), and hence this goal is unattainable. To overcome this difficulty, scatter P-estimators (Maronna *et al.*, 1992) were proposed as "best" approximations to (6.115), analogous to the approach in Section 5.11.2). Specifically, a scatter P-estimator is a matrix $\widehat{\mathbf{\Sigma}}$ that satisfies

$$\sup_{\mathbf{a} \neq \mathbf{0}} \left| \log \left(\frac{\widehat{\sigma}(\mathbf{a}'\mathbf{x})^2}{\mathbf{a}'\widehat{\mathbf{\Sigma}}\mathbf{a}} \right) \right| = \min. \tag{6.116}$$

A similar idea for location was proposed by Tyler (1994). If $\widehat{\mu}(.)$ denotes the sample mean, then the sample mean of $\mathbf{x}$ may be characterized as a vector $\boldsymbol{\nu}$ satisfying

$$\widehat{\mu}(\mathbf{a}'(\mathbf{x}-\boldsymbol{\nu})) = \mathbf{0} \quad \forall \ \mathbf{a} \in R^p. \tag{6.117}$$

Now let $\widehat{\mu}$ be a robust univariate location statistic. It would be desirable to find $\boldsymbol{\nu}$ satisfying (6.117); this unfeasible goal is avoided by defining a location P-estimator as a vector $\boldsymbol{\nu}$ such that

$$\max_{\|\mathbf{a}\|=1} \frac{|\widehat{\mu}(\mathbf{a}'(\mathbf{x}-\boldsymbol{\nu}))|}{\widehat{\sigma}(\mathbf{a}'\mathbf{x})} = \min, \tag{6.118}$$

where $\widehat{\sigma}$ is a robust scatter (the condition $\|\mathbf{a}\| = 1$ is equivalent to $\mathbf{a} \neq \mathbf{0}$). Note that $\boldsymbol{\nu}$ is the point minimizing the outlyingness measure (6.49). The estimator with $\widehat{\mu}$ and $\widehat{\sigma}$ equal to the median and MAD respectively, is called the *MP-estimator* by Adrover and Yohai (2002).

It is easy to verify that both location and scatter P-estimators are equivariant. It can be shown that their maximum asymptotic biases for the normal model do not depend on p. Adrover and Yohai (2002) computed the maximum biases of several estimators and concluded that MP has the smallest. Unfortunately, MP is not fast enough to be useful as a starting estimator.

6.16.2 Constrained M-estimators

Kent and Tyler (1996) define robust efficient estimators, called *constrained M-estimators* (CM estimators for short), as in Section 5.11.3:

$$(\widehat{\boldsymbol{\mu}}, \widehat{\boldsymbol{\Sigma}}) = \arg\min_{\boldsymbol{\mu},\boldsymbol{\Sigma}} \left\{ \frac{1}{n} \sum_{i=1}^{n} \rho(d_i) + \frac{1}{2} \ln |\boldsymbol{\Sigma}| \right\},$$

with the constraint

$$\frac{1}{n} \sum_{i=1}^{n} \rho(d_i) \leq \varepsilon,$$

where $d_i = (\mathbf{x}_i - \widehat{\mu})'\widehat{\boldsymbol{\Sigma}}^{-1}(\mathbf{x}_i - \widehat{\mu})$, ρ is a bounded ρ-function and $\boldsymbol{\Sigma}$ ranges over the set of symmetric positive-definite $p \times p$ matrices.

They show the FBP for data in general position to be

$$\varepsilon^* = \frac{1}{n} \min([n\varepsilon], [n(1-\varepsilon) - p]),$$

and hence the bound (6.27) is attained when $[n\varepsilon] = (n-p)/2$.

These estimators satisfy M-estimating equations (6.11)–(6.12). By a suitable choice of ρ, they can be tuned to attain a desired efficiency.

6.16.3 Multivariate depth

Another approach for location is based on extending the notion of order statistics to multivariate data, and then defining μ as a "multivariate median" or, more generally, a multivariate L-estimator. Among the large amount of literature on the subject, we note the work of Tukey (1975b), Liu (1990), Zuo and Serfling (2000), and Bai and He (1999). The maximum BP of this type of estimator is 1/3, which is much lower than the maximum BP for equivariant estimators given by (6.27); see Donoho and Gasko (1992) and Chen and Tyler (2002).

6.17 Appendix: proofs and complements

6.17.1 Why affine equivariance?

Let $\mathbf{x}$ have an elliptical density $f(\mathbf{x}, \mu, \Sigma)$ of the form (6.7). Here μ and Σ are the distribution parameters. Then if $\mathbf{A}$ is nonsingular, the usual formula for the density of transformed variables yields that $\mathbf{y} = \mathbf{Ax} + \mathbf{b}$ has density

$$f(\mathbf{A}^{-1}(\mathbf{y} - \mathbf{b}), \mu, \Sigma) = f(\mathbf{y}, \mathbf{A}\mu + \mathbf{b}, \mathbf{A}\Sigma\mathbf{A}'), \tag{6.119}$$

and hence the location and scatter parameters of $\mathbf{y}$ are $\mathbf{A}\mu + \mathbf{b}$ and $\mathbf{A}\Sigma\mathbf{A}'$ respectively.

Denote by $(\hat{\mu}(X), \hat{\Sigma}(X))$ the values of the estimators corresponding to a sample $X = \{\mathbf{x}_1, ..., \mathbf{x}_n\}$. Then it is desirable that the estimators $(\hat{\mu}(Y), \hat{\Sigma}(Y))$ corresponding to $Y = \{\mathbf{y}_1, .., \mathbf{y}_n\}$, with $\mathbf{y}_i = \mathbf{Ax}_i + \mathbf{b}$, transform in the same manner as the parameters do in (6.119); that is:

$$\hat{\mu}(Y) = \mathbf{A}\hat{\mu}(X) + \mathbf{b}, \quad \hat{\Sigma}(Y) = \mathbf{A}\hat{\Sigma}(X)\mathbf{A}', \tag{6.120}$$

which corresponds to (6.3).

Affine equivariance is natural in those situations where it is desirable that the result remains essentially unchanged under *any* nonsingular linear transformation, such as linear discriminant analysis, canonical correlations and factor analysis. This does not happen in PCA, since it is based on a fixed metric that is invariant only under orthogonal transformations.

6.17.2 Consistency of equivariant estimators

We shall show that affine equivariant estimators are consistent for elliptical distributions, in the sense that if $\mathbf{x} \sim f(\mathbf{x}, \mu, \Sigma)$ and $\hat{\mu}_\infty$ and $\hat{\Sigma}_\infty$, then

$$\hat{\mu}_\infty = \mu, \quad \hat{\Sigma}_\infty = c\Sigma, \tag{6.121}$$

where c is a constant.

Denote again for simplicity $(\hat{\mu}_\infty(\mathbf{x}), \hat{\Sigma}_\infty(\mathbf{x}))$ as the asymptotic values of the estimators corresponding to the distribution of $\mathbf{x}$. Note that the asymptotic values share the affine equivariance of the estimators; that is (6.120) holds also for $\hat{\boldsymbol{\mu}}$ and $\hat{\boldsymbol{\Sigma}}$ replaced by $\hat{\boldsymbol{\mu}}_\infty$ and $\hat{\boldsymbol{\Sigma}}_\infty$.

We first prove (6.121) for the case $\boldsymbol{\mu} = \mathbf{0}, \boldsymbol{\Sigma} = \mathrm{I}$. Then the distribution is spherical, and so $\mathcal{D}(\mathbf{Tx}) = \mathcal{D}(\mathbf{x})$ for any orthogonal matrix $\mathbf{T}$. In particular, for $\mathbf{T} = -\mathbf{I}$ we have

$$\hat{\boldsymbol{\mu}}_\infty(\mathbf{Tx}) = \hat{\boldsymbol{\mu}}_\infty(-\mathbf{x}) = -\hat{\boldsymbol{\mu}}_\infty(\mathbf{x}) = \hat{\boldsymbol{\mu}}_\infty(\mathbf{x}),$$

which implies $\hat{\boldsymbol{\mu}}_\infty = \mathbf{0}$. At the same time we have

$$\hat{\boldsymbol{\Sigma}}_\infty(\mathbf{Tx}) = \mathbf{T}\hat{\boldsymbol{\Sigma}}_\infty(\mathbf{x})\mathbf{T}' = \hat{\boldsymbol{\Sigma}}_\infty(\mathbf{x}) \tag{6.122}$$

for all orthogonal $\mathbf{T}$. Write $\hat{\boldsymbol{\Sigma}}_\infty = \mathbf{U}\boldsymbol{\Lambda}\mathbf{U}'$, where $\mathbf{U}$ is orthogonal and $\boldsymbol{\Lambda} = \mathrm{diag}(\lambda_1, .., \lambda_p)$. Putting $\mathbf{T} = \mathbf{U}^{-1}$ in (6.122) yields $\boldsymbol{\Lambda} = \hat{\boldsymbol{\Sigma}}_\infty(\mathbf{x})$, so that $\hat{\boldsymbol{\Sigma}}_\infty(\mathbf{x})$ is diagonal. Now let $\mathbf{T}$ be the transformation that interchanges the first two coordinate axes. Then $\mathbf{T}\hat{\boldsymbol{\Sigma}}_\infty(\mathbf{x})\mathbf{T}' = \hat{\boldsymbol{\Sigma}}_\infty(\mathbf{x})$ implies that $\lambda_1 = \lambda_2$, and the same procedure shows that $\lambda_1 = \ldots = \lambda_p$. Thus $\hat{\boldsymbol{\Sigma}}_\infty(\mathbf{x})$ is diagonal with all diagonal elements equal; that is, $\hat{\boldsymbol{\Sigma}}_\infty(\mathbf{x}) = c\mathbf{I}$.

To complete the proof of (6.121), put $\mathbf{y} = \boldsymbol{\mu} + \mathbf{Ax}$, where $\mathbf{x}$ is as before and $\mathbf{AA}' = \boldsymbol{\Sigma}$, so that $\mathbf{y}$ has distribution (6.7). Then the equivariance implies that

$$\hat{\boldsymbol{\mu}}_\infty(\mathbf{y}) = \boldsymbol{\mu} + \mathbf{A}\hat{\boldsymbol{\mu}}_\infty(\mathbf{x}) = \boldsymbol{\mu}, \;\; \hat{\boldsymbol{\Sigma}}_\infty(\mathbf{y}) = \mathbf{A}\hat{\boldsymbol{\Sigma}}_\infty(\mathbf{x})\mathbf{A}' = c\boldsymbol{\Sigma}.$$

The same approach can be used to show that the asymptotic covariance matrix of $\hat{\boldsymbol{\mu}}$ verifies (6.6), noting that if $\hat{\boldsymbol{\mu}}$ has asymptotic covariance matrix $\mathbf{V}$, then $\mathbf{A}\hat{\boldsymbol{\mu}}$ has asymptotic covariance matrix $\mathbf{AVA}'$ (see Section 6.17.7 for details).

6.17.3 The estimating equations of the MLE

We shall prove (6.11)–(6.12). As a generic notation, if $g(\mathbf{T})$ is a function of the $p \times q$ matrix $\mathbf{T} = [t_{ij}]$, then $\partial g/\partial \mathbf{T}$ will denote the $p \times q$ matrix with elements $\partial g/\partial t_{ij}$; a vector argument corresponds to $q = 1$. It is well known (see Seber, 1984) that

$$\frac{\partial|\mathbf{A}|}{\partial \mathbf{A}} = |\mathbf{A}|\mathbf{A}^{-1}, \tag{6.123}$$

and the reader can easily verify that

$$\frac{\partial \mathbf{b}'\mathbf{Ab}}{\partial \mathbf{b}} = (\mathbf{A} + \mathbf{A}')\mathbf{b} \;\text{ and }\; \frac{\partial \mathbf{b}'\mathbf{Ab}}{\partial \mathbf{A}} = \mathbf{bb}'. \tag{6.124}$$

Put $\mathbf{V} = \boldsymbol{\Sigma}^{-1}$. Then (6.9) becomes

$$\mathrm{ave}_i(\rho(d_i)) - \ln|\mathbf{V}| = \min, \tag{6.125}$$

with $d_i = (\mathbf{x}_i - \boldsymbol{\mu})'\,\mathbf{V}(\mathbf{x}_i - \boldsymbol{\mu})$. It follows from (6.124) that

$$\frac{\partial d_i}{\partial \boldsymbol{\mu}} = 2\mathbf{V}(\mathbf{x}_i - \boldsymbol{\mu}) \text{ and } \frac{\partial d_i}{\partial \mathbf{V}} = (\mathbf{x}_i - \boldsymbol{\mu})(\mathbf{x}_i - \boldsymbol{\mu})'. \tag{6.126}$$

Differentiating (6.125) yields

$$2\mathbf{V}\text{ave}_i\{W(d_i)(\mathbf{x}_i-\boldsymbol{\mu})\}=\mathbf{0}\ \text{ and }\ \text{ave}_i\{W(d_i)(\mathbf{x}_i-\boldsymbol{\mu})(\mathbf{x}_i-\boldsymbol{\mu})'\}-\mathbf{V}^{-1}=\mathbf{0},$$

which are equivalent to (6.11)–(6.12).

6.17.4 Asymptotic BP of monotone M-estimators

6.17.4.1 Breakdown point of a location estimator with $\boldsymbol{\Sigma}$ known

It will be shown that the BP of the location estimator given by (6.17) with $\boldsymbol{\Sigma}$ known is $\varepsilon^*=0.5$. It may be supposed, without loss of generality, that $\boldsymbol{\Sigma}=\mathbf{I}$ (Problem 6.4), so that $d(\mathbf{x},\boldsymbol{\mu},\boldsymbol{\Sigma})=\|\mathbf{x}-\boldsymbol{\mu}\|^2$. Let $v(d)=\sqrt{d}W_1(d)$. It is assumed that for all d

$$v(d)\le K=\lim_{s\to\infty}v(s)<\infty. \tag{6.127}$$

For a given ε and a contaminating sequence G_m, call $\boldsymbol{\mu}_m$ the solution of (6.17) corresponding to the mixture $(1-\varepsilon)F+\varepsilon G_m$. Then the scalar product of (6.17) with $\boldsymbol{\mu}_m$ yields

$$(1-\varepsilon)\mathrm{E}_F v(\|\mathbf{x}-\boldsymbol{\mu}_m\|^2)\frac{(\mathbf{x}-\boldsymbol{\mu}_m)'\boldsymbol{\mu}_m}{\|\mathbf{x}-\boldsymbol{\mu}_m\|\|\boldsymbol{\mu}_m\|}$$
$$+\varepsilon\mathrm{E}_{G_m}v(\|\mathbf{x}-\boldsymbol{\mu}_m\|^2)\frac{(\mathbf{x}-\boldsymbol{\mu}_m)'\boldsymbol{\mu}_m}{\|\mathbf{x}-\boldsymbol{\mu}_m\|\|\boldsymbol{\mu}_m\|}=0. \tag{6.128}$$

Assume that $\|\boldsymbol{\mu}_m\|\to\infty$. Then, since for each $\mathbf{x}$ we have

$$\lim_{m\to\infty}\|\mathbf{x}-\boldsymbol{\mu}_m\|^2=\infty,\quad \lim_{m\to\infty}\frac{(\mathbf{x}-\boldsymbol{\mu}_m)'\boldsymbol{\mu}_m}{\|\mathbf{x}-\boldsymbol{\mu}_m\|\|\boldsymbol{\mu}_m\|}=-1,$$

(6.128) yields

$$0\le(1-\varepsilon)K(-1)+\varepsilon K,$$

which implies $\varepsilon\ge 0.5$.

The assumption (6.127) holds in particular if v is monotone. The case with v not monotone has the complications already described for univariate location in Section 3.2.3.

6.17.4.2 Breakdown point of a scatter estimator

To prove (6.25), we deal only with $\boldsymbol{\Sigma}$, and hence assume $\boldsymbol{\mu}$ is known and equal to $\mathbf{0}$. Thus $\widehat{\boldsymbol{\Sigma}}_\infty$ is defined by an equation of the form

$$\mathrm{E}W(\mathbf{x}'\widehat{\boldsymbol{\Sigma}}_\infty^{-1}\mathbf{x})\mathbf{x}\mathbf{x}'=\widehat{\boldsymbol{\Sigma}}_\infty\,. \tag{6.129}$$

Let $\alpha=\mathrm{P}(\mathbf{x}=\mathbf{0})$. It will first be shown that in order for (6.129) to have a solution, it is necessary that

$$K(1-\alpha)\ge p. \tag{6.130}$$

Let $\mathbf{A}$ be any matrix such that $\widehat{\boldsymbol{\Sigma}}_\infty = \mathbf{AA}'$. Multiplying (6.129) by $\mathbf{A}^{-1}$ on the left and by $\mathbf{A}^{-1\prime}$ on the right yields

$$\mathrm{E}W(\mathbf{y}'\mathbf{y})\mathbf{yy}' = \mathbf{I}, \tag{6.131}$$

where $\mathbf{y} = \mathbf{A}^{-1}\mathbf{x}$. Taking the trace in (6.131) yields

$$p = \mathrm{E}W(\|\mathbf{y}\|^2)\|\mathbf{y}\|^2 \mathrm{I}(\mathbf{y} \neq \mathbf{0}) \leq \mathrm{P}(\mathbf{y} \neq \mathbf{0}) \sup_d (dW(d)) = K(1-\alpha), \tag{6.132}$$

which proves (6.130).

Now let F attribute zero mass to the origin, and consider a proportion ε of contamination with distribution G. Then (6.129) becomes

$$(1-\varepsilon)\mathrm{E}_F W(\mathbf{x}'\widehat{\boldsymbol{\Sigma}}_\infty^{-1}\mathbf{x})\mathbf{xx}' + \varepsilon \mathrm{E}_G W(\mathbf{x}'\widehat{\boldsymbol{\Sigma}}_\infty^{-1}\mathbf{x})\mathbf{xx}' = \widehat{\boldsymbol{\Sigma}}_\infty . \tag{6.133}$$

Assume $\varepsilon < \varepsilon^*$. Take G concentrated at $\mathbf{x}_0$:

$$(1-\varepsilon)\mathrm{E}_F W(\mathbf{x}'\widehat{\boldsymbol{\Sigma}}_\infty^{-1}\mathbf{x})\mathbf{xx}' + \varepsilon W(\mathbf{x}_0'\widehat{\boldsymbol{\Sigma}}_\infty^{-1}\mathbf{x}_0)\mathbf{x}_0\mathbf{x}_0' = \widehat{\boldsymbol{\Sigma}}_\infty . \tag{6.134}$$

Put $\mathbf{x}_0 = \mathbf{0}$ first. Then the distribution $(1-\varepsilon)F + \varepsilon G$ attributes mass ε to $\mathbf{0}$, and hence in order for a solution to exist, we must have (6.130); that is, $K(1-\varepsilon) \geq p$, and hence $\varepsilon^* \leq 1 - p/K$.

Let $\mathbf{x}_0$ now be arbitrary and again let $\mathbf{A}$ be any matrix such that $\widehat{\boldsymbol{\Sigma}}_\infty = \mathbf{AA}'$; then

$$(1-\varepsilon)\mathrm{E}_F W(\mathbf{y}'\mathbf{y})\mathbf{yy}' + \varepsilon W(\mathbf{y}_0'\mathbf{y}_0)\mathbf{y}_0\mathbf{y}_0' = \mathbf{I}, \tag{6.135}$$

where $\mathbf{y} = \mathbf{A}^{-1}\mathbf{x}$ and $\mathbf{y}_0 = \mathbf{A}^{-1}\mathbf{x}_0$. Let $\mathbf{a} = \mathbf{y}_0/\|\mathbf{y}_0\|$. Then multiplying in (6.135) by $\mathbf{a}'$ on the left and by $\mathbf{a}$ on the right yields

$$(1-\varepsilon)\mathrm{E}_F W(\|\mathbf{y}\|^2)(\mathbf{y}'\mathbf{a})^2 + \varepsilon W(\|\mathbf{y}_0\|^2)\|\mathbf{y}_0\|^2 = 1 \geq \varepsilon W(\|\mathbf{y}_0\|^2)\|\mathbf{y}_0\|^2. \tag{6.136}$$

Call λ_p and λ_1 the smallest and largest eigenvalues of $\widehat{\boldsymbol{\Sigma}}_\infty$. Let $\mathbf{x}_0$ now tend to infinity. Since $\varepsilon < \varepsilon^*$, the eigenvalues of $\widehat{\boldsymbol{\Sigma}}_\infty$ are bounded away from zero and infinity. Since

$$\|\mathbf{y}_0\|^2 = \mathbf{x}_0'\widehat{\boldsymbol{\Sigma}}_\infty^{-1}\mathbf{x}_0 \geq \frac{\|\mathbf{x}_0\|^2}{\lambda_p},$$

it follows that $\mathbf{y}_0$ tends to infinity. Hence the right-hand side of (6.136) tends to εK, and this implies $\varepsilon \leq 1/K$.

Now let $\varepsilon > \varepsilon^*$. Then either $\lambda_p \to 0$ or $\lambda_1 \to \infty$. Call $\mathbf{a}_1$ and $\mathbf{a}_p$ the unit eigenvectors corresponding to λ_1 and λ_p. Multiplying (6.133) by $\mathbf{a}_1'$ on the left and by $\mathbf{a}_1$ on the right yields

$$(1-\varepsilon)\mathrm{E}_F W(\mathbf{x}'\widehat{\boldsymbol{\Sigma}}_\infty^{-1}\mathbf{x})(\mathbf{x}'\mathbf{a}_1)^2 + \varepsilon \mathrm{E}_G W(\mathbf{x}'\widehat{\boldsymbol{\Sigma}}_\infty^{-1}\mathbf{x})(\mathbf{x}'\mathbf{a}_1)^2 = \lambda_1.$$

Suppose that $\lambda_1 \to \infty$. Divide the above expression by λ_1; recall that the first expectation is bounded and that

$$\frac{(\mathbf{x}'\mathbf{a}_1)^2}{\lambda_1} \leq \mathbf{x}'\widehat{\boldsymbol{\Sigma}}_\infty^{-1}\mathbf{x}.$$

Then in the limit we have

$$1 = \varepsilon \mathrm{E}_G W(\mathbf{x}'\widehat{\boldsymbol{\Sigma}}_{\infty}^{-1}\mathbf{x})\frac{(\mathbf{x}'\mathbf{a}_1)^2}{\lambda_1} \leq \varepsilon K,$$

and hence $\varepsilon \geq 1/K$.

On the other hand, taking the trace in (6.136) and proceeding as in the proof of (6.132) yields

$$p = (1-\varepsilon)\mathrm{E}_F W(\|\mathbf{y}\|^2)\|\mathbf{y}\|^2 + \varepsilon \mathrm{E}_G W(\|\mathbf{y}\|^2)\|\mathbf{y}\|^2 \geq (1-\varepsilon)\mathrm{E}_F W(\|\mathbf{y}\|^2)\|\mathbf{y}\|^2.$$

Note that

$$\|\mathbf{y}\|^2 \geq \frac{(\mathbf{x}'\mathbf{a}_p)^2}{\lambda_p}.$$

Hence $\lambda_p \to 0$ implies $\|\mathbf{y}\|^2 \to \infty$ and thus the right-hand side of the equation above tends to $(1-\varepsilon)K$, which implies $\varepsilon \geq 1 - p/K$.

6.17.5 The estimating equations for S-estimators

We are going to prove (6.31)–(6.32). Put, for simplicity,

$$\mathbf{V} = \boldsymbol{\Sigma} \text{ and } d_i = d(\mathbf{x}_i, \boldsymbol{\mu}, \mathbf{V}) = (\mathbf{x}_i - \boldsymbol{\mu})'\mathbf{V}(\mathbf{x}_i - \boldsymbol{\mu})$$

and call $\sigma(\boldsymbol{\mu}, \mathbf{V})$ the solution of

$$\mathrm{ave}_i\left\{\rho\left(\frac{d_i}{\sigma}\right)\right\} = \delta. \tag{6.137}$$

Then (6.28) amounts to minimizing $\sigma(\boldsymbol{\mu}, \mathbf{V})$ with $|\mathbf{V}| = 1$. Solving this problem by the method of Lagrange's multipliers becomes

$$g(\boldsymbol{\mu}, \mathbf{V}, \lambda) = \sigma(\boldsymbol{\mu}, \mathbf{V}) + \lambda(|\mathbf{V}| - 1) = \min$$

Differentiating g with respect to λ, $\boldsymbol{\mu}$ and $\mathbf{V}$, and recalling (6.123), we have $|\mathbf{V}| = 1$ and

$$\frac{\partial\sigma}{\partial\boldsymbol{\mu}} = \mathbf{0}, \quad \frac{\partial\sigma}{\partial\mathbf{V}} + \lambda\mathbf{V}^{-1} = \mathbf{0}. \tag{6.138}$$

Differentiating (6.137) and recalling the first equations in (6.138) and in (6.126) yields

$$\mathrm{ave}_i\left\{W\left(\frac{d_i}{\sigma}\right)\left(\sigma\frac{\partial d_i}{\partial\boldsymbol{\mu}} - d_i\frac{\partial\sigma}{\partial\boldsymbol{\mu}}\right)\right\} = 2\sigma\mathrm{ave}_i\left\{W\left(\frac{d_i}{\sigma}\right)(\mathbf{x}_i - \boldsymbol{\mu})\right\} = \mathbf{0},$$

which implies (6.31). Proceeding similarly with respect to $\mathbf{V}$ we have

$$\mathrm{ave}_i\left\{W\left(\frac{d_i}{\sigma}\right)\left(\sigma\frac{\partial d_i}{\partial\mathbf{V}} + d_i\lambda\mathbf{V}^{-1}\right)\right\}$$

$$= \sigma\mathrm{ave}_i\left\{W\left(\frac{d_i}{\sigma}\right)(\mathbf{x}_i - \boldsymbol{\mu})(\mathbf{x}_i - \boldsymbol{\mu})'\right\} + b\mathbf{V}^{-1} = \mathbf{0},$$

with

$$b = \lambda \quad \text{ave}_i \left\{ W\left(\frac{d_i}{\sigma}\right) d_i \right\};$$

and this implies (6.32) with $c = -b/\sigma$.

6.17.6 Behavior of S-estimators for high p

It will be shown that an S-estimator with continuous ρ becomes increasingly similar to the classical estimator when $p \to \infty$. For simplicity, this property is proved here only for normal data. However, it can be proved under more general conditions which include finite fourth moments.

Because of the equivariance of S-estimators, we may assume that the true parameters are $\mathbf{\Sigma} = \mathbf{I}$ and $\boldsymbol{\mu} = \mathbf{0}$. Then, since the estimator is consistent, its asymptotic values are $\widehat{\boldsymbol{\mu}}_\infty = \mathbf{0}$, $\widehat{\mathbf{\Sigma}}_\infty = \mathbf{I}$.

For each p, let $d^{(p)} = d(\mathbf{x}, \widehat{\boldsymbol{\mu}}_\infty, \widehat{\mathbf{\Sigma}}_\infty)$. Then $d^{(p)} = ||\mathbf{x}||^2 \sim \chi^2_p$ and hence

$$\text{E}\left(\frac{d^{(p)}}{p}\right) = 1, \quad \text{SD}\left(\frac{d^{(p)}}{p}\right) = \sqrt{\frac{2}{p}}, \tag{6.139}$$

which implies that the distribution of $d^{(p)}/p$ is increasingly concentrated around 1, and $d^{(p)}/p \to 1$ in probability when $p \to \infty$.

Since ρ is continuous, there exists $a > 0$ such that $\rho(a) = 0.5$. Call σ_p the scale corresponding to $d^{(p)}$

$$0.5 = \text{E}\rho\left(\frac{d^{(p)}}{\sigma_p}\right). \tag{6.140}$$

We shall show that

$$\frac{d^{(p)}}{\sigma_p} \to_p a. \tag{6.141}$$

Since $d^{(p)}/p \to_p 1$, we have for any $\varepsilon > 0$

$$\text{E}\rho\left(\frac{d^{(p)}}{p/(a(1+\varepsilon))}\right) \to \rho(a(1+\varepsilon)) > 0.5$$

and

$$\text{E}\rho\left(\frac{d^{(p)}}{p/(a(1-\varepsilon))}\right) \to \rho(a(1-\varepsilon)) < 0.5.$$

Then (6.140) implies that for large enough p

$$\frac{p}{a(1+\varepsilon)} \le \sigma_p \le \frac{p}{a(1-\varepsilon)}$$

and hence $\lim_{p\to\infty}(a\sigma_p/p) = 1$, which implies

$$\frac{d^{(p)}}{\sigma_p} = a\frac{d^{(p)}/p}{a\sigma_p/p} \to_p a$$

as stated. This implies that for large n and p the weights of the observations $W(d_i/\widehat{\sigma})$ are

$$W\left(\frac{d_i}{\widehat{\sigma}}\right) = W\left(\frac{d(\mathbf{x}_i, \widehat{\boldsymbol{\mu}}, \widehat{\boldsymbol{\Sigma}})}{\widehat{\sigma}}\right) \simeq W\left(\frac{d(\mathbf{x}_i, \widehat{\boldsymbol{\mu}}_\infty, \widehat{\boldsymbol{\Sigma}}_\infty)}{\sigma_p}\right) \approx W(a);$$

that is, they are practically constant. Hence $\widehat{\boldsymbol{\mu}}$ and $\widehat{\boldsymbol{\Sigma}}$, which are weighted means and covariances, will be very similar to E$\mathbf{x}$ and $\mathbf{Var}(\mathbf{x})$, and hence very efficient for normal data.

6.17.7 Calculating the asymptotic covariance matrix of location M-estimators

Recall that the covariance matrix of the classical location estimator is a constant multiple of the covariance matrix of the observations, since $\mathbf{Var}(\overline{\mathbf{x}}) = n^{-1}\mathbf{Var}(\mathbf{x})$. We shall show a similar result for M-estimators for elliptically distributed data. Let the estimators be defined by (6.14)–(6.15). As explained at the end of Section 6.12.2, it can be shown that if $\mathbf{x}_i$ has an elliptical distribution (6.7), the asymptotic covariance matrix of $\widehat{\boldsymbol{\mu}}$ has the form (6.6): $\mathbf{V} = v\boldsymbol{\Sigma}$, where v is a constant that we shall now calculate.

It can be shown that in the elliptical case the asymptotic distribution of $\widehat{\boldsymbol{\mu}}$ is the same as if $\boldsymbol{\Sigma}$ were assumed known. In view of the equivariance of the estimator, we may consider only the case $\boldsymbol{\mu} = \mathbf{0}$, $\boldsymbol{\Sigma} = \mathbf{I}$. Then it follows from (6.121) that $\widehat{\boldsymbol{\Sigma}}_\infty = c\mathbf{I}$, and taking the trace in (6.15) we have that c is the solution of (6.21).

It will be shown that

$$v = \frac{pa}{b^2},$$

where (writing $z = ||\mathbf{x}||^2$)

$$a = \mathrm{E}W_1\left(\frac{z}{c}\right)^2 z, \quad b = 2\mathrm{E}W_1'\left(\frac{z}{c}\right)\frac{z}{c} + p\mathrm{E}W_1\left(\frac{z}{c}\right). \tag{6.142}$$

We may write (6.14) as

$$\sum_{i=1}^{n} \boldsymbol{\Psi}(\mathbf{x}_i, \boldsymbol{\mu}) = \mathbf{0},$$

with

$$\boldsymbol{\Psi}(\mathbf{x}, \boldsymbol{\mu}) = W_1\left(\frac{\|\mathbf{x} - \boldsymbol{\mu}\|^2}{c}\right)(\mathbf{x} - \boldsymbol{\mu}).$$

It follows from (3.49) that $\mathbf{V} = \mathbf{B}^{-1}\mathbf{A}\mathbf{B}'^{-1}$, where

$$\mathbf{A} = \mathrm{E}\boldsymbol{\Psi}(x, 0)\boldsymbol{\Psi}(x, 0)', \quad \mathbf{B} = \mathrm{E}\dot{\boldsymbol{\Psi}}(\mathbf{x}_i, \mathbf{0}),$$

where $\dot{\boldsymbol{\Psi}}$ is the derivative of $\boldsymbol{\Psi}$ with respect to $\boldsymbol{\mu}$; that is, the matrix $\dot{\boldsymbol{\Psi}}$ with elements $\dot{\Psi}_{jk} = \partial\Psi_j/\partial\mu_k$.

We have

$$\mathbf{A} = \mathrm{E}W_1\left(\frac{\|\mathbf{x}\|^2}{c}\right)^2 \mathbf{x}\mathbf{x}'.$$

Since $\mathcal{D}(\mathbf{x})$ is spherical, $\mathbf{A}$ is a multiple of the identity: $\mathbf{A} = t\mathbf{I}$. Taking the trace and recalling that $\mathrm{tr}(\mathbf{x}\mathbf{x}') = \mathbf{x}'\mathbf{x}$, we have

$$\mathrm{tr}(\mathbf{A}) = \mathrm{E}W_1\left(\frac{\|\mathbf{x}\|^2}{c}\right)^2 \|\mathbf{x}\|^2 = a = tp,$$

and hence $\mathbf{A} = (a/p)\mathbf{I}$.

To calculate $\mathbf{B}$, recall that

$$\frac{\partial\|\mathbf{a}\|^2}{\partial\mathbf{a}} = 2\mathbf{a}, \quad \frac{\partial\mathbf{a}}{\partial\mathbf{a}} = \mathbf{I},$$

and hence

$$\dot{\boldsymbol{\Psi}}(\mathbf{x}_i, \boldsymbol{\mu}) = -\left\{2W_1'\left(\frac{\|\mathbf{x}-\boldsymbol{\mu}\|^2}{c}\right)\frac{(\mathbf{x}-\boldsymbol{\mu})}{c}(\mathbf{x}-\boldsymbol{\mu})' + W_1\left(\frac{\|\mathbf{x}-\boldsymbol{\mu}\|^2}{c}\right)\mathbf{I}\right\}.$$

Then the same reasoning yields $\mathbf{B} = -(b/p)\mathbf{I}$, which implies

$$\mathbf{V} = \frac{pa}{b^2}\mathbf{I},$$

as stated.

To compute c for normal data, note that it depends only on the distribution of $\|\mathbf{x}\|^2$, which is χ^2_p.

This approach can also be used to calculate the efficiency of location S-estimators.

6.17.8 The exact fit property

Let the dataset X contain $q \geq n - m^*$ points on the hyperplane $H = \{\mathbf{x} : \boldsymbol{\beta}'\mathbf{x} = \gamma\}$. It will be shown that $\widehat{\boldsymbol{\mu}}(X) \in H$ and $\widehat{\boldsymbol{\Sigma}}(X)\boldsymbol{\beta} = \mathbf{0}$.

Without loss of generality we can take $\|\boldsymbol{\beta}\| = 1$ and $\gamma = 0$. In fact, the equation defining H does not change if we divide both sides by $\|\boldsymbol{\beta}\|$, and since

$$\widehat{\boldsymbol{\mu}}(X + \mathbf{a}) = \widehat{\boldsymbol{\mu}}(X) + \mathbf{a}, \quad \widehat{\boldsymbol{\Sigma}}(X + \mathbf{a}) = \widehat{\boldsymbol{\Sigma}}(X),$$

we may replace $\mathbf{x}$ by $\mathbf{x} + \mathbf{a}$ where $\boldsymbol{\beta}'\mathbf{a} = 0$.

Now $H = \{\mathbf{x} : \boldsymbol{\beta}'\mathbf{x} = 0\}$ is a subspace. Call $\mathbf{P}$ the matrix corresponding to the orthogonal projection on the subspace orthogonal to H; that is, $\mathbf{P} = \boldsymbol{\beta}\boldsymbol{\beta}'$. Define for $t \in R$

$$\mathbf{y}_i = \mathbf{x}_i + t\boldsymbol{\beta}\boldsymbol{\beta}'\mathbf{x}_i = (\mathbf{I} + t\mathbf{P})\mathbf{x}_i.$$

Then $Y = \{\mathbf{y}_1, .., \mathbf{y}_n\}$ has at least q elements in common with $\mathbf{x}$, since $\mathbf{P}\mathbf{z} = \mathbf{0}$ for $\mathbf{z} \in H$. Hence by the definition of BP, $\widehat{\boldsymbol{\mu}}(Y)$ remains bounded for all t. Since

$$\widehat{\boldsymbol{\mu}}(Y) = \widehat{\boldsymbol{\mu}}(X) + t\boldsymbol{\beta}\boldsymbol{\beta}'\widehat{\boldsymbol{\mu}}(X)$$

the left-hand side is a bounded function of t, while the right-hand side tends to infinity with t unless $\boldsymbol{\beta}'\widehat{\boldsymbol{\mu}}(X) = 0$; that is, $\widehat{\boldsymbol{\mu}}(X) \in H$.

In the same way

$$\widehat{\boldsymbol{\Sigma}}(Y) = (\mathbf{I} + t\mathbf{P})\widehat{\boldsymbol{\Sigma}}(X)(\mathbf{I} + t\mathbf{P}) = \widehat{\boldsymbol{\Sigma}}(X)$$
$$+ t^2(\boldsymbol{\beta}'\widehat{\boldsymbol{\Sigma}}(X)\boldsymbol{\beta})\boldsymbol{\beta}\boldsymbol{\beta}' + t(\mathbf{P}\widehat{\boldsymbol{\Sigma}}(X) + \widehat{\boldsymbol{\Sigma}}(X)\mathbf{P})$$

is a bounded function of t, which implies that $\widehat{\boldsymbol{\Sigma}}(X)\boldsymbol{\beta} = \mathbf{0}$.

6.17.9 Elliptical distributions

A random vector $\mathbf{r} \in R^p$ is said to have a spherical distribution if its density f depends only on $\|\mathbf{r}\|$; that is, it has the form

$$f(\mathbf{r}) = h(\|\mathbf{r}\|) \tag{6.143}$$

for some nonnegative function h. It follows that for any orthogonal matrix $\mathbf{T}$:

$$\mathcal{D}(\mathbf{Tr}) = \mathcal{D}(\mathbf{r}). \tag{6.144}$$

In fact, (6.144) may be taken as the general definition of a spherical distribution, without requiring the existence of a density. However, we prefer the definition here for reasons of simplicity.

The random vector $\mathbf{x}$ will be said to have an elliptical distribution if

$$\mathbf{x} = \boldsymbol{\mu} + \mathbf{Ar} \tag{6.145}$$

where $\boldsymbol{\mu} \in R^p$, $\mathbf{A} \in R^{p \times p}$ is nonsingular and $\mathbf{r}$ has a spherical distribution. Let

$$\boldsymbol{\Sigma} = \mathbf{AA}'.$$

We shall call $\boldsymbol{\mu}$ and $\boldsymbol{\Sigma}$ the location vector and the scatter matrix of $\mathbf{x}$, respectively. We now state the most relevant properties of elliptical distributions. If $\mathbf{x}$ is given by (6.145), then:

1. The distribution of $\mathbf{Bx} + \mathbf{c}$ is also elliptical, with location vector $\mathbf{B}\boldsymbol{\mu} + \mathbf{c}$ and scatter matrix $\mathbf{B}\boldsymbol{\Sigma}\mathbf{B}'$
2. If the mean and variances of $\mathbf{x}$ exist, then

   $$\mathrm{E}\mathbf{x} = \boldsymbol{\mu}, \quad \mathbf{Var}(\mathbf{x}) = c\boldsymbol{\Sigma},$$

   where c is a constant.
3. The density of $\mathbf{x}$ is

   $$h((\mathbf{x} - \boldsymbol{\mu})'\boldsymbol{\Sigma}^{-1}(\mathbf{x} - \boldsymbol{\mu})).$$

4. The distributions of linear combinations of $\mathbf{x}$ belong to the same location-scale family; more precisely, for any $\mathbf{a} \in R^p$

$$\mathcal{D}(\mathbf{a}'\mathbf{x}) = \mathcal{D}(\mathbf{a}'\boldsymbol{\mu} + \sqrt{\mathbf{a}'\boldsymbol{\Sigma}a}r_1), \tag{6.146}$$

where r_1 is the first coordinate of $\mathbf{r}$.

The proofs of (1) and (3) are immediate. The proof of (2) follows from the fact that, if the mean and variances of $\mathbf{r}$ exist, then

$$\mathrm{E}\mathbf{r} = \mathbf{0}, \quad \mathbf{Var}(\mathbf{r}) = c\mathbf{I}$$

for some constant c.

Proof of (4): It will be shown that the distribution of a linear combination of $\mathbf{r}$ does not depend on its direction. More precisely, for all $\mathbf{a} \in R^p$

$$\mathcal{D}(\mathbf{a}'\mathbf{r}) = \mathcal{D}(\|\mathbf{a}\|r_1). \tag{6.147}$$

In fact, let $\mathbf{T}$ be an orthogonal matrix with columns $\mathbf{t}_1, ..., \mathbf{t}_p$ such that $\mathbf{t}_1 = \mathbf{a}/\|\mathbf{a}\|$. Then $\mathbf{Ta} = (\|\mathbf{a}\|, 0, 0, ..., 0)'$.

Then by (6.144):

$$\mathcal{D}(\mathbf{a}'\mathbf{r}) = \mathcal{D}(\mathbf{a}'\mathbf{T}'\mathbf{r}) = \mathcal{D}((\mathbf{Ta})'\mathbf{r}) = \mathcal{D}(\|\mathbf{a}\|r_1)$$

as stated; and (6.146) follows from (6.147) and (6.145).

6.17.10 Consistency of Gnanadesikan–Kettenring correlations

Let the random vector $\mathbf{x} = (x_1, \ldots, x_p)$ have an elliptical distribution: that is, $\mathbf{x} = \mathbf{Az}$ where $\mathbf{z} = (z_1, \ldots, z_p)$ has a spherical distribution. This implies that for any $\mathbf{u} \in R^p$,

$$\mathcal{D}(\mathbf{u}'\mathbf{x}) = \mathcal{D}(\|\mathbf{b}\|z_1) \text{ with } \mathbf{b} = \mathbf{A}'\mathbf{u},$$

and hence

$$\sigma(\mathbf{u}'\mathbf{x}) = \sigma_0\|\mathbf{b}\| \text{ with } \sigma_0 = \sigma(z_1). \tag{6.148}$$

Let $U_j = \mathbf{u}_j'\mathbf{x}$ ($j = 1, 2$) be two linear combinations of $\mathbf{x}$. It will be shown that their robust correlation (6.69) coincides with the ordinary one.

Assume that $\mathbf{z}$ has finite second moments. We may assume that $\mathbf{Var}(\mathbf{z}) = \mathbf{I}$. Then

$$\mathrm{Cov}(U_1, U_2) = \mathbf{b}_1'\mathbf{b}_2, \quad \mathrm{Var}(U_j) = \|\mathbf{b}_j\|^2,$$

where $\mathbf{b}_j = \mathbf{A}'\mathbf{u}_j$, and hence the ordinary correlation is

$$\mathrm{Corr}(U_1, U_2) = \frac{\mathbf{b}_1'\mathbf{b}_2}{\|\mathbf{b}_1\|\|\mathbf{b}_2\|}.$$

Put, for brevity, $\sigma_j = \sigma(U_j)$. It follows from (6.148) that

$$\sigma_j = \|\mathbf{b}_j\|\sigma_0, \quad \sigma\left(\frac{U_1}{\sigma_1} \pm \frac{U_2}{\sigma_2}\right) = \left\|\frac{\mathbf{b}_1}{\|\mathbf{b}_1\|} \pm \frac{\mathbf{b}_2}{\|\mathbf{b}_2\|}\right\|,$$

and hence (6.69) yields

$$\mathrm{RCorr}(U_1, U_2) = \frac{\mathbf{b}_1'\mathbf{b}_2}{\|\mathbf{b}_1\|\|\mathbf{b}_2\|} = \mathrm{Corr}(U_1, U_2).$$

6.17.11 Spherical principal components

We may assume, without loss of generality, that $\boldsymbol{\mu} = \mathbf{0}$. The covariance matrix of $\mathbf{x}/\|\mathbf{x}\|$ is

$$\mathbf{U} = \mathrm{E}\frac{\mathbf{x}\mathbf{x}'}{\|\mathbf{x}\|^2}. \tag{6.149}$$

It will be shown that $\mathbf{U}$ and $\boldsymbol{\Sigma}$ have the same eigenvectors.

It will be first assumed that $\boldsymbol{\Sigma}$ is diagonal: $\boldsymbol{\Sigma} = \mathrm{diag}\{\lambda_1, ..., \lambda_p\}$, where $\lambda_1, ..., \lambda_p$ are its eigenvalues. Then the eigenvectors of $\boldsymbol{\Sigma}$ are the vectors of the canonical basis $\mathbf{b}_1, ..., \mathbf{b}_p$ with $b_{jk} = \delta_{jk}$. It will be shown that the $\mathbf{b}_j$s are also the eigenvectors of $\mathbf{U}$; that is,

$$\mathbf{U}\mathbf{b}_j = \alpha_j\mathbf{b}_j \tag{6.150}$$

for some α_j. For a given j, put $\mathbf{u} = \mathbf{U}\mathbf{b}_j$. Then we must show that $k \neq j$ implies $u_k = 0$. In fact, for $k \neq j$,

$$u_k = \mathrm{E}\frac{x_j x_k}{\|\mathbf{x}\|^2},$$

where x_j $(j = 1, .., p)$ are the coordinates of $\mathbf{x}$. The symmetry of the distribution implies that $D(x_j, x_k) = D(x_j, -x_k)$, which implies

$$u_k = \mathrm{E}\frac{x_j(-x_k)}{\|\mathbf{x}\|^2} = -u_k$$

and hence $u_k = 0$. This proves (6.150). It follows from (6.150) that $\mathbf{U}$ is diagonal.

Now let $\boldsymbol{\Sigma}$ have arbitrary eigenvectors $\mathbf{t}_1, ..., \mathbf{t}_p$. Call λ_j $(j = 1, .., p)$ its eigenvalues, and let $\mathbf{T}$ be the orthogonal matrix with columns $\mathbf{t}_1, ..., \mathbf{t}_p$, so that

$$\boldsymbol{\Sigma} = \mathbf{T}\boldsymbol{\Lambda}\mathbf{T}',$$

where $\boldsymbol{\Lambda} = \mathrm{diag}\{\lambda_1, ..., \lambda_p\}$. We must show that the eigenvectors of $\mathbf{U}$ in (6.149) are the $\mathbf{t}_j$s.

Let $\mathbf{z} = \mathbf{T}'\mathbf{x}$. Then $\mathbf{z}$ has an elliptical distribution with location vector $\mathbf{0}$ and scatter matrix $\boldsymbol{\Lambda}$. The orthogonality of $\mathbf{T}$ implies that $\|\mathbf{z}\| = \|\mathbf{x}\|$. Let

$$\mathbf{V} = \mathrm{E}\frac{\mathbf{z}\mathbf{z}'}{\|\mathbf{z}\|^2} = \mathbf{T}'\mathbf{U}\mathbf{T}. \tag{6.151}$$

It follows from (6.150) that the $\mathbf{b}_j$s are the eigenvectors of $\mathbf{V}$, and hence that $\mathbf{V}$ is diagonal. Then (6.151) implies that $\mathbf{U} = \mathbf{T}\mathbf{V}\mathbf{T}'$, which implies that the eigenvectors of $\mathbf{U}$ are the columns of $\mathbf{T}$, which are the eigenvectors of $\boldsymbol{\Sigma}$. This completes the proof of the equality of the eigenvector of $\mathbf{U}$ and $\boldsymbol{\Sigma}$.

Now let $\sigma(.)$ be a scatter statistic. We shall show that the values of $\sigma(\mathbf{x}'\mathbf{t}_j)^2$ are proportional to the eigenvalues of $\boldsymbol{\Sigma}$. In fact, it follows from (6.146) and $\mathbf{t}_j'\boldsymbol{\Sigma}t_j = \lambda_j$ that for all j,

$$\mathcal{D}(\mathbf{t}_j'\mathbf{x}) = \mathcal{D}\left(\mathbf{t}_j'\boldsymbol{\mu} + \sqrt{\lambda_j}r_1\right),$$

and hence

$$\sigma(\mathbf{t}_j'\mathbf{x}) = \sigma\left(\sqrt{\lambda_j}r_1\right) = \sqrt{\lambda_j}d,$$

with $d = \sigma(r_1)$.

6.17.12 Fixed point estimating equations and computing algorithm for the GS estimator

Let $\mathbf{x} \in R^p$ be a possible observation and $\mathbf{u}$ the vector of zeros and ones indicating the missing observations (see Section 6.12.1). Call $\mathbf{m} \in R^p$ and Σ the mean vector and covariance matrix of $\mathbf{x}$. Then we define $\widehat{\mathbf{x}}(\mathbf{u}, \mathbf{x}^{(\mathbf{u})}, \mathbf{m}, \boldsymbol{\Sigma})$ as the best linear predictor of $\mathbf{x}$ given $\mathbf{x}^{(\mathbf{u})}$, and $\mathbf{C}(\mathbf{u}, \boldsymbol{\Sigma})$ as the covariance matrix for the prediction error $\mathbf{x} - \widehat{\mathbf{x}}(\mathbf{u}, \mathbf{x}^{(\mathbf{u})}, \mathbf{m}, \boldsymbol{\Sigma})$. In particular, if $\mathbf{u}$ has the first $q = p\ (\mathbf{u})$ entries equal to one and the remaining entries equal to zero, we have the following simple formulae. Let $\mathbf{v} = (v_1, ..., v_p)' \in \mathbf{A}_p$ such that $v_1 = \cdots = v_q = 0$ and $v_{q+1} = \cdots = v_p = 1$ and put

$$\mathbf{m} = \begin{pmatrix} \mathbf{m}^{(\mathbf{u})} \\ \mathbf{m}^{(\mathbf{v})} \end{pmatrix}, \ \boldsymbol{\Sigma} = \begin{pmatrix} \boldsymbol{\Sigma}_{\mathbf{uu}} & \boldsymbol{\Sigma}_{\mathbf{uv}} \\ \boldsymbol{\Sigma}_{\mathbf{vu}} & \boldsymbol{\Sigma}_{\mathbf{vv}} \end{pmatrix}.$$

Then,

$$\widehat{\mathbf{x}}(\mathbf{u}, \mathbf{x}^{(\mathbf{u})}, \mathbf{m}, \boldsymbol{\Sigma}) = \begin{pmatrix} \mathbf{x}^{(\mathbf{u})} \\ \\ \mathbf{m}^{(\mathbf{v})} + \boldsymbol{\Sigma}_{\mathbf{vu}}\boldsymbol{\Sigma}_{\mathbf{uu}}^{-1}(\mathbf{x}^{(\mathbf{u})} - \mathbf{m}^{(\mathbf{u})}) \end{pmatrix}, \tag{6.152}$$

$$\mathbf{C}(\mathbf{u}, \boldsymbol{\Sigma}) = \begin{pmatrix} \mathbf{0} & \mathbf{0} \\ \mathbf{0} & \boldsymbol{\Sigma}_{\mathbf{vv}} - \boldsymbol{\Sigma}_{\mathbf{vu}}\boldsymbol{\Sigma}_{\mathbf{uu}}^{-1}\boldsymbol{\Sigma}_{\mathbf{uv}} \end{pmatrix}. \tag{6.153}$$

Then the fixed-point estimating equations for the GS-estimators $\widehat{\mathbf{m}}_n$ $\tilde{\boldsymbol{\Sigma}}_n$ and $\boldsymbol{\Sigma}_n$ are the following:

$$\widehat{\mathbf{m}}_n = \frac{\sum_{i=1}^n w_i\widehat{\mathbf{x}}_i}{\sum_{i=1}^n w_i} \tag{6.154}$$

and

$$\tilde{\boldsymbol{\Sigma}}_n = \frac{\sum_{i=1}^n [w_i\,(\widehat{\mathbf{x}}_i - \widehat{\mathbf{m}}_n)\,(\widehat{\mathbf{x}}_i - \widehat{\mathbf{m}}_n)' + w_i w_i^* \mathbf{C}_i]}{\sum_{i=1}^n w_i w_i^*}, \tag{6.155}$$

where $\widehat{\mathbf{x}}_i = \widehat{\mathbf{x}}(\mathbf{u}_i, \mathbf{x}_i^{(\mathbf{u}_i)}, \widehat{\mathbf{m}}_n, \widehat{\boldsymbol{\Sigma}}_n)$, $\mathbf{C}_i = \mathbf{C}(\mathbf{u}_i, \widehat{\boldsymbol{\Sigma}}_n)$, $w_i = w(\mathbf{u}_i, \mathbf{x}_i^{(\mathbf{u}_i)}, \widehat{\mathbf{m}}_n, \widehat{\boldsymbol{\Sigma}}_n, \tilde{s}_n)$, $w_i^* = w^*(\mathbf{u}_i, \mathbf{x}_i^{(\mathbf{u}_i)}, \widehat{\mathbf{m}}_n, \widehat{\boldsymbol{\Sigma}}_n)$ with

$$w(\mathbf{u}, \mathbf{x}^{(\mathbf{u})}, \mathbf{m}, \boldsymbol{\Sigma}, s) = \frac{|\boldsymbol{\Sigma}^{(\mathbf{u})}|^{1/p(\mathbf{u})}}{|\widehat{\boldsymbol{\Omega}}_n^{(\mathbf{u})}|^{1/p(\mathbf{u})}} \rho' \left(\frac{d(\mathbf{x}^{(\mathbf{u})}, \mathbf{m}^{(\mathbf{u})}, \boldsymbol{\Sigma}^{(\mathbf{u})})}{c_{p(\mathbf{u})} \; s} \frac{|\boldsymbol{\Sigma}^{(\mathbf{u})}|^{1/p(\mathbf{u})}}{|\widehat{\boldsymbol{\Omega}}_n^{(\mathbf{u})}|^{1/p(\mathbf{u})}} \right), \tag{6.156}$$

$$w^*(\mathbf{u}, \mathbf{x}^{(\mathbf{u})}, \mathbf{m}, \boldsymbol{\Sigma}) = \frac{d(\mathbf{x}^{(\mathbf{u}_i)}, \mathbf{m}^{(\mathbf{u})}, \boldsymbol{\Sigma}^{(\mathbf{u})})}{p(\mathbf{u})}, \tag{6.157}$$

$\tilde{s}_n = S_n^*(\widehat{\mathbf{m}}_n, \tilde{\boldsymbol{\Sigma}}_n)$ and $\widehat{\boldsymbol{\Sigma}}_n = \widehat{s}_n \tilde{\boldsymbol{\Sigma}}_n$, where $\widehat{s}_n$ satisfies (6.98).

These equations show that the GS-estimators of location and scatter shape are a weighted mean and a weighted/corrected sample covariance matrix. If $w_i = w_i^* = 1$, $1 \le i \le n$ we obtain the equations of the MLE for Gaussian data.

6.17.12.1 Computing algorithm

Based on the fixed points equations, a following natural algorithm can be used to compute the GS-estimators. Given initial estimators $(\widehat{\mathbf{m}}_n^{(0)}, \widehat{\boldsymbol{\Sigma}}_n^{(0)}, \tilde{s}_n^{(0)})$, put $\widehat{\boldsymbol{\Omega}}_n = \widehat{\boldsymbol{\Sigma}}_n^{(0)}$ and define the sequence $(\widehat{\mathbf{m}}_n^{(k)}, \widehat{\boldsymbol{\Sigma}}_n^{(k)}, \tilde{s}_n^{(k)})$, $k \ge 0$, using the recursion below. A procedure to compute the initial estimators $(\widehat{\mathbf{m}}_n^{(0)}, \widehat{\boldsymbol{\Sigma}}_n^{(0)}, \tilde{s}_n^{(0)})$ can be found in Danilov *et al.* (2012).

Given $(\widehat{\mathbf{m}}_n^{(k)}, \widehat{\boldsymbol{\Sigma}}_n^{(k)}, \tilde{s}_n^{(k)})$, compute $(\widehat{\mathbf{m}}_n^{(k+1)}, \widehat{\boldsymbol{\Sigma}}_n^{(k+1)}, \tilde{s}_n^{(k+1)})$ as follows:

$$\widehat{\mathbf{m}}_n^{(k+1)} = \frac{\sum_{i=1}^n w_i^{(k)} \widehat{\mathbf{x}}_i^{(k)}}{\sum_{i=1}^n w_i^{(k)}} \tag{6.158}$$

and

$$\tilde{\boldsymbol{\Sigma}}_n^{(k+1)} = \frac{\sum_{i=1}^n [w_i^{(k)} (\widehat{\mathbf{x}}_i^{(k)} - \widehat{\mathbf{m}}_n^{(k)})(\widehat{\mathbf{x}}_i^{(k)} - \widehat{\mathbf{m}}_n^{(k)})' + w_i^{(k)} w_i^{*(k)} \mathbf{C}_i^{(k)}]}{\sum_{i=1}^n w_i^{(k)} w_i^{*(k)}}, \tag{6.159}$$

where $\widehat{\mathbf{x}}_i^{(k)} = \widehat{\mathbf{x}}(\mathbf{u}_i, \mathbf{x}_i^{(\mathbf{u}_i)}, \widehat{\mathbf{m}}_n^{(k)}, \widehat{\boldsymbol{\Sigma}}_n^{(k)})$, $\mathbf{C}_i^{(k)} = \mathbf{C}(\mathbf{u}_i, \widehat{\boldsymbol{\Sigma}}_n^{(k)})$, $w_i^{(k)} = w(\mathbf{u}_i, \mathbf{x}_i^{(\mathbf{u}_i)}, \widehat{\mathbf{m}}_n^{(k)}, \widehat{\boldsymbol{\Sigma}}_n^{(k)}, \tilde{s}_n^{(k)})$ and $w_i^* = w^*(\mathbf{u}_i, \mathbf{x}_i^{(\mathbf{u}_i)}, \widehat{\mathbf{m}}_n^{(k)}, \widehat{\boldsymbol{\Sigma}}_n^{(k)})$, where w and w^* are defined in (6.156) and (6.157) respectively. Set $\tilde{s}_n^{(k+1)} = S_n^*(\widehat{\mathbf{m}}_n^{(k+1)}, \tilde{\boldsymbol{\Sigma}}_n^{(k+1)})$ and $\widehat{\boldsymbol{\Sigma}}_n^{(k+1)} = \widehat{s}_n^{(k+1)} \tilde{\boldsymbol{\Sigma}}_n^{(k+1)}$, where $\widehat{s}_n^{(k+1)}$ is the solution to (6.98) with $\widehat{\mathbf{m}}_n = \widehat{\mathbf{m}}_n^{(k+1)}$ and $\tilde{\boldsymbol{\Sigma}}_n = \tilde{\boldsymbol{\Sigma}}_n^{(k+1)}$. The iteration stops when $|\tilde{s}_n^{(k+1)} / \tilde{s}_n^{(k)} - 1| < \varepsilon$ for some appropriately chosen $\varepsilon > 0$.

Note that the recursion equations for the classical EM algorithm for Gaussian data can be obtained from (6.158) and (6.159) by setting $w_i^{(k)} = w_i^{*(k)} = 1$ for all i.

6.18 Recommendations and software

For multivariate location and scatter we recommend the S-estimator with Rocke ρ-function (Section 6.4.4) and the MM-estimator with SHR ρ-function (Section 6.5), both starting from the Peña–Prieto estimator (Section 6.9.2). They are implemented together in **covRob** (RobStatTM). These functions use as initial estimator the function **initPP** (RobStatTM), which implements the Peña–Prieto estimator. As is explained in Section 6.10.5, we prefer the MM-estimator for $p < 10$ and the Rocke S-estimator for $p \geq 10$. Both are also available separately in **covRobMM** and **covRobRocke**.

For missing data we recommend the generalized S-estimator with Rocke ρ-function (Section 6.12.2), which is implemented in **GSE** (GSE).

For data with cell-wise contamination we recommend the two-step generalized S-estimator (Section 6.13),implemented in **TSGS** (2SGS).

For principal components we recommend the M-S estimator of Section 6.11.2, implemented in **pcaRobS** (RobStatTM).

Another option are the spherical principal components (Section 6.11.1), implemented in **PcaLocantore** (rrcov).

For mixed linear models we recommend the composite τ-estimator (Section 6.15.4). The function **varComprob** (RobustvarComp) computes this and other robust estimators for these models.

6.19 Problems

6.1. Show that if $\mathbf{x}$ has distribution (6.7), then E$\mathbf{x} = \boldsymbol{\mu}$ and Var($\mathbf{x}$) $= c\boldsymbol{\Sigma}$.

6.2. Prove that M-estimators (6.14)–(6.15) are affine equivariant.

6.3. Show that the asymptotic value of an equivariant $\widehat{\boldsymbol{\Sigma}}$ for a spherical distribution is a scalar multiple of $\mathbf{I}$.

6.4. Show that the result of the first part of Section 6.17.4.1 is valid for any $\boldsymbol{\Sigma}$.

6.5. Prove that if $\rho(t)$ is a bounded nondecreasing function, then $t\rho'(t)$ cannot be nondecreasing.

6.6. Let $\widehat{\mu}$ and $\widehat{\Sigma}$ be S-estimators of location and scatter based on the scale $\widehat{\sigma}$, and let $\widehat{\sigma}_0 = \widehat{\sigma}(\mathbf{d}(X, \widehat{\mu}, \widehat{\Sigma}))$. Given a constant σ_0, define $\widehat{\mu}^*$ and $\widehat{\Sigma}^*$ as the values μ and Σ that minimize $|\boldsymbol{\Sigma}|$ subject to $\widehat{\sigma}(\mathbf{d}(\mathbf{x}, \mu, \Sigma)) = \sigma_0$. Prove that $\widehat{\mu}^* = \widehat{\mu}$ and $\widehat{\Sigma}^* = (\widehat{\sigma}_0/\sigma_0)\widehat{\Sigma}$

6.7. Show that $\overline{\mathbf{x}}$ and $\mathbf{Var}(X)$ are the values of $\widehat{\boldsymbol{\mu}}$ and $\widehat{\Sigma}$ minimizing $|\boldsymbol{\Sigma}|$ subject to $(1/n)\sum_{i=1}^{n} d(\mathbf{x}_i, \boldsymbol{\mu}, \boldsymbol{\Sigma}) = p$

6.8. Let $\widehat{\boldsymbol{\mu}}$ and $\widehat{\boldsymbol{\Sigma}}$ be the MCD estimators of location and scatter, which minimize the scale $\widehat{\sigma}(d_1, ...d_n) = \sum_{i=1}^{h} d_{(i)}$. For each subsample $A = \{\mathbf{x}_{i_1}, .., \mathbf{x}_{i_h}\}$ of size h call $\overline{\mathbf{x}}_A$ and $\mathbf{C}_A$ the sample mean and covariance matrix corresponding to A. Let A^* be a subsample of size h that minimizes $|\mathbf{C}_A|$. Show that A^* is the set of observations corresponding to the h smallest values $\mathbf{d}(\mathbf{x}_i, \widehat{\boldsymbol{\mu}}, \widehat{\boldsymbol{\Sigma}})$, and that $\widehat{\boldsymbol{\mu}} = \overline{\mathbf{x}}_{A*}$ and $\widehat{\boldsymbol{\Sigma}} = |\mathbf{C}_{A*}|^{-1/p}\mathbf{C}_{A*}$.

6.9. Let $\widehat{\boldsymbol{\mu}}$ and $\widehat{\boldsymbol{\Sigma}}$ be the MVE estimators of location and scatter. Let $\widehat{\boldsymbol{\mu}}^*$, $\widehat{\boldsymbol{\Sigma}}^*$ be the values of $\boldsymbol{\mu}$ and $\boldsymbol{\Sigma}$ minimizing $|\boldsymbol{\Sigma}|$ under the constraint that the ellipsoid

$$\{\mathbf{x} \in R^p : (\mathbf{x} - \boldsymbol{\mu})'\boldsymbol{\Sigma}^{-1}(\mathbf{x} - \boldsymbol{\mu}) \leq 1\}$$

of volume $|\boldsymbol{\Sigma}|$ contains at least $n/2$ sample points. Show that $\widehat{\mu}^* = \widehat{\boldsymbol{\mu}}$ and $\widehat{\Sigma}^* = \lambda\widehat{\Sigma}$ where $\lambda = \text{Med}\,\{\mathbf{d}(X, \widehat{\boldsymbol{\mu}}, \widehat{\boldsymbol{\Sigma}})\}$.

6.10. Prove (6.30).

6.11. Let (x, y) be bivariate normal with zero means, unit variances and correlation ρ, and let ψ be a monotone ψ -function. Show that $\mathrm{E}(\psi(x)\psi(y))$ is an increasing function of ρ (Hint: $y = \rho x + \sqrt{1-\rho^2}z$ with $z \sim (0, 1)$ independent of x).

6.12. Prove (6.19).

6.13. The dataset **glass** from (Hettich and Bay, 1999) contains measurements of the presence of 7 chemical constituents in 76 pieces of glass from nonfloat windows. Compute the classical and robust estimators of location and scatter and the respective squared distances. For both, make the QQ plots of distances and the plots of distances against index numbers, and compare the results.

6.14. The first principal component is often used to represent multispectral images. The dataset **image** (Frery, 2005) contains the values corresponding to three frequency bands for each of the 1573 pixels of a radar image. Compute the classical and robust principal components and compare the directions of the respective eigenvectors. and the fits given by the first component.

7

Generalized Linear Models

In Chapter 4 we considered regression models where the response variable y depends linearly on several explanatory variables $x_1, \ldots, x_p$. In this case, y was a quantitative variable; that is, it could take on any real value, and the regressors – which could be quantitative or qualitative – affected only its mean.

In this chapter we shall consider more general situations in which the regressors affect the distribution function of y. However, to retain parsimony, it is assumed that this distribution depends on them only through a linear combination $\sum_j \beta_j x_j$ where the β_js are unknown.

We shall first consider the situation when y is a 0–1 variable.

7.1 Binary response regression

Let y be a 0–1 variable representing the death or survival of a patient after heart surgery. Here $y = 1$ and $y = 0$ represent death and survival, respectively. We want to predict this outcome by means of different regressors, such as x_1 = age, x_2 = diastolic pressure, and so on.

We observe $(\mathbf{x}, y)$ where $\mathbf{x} = (x_1, \ldots, x_p)'$ is the vector of explanatory variables. Assume first that $\mathbf{x}$ is fixed (i.e., nonrandom). To model the dependency of y on $\mathbf{x}$, we assume that $\mathrm{P}(y = 1)$ depends on $\boldsymbol{\beta}'\mathbf{x}$ for some unknown $\boldsymbol{\beta} \in R^p$. Since $\mathrm{P}(y = 1) \in [0, 1]$ and $\boldsymbol{\beta}'\mathbf{x}$ may take on any real value, we make the further assumption that

$$\mathrm{P}(y = 1) = F(\boldsymbol{\beta}'\mathbf{x}), \tag{7.1}$$

where F is any continuous distribution function. The function F^{-1} is called the *link function*. If instead $\mathbf{x}$ is random, it will be assumed that the probabilities are conditional; that is,

$$\mathrm{P}(y = 1|\mathbf{x}) = F(\boldsymbol{\beta}'\mathbf{x}). \tag{7.2}$$

Robust Statistics: Theory and Methods (with R), Second Edition.
Ricardo A. Maronna, R. Douglas Martin, Victor J. Yohai and Matías Salibián-Barrera.

Companion website: www.wiley.com/go/maronna/robust

In the common case of a model with an intercept, the first coordinate of each $\mathbf{x}_i$ is one, and the prediction may be written as

$$\boldsymbol{\beta}'\mathbf{x}_i = \beta_0 + \underline{\mathbf{x}}_i\boldsymbol{\beta}_1, \tag{7.3}$$

with $\underline{\mathbf{x}}_i$ and $\boldsymbol{\beta}_1$ as in (4.6).

The most popular functions F are those corresponding to the logistic distribution

$$F(y) = \frac{e^y}{1+e^y} \tag{7.4}$$

("logistic model") and to the standard normal distribution $F(y) = \Phi(y)$ ("probit model"). For the logistic model we have

$$\log \frac{\mathrm{P}(y=1)}{1-\mathrm{P}(y=1)} = \boldsymbol{\beta}'\mathbf{x}.$$

The left-hand side is called the *log odds ratio*, and is seen to be a linear function of $\mathbf{x}$.

Now let $(\mathbf{x}_1, y_1), \ldots, (\mathbf{x}_n, y_n)$ be a sample from model (7.1), where $\mathbf{x}_1, \ldots, \mathbf{x}_n$ are fixed. From now on we shall write for simplicity

$$p_i(\boldsymbol{\beta}) = F(\boldsymbol{\beta}'\mathbf{x}_i).$$

Then $y_1, \ldots, y_n$ are response random variables that take on values 1 and 0 with probabilities $p_i(\boldsymbol{\beta})$ and $1 - p_i(\boldsymbol{\beta})$ respectively, and hence their frequency function is

$$p(y_{i,}\boldsymbol{\beta}) = p_i^{y_i}(\boldsymbol{\beta})(1-p_i(\boldsymbol{\beta}))^{1-y_i}.$$

Hence the log-likelihood function of the sample $L(\boldsymbol{\beta})$ is given by

$$\log L(\boldsymbol{\beta}) = \sum_{i=1}^{n} [y_i \log p_i(\boldsymbol{\beta}) + (1-y_i)\log(1-p_i(\boldsymbol{\beta}))]. \tag{7.5}$$

Differentiating (7.5) yields the estimating equations for the maximum likelihood estimator (MLE):

$$\sum_{i=1}^{n} \frac{y_i - p_i(\boldsymbol{\beta})}{p_i(\boldsymbol{\beta})(1-p_i(\boldsymbol{\beta}))} F'(\boldsymbol{\beta}'\mathbf{x}_i)\mathbf{x}_i = \mathbf{0}. \tag{7.6}$$

In the case of random $\mathbf{x}_i$, (7.2) yields

$$\log L(\boldsymbol{\beta}) = \sum_{i=1}^{n} [y_i \log p_i(\boldsymbol{\beta}) + (1-y_i)\log(1-p_i(\boldsymbol{\beta}))] + \sum_{i=1}^{n} \log g(\mathbf{x}_i), \tag{7.7}$$

where g is the density of the $\mathbf{x}_i$. Differentiating this log likelihood again yields (7.6).

For predicting the values y_i from the corresponding regressor vector $\mathbf{x}_i$, the ideal situation would be that of "perfect separation"; that is, when there exist $\boldsymbol{\gamma} \in R^p$ and $\alpha \in R$ such that

$$\begin{aligned} \boldsymbol{\gamma}'\mathbf{x}_i &> \alpha \quad \text{if} \quad y_i = 1 \\ \boldsymbol{\gamma}'\mathbf{x}_i &< \alpha \quad \text{if} \quad y_i = 0, \end{aligned} \tag{7.8}$$

and therefore $\boldsymbol{\gamma}'\mathbf{x} = \alpha$ is a *separating hyperplane*. It is intuitively clear that if one such hyperplane exists, there must be an infinite number of them. However, this has the consequence that the MLE becomes undetermined. More precisely, let $\boldsymbol{\beta}(k) = k\boldsymbol{\gamma}$. Then

$$\lim_{k\to+\infty} p_i(\boldsymbol{\beta}(k)) = \lim_{u\to+\infty} F(u) = 1 \text{ if } y = 1$$

and

$$\lim_{k\to+\infty} p_i(\boldsymbol{\beta}(k)) = \lim_{u\to-\infty} F(u) = 1 \text{ if } y = 0.$$

Therefore

$$\lim_{k\to\infty} \sum_{i=1}^{n} [y_i \log p_i(\boldsymbol{\beta}(k)) + (1 - y_i)\log((1 - p_i(\boldsymbol{\beta}(k)))] = 0.$$

Since for all finite $\boldsymbol{\beta}$

$$\sum_{i=1}^{n} y_i \log p_i(\boldsymbol{\beta}(k)) + (1 - y_i)\log(1 - p_i(\boldsymbol{\beta}(k))) < 0,$$

according to (7.5)–(7.7), the MLE does not exist for either fixed or random $\mathbf{x}_i$.

Albert and Anderson (1984) showed that the MLE is unique and finite if and only if no $\boldsymbol{\gamma} \in R^p$ and $\alpha \in R$ exist such that

$$\begin{aligned} \boldsymbol{\gamma}'\mathbf{x}_i &\geq \alpha \quad \text{if} \quad y_i = 1 \\ \boldsymbol{\gamma}'\mathbf{x}_i &\leq \alpha \quad \text{if} \quad y_i = 0. \end{aligned}$$

For $\boldsymbol{\gamma} \in R^p$ and $\alpha \in R$, call $K(\gamma, \alpha)$ the number of points in the sample which do not satisfy (7.8), and define

$$k_0 = \min_{\boldsymbol{\gamma} \in R^p, \alpha \in R} K(\boldsymbol{\gamma}, \boldsymbol{\alpha}), \quad (\boldsymbol{\gamma}_0, \alpha_0) = \arg \min_{\boldsymbol{\gamma} \in R^p, \alpha \in R} K(\boldsymbol{\gamma}, \boldsymbol{\alpha}). \tag{7.9}$$

Then, replacing the k_0 points that do not satisfy (7.8) for $\boldsymbol{\gamma} = \boldsymbol{\gamma}_0$ and $\alpha = \alpha_0$ (called *overlapping points*) with other k_0 points lying on the correct side of the hyperplane $\boldsymbol{\gamma}_0'\mathbf{x} = \alpha$, the MLE goes to infinity. Then we can say that the breakdown point of the MLE in this case is k_0/n. Observe that the points that replace the k_0 misclassified points are not "atypical". They follow the pattern of the majority: those with $\boldsymbol{\gamma}_0'\mathbf{x}_i > \alpha$ have $y_i = 1$ and those with $\boldsymbol{\gamma}_0'\mathbf{x}_i < \alpha$ have $y_i = 0$. The fact that the points that produce breakdown to infinity are not outliers was observed for the first time by Croux *et al.* (2002), who also showed that the effect produced by outliers on the MLE is quite different; it will be described later in this section.

It is easy to show that the function (7.4) verifies $F'(y) = F(y)(1 - F(y))$. Hence in the logistic case, (7.6) simplifies to

$$\sum_{i=1}^{n} (y_i - p_i(\boldsymbol{\beta}))\mathbf{x}_i = \mathbf{0}. \tag{7.10}$$

We shall henceforth consider only the logistic case, which is probably the most commonly used, and which, as we shall now see, is easier to robustify.

According to (3.48), the influence function of the MLE for the logistic model is

$$\mathrm{IF}(y, \mathbf{x}, \boldsymbol{\beta}) = \mathbf{M}^{-1}(y - F(\boldsymbol{\beta}'\mathbf{x}))\mathbf{x},$$

where $\mathbf{M} = \mathrm{E}(F'(\boldsymbol{\beta}'\mathbf{x})\mathbf{x}\mathbf{x}')$. Since the factor $(y - F(\boldsymbol{\beta}'\mathbf{x}))$ is bounded, the only outliers that make this influence large are those such that $||\mathbf{x}_i|| \to \infty$, $y_i = 1$ and $\boldsymbol{\beta}'\mathbf{x}_i$ is bounded away from ∞, or those such that $||\mathbf{x}_i|| \to \infty$, $y_i = 0$ and $\boldsymbol{\beta}'\mathbf{x}_i$ is bounded away from $-\infty$. Croux *et al.* (2002) showed that if the model has an intercept (see (7.3)), then, unlike the case of ordinary linear regression, this kind of outlier makes the MLE of $\boldsymbol{\beta}_1$ tend to zero and not to infinity. More precisely, they show that by conveniently choosing not more than $2(p-1)$ outliers, the MLE $\hat{\boldsymbol{\beta}}_1$ of $\boldsymbol{\beta}_1$ can be made as close to zero as desired. This is a situation where, although the estimator remains bounded, we may say that it breaks down since its values are determined by the outliers rather than by the bulk of the data, and in this sense the breakdown point to zero of the MLE is $\leq 2(p-1)/n$.

To exemplify this lack of robustness, we consider a sample of size 100 from the model

$$\log \frac{\mathrm{P}(y=1)}{1-\mathrm{P}(y=1)} = \beta_0 + \beta_1 x,$$

where $\beta_0 = -2$, $\beta_1 = 3$ and x is uniform in the interval $[0, 1]$. Figure 7.1 shows the sample, and we find, as expected, that for low values of x, a majority of the ys are zero and the opposite occurs for large values of x. The MLE is $\beta_0 = -1.72, \beta_1 = 2.76$.

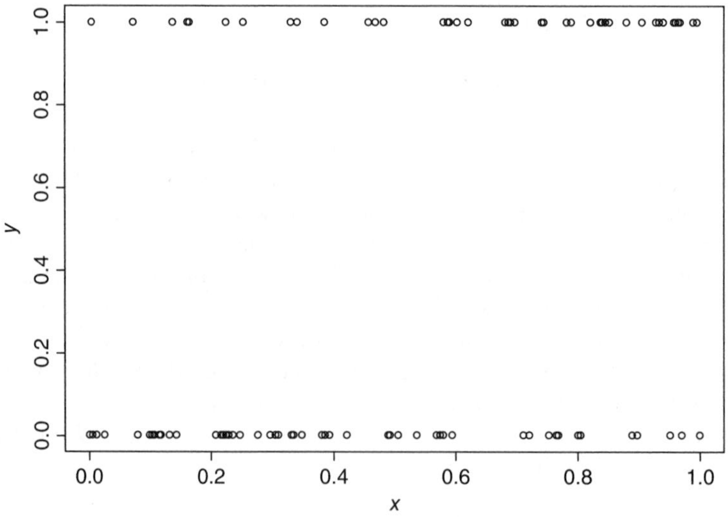

Figure 7.1 Simulated data: plot of y against x

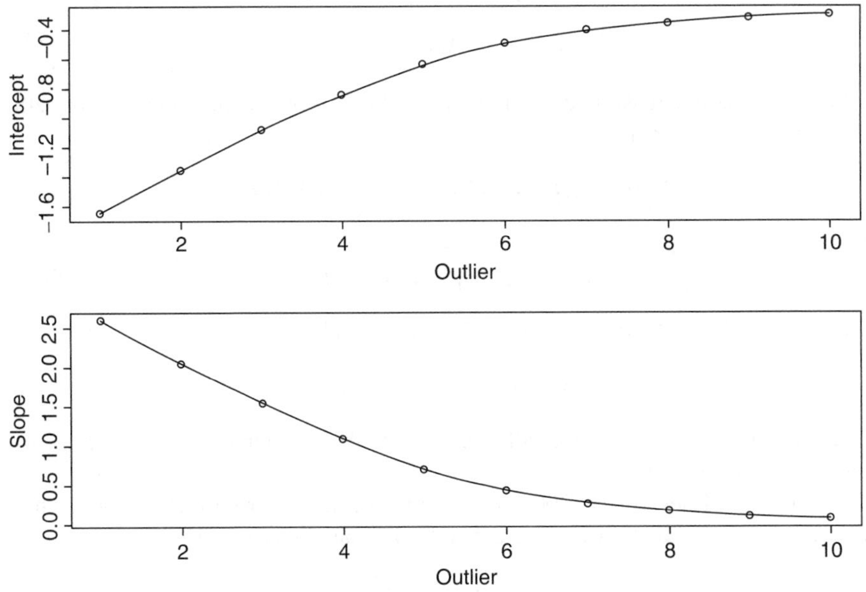

Figure 7.2 Simulated data: effect of one outlier

Now we add to this sample one outlier of the form $x_0 = i$ and $y_0 = 0$ for $i = 1, \ldots, 10$, and we plot in Figure 7.2 the values of β_0 and β_1. Observe that the value of β_1 tends to zero and β_0 converges to $\log(\alpha/(1-\alpha))$, where $\alpha = 45/101 \approx 0.45$ is the frequency of ones in the contaminated sample.

7.2 Robust estimators for the logistic model

7.2.1 Weighted MLEs

Carroll and Pederson (1993) proposed a simple way to turn the MLE into an estimator with bounded influence, namely by downweighting high-leverage observations. A measure of the leverage of observation $\mathbf{x}$ similar to (5.87) is defined as

$$h_n(\mathbf{x}) = ((\mathbf{x} - \hat{\boldsymbol{\mu}}_n)'\hat{\Sigma}_n^{-1}(\mathbf{x} - \hat{\boldsymbol{\mu}}_n))^{1/2},$$

where $\hat{\boldsymbol{\mu}}_n$ and $\hat{\Sigma}_n$ are respectively a robust location vector and a robust scatter matrix estimator of $\mathbf{x}$. Note that if $\hat{\mu}_n$ and $\hat{\Sigma}_n$ are affine equivariant, this measure is invariant under affine transformations.

Then robust estimators can be obtained by minimizing

$$\sum_{i=1}^{n} w_i[y_i \log p_i(\boldsymbol{\beta}) + (1 - y_i)\log(1 - p_i(\boldsymbol{\beta}))],$$

with

$$w_i = W(h_n(\mathbf{x}_i)), \tag{7.11}$$

where W is a nonincreasing function such that $W(u)u$ is bounded. According to (3.48), its influence function is

$$\mathrm{IF}(y, \mathbf{x}, \boldsymbol{\beta}) = \mathbf{M}^{-1}(y - F(\boldsymbol{\beta}'\mathbf{x}))\mathbf{x}W(h(\mathbf{x})),$$

with

$$h(\mathbf{x}) = ((\mathbf{x} - \boldsymbol{\mu})'\Sigma^{-1}(\mathbf{x} - \boldsymbol{\mu}))^{1/2}, \tag{7.12}$$

where $\boldsymbol{\mu}$ and $\boldsymbol{\Sigma}$ are the limit values of $\hat{\boldsymbol{\mu}}_n$ and $\hat{\Sigma}_n$, and

$$\mathbf{M} = \mathrm{E}(W(h(\mathbf{x}))F'(\beta\mathbf{x})\mathbf{x}\mathbf{x}').$$

These estimators are asymptotically normal and their asymptotic covariance matrix can be found using (3.50).

Croux and Haesbroeck (2003) proposed choosing W from the family of "hard rejection" weight functions, which depends on a parameter $c > 0$:

$$W(u) = \begin{cases} 1 & \text{if } 0 \leq u \leq c \\ 0 & \text{if } \quad u > c, \end{cases} \tag{7.13}$$

and as $\hat{\boldsymbol{\mu}}_n$ and $\hat{\Sigma}_n$ MCD estimators.

7.2.2 Redescending M-estimators

The MLE for the model can be also defined as minimizing the total *deviance*

$$D(\beta) = \sum_{i=1}^{n} d^2(p_i(\boldsymbol{\beta}), y_i),$$

where $d(\mathbf{u}, y)$ is given by

$$d(u, y) = \{-2[y\log(u) + (1 - y)\log(1 - u))]\}^{1/2} \operatorname{sgn}(y - u) \tag{7.14}$$

and is a signed measure of the discrepancy between a Bernoulli variable y and its expected value u. Observe that

$$d(u, y) = \begin{cases} 0 & \text{if } \; u = y \\ -\infty & \text{if } \; u = 1, \; y = 0 \\ \infty & \text{if } \; u = 0, \; y = 1. \end{cases}$$

In the logistic model, the values $d(p_i(\boldsymbol{\beta}), y_i)$ are called *deviance residuals*, and they measure the discrepancies between the probabilities fitted using the regression coefficients $\boldsymbol{\beta}$ and the observed values. In Section 7.3 we define the deviance residuals for a larger family of models.

Pregibon (1981) proposed robust M-estimators for the logistic model based on minimizing

$$M(\beta) = \sum_{i=1}^{n} \rho(d^2(p_i(\beta), y_i)),$$

where $\rho(u)$ is a function that increases more slowly than the identity function. Bianco and Yohai (1996) observed that for random $\mathbf{x}_i$ these estimators are not Fisher-consistent; that is, the respective score function does not satisfy (3.32). They found that this difficulty may be overcome by using a correction term. They proposed estimating $\boldsymbol{\beta}$ by minimizing

$$M(\boldsymbol{\beta}) = \sum_{i=1}^{n} [\rho(d^2(p_i(\boldsymbol{\beta}), y_i)) + q(p_i(\boldsymbol{\beta}))], \tag{7.15}$$

where $\rho(u)$ is nondecreasing and bounded and

$$q(u) = v(u) + v(1-u),$$

with

$$v(u) = 2\int_0^u \psi(-2\log t)dt$$

and $\psi = \rho'$.

Croux and Haesbroeck (2003) described sufficient conditions on ρ to guarantee a finite minimum of $M(\boldsymbol{\beta})$ for all samples with overlapping observations ($k_0 > 0$). They proposed choosing ψ in the family

$$\psi_c^{CH}(u) = \exp\left(-\sqrt{\max(u, c)}\right). \tag{7.16}$$

Differentiating (7.15) with respect to $\boldsymbol{\beta}$ and using the facts that

$$q'(u) = 2\psi(-2\log u) - 2\psi(-2\log(1-u))$$

and that in the logistic model $F' = F(1-F)$, we get

$$2\sum_{i=1}^{n} \psi(d_i^2(\boldsymbol{\beta}))(y_i - p_i(\boldsymbol{\beta}))\mathbf{x}_i$$
$$-2\sum_{i=1}^{n} p_i(\boldsymbol{\beta})(1-p_i(\boldsymbol{\beta}))[\psi(-2\log p_i(\boldsymbol{\beta})) - \psi(-2\log(1-p_i(\boldsymbol{\beta})))]\mathbf{x}_i = \mathbf{0},$$

where $d_i(\boldsymbol{\beta}) = d(p_i(\boldsymbol{\beta}), y_i)$ are the deviance residuals given in (7.14). This equation can also be written as

$$\sum_{i=1}^{n} [\psi(d_i^2(\boldsymbol{\beta}))(y_i - p_i(\boldsymbol{\beta})) - \mathrm{E}_{\boldsymbol{\beta}}(\psi(d_i^2(\boldsymbol{\beta}))(y_i - p_i(\boldsymbol{\beta}))|\mathbf{x}_i)]\mathbf{x}_i = \mathbf{0}, \tag{7.17}$$

where E_{β} denotes the expectation when $\mathrm{P}(y_i = 1|\mathbf{x}_i) = p_i(\boldsymbol{\beta})$. Putting

$$\Psi(y_i, \mathbf{x}_i, \boldsymbol{\beta}) = [\psi(d_i^2(\boldsymbol{\beta}))(y_i - p_i(\boldsymbol{\beta})) - \mathrm{E}_{\boldsymbol{\beta}}(\psi(d_i^2(\boldsymbol{\beta}))(y_i - p_i(\boldsymbol{\beta}))|\mathbf{x}_i)]\mathbf{x}_i, \qquad (7.18)$$

equation 7.17 can also be written as

$$\sum_{i=1}^{n} \Psi(y_i, \mathbf{x}_i, \boldsymbol{\beta}) = \mathbf{0}.$$

From (7.18) it is clear that $\mathrm{E}_{\boldsymbol{\beta}}(\Psi(y_i, \mathbf{x}_i, \boldsymbol{\beta})) = \mathbf{0}$, and therefore these estimators are Fisher-consistent. Their influence function can again be obtained from (3.48). Bianco and Yohai (1996) proved that under general conditions these estimators are asymptotically normal. The asymptotic covariance matrix can be obtained from (3.50).

In Figure 7.3 we repeat the same graph as in Figure 7.2 using both estimators: the MLE and a redescending M-estimator with $\psi_{0.5}^{CH}$. We observe that the changes in both the slope and intercept of the M-estimator are very small compared to those of the MLE.

Since the function $\Psi(y_i, \mathbf{x}_i, \boldsymbol{\beta})$ is not bounded, the M-estimator does not have bounded influence. To obtain bounded influence estimators, Croux and

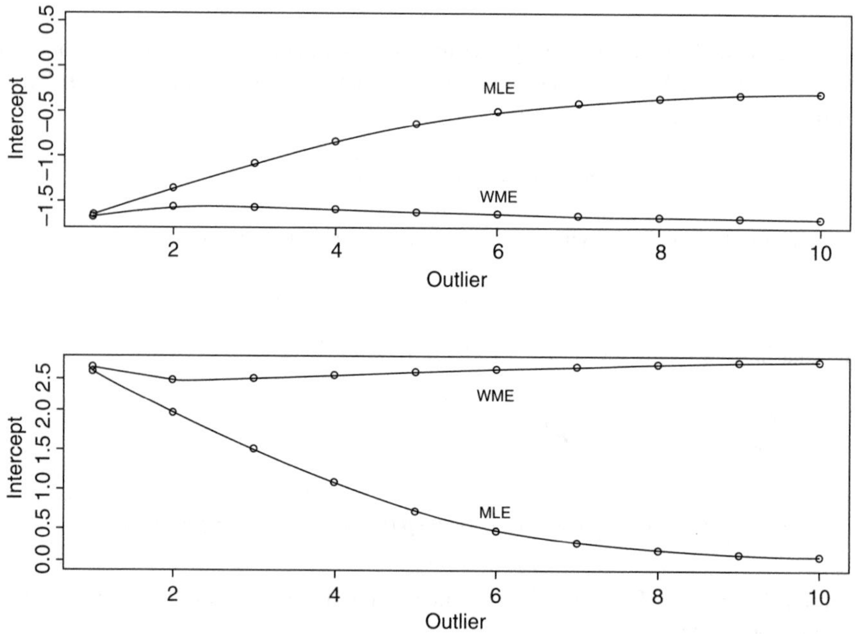

Figure 7.3 Effect of an oulier on the M-estimator of slope and intercept

Haesbroeck (2003) propose to downweight high-leverage observations. They define redescending weighted M- (WM-) estimators by minimizing

$$M(\beta) = \sum_{i=1}^{n} w_i[\rho(d^2(p_i(\boldsymbol{\beta}), y_i)) + q(p_i(\boldsymbol{\beta}))], \tag{7.19}$$

where the weights w_i are as in Section 7.2.1.

Example 7.1 *The following dataset was considered by Cook and Weisberg (1982, Ch. 5, p. 193) and Johnson (1985) to illustrate the identification of influential observations. The data are records for 33 leukemia patients. Tables and figures for this example are obtained with script* ***leukemia.R****.*

The response variable is 1 when the patient survives more than 52 weeks. Two covariates are considered: white blood cell count (WBC) and presence or absence of a certain morphological characteristic in the white cells (AG). The model also includes an intercept.

Cook and Weisberg detected an observation (#15) corresponding to a patient with WBC = 100.000, who survived for a long time. This observation was very influential on the MLE. They also noticed that after removing this observation a much better overall fit was obtained, and that the fitted survival probabilities of those observations corresponding to patients with small values of WBC increased significantly.

In Table 7.1 we give the estimated slopes and their asymptotic standard deviations corresponding to:

- the MLE with the complete sample (MLE);
- the MLE after removing the influential observation (MLE_{-15});
- the weighted MLE (WMLE);
- the redescending M-estimator (M) corresponding to the Croux and Haesbroeck family ψ_c^{CH} with $c = 0.5$;
- the redescending weighted M-estimator (WM).

Table 7.1 Estimators for Leukemia data and their standard errors

Estimate	Intercept	WBC($\times 10^{-4}$)	AG
MLE	−1.31 (0.81)	−0.32 (0.18)	2.26 (0.95)
MLE_{-15}	0.21 (1.08)	−2.35 (1.35)	2.56 (1.23)
WMLE	0.17 (1.08)	−2.25 (1.32)	2.52 (1.22)
M	0.16 (1.66)	−1.77 (2.33)	1.93 (1.16)
WM	0.20 (1.19)	−2.21 (0.98)	2.40 (1.30)
CUBIF	−1.04 (0.85)	−0.53 (0.30)	2.22 (0.98)

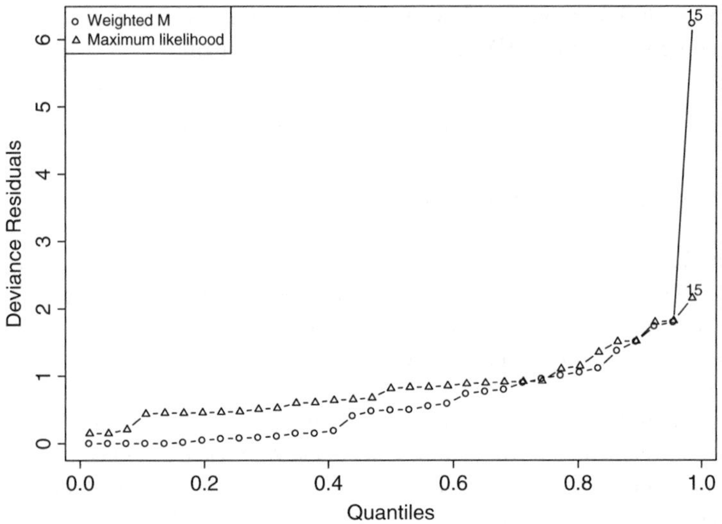

Figure 7.4 Leukemia data: ordered absolute deviances from the ML and WM-estimators

We also show the results for the optimal conditionally unbiased bounded influence estimator (CUBIF), which will be described for a more general family of models in Section 7.3.

We can observe that coefficients fitted with MLE_{-15} are very similar to those of the WMLE, M- and WM- estimators. The CUBIF estimator gives results intermediate between MLE and MLE_{-15}.

Figure 7.4 compares the ordered absolute deviances of the ML and WM-estimates. It is seen that the latter gives a better fit than the former, except for observation 15, which WM clearly pinpoints as atypical.

Example 7.2 *The following dataset was introduced by Finney (1947) and later studied by Pregibon (1982) and Croux and Haesbroeck (2003). The response is the presence or absence of vasoconstriction of the skin of the digits after air inspiration, and the explanatory variables are the logarithms of the volume of air inspired (log VOL) and of the inspiration rate (log RATE). Tables and figures for this example are obtained with script* ***skin.R****.*

Table 7.2 gives the estimated coefficients and standard errors for the MLE, WMLE, M-, WM- and CUBIF estimators. Since there are no outliers in the regressors, the weighted versions give similar results to the unweighted ones. This also explains in part why the CUBIF estimator is very similar to the MLE.

Table 7.2 Estimates for skin data

Estimator	Intercept	log VOL	log RATE
MLE	−9.53 (3.21)	3.88 (1.42)	2.65 (0.91)
WMLE	−9.51 (3.18)	3.87 (1.41)	2.64 (0.90)
M	−14.21 (10.88)	5.82 (4.52)	3.72 (2.70)
WM	−14.21 (10.88)	5.82 (4.53)	3.72 (2.70)
CUBIF	−9.47 (3.22)	3.85 (1.42)	2.63 (0.91)

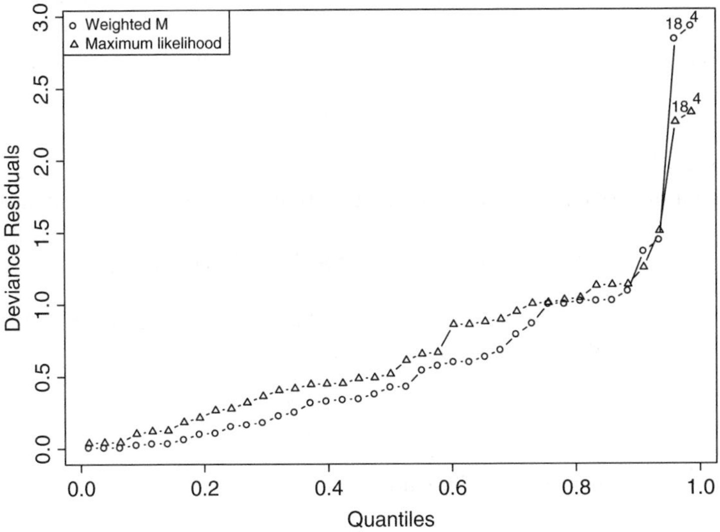

Figure 7.5 Skin data: ordered absolute deviances from the ML and WM-estimators

Figure 7.5 compares the ordered absolute deviances of the ML and WM-estimators. It is seen that the latter gives a better fit than the former for the majority of observations, and that it more clearly pinpoints observations 4 and 18 as outliers.

7.3 Generalized linear models

The binary response regression model is included in a more general class called *generalized linear models* (GLMs). If $\mathbf{x}$ is fixed, the distribution of y is given by a density

$f(y, \lambda)$ depending on a parameter λ, and there is a known one-to-one function l, called *the link function*, and an unknown vector $\boldsymbol{\beta}$, such that $l(\lambda) = \boldsymbol{\beta}'\mathbf{x}$. If $\mathbf{x}$ is random, the *conditional* distribution of y given $\mathbf{x}$ is given by $f(y, \lambda)$ with $l(\lambda) = \boldsymbol{\beta}'\mathbf{x}$. In the previous section we had $\lambda \in (0, 1)$ and

$$f(y, \lambda) = \lambda^y (1 - \lambda)^{1-y}, \tag{7.20}$$

and $l = F^{-1}$.

A convenient framework is the *exponential family* of distributions:

$$f(y, \lambda) = \exp[m(\lambda)y - G(m(\lambda)) - t(y)], \tag{7.21}$$

where m, G and t are given functions. When $l = m$, the link function is called *canonical*.

It is easy to show that if y has distribution (7.21), then

$$\mathrm{E}_\lambda(y) = g(m(\lambda)),$$

where $g = G'$.

This family contains the Bernoulli distribution (7.20), which corresponds to

$$m(\lambda) = \log \frac{\lambda}{1-\lambda}, \quad G(u) = \log(1 + e^u) \text{ and } t(y) = 0.$$

In this case, the canonical link corresponds to the logistic model and $\mathrm{E}_\lambda(y) = \lambda$.

Another example is the Poisson family with

$$f(y, \lambda) = \frac{\lambda^y e^{-\lambda}}{y!}.$$

This family corresponds to (7.21), with

$$m(\lambda) = \log \lambda, \ \ G(u) = e^u \text{ and } t(y) = \log y!$$

This yields $\mathrm{E}y = g(m(\lambda)) = \lambda$. The canonical link in this case is $l(\lambda) = \log \lambda$.

Define $\hat{\lambda}(y)$ as the value of λ that maximizes $f(y, \lambda)$ or, equivalently, that maximizes

$$\log f(y, \lambda) = m(\lambda)y - G(m(\lambda)) - t(y).$$

Differentiating, we obtain that $\lambda = \hat{\lambda}(y)$ should satisfy

$$m'(\lambda)y - g(m(\lambda))m'(\lambda) = 0$$

and therefore $g(m(\hat{\lambda}(y))) = y$. Define the deviance residual function by

$$\begin{aligned} d(y, \lambda) &= \{2 \log[f(y, \lambda)/f(y, \hat{\lambda}(y))]\}^{1/2} \operatorname{sgn}(y - g(m(\lambda)) \\ &= \{2[m(\lambda)y - G(m(\lambda)) - m(\hat{\lambda}(y))y + G(m(\hat{\lambda}(y)))]\}^{1/2} \operatorname{sgn}(y - g(m(\lambda))). \end{aligned}$$

It is easy to check that when y is a Bernoulli variable, this definition coincides with (7.14).

Consider now a sample $(\mathbf{x}_1, y_1), \ldots, (\mathbf{x}_n, y_n)$ from a GLM with the canonical link function and fixed $\mathbf{x}_i$. Then the log likelihood is

$$\log L(\beta) = \sum_{i=1}^{n} \log f(y_i, m^{-1}(\boldsymbol{\beta}'\mathbf{x}_i))$$
$$= \sum_{i=1}^{n} (\boldsymbol{\beta}'\mathbf{x}_i) y_i - \sum_{i=1}^{n} G(\boldsymbol{\beta}'\mathbf{x}_i) - \sum_{i=1}^{n} t(y_i). \tag{7.22}$$

The MLE maximizes $\log L(\beta)$ or equivalently

$$\sum_{i=1}^{n} 2(\log f(y_i, m^{-1}(\boldsymbol{\beta}'\mathbf{x}_i)) - \log f(y_i, \hat{\lambda}(y_i)))$$
$$= \sum_{i=1}^{n} d^2(y_i, m^{-1}(\boldsymbol{\beta}'\mathbf{x}_i)).$$

Differentiating (7.22) we get the equations for the MLE:

$$\sum_{i=1}^{n} (y_i - g(\boldsymbol{\beta}'\mathbf{x}_i))\mathbf{x}_i = \mathbf{0}. \tag{7.23}$$

For example, for the Poisson family this equation is

$$\sum_{i=1}^{n} (y_i - e^{\boldsymbol{\beta}'\mathbf{x}_i})\mathbf{x}_i = \mathbf{0}. \tag{7.24}$$

7.3.1 Conditionally unbiased bounded influence estimators

To robustify the MLE, Künsch *et al.* (1991) consider M-estimators of the form

$$\sum_{i=1}^{n} \Psi(y_i, \mathbf{x}_i, \boldsymbol{\beta}) = \mathbf{0},$$

where $\Psi : R^{1+p+p} \rightarrow R^p$ such that

$$\mathrm{E}(\Psi(y, \mathbf{x}, \boldsymbol{\beta})|x_i) = \mathbf{0}. \tag{7.25}$$

These estimators are referred to as *conditionally unbiased bounded influence* (CUBIF). Clearly these estimators are Fisher-consistent; that is, $\mathrm{E}(\Psi(y_i, \mathbf{x}_i, \boldsymbol{\beta})) = 0$. Künsch *et al.* (1989) found the estimator in this class that solves an optimization problem similar to Hampel's one, as studied in Section 3.5.4. This estimator minimizes a measure of efficiency – based on the asymptotic covariance matrix under the model – subject to a bound on a measure of infinitesimal sensitivity similar to the gross-error sensitivity. Since these measures are quite complicated and may be controversial, we do not give more details about their definition.

The optimal score function Ψ has the following form:

$$\Psi(y, \mathbf{x}, \boldsymbol{\beta}, b, \mathbf{B}) = W(\boldsymbol{\beta}, y, \mathbf{x}, b, \mathbf{B}) \left\{ y - g(\boldsymbol{\beta}'\mathbf{x}) - c\left(\boldsymbol{\beta}'\mathbf{x}, \frac{b}{h(\mathbf{x}, \mathbf{B})}\right) \right\} \mathbf{x},$$

where b is the bound on the measure of infinitesimal sensitivity, $\mathbf{B}$ is a dispersion matrix that will be defined below, and $h(\mathbf{x}, \mathbf{B}) = (\mathbf{x}'\mathbf{B}^{-1}\mathbf{x})^{1/2}$ is a leverage measure. Observe the similarity with (7.23). The function W downweights atypical observations and makes Ψ bounded, and therefore the corresponding M-estimator has bounded influence. The function $c(\boldsymbol{\beta}'\mathbf{x}, b/h(\mathbf{x}, \mathbf{B}))$ is a bias-correction term chosen so that (7.25) holds. Call $r(y, \mathbf{x}, \boldsymbol{\beta})$ the corrected residual:

$$r(y, \mathbf{x}, \boldsymbol{\beta}, b, \mathbf{B}) = y_i - g(\boldsymbol{\beta}'\mathbf{x}) - c\left(\boldsymbol{\beta}'\mathbf{x}, \frac{b}{h(\mathbf{x}, \mathbf{B})}\right). \tag{7.26}$$

Then the weights are of the form

$$W(\boldsymbol{\beta}, y, \mathbf{x}, b, \mathbf{B}) = W_b(r(y, \mathbf{x}, \boldsymbol{\beta})h(\mathbf{x}, \mathbf{B})),$$

where W_b is the Huber weight function (2.33) given by

$$W_b(x) = \min\left\{1, \frac{b}{|x|}\right\}.$$

Then, as in the Schweppe-type GM-estimators of Section 5.11.1, W downweights observations for which the product of corrected residuals and leverage has a high value.

Finally, the matrix $\mathbf{B}$ should satisfy

$$\mathrm{E}(\Psi(y, \mathbf{x}, \boldsymbol{\beta}, b, \mathbf{B})\Psi'(y, \mathbf{x}, \boldsymbol{\beta}, b, \mathbf{B})) = \mathbf{B}.$$

Details of how to implement these estimators, in particular of how to estimate $\mathbf{B}$, and a more precise description of their optimal properties can be found in Künsch *et al.* (1989). We shall call these estimators *optimal conditionally unbiased bounded influence*, or *optimal CUBIF estimators* for short.

7.4 Transformed M-estimators

7.4.1 Definition of transformed M-estimators

One source of difficulties for the development of robust estimators in the GLM is that – unlike in the linear model – the variability of the observations depends on the parameters, and this complicates assessing the outlyingness of an observation. There are, however, cases of GLMs in which a "variance stabilizing transformation" exists, and this fact greatly simplifies the problem. Let $f(y, \lambda)$ be a family of discrete or continuous densities with a real parameter λ, and let $t : R \rightarrow R$ be a function such that the variance of $t(y)$ is "almost" independent of λ. Given a ρ-function, a natural way to define to define M-estimators of λ is as follows. Let $m(\lambda)$ be defined by

$$m(\lambda) = \arg\min_m \mathrm{E}(\rho(t(y) - m)),$$

where y has density $f(y, \lambda)$. Then, given a sample $y_1, ..., y_n$, we can define a Fisher-consistent estimator of λ by

$$\hat{\lambda} = \arg\min_\lambda \sum_{i=1}^{n} \rho(t(y_i) - m(\lambda))$$

We call this estimator the *transformed M-estimator* (MT-estimator). The fact that the variability of $t(y)$ is almost constant makes scale estimation unnecessary.

Let $(\mathbf{x}_1, y_1), ..., (\mathbf{x}_n, y_n)$ be a sample from a GLM such that $y|\mathbf{x}$ has density $f(y, \lambda)$, and the link function is $l(\lambda) = \boldsymbol{\beta}_0'\mathbf{x}$. Then, Valdora and Yohai (2014) defined the MT-estimator of $\boldsymbol{\beta}$ by

$$\hat{\boldsymbol{\beta}} = \arg\min_\lambda \sum_{i=1}^{n} \rho(t(y_i) - m(l^{-1}(\boldsymbol{\beta}'\mathbf{x}))) \tag{7.27}$$

The MT-estimators are not applicable when $f(y, \lambda)$ is the Bernoulli family of distributions, since in this case these estimators coincide with the untransformed M-estimator.

Since ρ is a bounded function, the estimator defined in (7.27) is already robust. However, if high-leverage outliers are expected, penalizing high-leverage observations may increase the estimator's robustness. For this reason Valdora and Yohai (2014) defined weighted MT-estimators (WMT-estimators) as

$$\hat{\boldsymbol{\beta}} = \arg\min_\lambda \sum_{i=1}^{n} w_i\rho(t(y_i) - m(l^{-1}(\boldsymbol{\beta}'\mathbf{x}))) \tag{7.28}$$

where the weights w_i are defined as in Section 7.2.1; that is, $w_i = W(h(\mathbf{x}_i))$ with h defined as in (7.12).

Valdora and Yohai (2014) show that, under very general conditions, the WMT-estimators and in particular the MT-estimators are consistent and asymptotically normal. A family of ρ-functions satisfying these conditions is given by

$$\rho_c(u) = 1 - \left(1 - \left(\frac{u}{c}\right)^2\right)^4 \mathrm{I}(|u| \leq c), \tag{7.29}$$

where c is a positive tuning constant that should be chosen by trading-off between efficiency and robustness. Note the similarity with the bisquare function given in (2.38). The reason to use this function is that it has three bounded derivatives, as required for the asymptotic normality of the WMT-estimators. In contrast, the bisquare ρ-function has only two. Generally, while increasing c, we gain efficiency while the robustness of the estimator decreases.

Consider in particular the case of Poisson regression with $l(\lambda) = \log(\lambda)$. It is well-known that a variance-stabilizing transformation for this family of distributions is $t(y) = \sqrt{y}$. Monte Carlo simulations in Valdora and Yohai (2014) show that in this case the MT-estimator, with ρ in the family ρ_c given in (7.29), and with $c = 2.3$, compares favorably with other proposed robust estimators, such as optimal CUBIF or the robust quasi-likelihood estimators defined in Section 7.4.3.

7.4.2 Some examples of variance-stabilizing transformations

Consider a family of densities $f(y, \lambda)$ such that the corresponding means and variances are $\mu(\lambda)$ and $v(\lambda)$ respectively. A simple Taylor expansion shows that

$$t(y) = \int_0^y \frac{du}{v(\mu^{-1}(u))^{1/2}} \tag{7.30}$$

has approximately constant variance. From (7.30) we obtain that if $v(\lambda) = \mu(\lambda)^q$ then

$$t(y) = \begin{cases} y^{-(q/2)+1} & \text{if } q \neq 2 \\ \log(y) & \text{if } q = 2 \end{cases}. \tag{7.31}$$

In the Poisson case, we have $\mu(\lambda) = v(\lambda) = \lambda$, and this yields $q = 1$ and $t(y) = \sqrt{y}$.

In the case of an exponential distribution; that is, when

$$f(y, \lambda) = e^{-\lambda y} I(y \geq 0),$$

we have $\mu(\lambda) = 1/\lambda$ and $v(\lambda) = 1/\lambda^2$. Then $q = 2$ and $t(y) = \log(y)$.

In the case of a binomial distribution with k trials and success probability λ ($\text{Bi}(\lambda, k)$); that is, when

$$f(y, \lambda) = \binom{k}{y} \lambda^y (1-\lambda)^{k-y}, \ 0 \leq y \leq k$$

with k known, we have $\mu(\lambda) = k\lambda$ and $v(\lambda) = k\lambda(1-\lambda)$. Then in this case applying (7.30) we get $t(y) = \arcsin \sqrt{y/k}$.

7.4.3 Other estimators for GLMs

Bianco *et al.* (2013) extended the redescending M-estimators of Section 7.2.2 to other GLMs. Bianco *et al.* (2005) considered M-estimators for the case when the distribution of y is gamma and the link function is the logarithm. They showed that in this case no correction term is needed for Fisher-consistency.

Bergesio and Yohai (2011) extended the projection estimators of Section 5.11.2 to GLMs. These estimators are highly robust but not very efficient. For this reason they propose using a projection estimator followed by a one-step M-estimator. In this way a highly robust and efficient estimator is obtained. A drawback of these estimators is their computational complexity.

Cantoni and Ronchetti (2001) robustified the quasi-likelihood approach to estimate GLMs. The quasi-likelihood estimators proposed by Wedderburn (1974) are defined as solutions of the equation

$$\sum_{i=1}^{n} \frac{y_i - \mu(\boldsymbol{\beta}'\mathbf{x}_i)}{V(\boldsymbol{\beta}'\mathbf{x}_i)} \mu'(\boldsymbol{\beta}'\mathbf{x}_i)\mathbf{x}_i = 0,$$

where

$$\mu(\lambda) = \mathrm{E}_\lambda(y), \ \mathbf{V}(\lambda) = Var_\lambda(y).$$

The robustification proposed by Cantoni and Ronchetti is performed by bounding and centering the quasi-likelihood score function

$$\psi(y, \boldsymbol{\beta}) = \frac{y - \mu(\boldsymbol{\beta}'\mathbf{x})}{V(\boldsymbol{\beta}'\mathbf{x})} \mu'(\boldsymbol{\beta}'\mathbf{x})\mathbf{x},$$

similar to what was done with the maximum likelihood score function in Section 7.3.1. The purpose of centering is to obtain conditional Fisher-consistent estimators and that of bounding is to bound the IF.

To cope with high-leverage points, they propose giving weights to each observation, as in the definition of the weighted maximum likelihood estimators defined in Section 7.2.1.

Example 7.3 *Breslow (1996) used a Poisson GLM to study the effect of a drug in epilepsy patients using a sample of size 59. Tables and figures for this example are obtained with script* ***epilepsy.R****.*

The response variable is the number of attacks during four weeks (sumY) in a given time interval and the explanatory variables are: patient age divided by 10 (Age10), the number of attacks in the four weeks previous to the study (Base4), a dummy variable that takes values one or zero if the patient received the drug or a placebo respectively (Trt) and an interaction term (Base4*Trt). We fit a Poisson GLM with log link using five estimators: the MLE, the optimal CUBIF, a robustified quasi likelihood (RQL) estimator, the projection estimator followed by a one-step estimator (MP) and the MT-estimator with ρ_c given in (7.29) and $c = 2.3$. Figure 7.6 shows boxplots of the absolute values of the respective deviance residuals. The left-hand plot shows the residuals corresponding to all the observations. It is seen that all robust estimators identify a large outlier (observation 49), while the MLE identifies none. In order to make the boxes more clearly visible, the right-hand plot shows the boxplots

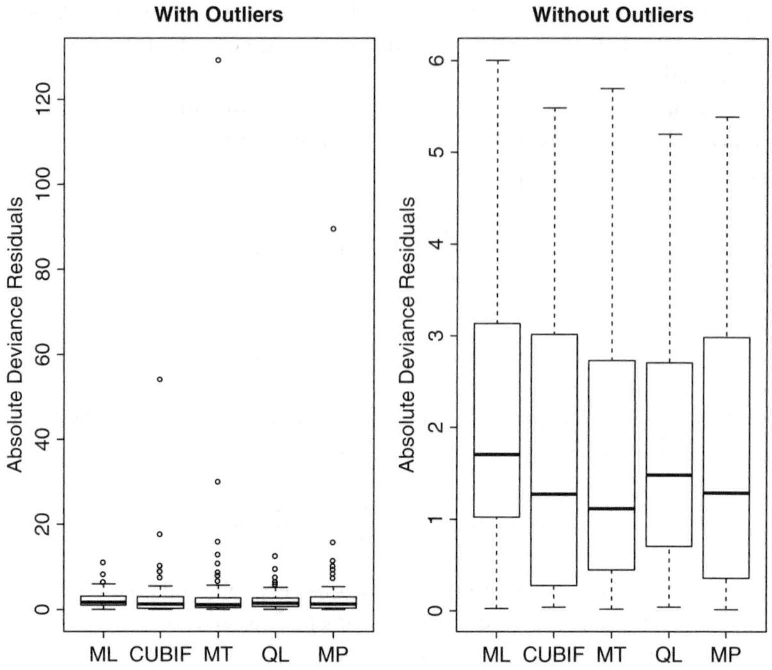

Figure 7.6 Epilepsy data: boxplots of deviances

Table 7.3 Estimates for epilepsy data

Estimator	Intercept	Age10	Base4	Trt	Base4*Trt
MLE	1.84 (0.13)	0.24 (0.04)	0.09 (0.002)	−0.13 (0.04)	0.004 (0.002)
MLE_{-49}	1.60 (0.15)	0.29 (0.04)	0.10 (0.004)	−0.24 (0.05)	0.018 (0.004)
CUBIF	1.84 (0.69)	0.12 (0.12)	0.14 (0.24)	−0.41 (0.12)	0.022 (0.022)
MT	1.62 (0.26)	0.15 (0.091)	0.17 (0.013)	−0.60 (0.29)	0.042 (0.03)
MP	2.00	0.071	0.13	−0.49	0.0476
RQL	2.04 (0.15)	0.16 (0.047)	0.084 (0.004)	−0.33 (0.86)	0.012 (0.0049)

without the outliers. It is seen that the MT-estimator gives the best fit for the bulk of the data.

The coefficient estimates and their standard errors are shown in Table 7.3. Figure 7.7 compares the ordered absolute deviances corresponding to the MT- and ML estimators. It is seen that MT- gives a much better fit to all the data, except for observation 49, which is pinpointed as an outlier.

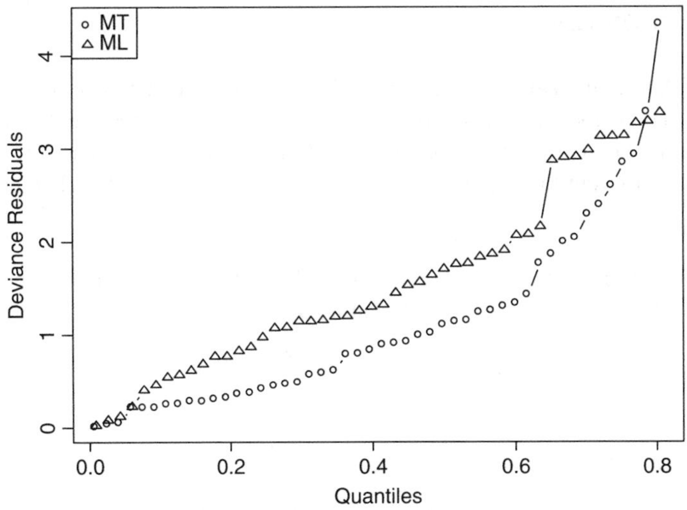

Figure 7.7 Epilepsy data: ordered absolute deviances from the ML and MT-estimators

7.5 Recommendations and software

For logistic regression, we recommend the redescending weighted M-estimator defined in Section 7.2.2, which is implemented in **logregWBY** (`RobStatTM`).

Other options are:

- to use an M-estimator (Section 7.2.2) that can be computed with **logregBY** (`RobStatTM`);
- a weighted maximum likelihood estimator (Section 7.2.1) implemented in **logreg-WML** (`RobStatTM`);
- the conditional unbiased bounded influence estimators that can be computed with **glmRob** (`robust`) with the parameter "method" equal to "cubif";
- the robust quasi-likelihood estimator (Section 7.4.3) that can be computed with **glmrob**(`robustbase`) with the parameter "method" equal to "Mqle".

For Poisson regression we recommend the MT-estimator defined in Section 7.4, computed with **glmrob** (`robustbase`) with the parameter "method" equal to "MT".

Another option is the robust quasi likelihood estimator (Section 7.4.3), which can be computed with **glmrob** (`robustbase`) with the parameter "method" equal to "Mqle".

7.6 Problems

7.1. The dataset **neuralgia** (Piegorsch, 1992) contains the values of four predictors for 18 patients, the outcome being whether the patient experimented pain relief after a treatment. The data are described in Chapter 11. Compare the fits given by the different logistic regression estimators discussed in this chapter.

7.2. Prove (7.10).

7.3. Consider the univariate logistic model without intercept for the sample $(x_1, y_1), \ldots, (x_n, y_n)$ with $x_i \in R$; that is, $\boldsymbol{\beta}'\mathbf{x} = \beta x$. Let

$$p(x, \beta) = e^{\beta x}/(1 + e^{\beta x}) = P(y = 1).$$

(a) Show that

$$A_n(\beta) = \sum_{i=1}^{n} (y_i - p(x_i, \beta))x_i$$

is decreasing in β.

(b) Call $\hat{\beta}_n$ the ML estimator. Assume $\hat{\beta}_n > 0$. Add one outlier $(K, 0)$ where $K > 0$; call $\hat{\beta}_{n+1}(K)$ the MLE computed with the enlarged sample. Show that $\lim_{K\to\infty}\hat{\beta}_{n+1}(K) = 0$. State a similar result when $\hat{\beta}_n < 0$.

7.4. Let $Z_n = \{(\mathbf{x}_1, y_1), \ldots, (\mathbf{x}_n, y_n)\}$, be a sample for the logistic model, where the first coordinate of each $\mathbf{x}_i$ is 1 if the model contains an intercept. Consider a new sample $Z_n^* = \{(-\mathbf{x}_1, 1 - y_1), \ldots, (-\mathbf{x}_n, 1 - y_n)\}$.

(a) Explain why is desirable that an estimator $\hat{\boldsymbol{\beta}}$ satisfies the equivariance property $\hat{\boldsymbol{\beta}}(Z_n^*) = \hat{\boldsymbol{\beta}}(Z_n)$.

(b) Show that M-, WM- and CUBIF estimators satisfy this property.

7.5. For the model in Problem 7.3, define an estimator by the equation

$$\sum_{i=1}^{n} (y_i - p(x_i, \beta))\text{sgn}(x_i) = 0.$$

Since deleting all $x_i = 0$ yields the same estimate, it will be henceforth assumed that $x_i \neq 0$ for all i.

(a) Show that this estimator is Fisher-consistent.

(b) Show that the estimator is a weighted ML estimator.

(c) Given the sample $Z_n = \{(x_i, y_i), i = 1, \ldots, n\}$, define the sample $Z_n^* = \{(x_i^*, y_i^*), i = 1, \ldots, n\}$, where $(x_i^*, y_i^*) = (x_i, y_i)$ if $x_i > 0$ and $(x_i^*, y_i^*) = (-x_i, 1 - y_i)$ if $x_i < 0$. Show that $\hat{\beta}_n(Z_n) = \hat{\beta}_n(Z_n^*)$.

(d) Show that $\hat{\beta}_n(Z_n^*)$ fulfills the equation $\sum_{i=1}^n y_i^* = \sum_{i=1}^n p(x_i^*, \beta)$, and hence $\hat{\beta}_n(Z_n^*)$ is the value of β that matches the empirical frequency of $y_i = 1$ with the theoretical one.
(e) Prove that $\sum_i p(x_i^*, 0) = n/2$.
(f) Show that if n is even, then the minimum number of observations that it is necessary to change in order to make $\hat{\boldsymbol{\beta}}_n = 0$ is $|n/2 - \sum_{i=1}^n y_i^*|$.
(g) Discuss the analogue property for odd n.
(h) Show that the influence function of this estimator is

$$\mathrm{IF}(y, x, \beta) = \frac{(y - p(x, \beta))\mathrm{sgn}(x)}{A}$$

where $A = \mathrm{E}(p(x, \beta)(1 - p(x, \beta))|x|)$, and hence GES $= 1/A$.

7.6. Consider the CUBIF estimator defined in Section 7.3.

(a) Show that he correction term $c(a, b)$ defined above (7.26) is a solution of the equation

$$\mathrm{E}_{m^{-1}(a)}(\psi_b^H(y - g(a) - c(a, b))) = 0.$$

(b) In the case of the logistic model for the Bernoulli family put $g(a) = e^a/(1 + e^a)$. Then prove that $c(a, b) = c^*(g(a), b)$, where

$$c^*(p, b) = \begin{cases} (1-p)(p-b)/p & \text{if } p > \max\left(\frac{1}{2}, b\right) \\ p(b-1+p)/(1-p) & \text{if } p < \min\left(\frac{1}{2}, b\right) \\ 0 & \text{elsewhere.} \end{cases}$$

(c) Show that the limit when $b \to 0$ of the CUBIF estimator for the model in Problem 7.3 satisfies the equation

$$\sum_{i=1}^{n} \frac{(y - p(x_i, \beta))}{\max(p(x_i, \beta), 1 - p(x_i, \beta))} \mathrm{sgn}(x_i) = 0.$$

Compare this estimator with the one of Problem 7.5.
(d) Show that the influence function of this estimator is

$$\mathrm{IF}(y, x, \beta) = \frac{1}{A} \frac{(y - p(x, \beta))\mathrm{sgn}(x_i)}{\max(p(x, \beta), 1 - p(x, \beta))}$$

with $A = \mathrm{E}(\min(p(x, \beta)(1 - p(x, \beta))|x|)$; and that the gross error sensitivity is $\mathrm{GES}(\beta) = 1/A$.
(e) Show that this GES is smaller than the GES of the estimator given in Problem 7.5. Explain why this may happen.

8

Time Series

Throughout this chapter we shall focus on time series in *discrete time*; those whose time index t is integer valued; that is, $t = 0, \pm 1, \pm 2, \ldots$. We shall typically label the observed values of time series as x_t or y_t, and so on.

We shall assume that our time series is either stationary in some sense or may be reduced to stationarity by a combination of elementary differencing operations and regression trend removal. Two types of stationarity are in common use, *second-order* stationarity and *strict* stationarity. The sequence is said to be second-order (or *wide-sense*) stationary if the first- and second-order moments $\mathrm{E}y_t$ and $\mathrm{E}(y_{t_1}y_{t_2})$ exist and are finite, with $\mathrm{E}y_t = \mu$ a constant independent of t, and the covariance of y_{t+l} and y_t depends only on the lag l:

$$\mathrm{Cov}(y_{t+l}, y_t) = C(l) \text{ for all } t, \tag{8.1}$$

where C is called the *covariance function*, or alternatively the *autocovariance function*.

The time series is said to be strictly stationary if, for every integer $k \geq 1$ and every subset of times $t_1, t_2, \ldots, t_k$, the joint distribution of $y_{t_1}, y_{t_2}, \ldots, y_{t_k}$ is invariant with respect to shifts in time; that is, for every positive integer k and every integer l we have

$$\mathcal{D}(y_{t_1+l}, y_{t_2+l}, \ldots, y_{t_k+l}) = \mathcal{D}(y_{t_1}, y_{t_2}, \ldots, y_{t_k}),$$

where $\mathcal{D}$ denotes the joint distribution. A strictly stationary time series with finite second moments is obviously second-order stationary, and we shall assume unless otherwise stated that our time series is at least second-order stationary.

Robust Statistics: Theory and Methods (with R), Second Edition.
Ricardo A. Maronna, R. Douglas Martin, Victor J. Yohai and Matías Salibián-Barrera.

Companion website: www.wiley.com/go/maronna/robust

8.1 Time series outliers and their impact

Outliers in time series are more complex than in the situations dealt with in the previous chapters, where there is no temporal dependence in the data. This is because in the time series setting we encounter several different types of outliers, as well as other important behaviors that are characterized by their temporal structure. Specifically, in fitting time series models we may have to deal with one or more of the following:

- isolated outliers
- patchy outliers
- level shifts in mean value.

While level shifts have a different character than outliers, they are a frequently occurring phenomenon that must be dealt with in the context of robust model fitting, and so we include them in our discussion of robust methods for time series. The following figures display time series which exhibit each of these types of behavior. Figure 8.1 shows a time series of stock returns for a company, with stock ticker NHC, which contains an isolated outlier. Here we define stock returns r_t as the relative change in price $r_t = (p_t - p_{t-1})/p_{t-1}$.

Figure 8.2 shows a time series of stock prices (for a company with stock ticker WYE) which has a patch outlier of length four with roughly constant size. Patch outliers can have different shapes or "configurations". For example, the stock returns for the company with ticker GHI in Figure 8.3 have a "doublet" patch outlier.

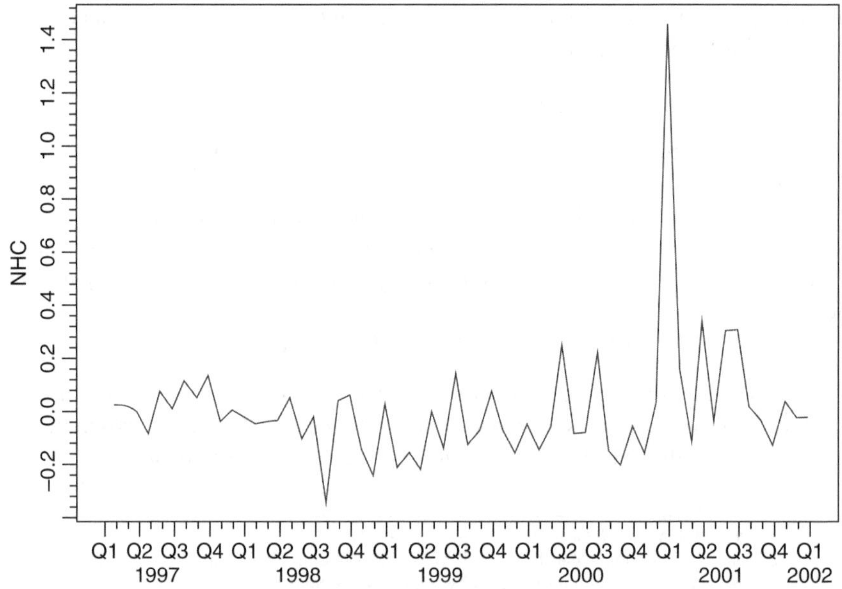

Figure 8.1 Stock returns (NHC) with isolated outlier

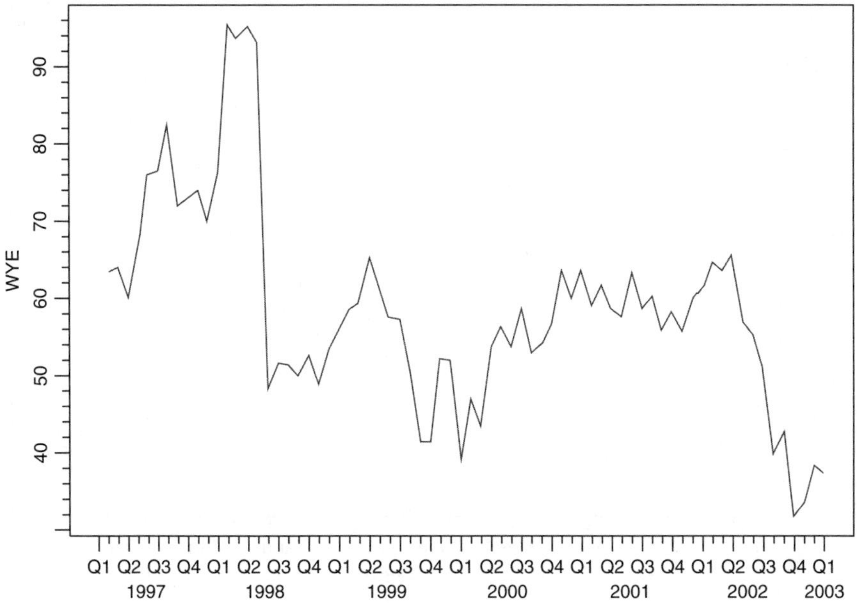

Figure 8.2 Stock prices (WYE) with patch outliers

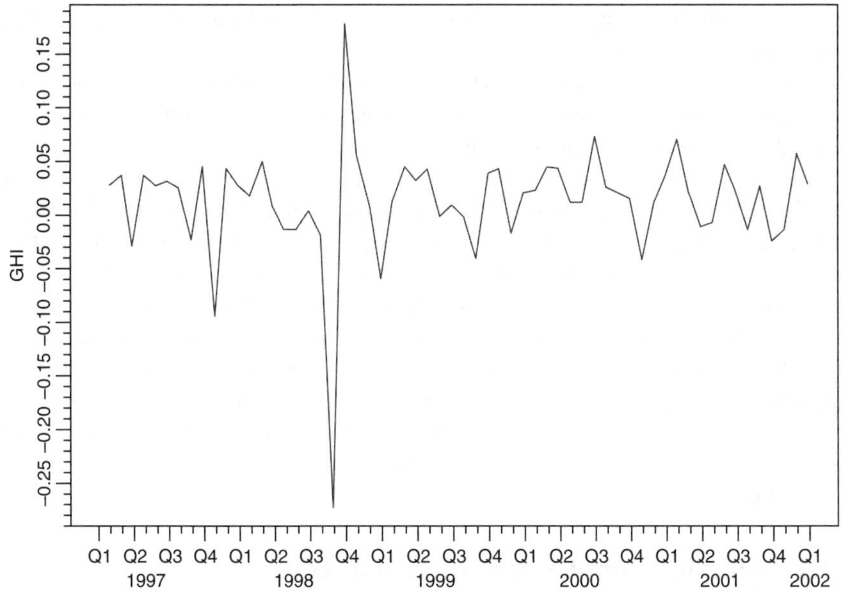

Figure 8.3 Stock returns (GHI) with doublet outlier

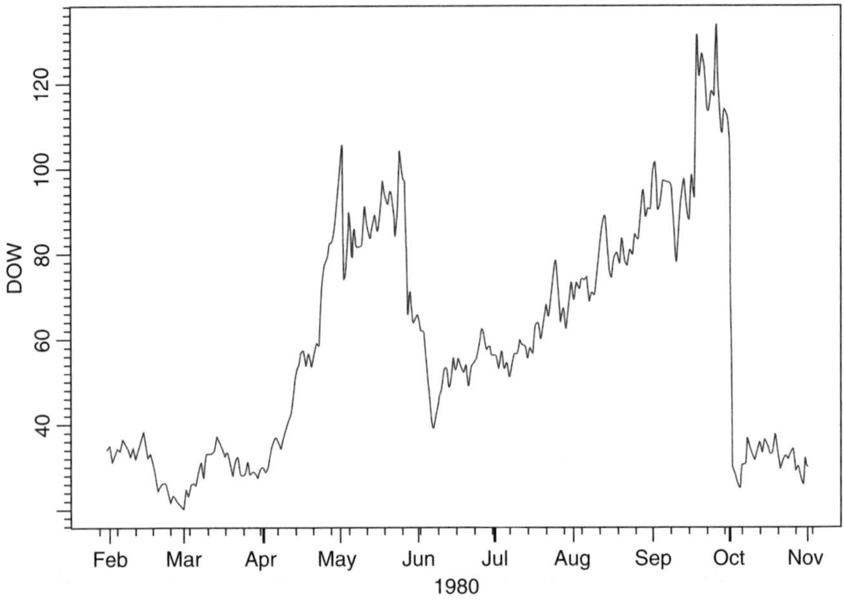

Figure 8.4 Stock prices with level shift

The doublet outlier in the GHI returns arises because of the differencing operation in the two returns computations. This involves the isolated outlier in the GHI price series.

Figure 8.4 shows a price time series for Dow Jones (ticker DOW), which has a large level shift at the beginning of October. Note that this level shift will produce an isolated outlier in the Dow Jones returns series.

Finally, Figure 8.5 shows a time series of tobacco sales in the UK (West and Harrison, 1997), which contains both an isolated outlier and two or three level shifts. The series also appears to contain trend segments at the beginning and end of the series. It is important to note that since isolated outliers, patch outliers and level shifts can all occur in a single time series, it will not suffice to discuss robustness toward outliers without taking into consideration handling of patch outliers and level shifts. Note also that when one first encounters an outlier – that is, as the most recent observation in a time series – then lacking side information we do not know whether it is an isolated outlier, or a level shift or a short patch outlier. Consequently it will take some amount of future data beyond the time of occurrence of the outlier in order to resolve this uncertainty.

8.1.1 Simple examples of outliers influence

Time series outliers can have an arbitrarily adverse influence on parameter estimators for time series models, and the nature of this influence depends on the type of outlier.

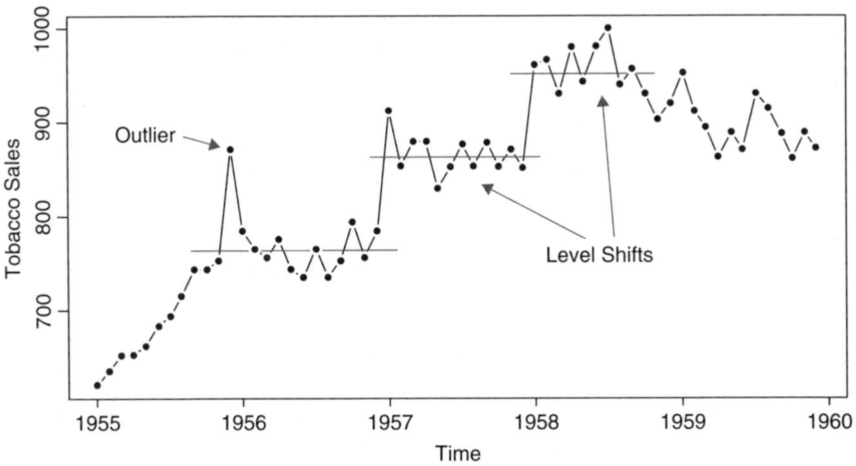

Figure 8.5 Tobacco and related sales in the UK

We focus on the lag-k autocorrelation

$$\rho(k) = \frac{\mathrm{Cov}(y_{t+k}, y_t)}{\mathrm{Var}(y_t)} = \frac{C(k)}{C(0)}. \tag{8.2}$$

Here we take a simple first look at the impact of time series outliers of different types by focusing on the special case of the estimation of $\rho(1)$. Let $y_1, y_2, \ldots, y_T$ be the observed values of the series. We initially assume for simplicity that $\mu = \mathrm{E}y = 0$. In that case, a natural estimator $\widehat{\rho}(1)$ of $\rho(1)$ is given by the lag-1 sample autocorrelation coefficient

$$\widehat{\rho}(1) = \frac{\sum_{t=1}^{T-1} y_t y_{t+1}}{\sum_{t=1}^{T} y_t^2}. \tag{8.3}$$

It may be shown that $|\widehat{\rho}(1)| \leq 1$, which is certainly a reasonable property for such an estimator (see Problem 8.1).

Now suppose that for some t_0, the true value y_{t_0} is replaced by an arbitrary value A, where $2 \leqslant t_0 \leqslant T-1$. In this case the estimator becomes

$$\widehat{\rho}(1) = \frac{\sum_{t=1}^{T-1} y_t y_{t+1} \mathrm{I}(t \notin \{t_0 - 1, t_0\})}{\sum_{t=1}^{T} y_t^2 \mathrm{I}(t \neq t_0) + A^2} + \frac{y_{t_0-1}\, A + A\, y_{t_0+1}}{\sum_{t=1}^{T} y_t^2 \mathrm{I}(t \neq t_0) + A^2}.$$

Since A appears quadratically in the denominator and only linearly in the numerator of the above estimator, $\widehat{\rho}(1)$ goes to zero as $A \to \infty$, with all other values of y_t for $t \neq t_0$ held fixed. So, whatever the original value of $\widehat{\rho}(1)$, the alteration of an original value y_{t_0} to an outlying value $y_{t_0} = A$ results in a "bias" of $\widehat{\rho}(1)$ toward zero, the more so the larger the magnitude of the outlier.

Now consider the case of a patch outlier of constant value A and patch length k, where the values y_i for $i = t_0, \ldots, t_0 + k - 1$ are replaced by A. In this case the above estimator has the form

$$\hat{\rho}(1) = \frac{\sum_{t=1}^{T-1} y_t y_{t+1} \mathrm{I}\{t \notin [t_0 - 1, t_0 + k - 1]\}}{\sum_{t=1}^{T} y_t^2 \mathrm{I}\{t \notin [t_0, t_0 + k - 1]\} + k\, A^2} + \frac{y_{t_0 - 1}\, A + (k-1)A^2 + A\, y_{t_0 + k}}{\sum_{t=1}^{T} y_t^2 \mathrm{I}\{t \notin [t_0, t_0 + k - 1]\} + k\, A^2},$$

and therefore

$$\lim_{A \to \infty} \hat{\rho}(1) = \frac{k-1}{k}.$$

Hence, the limiting value of $\hat{\rho}(1)$ with the patch outlier can either increase or decrease relative to the original value, depending on the value of k and the value of $\hat{\rho}(1)$ without the patch outlier. For example, if $k = 10$ with $\hat{\rho}(1) = 0.5$ without the patch outlier, then $\hat{\rho}(1)$ increases to the value 0.9 as $A \to \infty$.

In some applications, one may find that outliers come in pairs of opposite signs. For example, when computing first differences of a time series that has isolated outliers, we get a *doublet* outlier, as shown in Figure 8.3. We leave it to the reader to show that for a doublet outlier with adjacent values having equal magnitude but opposite signs – that is, values $\pm A$ – the limiting value of $\hat{\rho}(1)$ as $A \to \infty$ is $\hat{\rho}(1) = -0.5$ (Problem 8.2).

Of course, one can seldom make the assumption that the time series has zero mean, so one usually defines the lag-1 sample autocorrelation coefficient using the definition

$$\hat{\rho}(1) = \frac{\sum_{t=1}^{T-1} (y_t - \bar{y})(y_{t+1} - \bar{y})}{\sum_{t=1}^{T} (y_t - \bar{y})^2}. \tag{8.4}$$

Determining the influence of outliers in this more realistic case is often algebraically quite messy, but achievable. For example, in the case of an isolated outlier of size A, it may be shown that the limiting value of $\hat{\rho}(1)$ as $A \to \infty$ is approximately $-1/T$ for large T (Problem 8.3). However, it is usually easier to resort to some type of influence function calculation, as described in Section 8.11.1.

8.1.2 Probability models for time series outliers

In this section we describe several probability models for time series outliers, including *additive* outliers (AOs), *replacement* outliers (ROs) and *innovations* outliers (IOs). Let x_t be a wide-sense stationary "core" process of interest, and let v_t be a stationary outlier process that is non-zero a fraction ε of the time; that is, $\mathrm{P}(v_t = 0) = 1 - \varepsilon$. In practice the fraction ε is often positive but small.

Under an AO model, instead of x_t one actually observes

$$y_t = x_t + v_t \tag{8.5}$$

where the processes x_t and υ_t are assumed to be independent of one another. A special case of the AO model was originally introduced by Fox (1972), who called them Type I outliers. Fox attributed such outliers to a "gross-error of observation or a recording error that affects a single observation". The AO model will generate mostly isolated outliers if υ_t is an independent and identically distributed (i.i.d.) process, with standard deviation (or scale) much larger than that of x_t. For example, suppose that x_t is a zero-mean normally distributed process with $\text{Var}(x_t) = \sigma_x^2$, and υ_t has a normal mixture distribution with degenerate central component

$$\upsilon_t \sim (1 - \varepsilon)\delta_0 + \varepsilon \text{N}(\mu_\upsilon, \sigma_\upsilon^2). \tag{8.6}$$

Here δ_0 is a point-mass distribution located at zero, and we assume that the normal component $\text{N}(\mu_\upsilon, \sigma_\upsilon^2)$ has variance $\sigma_\upsilon^2 \gg \sigma_x^2$. In this case y_t will contain an outlier at any fixed time t with probability ε, and the probability of getting two outliers in a row is the much smaller ε^2. It will be assumed that $\mu_\upsilon = 0$ unless otherwise stated.

Additive patch outliers can be obtained by specifying that at any given t, $\upsilon_t = 0$ with probability $1 - \varepsilon$; and with probability ε, υ_t is the first observation of a patch outlier having a particular structure that persists for k time periods. We leave it for the reader (Problem 8.4) to construct a probability model to generate patch outliers.

RO models have the form

$$y_t = (1 - z_t)x_t + z_t w_t \tag{8.7}$$

where z_t is a 0–1 process with $\text{P}(z_t = 0) = 1 - \varepsilon$, and w_t is a "replacement" process that is not necessarily independent of x_t. In fact, RO models contain AO models as a special case in which $w_t = x_t + \upsilon_t$ and z_t is a Bernoulli process; that is, z_t and z_u are independent for $t \neq u$. Outliers that are mostly isolated are obtained, for example, when z_t is a Bernoulli process and x_t and w_t are zero-mean normal processes with $\text{Var}(w_t) = \sigma_w^2 \gg \sigma_x^2$. For the reader familiar with Markov chains, we can say that patch outliers may be obtained by letting z_t be a Markov process that remains in the "one" state for more than one time period (of fixed or random duration), and w_t has an appropriately specified probability model.

IOs are a highly specialized form of outlier that occur in linear processes such as AR, ARMA and ARIMA models, which will be discussed in subsequent sections. IO models were first introduced by Fox (1972), who termed them Type II outliers, and noted that an IO "will affect not only the current observation but also subsequent observations". For the sake of simplicity, we illustrate IOs here in the special case of a first-order autoregression model, which is adequate to reveal the character of this type of outlier. A stationary first-order AR model is given by

$$x_t = \phi x_{t-1} + u_t \tag{8.8}$$

where the *innovations* process u_t is i.i.d. with zero mean and finite variance, and $|\phi| < 1$. An IO is an outlier in the u_t process. IOs are obtained, for example, when the innovations process has a zero-mean normal mixture distribution

$$(1 - \varepsilon)\text{N}(0, \sigma_0^2) + \varepsilon \text{N}(0, \sigma_1^2) \tag{8.9}$$

where $\sigma_1^2 \gg \sigma_0^2$. More generally, we may say that the process has IOs when the distribution of u_t is heavy-tailed (e.g., a Student t-distribution).

Example 8.1 *AR(1) with innovation outlier and additive outliers*

To illustrate the differences between AOs and IOs, we display in the first row of Figure 8.6 a Gaussian first-order AR series x_t of length 100, with parameter $\phi = 0.9$, and free of outliers. The second shows the same series with ten additive outliers (marked with circles) obtained by adding the value 4 at ten equidistant positions. The third row displays the same series with one innovation outlier at position 50, also marked with a circle. This IO was created by replacing (only) the normal innovation u_{50} by an atypical innovation with value $u_{50} = 10$. The persistent effect of the IO at $t = 50$ on subsequent observations is quite clear. The effect of this outlier decays roughly as ϕ^{t-50} for times $t > 50$.

One may think of an IO as an "impulse" input to a dynamic system driven by a background of uncorrelated or i.i.d. white noise. Consequently, the output of the system behaves transitorily like an *impulse response* – a concept widely used in linear

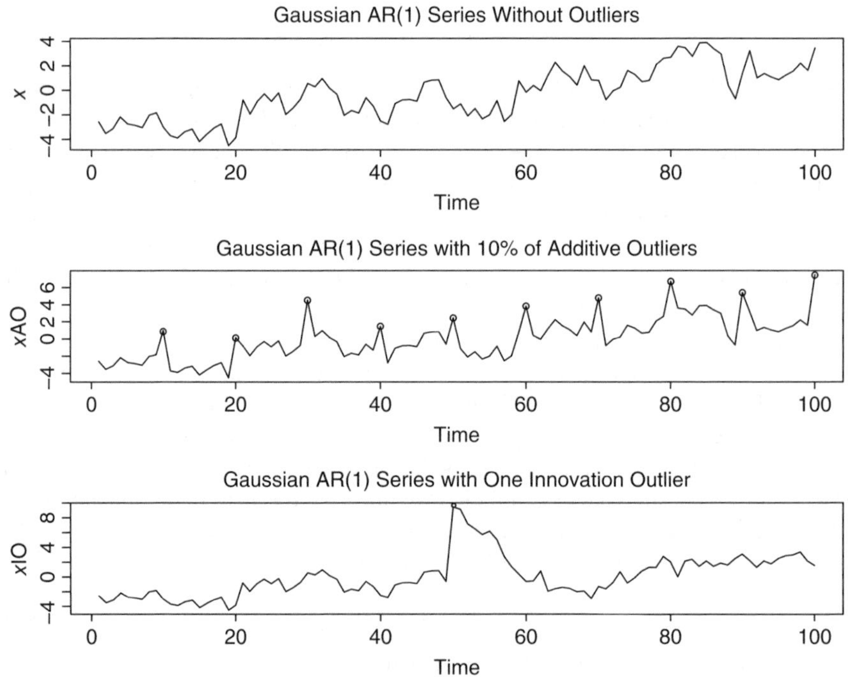

Figure 8.6 Top, a Gaussian AR(1) series with $\rho = 0.9$; middle, the same series with 10 additive outliers at equidistant locations; bottom, the same series with one additive outlier at location 50

systems theory – immediately after the occurrence of the outlier. It will be seen in Section 8.4.3 that IOs are "good" outliers, in the sense that they can improve the precision of the estimation of the parameters in AR and ARMA models, for example the parameter ϕ in the AR(1) model.

8.1.3 Bias impact of AOs

In this section we provide a simple illustrative example of the bias impact of AOs on the estimation of a first-order zero-mean AR model.

The reader may easily check that the lag-1 autocorrelation coefficient $\rho(1)$ for the AR(1) model (8.8) is equal to ϕ (see Problem 8.6). Furthermore, a natural least-squares (LS) estimator $\widehat{\phi}$ of ϕ in the case of perfect observations $y_t \equiv x_t$ is obtained by solving the minimization problem

$$\min_{\phi} \sum_{t=2}^{T} (y_t - \phi y_{t-1})^2. \tag{8.10}$$

Differentiation with respect to ϕ gives the estimating equation

$$\sum_{t=2}^{T} y_{t-1}(y_t - \widehat{\phi} y_{t-1}) = 0$$

and solving for $\widehat{\phi}$ gives the LS estimator

$$\widehat{\phi} = \frac{\sum_{t=2}^{T} y_t y_{t-1}}{\sum_{t=1}^{T-1} y_t^2}. \tag{8.11}$$

A slightly different estimator is

$$\phi^* = \frac{\sum_{t=2}^{T} y_t y_{t-1}}{\sum_{t=1}^{T} y_t^2}, \tag{8.12}$$

which coincides with (8.3). The main difference between these two estimators is that $|\phi^*|$ is bounded by one, while $|\widehat{\phi}|$ can take on values larger than one. Since the true autocorrelation coefficient ϕ has $|\phi| \leq 1$, and actually $|\phi| < 1$ except in the case of perfect linear dependence, the estimator ϕ^* is usually preferred.

Let $\rho_y(1)$ be the lag-1 autocorrelation coefficient for the AO observations $y_t = x_t + v_t$ where x_t is given by (8.8). It may be shown that when $T \to \infty$, ϕ^* converges to $\rho_y(1)$ under mild regularity conditions, and the same is true of $\widehat{\phi}$ (Brockwell and Davis, 1991). If we assume that v_t is independent of x_t, and that v_t has lag-1 autocorrelation coefficient $\rho_v(1)$, then

$$\rho_y(1) = \frac{\mathrm{Cov}(y_t, y_{t-1})}{\mathrm{Var}(y_t)} = \frac{\mathrm{Cov}(x_t, x_{t-1}) + \mathrm{Cov}(v_t, v_{t-1})}{\sigma_x^2 + \sigma_v^2}$$

$$= \frac{\sigma_x^2 \phi + \sigma_v^2 \rho_v(1)}{\sigma_x^2 + \sigma_v^2}.$$

The large-sample bias of ϕ^* is

$$\text{Bias}(\phi^*) = \rho_y(1) - \phi = \frac{\sigma_v^2}{\sigma_x^2 + \sigma_v^2}(\rho_v(1) - \phi)$$

$$= \frac{R}{1+R}(\rho_v(1) - \phi) \tag{8.13}$$

where $R = \sigma_v^2/\sigma_x^2$ is the "noise-to-signal" ratio. We see that the bias is zero when R is zero; that is, when the AOs have zero variance. The bias is bounded in magnitude by $|\rho_v(1) - \phi|$ and approaches $|\rho_v(1) - \phi|$ as R approaches infinity. When the AOs have lag-1 autocorrelation, $\rho_v(1) = 0$ and R is very large, the bias is approximately $-\phi$ and correspondingly the estimator ϕ^* has a value close to zero. As an intermediate example, suppose that $\rho_v(1) = 0$, $\phi = 0.5$, $\sigma_x^2 = 1$ and that v_t has distribution (8.6) with $\varepsilon = 0.1$, $\mu_v = 0$ and $\sigma_v^2 = 0.9$. Then the bias is negative and equal to −0.24. On the other hand, if the values of $\rho_v(1)$ and ϕ are interchanged, with the other parameter values remaining fixed, then the bias is positive and has value +0.24.

8.2 Classical estimators for AR models

In this section we describe the properties of classical estimators of the parameters of an autoregression model. In particular we describe the form of these estimators, state the form of their limiting multivariate normal distribution, and describe their efficiency and robustness in the absence of outliers.

An autoregression model of order p, called an AR(p) model for short, generates a time series according to the stochastic difference equation

$$y_t = \gamma + \phi_1 y_{t-1} + \phi_2 y_{t-2} + \ldots + \phi_p y_{t-p} + u_t \tag{8.14}$$

where the innovations u_t are an i.i.d. sequence of random variables with mean 0 and finite variance σ_u^2. The innovations are assumed to be independent of past values of the y_ts. It is known that the time series y_t is stationary if all the roots of the characteristic polynomial

$$\phi(z) = 1 - \phi_1 z - \phi_2 z^2 - \ldots - \phi_p z^p \tag{8.15}$$

lie outside the unit circle in the complex plane, and the sum of the ϕ_i is less than one. When y_t is stationary, it has a mean value $\mu = \mathrm{E}(y_t)$ that is determined by taking the mean value of both sides of (8.14), giving

$$\mu = \gamma + \phi_1\mu + \phi_2\mu + \ldots + \phi_p\mu + 0,$$

which implies

$$\mu\left(1 - \sum_{i=1}^{p}\phi_i\right) = \gamma, \tag{8.16}$$

and hence

$$\mu = \frac{\gamma}{1 - \sum_{i=1}^{p} \phi_i}. \tag{8.17}$$

In view of (8.16), the AR(p) model may also be written in the form

$$y_t - \mu = \phi_1(y_{t-1} - \mu) + \phi_2(y_{t-2} - \mu) + \ldots + \phi_p(y_{t-p} - \mu) + u_t. \tag{8.18}$$

There are several asymptotically equivalent forms of LS estimators of the AR(p) model parameters that are asymptotically efficient when the distribution of u_t is normal. Given a sample of observations $y_1, y_2, \ldots, y_T$, it seems natural at first glance to compute LS estimators of the parameters by choosing $\gamma, \phi_1, \phi_2, \ldots, \phi_p$ to minimize the sum of squares:

$$\sum_{t=p+1}^{T} \widehat{u}_t^2(\boldsymbol{\phi}, \gamma) \tag{8.19}$$

where $\widehat{u}_t$ are the prediction residuals defined by

$$\widehat{u}_t = \widehat{u}_t(\boldsymbol{\phi}, \gamma) = y_t - \gamma - \phi_1 y_{t-1} - \phi_2 y_{t-2} - \ldots - \phi_p y_{t-p}. \tag{8.20}$$

This is equivalent to applying ordinary LS to the linear regression model

$$\mathbf{y} = \mathbf{G}\boldsymbol{\beta} + \mathbf{u} \tag{8.21}$$

where

$$\begin{aligned} \mathbf{y}' &= (y_{p+1}, y_{p+2}, \ldots, y_T) \\ \mathbf{u}' &= (u_{p+1}, u_{p+2}, \ldots, u_T) \\ \boldsymbol{\beta}' &= (\boldsymbol{\phi}', \gamma) = (\phi_1, \phi_2, \ldots, \phi_p, \gamma) \end{aligned} \tag{8.22}$$

and

$$\mathbf{G} = \begin{bmatrix} y_p & y_{p-1} & \cdots & y_1 & 1 \\ y_{p+1} & y_p & & y_2 & 1 \\ \vdots & \vdots & \vdots & \vdots & \vdots \\ y_{T-1} & y_{T-2} & \cdots & y_{T-p} & 1 \end{bmatrix}. \tag{8.23}$$

This form of LS estimator may also be written as

$$\widehat{\boldsymbol{\beta}} = (\widehat{\phi}_1, \widehat{\phi}_2, \ldots, \widehat{\phi}_p, \widehat{\gamma})' = (\mathbf{G}'\mathbf{G})^{-1}\mathbf{G}'\mathbf{y} \tag{8.24}$$

and the mean value estimator can be computed as

$$\widehat{\mu} = \frac{\widehat{\gamma}}{1 - \sum_{i=1}^{p} \widehat{\phi}_i}. \tag{8.25}$$

An alternative approach is to estimate μ by the sample mean $\overline{y}$ and compute the LS estimator of $\widehat{\boldsymbol{\phi}}$ as

$$\hat{\boldsymbol{\phi}} = (\mathbf{G}^{*\prime}\mathbf{G}^*)^{-1}\mathbf{G}^{*\prime}\mathbf{y}^*, \tag{8.26}$$

where $\mathbf{y}^*$ is now the vector of centered observations $y_t - \bar{y}$, and $\mathbf{G}^*$ is defined as in (8.23), but replacing the y_t's by the y_t^* and omitting the last column of ones.

Unfortunately, the above forms of the LS estimator do not ensure that the estimators $\hat{\boldsymbol{\phi}} = (\hat{\phi}_1, \hat{\phi}_2, \ldots, \hat{\phi}_p)'$ correspond to a stationary autoregression; that is, it can happen that one or more of the roots of the estimated characteristic polynomial

$$\hat{\phi}(z) = 1 - \hat{\phi}_1 z - \hat{\phi}_2 z^2 - \ldots - \hat{\phi}_p z^p$$

lie inside the unit circle. A common way around this is to use the so-called Yule–Walker equations to estimate $\boldsymbol{\phi}$. Let $C(l)$ be the autocovariance (8.1) at lag l. The Yule–Walker equations relating the autocovariances and the parameters of an AR(p) process are obtained from (8.18) as

$$C(k) = \sum_{i=1}^{p} \phi_i C(k-i) \quad (k \geq 1). \tag{8.27}$$

For $1 \leq k \leq p$, (8.27) may be expressed in matrix equation form as

$$\mathbf{C}\boldsymbol{\phi} = \mathbf{g} \tag{8.28}$$

where $\mathbf{g}' = (C(1), C(2), \ldots, C(p))$ and the $p \times p$ matrix $\mathbf{C}$ has elements $C_{ij} = C(i-j)$. It is left for the reader (Problem 8.5) to verify the above equations for an AR(p) model.

The Yule–Walker equations can also be written in terms of the autocorrelations as

$$\rho(k) = \sum_{i=1}^{p} \phi_i \rho(k-i) \quad (k \geq 1). \tag{8.29}$$

The Yule–Walker estimator $\hat{\boldsymbol{\phi}}_{YW}$ of $\boldsymbol{\phi}$ is obtained by replacing the unknown lag-l covariances in $\mathbf{C}$ and $\mathbf{g}$ of (8.28) by the covariance estimators

$$\hat{C}(l) = \frac{1}{T} \sum_{t=1}^{T-|l|} (y_{t+l} - \bar{y})(y_t - \bar{y}), \tag{8.30}$$

and solving for $\hat{\boldsymbol{\phi}}_{YW}$. It can be shown that the above lag-l covariance estimators are biased and that unbiased estimators can be obtained under normality by replacing the denominator T by $T - |l| - 1$. However, the covariance matrix estimator $\hat{\mathbf{C}}$ based on the above biased lag-l estimators is preferred since it is known to be positive definite (with probability 1), and furthermore the resulting Yule–Walker parameter estimator $\hat{\boldsymbol{\phi}}_{YW}$ corresponds to a stationary autoregression (see, for example, Brockwell and Davis, 1991).

The Durbin–Levinson algorithm, to be described in Section 8.2.1, is a convenient recursive procedure to compute the sequence of Yule–Walker estimators for AR(1), AR(2), and so on.

8.2.1 The Durbin–Levinson algorithm

We shall describe the Durbin–Levinson algorithm, which is a recursive method to derive the best memory-$(m-1)$ linear predictor from the best memory-$(m-1)$ predictor.

Let y_t be a second-order stationary process. It will be assumed that $\mathrm{E}y_t = 0$. Otherwise, we apply the procedure to the centered process $y_t - \mathrm{E}y_t$ rather than y_t.

Let

$$\hat{y}_{t,m} = \phi_{m,1}y_{t-1} + \ldots + \phi_{m,m}y_{t-m} \tag{8.31}$$

be the minimum mean-square error (MMSE) memory-m linear predictor of y_t based on $y_{t-1}, \ldots, y_{t-m}$. The "diagonal" coefficients $\phi_{m,m}$ are the so-called *partial autocorrelations*, which are very useful for the identification of AR models, as will be seen in Section 8.7.

The $\phi_{m,m}$ satisfy

$$|\phi_{m,m}| < 1, \tag{8.32}$$

except when the process is deterministic.

The MMSE *forward* prediction residuals are defined as

$$\hat{u}_{t,\,m} = y_t - \phi_{m,1}y_{t-1} - \ldots - \phi_{m,m}y_{t-m}. \tag{8.33}$$

The memory-m backward linear predictor of y_t – that is, the MMSE predictor of y_t as a linear function of $y_{t+1}, \ldots, y_{t+m}$ – can be shown to be

$$\hat{y}^*_{t,m} = \phi_{m,1}y_{t+1} + \ldots + \phi_{m,m}y_{t+m},$$

and the *backward* MMSE prediction residuals are

$$u^*_{t,m} = y_t - \phi_{m,1}y_{t+1} - \ldots - \phi_{m,m}y_{t+m}. \tag{8.34}$$

Note that $\hat{u}_{t,m-1}$ and $u^*_{t-m,m-1}$ are both orthogonal to the linear space spanned by $y_{t-1}, \ldots, y_{t-m+1}$, with respect to the expectation inner product; that is,

$$\mathrm{E}\hat{u}_{t,m-1}y_{t-k} = \mathrm{E}u^*_{t-m,m-1}y_{t-k} = 0, \quad k = 1, \ldots, m-1.$$

We shall first derive the form of the memory-m predictor assuming that the true memory-$(m-1)$ predictor and the true values of the autocorrelations $\rho(k)$, $k = 1, \ldots, m$; are known.

Let $\zeta^* u^*_{t-m,m-1}$ be the MMSE linear predictor of $\hat{u}_{t,m-1}$ based on $u^*_{t-m,m-1}$. Then

$$\mathrm{E}(\hat{u}_{t,m-1} - \zeta^* u^*_{t-m,m-1})^2 = \min_{\zeta} \mathrm{E}(\hat{u}_{t,m-1} - \zeta u^*_{t-m,m-1})^2. \tag{8.35}$$

It can be proved that the MMSE memory-m predictor is given by

$$\begin{aligned}\hat{y}_{t,m} &= \hat{y}_{t,m-1} + \zeta^* u^*_{t-m,m-1} \\ &= (\phi_{m-1,1} - \zeta^*\phi_{m-1,m-1})y_{t-1} + \ldots \\ &\quad + (\phi_{m-1,i} - \zeta^*\phi_{m-1,1})y_{t-m+1} + \zeta^* y_{t-m}.\end{aligned} \tag{8.36}$$

To show (8.36) it suffices to prove that

$$\mathrm{E}[(y_t - \widehat{y}_{t,m} - \zeta^* u^*_{t-m,m})y_{t-i}] = 0,\ i = 1, \ldots, m, \tag{8.37}$$

which we leave to the reader as Problem 8.8. Then from (8.36) we have

$$\phi_{m,i} = \begin{cases} \zeta^* & \text{if} \quad i = m \\ \phi_{m-1,i} - \zeta^* \phi_{m-1,m-i} & \text{if} \quad 1 \le i \le m-1. \end{cases} \tag{8.38}$$

According to (8.38), if we already know $\phi_{m-1,i}$, $1 \le i \le m-1$, then to compute all the $\phi_{m,i}$, we only need $\zeta^* = \phi_{m,m}$.

It is easy to show that

$$\begin{aligned} \phi_{m,m} &= \mathrm{Corr}(\widehat{u}_{t,m-1}, u^*_{t-m,m-1}) \\ &= \frac{\rho(m) - \sum_{i=1}^{m-1} \rho(m-i)\phi_{m-1,i}}{1 - \sum_{i=1}^{m-1} \rho(i)\phi_{m-1,i}}. \end{aligned} \tag{8.39}$$

The first equality above justifies the term "partial autocorrelation": it is the correlation between y_t and y_{t-m} after the linear contribution of y_i $(i = t-1, \ldots, t-m+1)$ has been subtracted out.

If y_t is a stationary AR(p) process with parameters $\phi_1, \ldots, \phi_p$, it may be shown (Problem 8.11) that

$$\phi_{p,i} = \phi_i, \quad 1 \le i \le p, \tag{8.40}$$

and

$$\phi_{m,m} = 0, \quad m > p. \tag{8.41}$$

In the case that we have only a sample from a process, the unknown autocorrelations $\rho(k)$ can be estimated by their empirical counterparts, and the Durbin–Levinson algorithm can be used to estimate the predictor coefficients $\phi_{m,j}$ in a recursive way. In particular, if the process is assumed to be AR(p), then the AR coefficient estimators are obtained by substituting estimators in (8.40).

It is easy to show that $\phi_{1,1} = \rho(1)$ or, equivalently,

$$\phi_{1,1} = \arg\min_{\zeta} \mathrm{E}(y_t - \zeta y_{t-1})^2. \tag{8.42}$$

We shall now describe the classical Durbin–Levinson algorithm in such a way as to clarify the basis for its robust version, as given in Section 8.6.4.

The first step is to set $\widehat{\phi}_{1,1} = \widehat{\rho}(1)$, which is equivalent to

$$\widehat{\phi}_{1,1} = \arg\min_{\zeta} \sum_{t=2}^{T} (y_t - \zeta y_{t-1})^2. \tag{8.43}$$

Assuming that estimators $\widehat{\phi}_{m-1,i}$ of $\phi_{m-1,i}$, for $1 \le i \le m-1$, have already been computed, $\widehat{\phi}_{m,m}$ can be computed from (8.39), where the ρs and ϕs are replaced by their estimators. Alternatively, $\widehat{\phi}_{m,m}$ is obtained as

$$\widehat{\phi}_{m,m} = \arg\min_{\zeta} \sum_{t=m+1}^{T} \widehat{u}_{t,m}^2(\zeta), \tag{8.44}$$

with

$$\begin{aligned}\widehat{u}_{t,m}(\zeta) &= y_t - \widehat{y}_{t,m-1} - \zeta u^*_{t-m,m-1} \\ &= y_t - (\widehat{\phi}_{m-1,1} - \zeta \widehat{\phi}_{m-1,m-1}) y_{t-1} - \dots \\ &\quad - (\widehat{\phi}_{m-1,m-1} - \zeta \widehat{\phi}_{m-1,1}) y_{t-m+1} - \zeta y_{t-m},\end{aligned} \tag{8.45}$$

and where the backward residuals $u_t^* - m, m-1$ are computed here by

$$u^*_{t-m,m-1} = y_{t-m} - \widehat{\phi}_{m-1,1} y_{t-m+1} - \dots - \widehat{\phi}_{m-1,m-1} y_{t-1}.$$

The remaining $\widehat{\phi}_{m,i}$ are computed using the recursion (8.38). It is easy to verify that this sample Durbin–Levinson method is essentially equivalent to obtaining the AR(m) estimator $\widehat{\boldsymbol{\phi}}_m$ by solving the Yule–Walker equations for $m = 1, 2, \dots, p$.

8.2.2 Asymptotic distribution of classical estimators

The LS and Yule–Walker estimators have the same asymptotic distribution, which will be studied in this section. Call $\widehat{\lambda} = (\widehat{\phi}_1, \widehat{\phi}_2, \dots, \widehat{\phi}_p, \widehat{\mu})$ the LS or Yule–Walker estimator of $\lambda = (\phi_1, \phi_2, \dots, \phi_p, \mu)$ based on a sample of size T. Here $\widehat{\mu}$ can be either the sample mean estimator or the estimator of (8.25) based on the LS estimator $\widehat{\boldsymbol{\beta}}$ defined in (8.19). It is known that $\widehat{\lambda}$ converges in distribution to a $(p+1)$-dimensional multivariate normal distribution

$$\sqrt{T}(\widehat{\lambda} - \lambda) \to_d \mathrm{N}_{p+1}(\mathbf{0}, \mathbf{V}_{LS}) \tag{8.46}$$

where the asymptotic covariance matrix $\mathbf{V}_{LS}$ is given by

$$\mathbf{V}_{LS} = \begin{bmatrix} \mathbf{V}_{LS,\phi} & \mathbf{0}' \\ \mathbf{0} & V_{LS,\mu} \end{bmatrix} \tag{8.47}$$

with

$$V_{LS,\mu} = \frac{\sigma_u^2}{\left(1 - \sum_{i=1}^p \phi_i\right)^2} \tag{8.48}$$

and

$$\mathbf{V}_{LS,\phi} = \mathbf{V}_{LS}(\boldsymbol{\phi}) = \sigma_u^2 \mathbf{C}^{-1}, \tag{8.49}$$

where $\mathbf{C}$ is the $p \times p$ covariance matrix of $(y_{t-1}, \dots, y_{t-p})$, which does not depend on t (due to the stationarity of y_t), and $\sigma_u^2 \mathbf{C}^{-1}$ depends only on the AR parameters $\boldsymbol{\phi}$; see, for example, Anderson (1994) or Brockwell and Davis (1991).

In Section 8.15 we give a heuristic derivation of this result and the expression for $D = \mathbf{C}/\sigma_u{}^2$.

Remark: Note that if we apply formula (5.6) for the asymptotic covariance matrix of the LS estimator under a linear model with random predictors, to the regression model (8.21), then the result coincides with (8.49).

The block–diagonal structure of $\mathbf{V}$ shows that $\widehat{\mu}$ and $\widehat{\boldsymbol{\phi}} = (\widehat{\phi}_1, \widehat{\phi}_2, \cdots, \widehat{\phi}_p)'$ are asymptotically uncorrelated. The standard estimator of the innovations variance σ_u^2 is

$$\widehat{\sigma}_u^2 = \frac{1}{T-p} \sum_{t=p+1}^{T} (y_t - \widehat{\gamma} - \widehat{\phi}_1 y_{t-1} - \widehat{\phi}_2 y_{t-2} - \ldots - \widehat{\phi}_p y_{t-p})^2 \tag{8.50}$$

or alternatively

$$\widehat{\sigma}_u^2 = \frac{1}{T-p} \sum_{t=p+1}^{T} (\widetilde{y}_t - \widehat{\phi}_1 \widetilde{y}_{t-1} - \widehat{\phi}_2 \widetilde{y}_{t-2} - \ldots - \widehat{\phi}_p \widetilde{y}_{t-p})^2 \tag{8.51}$$

where $\widetilde{y}_{t-i} = y_{t-i} - \widehat{\mu}$, $i = 0, 1, \ldots, p$. It is known that $\widehat{\sigma}_u^2$ is asymptotically uncorrelated with $\widehat{\lambda}$ and has asymptotic variance

$$\text{AsVar}(\widehat{\sigma}_u^2) = \text{E}(u^4) - \sigma_u^4. \tag{8.52}$$

In the case of normally distributed u_t, this expression reduces to $\text{AsVar}(\widehat{\sigma}_u^2) = 2\sigma_u^4$.

What is particularly striking about the asymptotic covariance matrix $\mathbf{V}_{LS}$ is that the $p \times p$ submatrix $\mathbf{V}_{LS,\phi}$ is a constant that depends only on $\boldsymbol{\phi}$, and not at all on the distribution F_u of the innovations u_t (assuming finite variance innovations!). This distribution-free character of the estimator led Whittle (1962) to use the term *robust* to describe the LS estimators of AR parameters. With hindsight, this was a rather misleading use of this term because the constant character of $\mathbf{V}_{LS,\phi}$ holds only under *perfectly* observed autoregressions; that is, with no AOs or ROs. Furthermore it turns out that the LS estimator will not be efficiency robust with respect to heavy-tailed deviations of the IOs from normality, as we discuss in Section 8.4. It should also be noted that the variance V_μ is not constant with respect to changes in the variance of the innovations, and $\text{AsVar}(\widehat{\sigma}_u^2)$ depends upon the fourth moment as well as the variance of the innovations.

8.3 Classical estimators for ARMA models

A time series y_t is called an autoregressive moving-average model of orders p and q, or ARMA(p, q) for short, if it obeys the stochastic difference equation

$$(y_t - \mu) - \phi_1(y_{t-1} - \mu) - \ldots - \phi_p(y_{t-p} - \mu) = -\theta_1 u_{t-1} - \ldots - \theta_q u_{t-q} + u_t, \tag{8.53}$$

where the i.i.d. innovations u_t have mean 0 and finite variance σ_u^2. This equation may be written in more compact form as

$$\phi(B)(y_t - \mu) = \theta(B)u_t \tag{8.54}$$

where B is the back-shift operator; that is, $By_t = y_{t-1}$, and $\phi(B)$ and $\theta(B)$ are polynomial back-shift operators given by

$$\phi(B) = 1 - \phi_1 B - \phi_2 B^2 - \ldots - \phi_p B^p \tag{8.55}$$

and

$$\theta(B) = 1 - \theta_1 B - \theta_2 B^2 - \ldots - \theta_q B^q. \tag{8.56}$$

The process is called *invertible* if y_t can be expressed as an infinite linear combination of the y_s for $s < t$ plus the innovations:

$$y_t = u_t + \sum_{i=1}^{\infty} \eta_i y_{t-i} + \gamma.$$

It will henceforth be assumed that the ARMA process is stationary and invertible. The first assumption requires that all roots of the polynomial $\phi(B)$ lie outside the unit circle and the second requires the same of the roots of $\theta(B)$.

Let $\lambda = (\boldsymbol{\phi}, \boldsymbol{\theta}, \mu) = (\phi_1, \phi_2, \ldots, \phi_p, \theta_1, \theta_2, \ldots, \theta_q, \mu)$ and consider the sum of squared residuals

$$\sum_{t=p+1}^{T} \hat{u}_t^2(\lambda) \tag{8.57}$$

where the residuals $\hat{u}_t(\lambda)$ may be computed recursively as

$$\begin{aligned} \hat{u}_t(\lambda) &= (y_t - \mu) - \phi_1(y_{t-1} - \mu) - \ldots - \phi_p(y_{t-p} - \mu) \\ &\quad + \theta_1 \hat{u}_{t-1}(\lambda) + \ldots + \theta_q \hat{u}_{t-q}(\lambda) \end{aligned} \tag{8.58}$$

with the initial conditions

$$\hat{u}_p(\lambda) = \hat{u}_{p-1}(\lambda) = \ldots = \hat{u}_{p-q+1}(\lambda) = 0. \tag{8.59}$$

Minimizing the sum of squared residuals (8.57) with respect to λ produces an LS estimator $\hat{\lambda}_{LS} = (\hat{\boldsymbol{\phi}}, \hat{\boldsymbol{\theta}}, \hat{\mu})$. When the innovations u_t have a normal distribution with mean 0 and finite variance σ_u^2, this LS estimator is a *conditional* maximum likelihood estimator, conditioned on $y_1, y_2, \ldots, y_p$ and on

$$u_{p-q+1} = u_{p-q+2} = \ldots = u_{p-1} = u_p = 0.$$

See for example Harvey and Philips (1979), where it is also shown how to compute the exact Gaussian maximum likelihood estimator of ARMA model parameters.

It is known that under the above assumptions for the ARMA(p, q) process, the LS estimator, as well as the conditional and exact maximum likelihood estimators, converge asymptotically to a multivariate normal distribution:

$$\sqrt{T}(\hat{\lambda}_{LS} - \lambda) \to_d \mathrm{N}_{p+q+1}(\mathbf{0}, \mathbf{V}_{LS}) \tag{8.60}$$

where

$$\mathbf{V}_{LS} = \mathbf{V}_{LS}(\boldsymbol{\phi}, \boldsymbol{\theta}, \sigma_u^2) = \begin{bmatrix} \mathbf{D}^{-1}(\boldsymbol{\phi}, \boldsymbol{\theta}) & \mathbf{0}' \\ \mathbf{0} & V_{LS,\mu} \end{bmatrix}, \tag{8.61}$$

with $V_{LS,\mu}$ the asymptotic variance of the location estimator $\hat{\mu}$ and $\mathbf{D}(\boldsymbol{\phi}, \boldsymbol{\theta})$ the $(p+q) \times (p+q)$ asymptotic covariance matrix of $(\hat{\boldsymbol{\phi}}, \hat{\boldsymbol{\theta}})$. Expressions for the elements of $\mathbf{D}(\boldsymbol{\phi}, \boldsymbol{\theta})$ are given in Section 8.15. As the notation indicates, $\mathbf{D}(\boldsymbol{\phi}, \boldsymbol{\theta})$ depends only on $\boldsymbol{\phi}$ and $\boldsymbol{\theta}$, and so the LS estimator $(\hat{\boldsymbol{\phi}}, \hat{\boldsymbol{\theta}})$ has the same distribution-free property as in the AR case, described at the end of Section 8.2.2. The expression for $V_{LS,\mu}$ is

$$V_{LS,\mu} = \frac{\sigma_u^2}{\xi^2} \tag{8.62}$$

with

$$\xi = -\frac{1 - \phi_1 - \ldots - \phi_p}{1 - \theta_1 - \ldots - \theta_q}, \tag{8.63}$$

which depends upon the variance of the innovations σ_u^2 as well as on $\boldsymbol{\phi}$ and $\boldsymbol{\theta}$.

The conditional MLE of the innovations variance σ_u^2 for an ARMA(p, q) model is given by

$$\hat{\sigma}_u^2 = \frac{1}{T-p} \sum_{t=p+1}^{T} \hat{u}_t^2(\hat{\lambda}). \tag{8.64}$$

The estimator $\hat{\sigma}_u^2$ is asymptotically uncorrelated with $\hat{\lambda}_{LS}$ and has the same asymptotic distribution as in the AR case, namely $\mathrm{AsVar}(\hat{\sigma}_u^2) = \mathrm{E}(u_t^4) - \sigma_u^4$.

Note that the asymptotic distribution of $\hat{\lambda}$ does not depend on the distribution of the innovations, and hence the precision of the estimators does not depend on their variance, as long as it is finite.

A natural estimator of the variance of the estimator $\hat{\mu}$ is obtained by plugging parameter estimates into (8.62).

8.4 M-estimators of ARMA models

8.4.1 M-estimators and their asymptotic distribution

An M-estimator $\hat{\lambda}_M$ of the parameter vector λ for an ARMA(p, q) model is obtained by minimizing

$$\sum_{t=p+1}^{T} \rho\left(\frac{\hat{u}_t(\lambda)}{\hat{\sigma}}\right), \tag{8.65}$$

where ρ is a ρ-function already used for regression in (5.7). The residuals $\widehat{u}_t(\lambda)$ are defined as in the case of the LS estimator, and $\widehat{\sigma}$ is a robust scale estimator that is obtained either simultaneously with λ (e.g., as an M-scale of the $\widehat{u}_t$s as in Section 2.7.2) or previously as with MM-estimators in Section 5.5). We assume that when $T \to \infty$, $\widehat{\sigma}$ converges in probability to a value σ that is a scale parameter of the innovations. It is also assumed that σ is standardized so that if the innovations are normal, σ coincides with the standard deviation σ_u of u_t, as explained for the location case at the end of Section 2.5.

Let $\psi = \rho'$ and assume that $\widehat{\sigma}$ has a limit in probability σ and

$$\mathrm{E}\psi\left(\frac{u_t}{\sigma}\right) = 0, \tag{8.66}$$

Note that this condition is analogous to (4.41) used in regression. Under the assumptions concerning the ARMA process made in Section 8.3 and under reasonable regularity conditions, which include that $\sigma_u^2 = \mathrm{Var}(u_t) < \infty$, the M-estimator $\widehat{\lambda}_M$ has an asymptotic normal distribution given by

$$\sqrt{T}(\widehat{\lambda}_M - \lambda) \to_d \mathrm{N}_{p+q+1}(\mathbf{0}, \mathbf{V}_M), \tag{8.67}$$

with

$$\mathbf{V}_M = \mathbf{V}_M(\boldsymbol{\phi}, \boldsymbol{\theta}, \sigma^2) = a\mathbf{V}_{LS} \tag{8.68}$$

where a depends on the distribution F of the u_ts:

$$a = a(\psi, F) = \frac{\sigma^2 \mathrm{E}\psi(u_t/\sigma)^2}{\sigma_u^2(\mathrm{E}\psi'(u_t/\sigma))^2}. \tag{8.69}$$

A heuristic proof is given in Section 8.15. In the normal case, $\sigma = \sigma_u$ and a coincides with the reciprocal of the efficiency of a location or regression M-estimator (see (4.45)).

In the case that $\rho(t) = -\log f(t)$, where f is the density of the innovations, the M-estimator $\widehat{\lambda}_M$ is a conditional MLE, and in this case the M-estimator is asymptotically efficient.

8.4.2 The behavior of M-estimators in AR processes with additive outliers

We already know from the discussion in Sections 8.1.1 and 8.1.3 that LS estimators of AR models are not robust in the presence of AOs or ROs. Such outliers cause both bias and inflated variability of LS estimators.

The LS estimator (8.19) proceeds as an ordinary regression, where y_t is regressed on the "predictors" $y_{t-1}, \ldots, y_{t-p}$. Similarly, any robust regression estimator based on the minimization of a function of the residuals can be applied to the AR model, in particular the S-, M- and MM-estimators defined in Chapter 5. In order to obtain some degree of robustness, it is necessary, just as in the treatment of ordinary regression in that chapter, that ρ be bounded, and in addition a suitable algorithm must be used to help ensure a good local minimum.

This approach has the advantage that existing software for regression can be readily used. However, it has the drawback that if the observations y_t are actually an AR(p) process contaminated with an AO or RO, the robustness of the estimators decreases with increasing p. The reason for this is that in the estimation of AR(p) parameters, the observation y_t is used in computing the $p + 1$ residuals $\hat{u}_t(\gamma, \boldsymbol{\phi}), \hat{u}_{t+1}(\gamma, \boldsymbol{\phi}), \ldots, \hat{u}_{t+p}(\gamma, \boldsymbol{\phi})$. Each time an outlier appears in the series it may spoil $p + 1$ residuals. In an informal way, we can say that the BP of an M-estimator is not larger than $0.5/(p + 1)$. Correspondingly, the bias due to an AO can be quite high and one expects only a limited degree of robustness.

Example 8.2 *Simulated AR(3) data with AO. Tables for this example are obtained with script* ***ar3.R****.*

To demonstrate the effect of contamination on these estimators, we generated $T = 200$ observations x_t from a stationary normal AR(3) model with $\sigma_u = 1$, $\gamma = 0$ and $\boldsymbol{\phi} = (8/6, -5/6, 1/6)'$. We then modified k evenly spaced observations by adding four to each, for $k = 10$ and 20. Table 8.1 shows the results for LS and for the MM regression estimator with bisquare function and efficiency 0.85 (script **AR3**).

It is seen that the LS estimator is much affected by 10 outliers. The MM-estimator is similar to the LS estimator when there are no outliers. It is less biased and so better than LS when there are 10 outliers, but it is highly affected by them when there are 20 outliers. The reason is that in this case the proportion of outliers is 20/200 = 0.1, which is near the heuristic BP value of $0.125 = 0.5/(p + 1)$, as discussed in Section 8.4.2.

8.4.3 The behavior of LS and M-estimators for ARMA processes with infinite innovation variance

The asymptotic behavior of the LS and M-estimators for ARMA processes has been discussed in the previous sections under the assumption that the innovations u_t have

Table 8.1 LS and MM-estimators of the parameters of a AR(3) simulated process

Estimator	#(outliers)	ϕ_1	ϕ_2	ϕ_3	γ	σ_u
LS	0	1.41	−0.92	0.21	0.00	0.99
	10	0.74	−0.09	−0.14	0.10	1.68
	20	0.58	−0.01	−0.16	0.24	2.02
MM	0	1.39	−0.93	0.23	−0.02	0.91
	10	1.12	−0.51	0.04	−0.03	1.16
	20	0.56	0.02	−0.04	−0.19	1.61
	True values	1.333	−0.833	0.166	0.00	1.00

finite variance. When this is not true, it may be surprising to know that under certain conditions the LS estimator not only is still consistent, but also converges to the true value at a faster rate than it would under finite innovation variance, with the consistency rate depending on the rate at which $P(|u_t| > k)$ tends to zero as $k \to \infty$.

For the case of an M-estimator with bounded ψ, and assuming that a good robust scale estimator $\hat{\sigma}$ is used, a heavy-tailed f can lead to ultra-precise estimation of the ARMA parameters $(\boldsymbol{\phi}, \boldsymbol{\theta})$ (but not of μ), in the sense that $\sqrt{T}(\hat{\boldsymbol{\phi}} - \boldsymbol{\phi}) \to_p \mathbf{0}$ and $\sqrt{T}(\hat{\boldsymbol{\theta}} - \boldsymbol{\theta}) \to_p \mathbf{0}$. This fact can be understood by noting that if u_t has a heavy-tailed distribution, such as the Cauchy distribution, then the expectations in (8.69) and σ are finite, while σ_u is infinite.

To make this clear, consider fitting an AR(1) model. Estimating ϕ is equivalent to fitting a straight line to the lag-1 scatterplot of y_t against y_{t-1}. Each IO appears twice in the scatterplot: as y_{t-1} and as y_t. In the first case it is a "good" leverage point, and in the second it is an outlier. Both LS and M-estimators take advantage of the leverage point. But the LS estimator is affected by the outlier, while the M-estimator is not.

The main LS results were derived by Kanter and Steiger (1974), Yohai and Maronna (1977), Knight (1987) and Hannan and Kanter (1977) for AR processes, and by Mikosch *et al.* (1995), Davis (1996) and Rachev and Mittnik (2000) in the ARMA case. Results for monotone M-estimators were obtained by Davis *et al.* (1992) and Davis (1996).

The challenges of establishing results in time series with infinite-variance innovations has been of great interest to academics and has resulted in many papers, particularly in the econometrics and finance literature. See, for example, applications to unit root tests (Samarakoon and Knight, 2005), and references therein, and applications to GARCH models (Rachev and Mittnik, 2000), one of the most interesting of which is the application to option pricing by Menn and Rachev (2005).

8.5 Generalized M-estimators

One approach to curb the effect of "bad leverage points" due to outliers is to modify M-estimators in a way similar to ordinary regression. Note first that the estimating equation of an M-estimator, obtained by differentiating the objective function with respect to $(\hat{\gamma}, \hat{\boldsymbol{\phi}})$, is

$$\sum_{t=p+1}^{T} \mathbf{z}_{t-1} \psi \left(\frac{\hat{u}_t}{\hat{\sigma}_u} \right) = 0 \tag{8.70}$$

where $\psi = \rho'$ is bounded and $\mathbf{z}_t = (1, y_t, y_{t-1}, \ldots, y_{t-p+1})'$.

One way to improve the robustness of the estimators is to modify (8.70) by bounding the influence of outliers in $\mathbf{z}_{t-1}$ as well as in the residuals $\hat{u}_t(\hat{\gamma}, \hat{\boldsymbol{\phi}})$. This results in the class of *generalized M-estimators* (GM-estimators), similar to those defined for regression in Section 5.11.1. A GM-estimator $(\hat{\gamma}, \hat{\boldsymbol{\phi}})$ is obtained by solving

$$\sum_{t=p+1}^{T} \eta\left(d_T(\mathbf{y}_{t-1}), \frac{\hat{u}_t(\hat{\gamma}, \hat{\boldsymbol{\phi}})}{\hat{\sigma}_u}\right)\mathbf{z}_{t-1} = 0 \tag{8.71}$$

where the function $\eta(.,.)$ is bounded and continuous in both arguments (say, of Mallows or Schweppe type, as defined in Section 5.11.1) and $\hat{\sigma}$ is obtained from a simultaneous M-equation of the form

$$\frac{1}{n}\sum_{i=1}^{n} \rho\left(\frac{\hat{u}_t(\hat{\gamma}, \hat{\boldsymbol{\phi}})}{\hat{\sigma}_u}\right) = \delta. \tag{8.72}$$

Here

$$d_T(\mathbf{y}_{t-1}) = \frac{1}{p}\mathbf{y}_{t-1}'\hat{\mathbf{C}}^{-1}\mathbf{y}_{t-1}, \tag{8.73}$$

with $\hat{\mathbf{C}}$ an estimator of the $p \times p$ covariance matrix $\mathbf{C}$ of $\mathbf{y}_{t-1} = (y_{t-1}, y_{t-2}, \ldots, y_{t-p})'$.

In the remark above (8.50), it was pointed out that the asymptotic distribution of LS estimators for AR models coincides with that of LS estimators for the regression model (8.21). The same can be said of GM-estimators.

GM-estimators for AR models were introduced by Denby and Martin (1979) and Martin (1980, 1981), who called them bounded influence autoregressive (BIFAR) estimators. Bustos (1982) showed that GM-estimators for AR(p) models are asymptotically normal, with covariance matrix given by the analog of the regression case (5.90). Künsch (1984) derived Hampel-optimal GM-estimators.

There are two main possibilities for $\hat{\mathbf{C}}$. The first is to use the representation $\mathbf{C} = \sigma_u^2\mathbf{D}(\boldsymbol{\phi})$ given by the matrix $\mathbf{D}$ in Section 8.15, where $\boldsymbol{\phi}$ is the parameter vector of the pth order autoregression, and put $\hat{\mathbf{C}} = \sigma_u^2\mathbf{D}(\hat{\boldsymbol{\phi}})$ in (8.73). Then $\hat{\boldsymbol{\phi}}$ appears twice in (8.71): in d_T and in $\hat{u}_t$. This is a natural approach when fitting a single autoregression of given order p.

The second possibility is convenient in the commonly occurring situation where one fits a sequence of autoregressions of increasing order, with a view toward determining a "best" order p_{opt}.

Let $\phi_{k,1}, \ldots, \phi_{k,k}$ be the coefficients of the " best-fitting" autoregression of order k, given in Section 8.2.1. The autocorrelations $\rho(1), \ldots, \rho(p-1)$ depend only on $\phi_{p-1,1}, \ldots, \phi_{p-1,p-1}$ and can be obtained from the Yule–Walker equations by solving a linear system. Therefore the correlation matrix $\mathbf{R}$ of $\mathbf{y}_{t-1}$ also depends only on $\phi_{p-1,1}, \ldots, \phi_{p-1,p-1}$. We also have that

$$\mathbf{C} = C(0)\mathbf{R}.$$

Then we can estimate $\phi_{p,1}, \ldots, \phi_{p,p}$ recursively as follows. Suppose that we have already computed estimators $\hat{\phi}_{p-1,1}, \ldots, \hat{\phi}_{p-1,p-1}$. Then, we estimate $\phi_{p,1}, \ldots, \phi_{p,p}$ by solving (8.71) and (8.72) with $\hat{\mathbf{C}} = \hat{C}(0)\hat{\mathbf{R}}$, where $\hat{\gamma}(0)$ is a robust estimator of

Table 8.2 GM-estimators of the parameters of AR(3) simulated process

#(outliers)	ϕ_1	ϕ_2	ϕ_3	γ	σ_u
0	1.31	−0.79	0.10	0.11	0.97
10	1.15	−0.52	−0.03	0.17	1.06
20	0.74	−0.16	−0.09	0.27	1.46
True values	1.333	−0.833	0.166	0.00	1.00

the variance of the y_ts (say, the square of the MADN) and $\widehat{\mathbf{R}}$ is computed from the Yule–Walker equations using $\widehat{\phi}_{p-1,1}, \ldots, \widehat{\phi}_{p-1,p-1}$.

Table 8.2 shows the results of applying a Mallows-type GM-estimator to the data of Example 8.2 (script **AR3**). It is seen that the performance of the GM-estimator is no better than that of the MM-estimator shown in Table 8.1.

8.6 Robust AR estimation using robust filters

In this section we assume that the observations process y_t has the AO form $y_t = x_t + v_t$, with x_t an AR(p) process as given in (8.14), with parameters $\lambda = (\phi_1, \phi_2, \ldots, \phi_p, \gamma)'$. v_t is independent of x_t. An attractive approach is to define robust estimators by minimizing a robust scale of the prediction residuals, as with regression S-estimators in Section 5.4.1. It turns out that this approach is not sufficiently robust. A more robust method is obtained by minimizing a robust scale of prediction residuals obtained with a robust filter that curbs the effect of outliers. We begin by explaining why the simple approach of minimizing a robust scale of the prediction residuals is not adequate. Most of the remainder of this section is devoted to describing the robust filtering method, the scale minimization approach using prediction residuals from a robust filter, and the computational details for the whole procedure. The section concludes with an application example and an extension of the method to integrated AR(p) models.

8.6.1 Naive minimum robust scale autoregression estimators

In this section we deal with the robust estimation of the AR parameters by minimizing a robust scale estimator $\widehat{\sigma}$ of prediction residuals. Let y_t, $1 \le t \le T$, be observations corresponding to an AO model $y_t = x_t + v_t$, where x_t is an AR(p) process. For any $\lambda = (\phi_1, \phi_2, \ldots, \phi_p, \mu)' \in R^{p+1}$, define the residual vector as

$$\widehat{\mathbf{u}}(\lambda) = (\widehat{u}_{p+1}(\lambda), \ldots, \widehat{u}_T(\lambda))',$$

where

$$\hat{u}_t(\lambda) = (y_t - \mu) - \phi_1(y_{t-1} - \mu) - \ldots - \phi_p(y_{t-p} - \mu). \quad (8.74)$$

Given a scale estimator $\hat{\sigma}$, an estimator of λ can be defined by

$$\hat{\lambda} = \arg\min_{\lambda} \hat{\sigma}(\hat{\mathbf{u}}(\lambda)). \quad (8.75)$$

If $\hat{\sigma}$ is a high-BP M-estimator of scale we would have the AR analog of regression S-estimators. Boente *et al.* (1987) generalized the notion of qualitative robustness (Section 3.7) to time series, and proved that S-estimators for autoregression are qualitatively robust and have the same efficiency as regression S-estimators. As happens in the regression case (see (5.24), estimators based on the minimization of an M-scale are M-estimators, where the scale is the minimum scale, and therefore all the asymptotic theory of M-estimators applies under suitable regularity conditions.

If $\hat{\sigma}$ is a τ-estimator of scale (Section 5.4.3), it can be shown that, as in the regression case, the resulting AR estimators have a higher normal efficiency than that corresponding to an M-scale.

For the reasons given in Section 8.4.2, any estimator based on the prediction residuals has a BP not larger than $0.5/(p+1)$ for AR(p) models. Since invertible MA and ARMA models have infinite AR representations, the BP of estimators based on the prediction residuals will be zero for such models.

Section 8.6.2 shows how to obtain an improved S-estimator through the use of robust filtering.

8.6.2 The robust filter algorithm

Let y_t be an AO process (8.5), where x_t is a stationary AR(p) process with mean 0, and $\{v_t\}$ are i.i.d. and independent of $\{x_t\}$ with distribution (8.9). To avoid the propagation of outliers to many residuals, as described above, we shall replace the prediction residuals $\hat{u}_t(\lambda)$ in (8.74) by the *robust prediction residuals*

$$\tilde{u}_t(\lambda) = (y_t - \mu) - \phi_1(\hat{x}_{t-1|t-1} - \mu) - \ldots - \phi_p(\hat{x}_{t-p|t-1} - \mu) \quad (8.76)$$

obtained by replacing the AO observations $y_{t-i}, i = 1, \ldots, p$, in (8.74) by the *robust filtered values* $\hat{x}_{t-i|t-1} = \hat{x}_{t-i|t-1}(\lambda), i = 1, \ldots, p$, which are approximations to the values $\mathrm{E}(x_{t-i}|y_1, \ldots, y_t)$.

These approximated conditional expectations were derived by Masreliez (1975) and are obtained by means of a *robust filter*. To describe this filter we need the so-called *state-space representation* of the x_ts (see, for example, Brockwell and Davis, 1991), which for an AR(p) model is

$$\mathbf{x}_t = \boldsymbol{\mu} + \Phi(\mathbf{x}_{t-1} - \boldsymbol{\mu}) + \mathbf{d}u_t \quad (8.77)$$

where $\mathbf{x}_t = (x_t, x_{t-1}, \ldots, x_{t-p+1})'$ is called the *state vector*, $\mathbf{d}$ is defined by

$$\mathbf{d} = (1, 0, \ldots, 0)', \quad \boldsymbol{\mu} = (\mu, \ldots, \mu)', \quad (8.78)$$

and $\mathbf{\Phi}$ is the *state-transition matrix* given by

$$\mathbf{\Phi} = \begin{bmatrix} \phi_1 \ldots \phi_{p-1} & \phi_p \\ \mathbf{I}_{p-1} & \mathbf{0}_{p-1} \end{bmatrix}. \tag{8.79}$$

Here $\mathbf{I}_k$ is the $k \times k$ identity matrix and $\mathbf{0}_k$ the zero vector in R^k.

The following recursions compute robust filtered vectors $\widehat{\mathbf{x}}_{t|t}$, which are approximations of E($\mathbf{x}_t | \, y_1, y_2, \ldots, y_t$), and robust one-step-ahead predictions $\widehat{\mathbf{x}}_{t|t-1}$, which are approximations of E($\mathbf{x}_t | y_1, y_2, \ldots, y_{t-1}$).At each time $t-1$, the robust prediction vectors $\widehat{\mathbf{x}}_{t|t-1}$ are computed from the robustly filtered vectors $\widehat{\mathbf{x}}_{t-1|t-1}$ as

$$\hat{\mathbf{x}}_{t|t-1} = \boldsymbol{\mu} + \mathbf{\Phi}(\hat{\mathbf{x}}_{t-1|t-1} - \boldsymbol{\mu}). \tag{8.80}$$

Then the prediction vector $\widehat{x}_{t|t-1}(\lambda)$ and the AO observation y_t are used to compute the residual $\widetilde{u}_t(\lambda)$ and $\widehat{\mathbf{x}}_{t|t}$ using the recursions

$$\widetilde{u}_t(\lambda) = (y_t - \mu) - \boldsymbol{\phi}'(\hat{\mathbf{x}}_{t-1|t-1} - \boldsymbol{\mu}) \tag{8.81}$$

and

$$\widehat{\mathbf{x}}_{t|t} = \widehat{\mathbf{x}}_{t|t-1} + \frac{1}{s_t}\mathbf{m}_t \psi\left(\frac{\widetilde{u}_t(\lambda)}{s_t}\right), \tag{8.82}$$

where s_t is an estimator of the scale of the prediction residual $\widetilde{u}_t$ and $\mathbf{m}_t$ is a vector. Recursions for s_t and $\mathbf{m}_t$ are provided in Section 8.16. Here ψ is a bounded ψ-function that for some constants $a < b$ satisfies

$$\psi(u) = \begin{cases} u & \text{if} \quad |u| \le a \\ 0 & \text{if} \quad |u| > b. \end{cases} \tag{8.83}$$

It turns out that the first element of $\mathbf{m}_t$ is s_t^2, and hence the first coordinate of the vector recursion (8.82) gives the scalar version of the filter. Hence if $\widehat{\mathbf{x}}_{t|t} = (\widehat{x}_{t|t}, \ldots, \widehat{x}_{t-p+1|t})'$ and $\widehat{\mathbf{x}}_{t|t-1} = (\widehat{x}_{t|t-1}, \ldots, \widehat{x}_{t-p+1|t-1})'$, we have

$$\widehat{x}_{t|t} = \widehat{x}_{t|t-1} + s_t \psi\left(\frac{\widetilde{u}_t(\lambda)}{s_t}\right). \tag{8.84}$$

It follows that

$$\widehat{x}_{t|t} = \widehat{x}_{t|t-1} \quad \text{if} \quad |\widetilde{u}_t| > bs_t \tag{8.85}$$

and

$$\widehat{x}_{t|t} = y_t \quad \text{if} \quad |\widetilde{u}_t| \le as_t. \tag{8.86}$$

Equation (8.85) shows that the robust filter rejects observations with scaled absolute robust prediction residuals $|\widetilde{u}_t/s_t| \ge b$, and replaces them with predicted values based on previously filtered data. Equation (8.86) shows that observations with $|\widetilde{u}_t/s_t| \le a$ remain unaltered. Observations for which $|\widetilde{u}_t/s_t| \in (a, b)$ are modified depending on how close the values are to a or b. Consequently, the action of the robust filter is to "clean" the data of outliers by replacing them with predictions (one-sided interpolates) while leaving most of the remaining data unaltered. As such, the robust filter might well be called an "outlier-cleaner".

The above robust filter recursions have the same general form as the class of approximate conditional mean robust filters introduced by Masreliez (1975). See also Masreliez and Martin (1977), Kleiner *et al.* (1979), Martin and Thomson (1982), Martin *et al.* (1983), Martin and Yohai (1985), Brandt and Künsch (1988) and Meinhold and Singpurwalla (1989). In order that the filter $\widehat{x}_{t|t}$ be robust in a well-defined sense, it is sufficient that the functions ψ and $\psi(u)/u$ be bounded and continuous (Martin and Su, 1985).

The robust filtering algorithm, which we have just described for the case of a true AR(p) process x_t, can also be used for data cleaning and prediction based on cleaned data for a memory-l predictor, $1 \leq l < p$. Such use of the robust filter algorithm is central to the robustified Durbin–Levinson algorithm that we will describe shortly.

Remark 1: Note that the filter as described modifies all observations that are far enough from their predicted values, including innovations outliers. But this may damage the output of the filter, since altering one IO spoils the prediction of the ensuing values. The following modification of the above procedure deals with this problem. When a sufficiently large number of consecutive observations have been corrected – that is, $\widehat{x}_{t|t} \neq y_t$ for $t = t_0, ..., t_0 + h$ – the procedure goes back to t_0 and redefines $\widehat{x}_{t_0|t_0} = y_t$ and then goes on with the recursions.

Remark 2: Note that the robust filter algorithm replaces large outliers with predicted values based on the past, and as such produces "one-sided" interpolated values. One can improve the quality of the outlier treatment by using a two-sided interpolation at outlier positions by means of a robust smoother algorithm. One such algorithm is described by Martin (1979), who derives the robust smoother as an approximate conditional mean smoother analogous to Masreliez's approximate conditional mean filter.

8.6.3 Minimum robust scale estimators based on robust filtering

If $\widehat{\sigma}$ is a robust scale estimator, an estimator based on robust filtering may be defined as

$$\widehat{\lambda} = \arg\min_{\lambda} \widehat{\sigma}(\widetilde{u}(\lambda)) \tag{8.87}$$

where $\widetilde{u}\,(\lambda) = (\widetilde{u}_{p+1}(\lambda), \ldots, \widetilde{u}_T(\lambda))'$ is the vector of robust prediction residuals $\widetilde{u}_t$ given by (8.76). The use of these residuals in place of the raw prediction residuals (8.74) prevents the smearing effect of isolated outliers, and therefore will result in an estimator that is more robust than M-estimators or estimators based on a scale of the raw residuals $\widehat{u}_t$.

One problem with this approach is that the objective function $\widehat{\sigma}(\widetilde{u}\,(\lambda))$ in (8.87) typically has multiple local minima, making it difficult to find a global minimum.

Fortunately, there is a computational approach based on a different parameterization in which the optimization is performed one parameter at a time. This procedure amounts to a robustified Durbin–Levinson algorithm, as described in Section 8.6.4.

8.6.4 A robust Durbin–Levinson algorithm

There are two reasons why the Durbin–Levinson procedure is not robust:

- The quadratic loss function in (8.35) is unbounded.
- The residuals $\widehat{u}_{t,m}(\boldsymbol{\phi})$ defined in (8.45) are subject to an outlier "smearing" effect: if y_t is an isolated outlier, it spoils the $m+1$ residuals $\widehat{u}_{t,m}(\phi)$,$\widehat{u}_{t+1,m}(\phi), \ldots$, $\widehat{u}_{t+m,m}(\phi)$.

We now describe a modification of the standard sample-based Durbin–Levinson method that eliminates the preceding two sources of nonrobustness. The observations y_t are assumed to have been previously robustly centered by the subtraction of the median or another robust location estimator.

A robust version of (8.31) will be obtained in a recursive way, analogous to the classical Durbin–Levinson algorithm, as follows.

Let $\widetilde{\phi}_{m-1,1}, \ldots, \widetilde{\phi}_{m-1,m-1}$ be robust estimators of the coefficients $\phi_{m-1,1}, \ldots,$ $\phi_{m-1,m-1}$ of the memory-$(m-1)$ linear predictor. If we knew that $\phi_{m,m} = \zeta$, then according to (8.38), we could estimate the memory-m predictor coefficients as

$$\widetilde{\phi}_{m,i}(\zeta) = \widetilde{\phi}_{m-1,i} - \zeta \widetilde{\phi}_{\,m-1,m-i}, \quad i = 1, \ldots, m-1. \tag{8.88}$$

Therefore it would only remain to estimate ζ.

The robust memory-m prediction residuals $\widetilde{u}_{t,m}(\zeta)$ may be written in the form

$$\begin{aligned} \widetilde{u}_{t,m}(\zeta) = y_t - \widetilde{\phi}_{m,1}(\zeta)\, \widehat{x}^{(m)}_{t-1|t-1}(\zeta) - \ldots - \widetilde{\phi}_{m,m-1}(\zeta)\, \widehat{x}^{(m)}_{t-m+1|t-1}(\zeta) \\ - \zeta\, \widehat{x}^{(m)}_{t-m|t-1}(\zeta), \end{aligned} \tag{8.89}$$

where $\widehat{x}^{(m)}_{t-i|t-1}(\zeta)$, $i = 1, \ldots, m$, are the components of the robust state vector estimator $\widehat{\mathbf{x}}^{(m)}_{t-1|t-1}$ obtained using the robust filter (8.82), corresponding to an order-m autoregression with parameters

$$(\widetilde{\phi}_{m,1}(\zeta), \widetilde{\phi}_{m,2}(\zeta), \ldots, \widetilde{\phi}_{m,m-1}(\zeta), \zeta). \tag{8.90}$$

Observe that $\widetilde{u}_{t,m}(\zeta)$ is defined as $\widehat{u}_{t,m}(\zeta)$ in (8.45), except for the replacement of $y_{t-1}, \ldots, y_{t-m}$ with the robustly filtered values $\widehat{x}^{(m)}_{t-1|t-1}(\zeta), \ldots, \widehat{x}^{(m)}_{t-m|t-1}(\zeta)$. Now an outlier y_t may spoil only a single residual $\widetilde{u}_{t,m}(\zeta)$, rather than $p+1$ residuals, as in the case of the usual AR(p) residuals in (8.74).

The standard Durbin–Levinson algorithm computes $\widehat{\phi}_{m,m}$ by minimizing the sum of squares (8.44), which in the present context is equivalent to minimizing the sample standard deviation of the $\widetilde{u}_{t,m}(\zeta)$s defined by (8.89). Since the $\widetilde{u}_{t,m}(\zeta)$s may have outliers in the y_t term and the sample standard deviation is not robust, we replace it with a highly robust scale estimator $\widehat{\sigma} = \widehat{\sigma}(\widetilde{u}_{m+1,m}(\zeta), \ldots, \widetilde{u}_{T,m}(\zeta))$. We have thus eliminated the two sources of non-robustness of the standard Durbin–Levinson algorithm. Finally, the robust partial autocorrelation coefficient estimators $\widehat{\phi}_{m,m}$, $m = 1, 2, \ldots, p$, are obtained sequentially by solving

$$\widehat{\phi}_{m,m} = \arg\min_{\zeta} \widehat{\sigma}(\widetilde{u}_{m+1,m}(\zeta), \ldots, \widetilde{u}_{T,m}(\zeta)), \tag{8.91}$$

where for each m the values $\widehat{\phi}_{m,i}$, $i = 1, \ldots, m-1$, are obtained from (8.38). This minimization can be performed by a grid search on $(-1, 1)$.

The first step of the procedure is to compute a robust estimator $\widetilde{\phi}_{1,1}$ of $\phi_{1,1}$ by means of a robust version of (8.43), namely (8.91) with $m = 1$, where

$$\widetilde{u}_{t,1}(\zeta) = y_t - \zeta \widehat{x}_{t-1|t}(\zeta).$$

8.6.5 Choice of scale for the robust Durbin–Levinson procedure

One possibility for the choice of a robust scale in (8.91) is to use an M-scale with a BP of 0.5, in which case the resulting estimator is an S-estimator of autoregression using robustly filtered values. However, it was pointed out in Section 5.4.1 that Hössjer (1992) proved that an S-estimator of regression with a BP of 0.5 cannot have a large-sample efficiency greater than 0.33 when the errors have a normal distribution. This fact provided the motivation for using τ-estimators of regression, as defined in equations (5.26)–(5.28) of Section 5.4.3. These estimators can give high efficiency, say 95%, when the errors have a normal distribution, while at the same time having a high BP of 0.5. The relative performance of a τ-estimator versus an S-estimator with regard to BP and normal efficiency is expected to carry over to a large extent to the present case of robust AR model fitting using robustly filtered observations. Thus we recommend that the robust scale estimator $\widehat{\sigma}$ in (8.91) be a τ-scale defined as in (5.26)–(5.28), but with residuals given by (8.88) and (8.89). The examples we show for robust fitting of AR, ARMA, ARIMA and REGARIMA models in the remainder of this chapter are all computed with an algorithm that uses a τ-scale applied to robustly filtered residuals. We shall call such estimators *filtered* τ*- (or F*τ*-) estimators*. These estimators were studied by Bianco *et al.* (1996).

Table 8.3 shows the results of applying a filtered τ (Fτ) estimator to the data of Example 8.2. It is seen that the impact of outliers is slight, and comparison with Tables 8.1 and 8.2 shows the performance of the Fτ-estimator to be superior to that of the MM- and GM-estimators.

8.6.6 Robust identification of AR order

The classical approach based on Akaike's information criterion (AIC; Akaike, 1973, 1974a), when applied to the choice of the order of AR models, leads to

Table 8.3 Fτ-estimators of the parameters of AR(3) simulated process

#(outliers)	ϕ_1	ϕ_2	ϕ_3	γ	σ_u
0	1.37	−0.89	0.22	−0.01	0.87
10	1.43	−0.92	0.19	0.01	0.97
20	1.37	−0.89	0.15	0.00	1.00
True values	1.333	−0.833	0.166	0.00	1.00

the minimization of

$$\text{AIC}_p = \log\left(\frac{1}{T-p}\sum_{t=p+1}^{T} \widehat{u}_t^2(\widehat{\lambda}_{p,\text{LS}})\right) + \frac{2p}{T-p},$$

where $\widehat{\lambda}_{p,\text{LS}}$ is the LS estimator corresponding to an AR(p) model, and $\widehat{u}_t$ are the respective residuals. The robust implementation of this criterion that is used here is based on the minimization of

$$\text{RAIC}_p = \log(\tau^2(\widetilde{u}_{p+1}(\widehat{\lambda}_{p,\text{rob}}),\ldots,\widetilde{u}_T(\widehat{\lambda}_{p,\text{rob}}))) + \frac{2p}{T-p},$$

where $\widetilde{u}_i(\widehat{\lambda}_{p,\text{rob}})$ are the filtered residuals corresponding to the Fτ-estimator and τ is the respective scale.

As with the RFPE criterion in (5.39), we believe that it would be better to multiply the penalty term $2p/(T-p)$ by a factor depending on the choice of the scale and on the distribution of the residuals. This area requires further research.

8.7 Robust model identification

Time series autocorrelations are often computed for exploratory purposes without reference to a parametric model. In addition, autocorrelations are generally computed along with partial autocorrelations for use in identification of ARMA and ARIMA models; see for example Brockwell and Davis (1991). We know already from Sections 8.1.1 and 8.1.3 that additive outliers can have considerable influence and cause bias and inflated variability in the case of a lag-1 correlation estimator, and Section 8.2 indicates that additive outliers can have a similar adverse impact on partial autocorrelation estimators. Thus there is a need for robust estimators of autocorrelations and partial autocorrelations in the presence of AOs or IOs.

While methods for robust estimation of autocorrelations and partial autocorrelations have been discussed in the literature (see, for example, Ma and Genton, 2000), our recommendation is to use one of the following two procedures, based on robust fitting of an AR model of order p^* using the robustly Fτ-scale estimator, and where p^* was selected using the robust AIC criterion described in Subsection 8.6.6:

- Procedure A: Compute classical autocorrelations and partial autocorrelations based on the robustly filtered values $\widehat{x}_{t|t}$ for the AR(p^*) model.
- Procedure B: Compute the theoretical autocorrelations and partial autocorrelations corresponding to the fitted model AR(p^*). This can be done using Yule-Walker equations (8.29)

$$\rho(k) = \sum_{i=1}^{p^*} \phi_i \rho(k-i) \quad (k \geq 1) \tag{8.92}$$

  for the values of the unknown $\rho(k)$, where the unknown ϕ_i are replaced by the estimators $\widehat{\phi}_i$. Note that the first $p^* - 1$ equations of the above set suffice to determine $\rho(k), k = 1, \ldots, p^* - 1$, and that $\rho(k)$ for $k \geq p^*$ are obtained recursively from (8.92). Once the autocorrelations have been computed, the partial autocorrelations can be computed using the Durbin–Levinson algorithm to estimate an AR(p) models with $1 \leq p \leq p^{ast}$.

Example 8.3 *Identification of simulated AR(2) AO model. Tables and figures for this example are obtained with script* ***identAR2.R****.*

Consider an AO model $y_t = x_t + v_t$ where x_t is a zero-mean Gaussian AR(2) model with parameters $\boldsymbol{\phi} = (\phi_1, \phi_2)' = (4/3, -5/6)'$ and innovations variance $\sigma_u^2 = 1$; and $v_t = z_t w_t$, where z_t and w_t are independent, P($z_t = \pm 1$) = 0.5, and w_t has the mixture distribution (8.6) with $\varepsilon = 0.1$, $\sigma_v = 1$ and $\mu_v = 4$. The figures for this example were obtained using script **identAR2.R**. Figure 8.7 shows two series x_t and y_t of length 200 generated in this way.

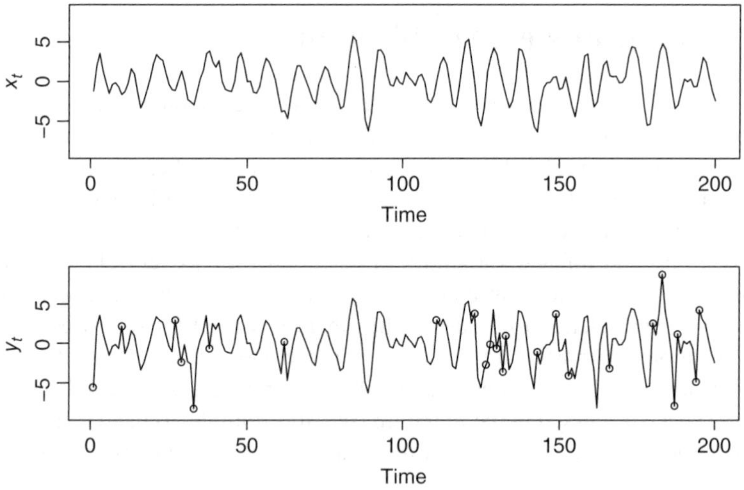

Figure 8.7 Above: Gaussian AR(2) series, below Gaussian AR(2) series with additive outliers

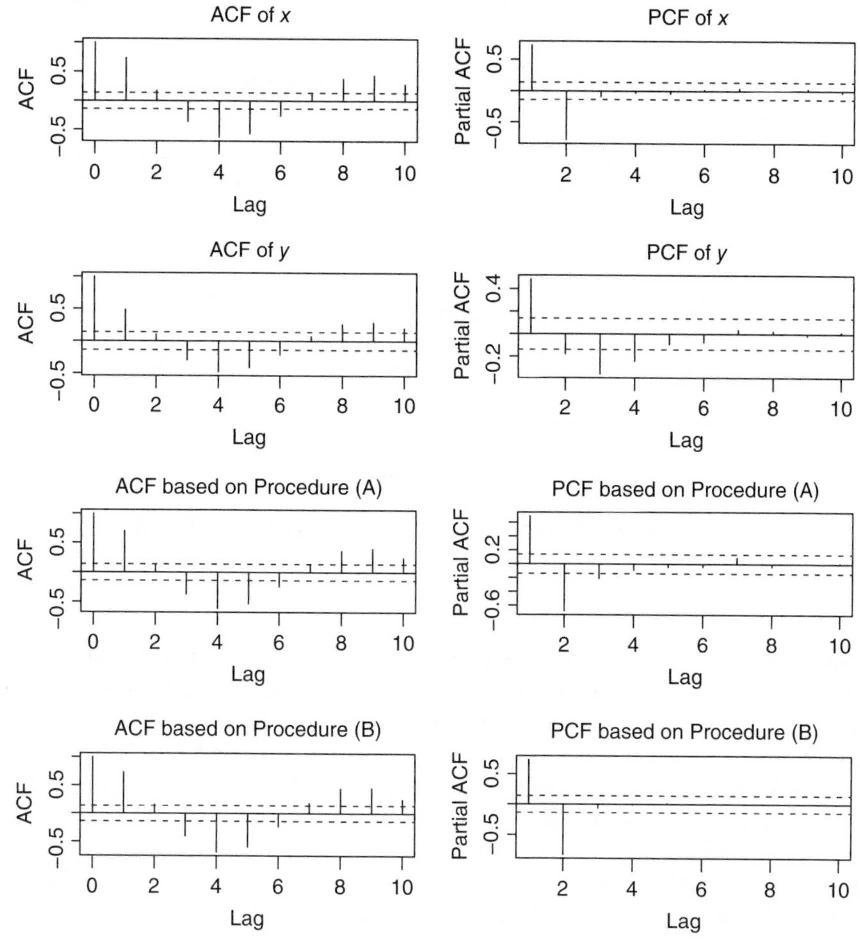

Figure 8.8 Estimated autocorrelations and partial autocorrelations for the AR(2) model

The first row of Figure 8.8 displays the autocorrelation function (ACF) and partial autocorrelation function (PCF) of x_t, the second those of y_t and the third and fourth rows the robust ACF and PCF corresponding to Procedures A and B respectively. We observe that while the ACF and PCF of x_t allow the correct identification of the AR(2) model, those of y_t do not. The robust ACF and PCF obtained according to the two procedures are similar to those of x_t and they also lead to correct model identification.

Example 8.4 *Identification of a simulated MA(1) AO model. Tables and figures for this example are obtained with script **identMA1.R**.*

Let us consider now an AO model $y_t = x_t + v_t$ where x_t is a Gaussian MA(1) process $x_t = u_t - \theta u_{t-1}$. It is easy to show that the lag-k autocorrelations $\rho(k)$ of x_t are zero,

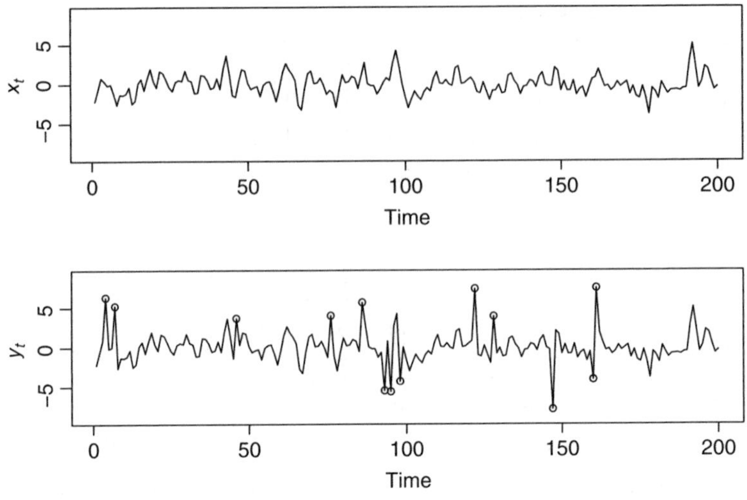

Figure 8.9 Top, Gaussian MA(1) series; bottom, Gaussian MA(1) series with additive outliers

except for $k = 1$ where $\rho(1) = -\theta/(1+\theta^2)$, and that $\rho(1) = \rho(1, \theta)$ is bounded in magnitude by $1/2$ for $-1 < \theta < 1$. Script **identMA1** generates the figures and other results for this example. We obtain by Monte Carlo simulation a series x_t of length 200 of an invertible Gaussian MA(1) process with $\theta = -0.8$ and $\sigma_u^2 = 1$. The series y_t with AO is generated as as in Example 8.3, except that now $\mu_v = 6$ instead of 4. Both series x_t and y_t are displayed in Figure 8.9.

Figure 8.10 displays the same four ACFs and PCFs as in Figure 8.8. The ACF and PCF of x_t and those obtained according to Procedures A and B identify a MA(1) model. The ACF and PCF of y_t lead instead to incorrect identification.

8.8 Robust ARMA model estimation using robust filters

In this section we assume that we observe $y_t = x_t + v_t$, where x_t follows an ARMA model given by equation (8.53). The parameters to be estimated are given by the vector $\boldsymbol{\lambda} = (\boldsymbol{\phi}, \boldsymbol{\theta}, \mu) = (\phi_1, \phi_2, \ldots, \phi_p, \theta_1, \theta_2, \ldots, \theta_q, \mu)$.

8.8.1 τ-estimators of ARMA models

In order to motivate the use of Fτ-estimators for fitting ARMA models, we first describe naive τ-estimators that do not involve the use of robust filters. Assume first there is no contamination; that is, $y_t = x_t$. For $t \geq 2$ call $\widehat{y}_{t|t-1}(\boldsymbol{\lambda})$ the optimal linear

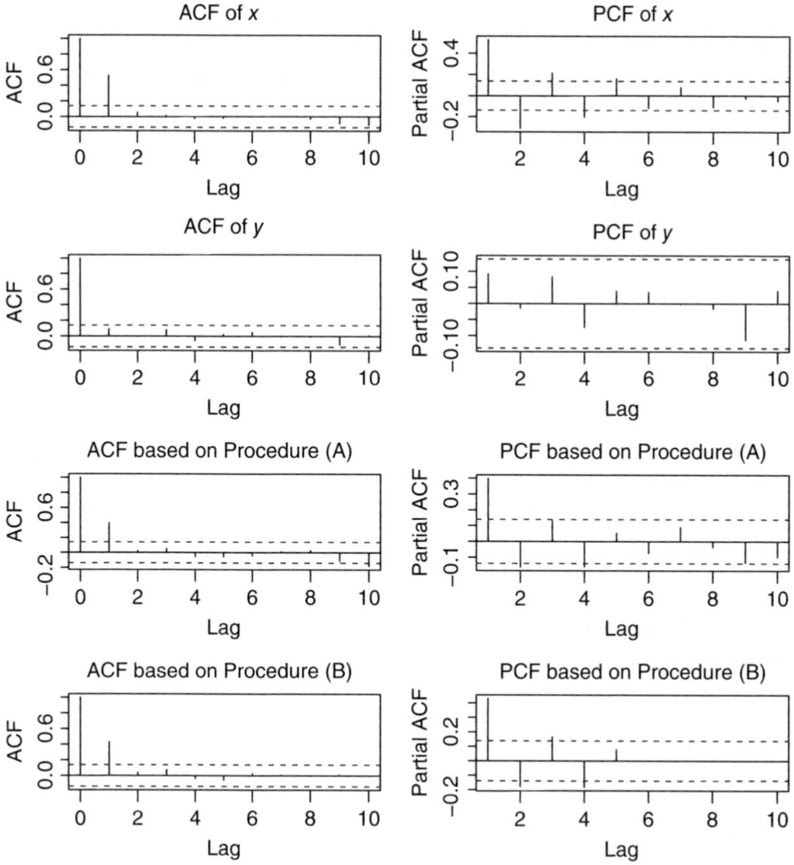

Figure 8.10 Estimated autocorrelations and partial autocorrelations for the MA(1) model

predictor of y_t based on $y_1, \ldots, y_{t-1}$ when the true parameter is λ, as described in Section 8.2.1. For $t = 1$ put $\widehat{y}_{t|t-1}(\lambda) = \mu = \mathrm{E}(y_t)$. Then if u_t are normal we also have

$$\widehat{y}_{t|t-1}(\lambda) = \mathrm{E}(y_t | y_1, \ldots, y_{t-1}), \ t > 1. \tag{8.93}$$

Define the prediction errors as

$$\widehat{u}_t(\lambda) = y_t - \widehat{y}_{t|t-1}(\lambda) \tag{8.94}$$

Note that these errors are not the same as $\widehat{u}_t(\lambda)$ defined by (8.58).

The variance of $\widehat{u}_t(\lambda)$, $\sigma_t^2(\lambda) = \mathrm{E}(y_t - \widehat{y}_{t|t-1}(\lambda))^2$ has the form

$$\sigma_t^2(\lambda) = a_t^2(\lambda)\sigma_u^2, \tag{8.95}$$

with $\lim_{t\to\infty} a_t^2(\lambda) = 1$ (see Brockwell and Davis, 1991). In the AR case we have $a = 1$ for $t \geq p + 1$.

Suppose that the innovations u_t have a $\mathrm{N}(0, \sigma_u^2)$ distribution. Let $L(y_1, \ldots, y_T, \lambda, \sigma_u)$ be the likelihood, and define

$$Q(\lambda) = -2 \max_{\sigma_u} \log L(y_1, \ldots, y_T, \lambda, \sigma_u). \tag{8.96}$$

Except for a constant, we have (see Brockwell and Davis, 1991)

$$Q(\lambda) = \sum_{t=1}^{T} \log a_t^2(\lambda) + T \log \left(\frac{1}{T} \sum_{t=1}^{T} \frac{\widehat{u}_t^2(\lambda)}{a_t^2(\lambda)} \right). \tag{8.97}$$

Then the MLE of λ is given by

$$\hat{\lambda} = \arg\min_{\lambda} Q(\lambda). \tag{8.98}$$

Observe that

$$\frac{1}{T} \sum_{t=1}^{T} \frac{\widehat{u}_t^2(\lambda)}{a_t^2(\lambda)}$$

is the square of an estimator of σ_u based on the values $\widehat{u}_t(\lambda)/a_t(\lambda)$, $t = 1, \ldots, T$. Then it seems natural to define a τ-estimator $\widehat{\lambda}$ of λ by minimizing

$$Q^*(\lambda) = \sum_{t=1}^{T} \log a_t^2(\lambda) + T \log \left(\tau^2 \left(\frac{\widehat{u}_1(\lambda)}{a_1(\lambda)}, \ldots, \frac{\widehat{u}_T(\lambda)}{a_T(\lambda)} \right) \right), \tag{8.99}$$

where, for any $\mathbf{u} = (u_1, \ldots, u_T)'$, a τ-scale estimator is defined by

$$\tau^2(\mathbf{u}) = s^2(\mathbf{u}) \sum_{t=1}^{T} \rho_2 \left(\frac{u_t}{s(\mathbf{u})} \right), \tag{8.100}$$

with $s(\mathbf{u})$ an M-scale estimator based on a bounded ρ-function ρ_1. See Section 5.4.3 for further details in the context of regression τ-estimators.

While regression τ-estimators can simultaneously have a high BP value of 0.5 and a high efficiency for the normal distribution, the τ-estimator $\widehat{\lambda}$ of λ has a BP of at most $0.5/(p + 1))$ in the AR(p) case, and is zero in the MA and ARMA cases, for the reasons given at the end of Section 8.6.1.

8.8.2 Robust filters for ARMA models

One way to achieve a positive (and hopefully reasonably high) BP for ARMA models with AO is to extend the AR robust filter method of Section 8.6.2 to ARMA models, by using a state-space representation of them.

The extension consists of modifying the state-space representation (8.77)–(8.79) as follows. Let x_t be an ARMA(p, q) process and $k = \max(p, q + 1)$. In

Section 8.17 we show that it is possible to define a k-dimensional state-space vector $\boldsymbol{\alpha}_t = (\alpha_{1,t}, \ldots, \alpha_{k,t})'$ with $\alpha_{1,t} = x_t - \mu$, so that the following representation holds:

$$\boldsymbol{\alpha}_t = \Phi\boldsymbol{\alpha}_{t-1} + \mathbf{d}u_t, \tag{8.101}$$

where

$$\mathbf{d} = (1, -\theta_1, \ldots, -\theta_{k-1})', \tag{8.102}$$

with $\theta_i = 0$ for $i > q$ in case $p > q$. The state-transition matrix Φ is now given by

$$\Phi = \begin{bmatrix} \boldsymbol{\phi}_{k-1} & \mathbf{I}_{k-1} \\ \phi_k & \mathbf{0}_{k-1} \end{bmatrix} \tag{8.103}$$

and where $\boldsymbol{\phi}_{k-1} = (\phi_1, \ldots, \phi_{k-1})$ and $\phi_i = 0$ for $i > p$.

Suppose now that the observations y_t follow the AO process (8.5). The Masreliez approximate robust filter can be derived in a way similar to the AR(p) case in Section 8.6.2. The filter yields approximations

$$\widehat{\boldsymbol{\alpha}}_{t|t} = (\widehat{\alpha}_{t,1|t}, \ldots, \widehat{\alpha}_{t,k|t}) \quad \text{and} \quad \widehat{\boldsymbol{\alpha}}_{t|t-1} = (\widehat{\alpha}_{t,1|t-1}, \ldots, \widehat{\alpha}_{t,k|t-1})$$

of $\mathrm{E}(\boldsymbol{\alpha}_t|y_1, \ldots, y_t)$ and $\mathrm{E}(\boldsymbol{\alpha}_t|y_1, \ldots, y_{t-1})$ respectively. Observe that

$$\widehat{x}_{t|t} = \widehat{x}_{t|t}(\boldsymbol{\lambda}) = \widehat{\boldsymbol{\alpha}}_{t,1|t}(\boldsymbol{\lambda}) + \mu$$

and

$$\widehat{x}_{t|t-1} = \widehat{x}_{t|t-1}(\boldsymbol{\lambda}) = \widehat{\alpha}_{t,1|t-1}(\boldsymbol{\lambda}) + \mu$$

approximate $\mathrm{E}(x_t|y_1, \ldots, y_t)$ and $\mathrm{E}(x_t|y_1, \ldots, y_{t-1})$, respectively.

The recursions to obtain $\widehat{\boldsymbol{\alpha}}_{t|t}$ and $\widehat{\boldsymbol{\alpha}}_{t|t-1}$ are as follows:

$$\begin{aligned} \widehat{\boldsymbol{\alpha}}_{t|t-1} &= \Phi\widehat{\boldsymbol{\alpha}}_{t-1|t-1}, \\ \widetilde{u}_t(\boldsymbol{\lambda}) &= y_t - \widehat{x}_{t|t-1} = y_t - \widehat{\alpha}_{t,1|t-1}(\boldsymbol{\lambda}) - \mu, \end{aligned} \tag{8.104}$$

and

$$\widehat{\boldsymbol{\alpha}}_{t|t} = \widehat{\boldsymbol{\alpha}}_{t|t-1} + \frac{1}{s_t}\mathbf{m}_t\psi\left(\frac{\widetilde{u}_t(\boldsymbol{\lambda})}{s_t}\right). \tag{8.105}$$

Taking the first component in the above equation, adding μ to each side, and using the fact that the first component of $\mathbf{m}_t$ is s_t^2 yields

$$\widehat{x}_{t|t} = \widehat{x}_{t|t-1} + s_t\psi\left(\frac{\widetilde{u}_t(\boldsymbol{\lambda})}{s_t}\right),$$

and therefore (8.85) and (8.86) hold.

Further details of the recursions of s_t and $\mathbf{m}_t$ are provided in Section 8.16. The recursions for this filter are the same as (8.80), (8.82) and the associated filter covariance recursions in Section 8.16, with $\widehat{\mathbf{x}}_{t|t}$ replaced with $\widehat{\boldsymbol{\alpha}}_{t|t}$ and $\widehat{\mathbf{x}}_{t|t-1}$ replaced with $\widehat{\boldsymbol{\alpha}}_{t|t-1}$. Further details are provided in Section 8.17.

As we shall see in Section 8.17, in order to implement the filter, a value for the ARMA innovation variance σ_u^2 is needed as well as a value for λ. We deal with this issue as in the case of an AR model by replacing this unknown variance with an estimator $\widehat{\sigma}_u^2$ in a manner described subsequently.

IOs are dealt with as described in the remark at the end of Section 8.6.2.

8.8.3 Robustly filtered τ-estimators

A τ-estimator $\widehat{\lambda}$ based on the robustly filtered observations y_t can now be obtained by replacing the raw unfiltered residuals (8.94) in $Q^*(\lambda)$ of (8.99) with the new robustly filtered residuals (8.104) and then minimizing $Q^*(\lambda)$. Then, by defining

$$Q^*(\lambda) = \sum_{t=1}^{T} \log a_t^2(\lambda) + T \log \left(\tau^2 \left(\frac{\widetilde{u}_1(\lambda)}{a_1(\lambda)}, \dots, \frac{\widetilde{u}_T(\lambda)}{a_T(\lambda)} \right) \right),$$

the Fτ-estimator is defined by

$$\hat{\lambda} = \arg\min_{\lambda} \; Q^*(\lambda). \tag{8.106}$$

Since the above $Q^*(\lambda)$ may have several local minima, a good robust initial estimator is required. Such an estimator is obtained by the following steps:

1. Fit an AR(p^*) model using the robust Fτ-estimator of Section 8.6.3, where p^* is selected by the robust order selection criterion – RAIC – described in Section 8.6.6. The value of p^* will almost always be larger than p, and sometimes considerably larger. This fit gives the required estimator $\widehat{\sigma}_u^2$, as well as robust parameter estimators $(\widehat{\phi}_1^o, ..., \widehat{\phi}_{p^*}^o)$ and robustly filtered values $\widehat{x}_{t|t}$.
2. Compute estimators of the first p autocorrelations of x_t and of η_i, $1 \le i \le q$ where
$$\eta_i = \frac{\mathrm{Cov}(x_t, u_{t-i})}{\sigma_u^2}. \tag{8.107}$$
using the estimators $(\widehat{\phi}_1^o, ..., \widehat{\phi}_{p^*}^o)$ and $\widehat{\sigma}_u^2$
3. Finally compute the initial parameter estimators of the ARMA(p, q) model by matching the first p autocorrelations and the q values η_i with those obtained in Step 2.

Example 8.5 *Estimation of a simulated MA(1) series with AO*

As an example, we generated an MA(1) series of 200 observations with 10 equally spaced additive outliers as follows (script **MA1-AO**):

$$y_t = \begin{cases} x_t + 4 & \text{if } t = 20i, \quad i = 1, ..., 10 \\ x_t & \text{otherwise} \end{cases}$$

where $x_t = 0.8u_{t-1} + u_t$ and the u_t are i.i.d. N(0,1) variables.

Table 8.4 Estimates of the parameters of MA(1) simulated process

	θ	μ	σ_u
Fτ	−0.80	−0.02	0.97
LS	−0.39	0.20	1.97
True values	−0.80	0.00	1.00

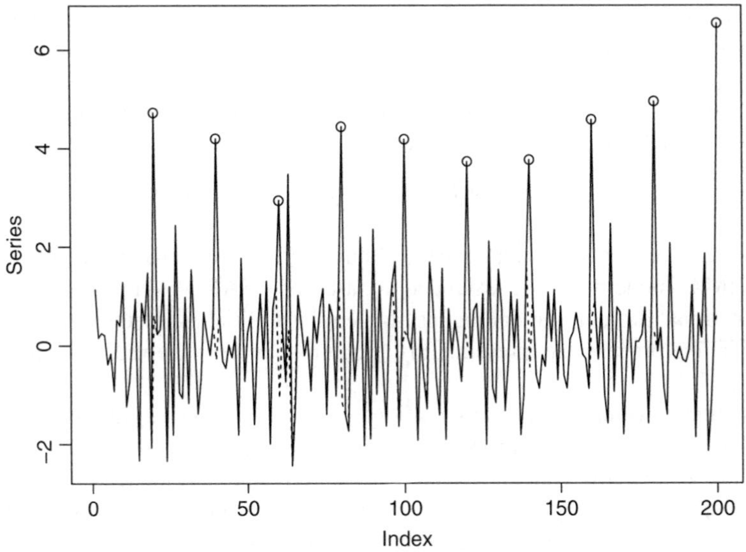

Figure 8.11 Simulated MA(1) series with 10 AOs: observed (solid line) and filtered (dashed line) data

The model parameters were estimated using the Fτ- and the LS estimators, and the results are shown in Table 8.4. We observe that the robust estimate is very close to the true value, while the LS estimate is very much influenced by the outliers. Figure 8.11 shows the observed series y_t and the filtered series $\widehat{x}_{t|t}$. It is seen that the latter is almost coincident with y_t except for the ten outliers, which are replaced by the predicted values.

8.9 ARIMA and SARIMA models

We define an autoregression integrated moving-average process y_t of orders p, d, q (ARIMA(p, d, q) for short) as one such that its order-d differences are a stationary

ARMA(p, q) process. It therefore satisfies

$$\phi(B)(1-B)^d y_t = \gamma + \theta(B)u_t, \tag{8.108}$$

where ϕ and θ are polynomials of order p and q, and u_t are the innovations.

A seasonal ARIMA process y_t of regular orders p, d, q, seasonal period s, and seasonal orders P, D, Q (SARIMA$(p, d, q) \times (P, D, Q)_s$ for short) fulfills the equation

$$\phi(B)\Phi(B^s)(1-B)^d(1-B^s)^D y_t = \gamma + \theta(B)\Theta(B^s)u_t, \tag{8.109}$$

where ϕ and θ are as above, and Φ and Θ are polynomials of order P and Q respectively. It is assumed that the roots of ϕ, θ, Φ and Θ lie outside the unit circle and then the differenced series $(1-B)^d(1-B^s)^D y_t$ is a stationary and invertible ARMA process.

In what follows, we shall restrict ourselves to the case $P = 0$ and $Q \leq 1$. Then $\Theta(B) = 1 - \Theta_s B$ and (8.109) reduces to

$$(1-B)^d(1-B^s)^D\phi(B)y_t = \gamma + \theta(B)(1-\Theta_s B^s)u_t. \tag{8.110}$$

The reason for this limitation is that, although the Fτ-estimators already defined for ARMA models can be extended to arbitrary SARIMA models, there is a computational difficulty in finding a suitable robust initial estimator for the iterative optimization process. At present, this problem has been solved only for $P = 0$ and $Q \leq 1$.

Assume now that we have observations $y_t = x_t + \upsilon_t$, where x_t fulfills an ARIMA model and υ_t is an outlier process. A naïve way to estimate the parameters is to difference y_t, thereby reducing the model to an ARMA(p, q) model, and then apply the Fτ-estimator already described. The problem with this approach is that the differencing operations will result in increasing the number of outliers. For example, with an ARIMA($p, 1, q$) model, the single regular difference operation will convert isolated outliers into two consecutive outliers of opposite sign (a so called "doublet"). However, one need not difference the data and may instead use the robust filter on the observations y_t as in the previous section, but based on the appropriate state-space model for the process (8.110).

The state-space representation is of the same form as (8.101), except that it uses a state-transition matrix Φ^* based on the coefficients of the polynomial operator of order $p^* = p + d + sD$

$$\phi^*(B) = (1-B)^d(1-B^s)^D\phi(B). \tag{8.111}$$

For example, in the case of an ARIMA($1, 1, q$) model with AR polynomial operator $\phi(B) = 1 - \phi_1 B$, we have

$$\phi^*(B) = 1 - \phi_1^* B - \phi_2^* B^2$$

with coefficients $\phi_1^* = 1 + \phi_1$ and $\phi_2^* = -\phi_1$. And for model (8.110) with $p = D = 1, d = q = Q = 0$ and seasonal period $s = 12$, we have

$$\phi^*(B) = 1 - \phi_1^* B - \phi_{12}^* B^{12} - \phi_{13}^* B^{13}$$

with

$$\phi_1^* = \phi_1, \quad \phi_{12}^* = 1, \quad \phi_{13}^* = -\phi_1.$$

Therefore, for each value of $\lambda = (\boldsymbol{\phi}, \boldsymbol{\theta}, \gamma, \Theta_s)$ (where Θ_s is the seasonal MA parameter when $Q = 1$), the filtered residuals corresponding to the operators ϕ^* and θ^* are computed, yielding the residuals $\widetilde{u}_t(\lambda)$. Then the Fτ-estimator is defined by the λ minimizing $Q^*(\lambda)$, with Q^* defined as in (8.99) but with $\phi^*(B)$ instead of $\phi(B)$.

More details can be found in Bianco *et al.* (1996).

Example 8.6 *Residential telephone extensions (RESEX) series. Tables and figures for this example are obtained with script* ***RESEX.R***.

This example deals with a monthly series of inward movement of residential telephone extensions in a fixed geographic area from January 1966 to May 1973 (RESEX). The series was analyzed by Brubacher (1974), who identified a SARIMA(2,0,0) $\times$ $(0,1,0)_{12}$ model, and by Martin *et al.* (1983).

Table 8.5 displays the LS, and Fτ-estimators of the parameters (script **RESEX**). We observe that they are quite different, and the estimation of the SD of the innovation corresponding to the LS estimator is much larger than the ones obtained with the Fτ-estimators.

Figure 8.12 shows the observed data y_t and the filtered values $\widehat{x}_{t|t}$,which are seen to be almost coincident with y_t except at outlier locations.

In Figure 8.13 we show the quantiles of the absolute values of the residuals of the three estimators. The three largest residuals are huge and hence were not included to improve the graph readability It is seen that the Fτ-estimator yields the smallest quantiles, and hence gives the best fit to the data.

Table 8.5 Estimates of the parameters of RESEX series

Estimators	ϕ_1	ϕ_2	γ	σ_u
Fτ	0.27	0.49	0.41	1.12
LS	0.48	−0.17	1.86	6.45

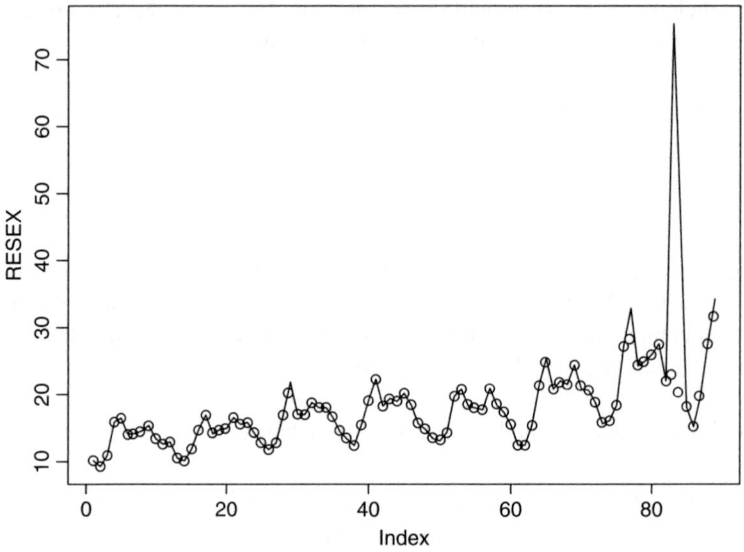

Figure 8.12 RESEX series: observed (solid line) and filtered (circles) values

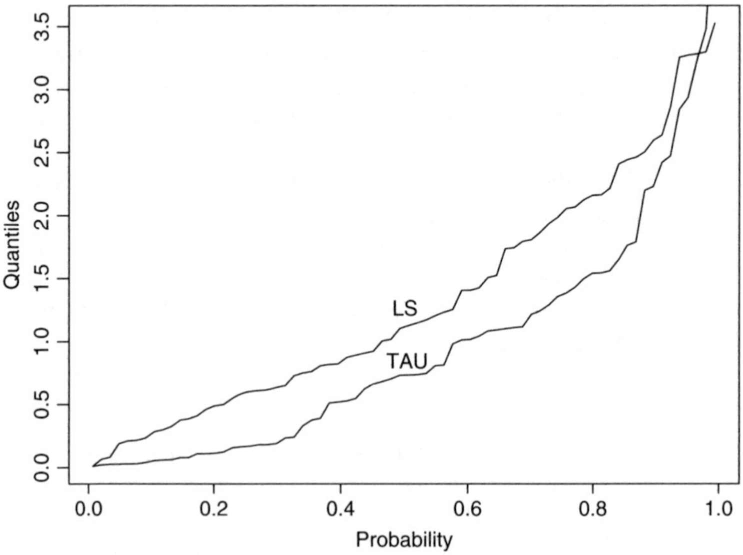

Figure 8.13 Quantiles of absolute residuals of estimators for RESEX series

8.10 Detecting time series outliers and level shifts

In many situations it is important to identify the type of perturbations that the series undergo. In this section we describe classical and robust diagnostic methods to detect outliers and level shifts in ARIMA models. As for the diagnostic procedures described in Chapter 4 for regression, the classical procedures are based on residuals obtained using nonrobust estimators. In general, these procedures succeed only when the proportion of outliers is very low and the outliers are not very large. Otherwise, due to masking effects, the outliers may not be detected.

Let y_t, $1 \leq t \leq T$, be an observed time series. We consider perturbed models of the form

$$y_t = x_t + \omega\, \xi_t^{(t_0)}, \tag{8.112}$$

where the unobservable series x_t is an ARIMA process satisfying

$$\phi(B)(1-B)^d x_t = \theta(B)u_t, \tag{8.113}$$

and the term $\omega\xi_t^{(t_0)}$ represents the effect on period t of the perturbation occurring at time t_0.

The value of ω in (8.112) measures the size of the perturbation at time t_0 and the form of $\xi_t^{(t_0)}$ depends on the type of the outlier. Let $o_t^{(t_0)}$ be an indicator variable for time t_0 ($o_t^{(t_0)} = 1$ for $t = t_0$ and 0 otherwise). Then an AO at time t_0 can be modeled by

$$\xi_t^{(t_0)} = o_t^{(t_0)} \tag{8.114}$$

and a level shift at time t_0 by

$$\xi_t^{(t_0)} = \begin{cases} 0 & \text{if } t < t_0 \\ 1 & \text{if } t \geq t_0. \end{cases}$$

To model an IO at time t_0, the observed series y_t is given by

$$\phi(B)(1-B)^d y_t = \theta(B)(u_t + \omega\, o_t^{(t_0)}).$$

Then, for an IO we get

$$\xi_t^{(t_0)} = \phi(B)^{-1}(1-B)^{-d}\theta(B)\, o_t^{(t_0)}. \tag{8.115}$$

We know that robust estimators are not very much influenced by a small fraction of atypical observations in the cases of IO or AO. The case of level shifts is different. A level shift at period t_0 modifies all the observations y_t with $t \geq t_0$. However, if the model includes a first-order difference, then differencing (8.112) gives

$$(1-B)y_t = (1-B)x_t + \omega(1-B)\xi_t^{(t_0)},$$

and since $(1-B)\xi_t^{(t_0)} = o_t^{(t_0)}$, the differenced series has an AO at time t_0. Then a robust estimator applied to the differenced series is not going to be influenced by a level shift. Therefore, the only case in which a robust procedure may be influenced by a level shift is when the model does not contain any difference. Note however that a second order difference converts a level shift to a doublet outlier.

8.10.1 Classical detection of time series outliers and level shifts

In this subsection, we shall describe the basic ideas of Chang *et al.* (1988) for outlier detection in ARIMA models. Similar approaches were considered by Tsay (1988) and Chen and Liu (1993). Procedures based on deletion diagnostics were proposed by Peña (1987, 1990), Abraham and Chuang (1989), Bruce and Martin (1989) and Ledolter (1991).

For the sake of simplicity, we start by assuming that the parameters of the ARIMA model, λ and σ_u^2, are known.

Let $\pi(B)$ be the filter defined by

$$\pi(B) = \theta(B)^{-1}\phi(B)(1-B)^d = 1 - \pi_1 B - \pi_2 B^2 - \ldots - \pi_k B^k - \ldots. \tag{8.116}$$

Then, from (8.113), $\pi(B)x_t = u_t$. Since $\pi(B)$ is a linear operator, we can apply it to both sides of (8.112), obtaining

$$\pi(B)\, y_t = u_t + \omega\, \pi(B)\, \xi_t^{(t_0)}, \tag{8.117}$$

which is a simple linear regression model with independent errors and regression coefficient ω.

Therefore, the LS estimator of ω is given by

$$\widehat{\omega} = \frac{\sum_{t=t_0}^{T} (\pi(B)\, y_t)\, (\pi(B)\, \xi_t^{(t_0)})}{\sum_{t=t_0}^{T} (\pi(B)\, \xi_t^{(t_0)})^2}, \tag{8.118}$$

with variance

$$\mathrm{Var}(\widehat{\omega}) = \frac{\sigma_u^2}{\sum_{t=t_0}^{T} (\pi(B)\, \xi_t^{(t_0)})^2}, \tag{8.119}$$

where σ_u^2 is the variance of u_t.

In practice, since the parameters of the ARIMA model are unknown, (8.118) and (8.119) are computed using LS or ML estimators of the ARIMA parameter. Let $\widehat{\pi}$ be defined as in (8.116) but using the estimators instead of the true parameters. Then (8.118) and (8.119) are replaced by

$$\widehat{\omega} = \frac{\sum_{t=t_0}^{T} \widehat{u}_t\, (\widehat{\pi}(B)\, \xi_t^{(t_0)})}{\sum_{t=t_0}^{T} (\widehat{\pi}(B)\, \xi_t^{(t_0)})^2}, \tag{8.120}$$

and

$$\widehat{\mathrm{Var}}(\widehat{\omega}) = \frac{\widehat{\sigma}_u^2}{\sum_{t=t_0}^{T} (\widehat{\pi}(B)\, \xi_t^{(t_0)})^2}, \tag{8.121}$$

where

$$\widehat{u}_t = \widehat{\pi}(B) y_t$$

and

$$\widehat{\sigma}_u^2 = \frac{1}{T - t_0} \sum_{t=t_0}^{T} (\widehat{u}_t - \widehat{\omega}(\widehat{\pi}(B)\, \xi_t^{(t_0)}))^2.$$

In the case of IO, the estimator of the outlier size given by (8.118) reduces to the innovation residual at t_0; that is, $\widehat{\omega} = \widehat{u}_{t_0}$.

A test to detect the presence of an outlier at a given t_0 can be based on the t-like statistic

$$U = \frac{|\widehat{\omega}|}{(\widehat{\mathrm{Var}}(\widehat{\omega}))^{1/2}}. \tag{8.122}$$

Since, in general, neither t_0 nor the type of outlier are known, in order to decide if there is an outlier at any position, the statistic

$$U_0 = \max_{t_0} \max\{U_{t_0,\mathrm{AO}}, U_{t_0,\mathrm{LvS}}, U_{t_0,\mathrm{IO}}\}$$

is used, where $U_{t_0,\mathrm{AO}}$, $U_{t_0,\mathrm{LvS}}$ and $U_{t_0,\mathrm{IO}}$ are the statistics defined by (8.122) corresponding to an AO, level shift (LvS) and IO at time t_0, respectively. If $U_0 > M$, where M is a conveniently chosen constant, one declares that there is an outlier or level shift. The time t_0 at which the outlier or level shift occurs and and whether the additive effect is an AO, IO or LvS is determined by where the double maximum is attained.

Since the values

$$\widehat{u}_t = \sum_{i=0}^{\infty} \widehat{\pi}_i y_{t-i}$$

can only be computed from a series extending into the infinite past, in practice, with data observed for $t = 1, \ldots, T$, they are approximated by

$$\widehat{u}_t = \sum_{i=0}^{t-1} \widehat{\pi}_i y_{t-i}.$$

As mentioned above, this type of procedure may fail in the presence of a large fraction of outliers and/or level shifts. This failure may be due to two facts. On one hand, the outliers or level shift may have a large influence on the MLE, and therefore the residuals may not reveal the outlying observations. This drawback may be overcome by using robust estimators of the ARMA coefficients. On the other hand, if y_{t_0} is an outlier or level shift, as noted before, not only is $\widehat{u}_{t_0}$ affected, but the effect of the outlier or level shift is propagated to the subsequent innovation residuals. Since the statistic U_0 is designed to detect the presence of an outlier or level shift at time t_0, it is desirable that U_0 be influenced by only an outlier at t_0. Outliers or level shifts at previous locations, however, may have a misleading influence on U_0. In the next subsection we show how to overcome this problem by replacing the innovation residuals $\widehat{u}_t$ by the filtered residuals studied in Section 8.8.3.

8.10.2 Robust detection of outliers and level shifts for ARIMA models

In this section we describe an iterative procedure introduced by Bianco (2001) for the detection of AO, level shifts and IO in an ARIMA model. The algorithm is similar to the one described in the previous subsection. The main difference is that the new method uses innovation residuals based on the Fτ-estimators of the ARIMA parameters instead of a Gaussion MLE, and uses a robust filter instead of the filter π to obtain an equation analogous to (8.117).

A detailed description of the procedure follows:

Step 1 Estimate the parameters λ and σ_u robustly using an Fτ–estimator. These estimators will be denoted by $\widehat{\lambda}$ and $\widehat{\sigma}_u$ respectively.

Step 2 Apply the robust filter described in Section 8.8.3 to y_t using the estimators computed in Step 1. This step yields the filtered residuals $\widetilde{u}_t$ and the scales s_t.

Step 3 In order to make the procedure less costly in terms of computing time, a preliminary set of outlier locations is determined in the following way:

- Declare that time t_0 is a candidate for an outlier or level shift location if

$$|\widetilde{u}_{t_0}| > M_1 s_{t_0}, \tag{8.123}$$

  where M_1 is a conveniently chosen constant.
- Denote by C the set of t_0s where (8.123) holds.

Step 4 For each $t_0 \in C$, let $\widehat{\pi}^*$ be a robust filter similar to the one applied in Step 2, but such that for $t_0 \leq t \leq t_0 + h$ the function ψ is replaced by the identity for a conveniently chosen value of h. Call $\widetilde{u}_t^* = \widehat{\pi}^*(B)y_t$ the residuals obtained with this filter. Since these residuals now have different variances, we estimate ω by weighted LS, with weights proportional to $1/s_t^2$. Then (8.120), (8.121) and (8.122) are now replaced by

$$\tilde{\omega} = \frac{\sum_{t=t_0}^{T} \widetilde{u}_t^* \ \widehat{\pi}^*(B) \ \xi_t^{(t_0)}/s_t^2}{\sum_{t=t_0}^{T} (\widehat{\pi}^*(B) \ \xi_t^{(t_0)})^2/s_t^2}, \tag{8.124}$$

$$\widehat{\mathrm{Var}}(\tilde{\omega}) = \frac{1}{\sum_{t=t_0}^{T} (\widehat{\pi}^*(B) \ \xi^{(t_0)})^2/s_t^2} \tag{8.125}$$

and

$$U^* = \frac{|\tilde{\omega}|}{(\widehat{\mathrm{Var}}(\tilde{\omega}))^{1/2}}. \tag{8.126}$$

The purpose of replacing $\widehat{\pi}$ by $\widehat{\pi}^*$ is to eliminate the effects of outliers at positions different from t_0. For this reason, the effect of those outliers before t_0 and after $t_0 + h$ is reduced by means of the robust filter. Since we want to detect a possible outlier at time t_0, and the effect of an outlier propagates to the subsequent observations, we do not downweight the effect of possible outliers between t_0 and $t_0 + h$.

Step 5 Compute

$$U_0^* = \max_{t_0 \in C} \max\{U^*_{\cdot t_0,\mathrm{AO}}, U^*_{\cdot t_0,\mathrm{LvS}}, U^*_{\cdot t_0,\mathrm{IO}}\},$$

where $U^*_{t_0,\mathrm{AO}}$, $U^*_{t_0,\mathrm{LvS}}$ and $U^*_{t_0,\mathrm{IO}}$ are the statistics defined by (8.126) corresponding to an AO, level shift and IO at time t_0, respectively. If $U_0^* \leq M_2$,where M_2 is a conveniently chosen constant, no new outliers are detected and the iterative procedure is stopped. Instead, if $U_0^* > M_2$, a new AO, level shift or IO is detected, depending on where the maximum is attained.

Step 6 Clean the series from the detected AO, level shifts or IO by replacing y_t by $y_t - \tilde{\omega}\xi_t^{(t_0)}$, where $\xi_t^{(t_0)}$ corresponds to the perturbation at t_0, pointed out by the test. Then the procedure is iterated, going back to Step 2 until no new perturbations are found.

The constant M_1 should be chosen rather small (for example $M_1 = 2$) to increase the power of the procedure for the detection of outliers. Based on simulations we recommend using $M_2 = 3$.

As already mentioned, this procedure will be reliable to detect level shifts only if the ARIMA model includes at least an ordinary difference ($d > 0$).

Example 8.7 *Continuation of Example 8.5. Tables and figures for this example are obtained with script **identMA1.R**.*

On applying the robust procedure just described to the data, all the outliers were detected. Table 8.6 shows the outliers found with this procedure as well as their corresponding type and size and the value of the test statistic.

Table 8.6 Outliers detected with the robust procedure in simulated MA(1) series

Index	Type	Size	U_0^*
20	AO	4.72	7.47
40	AO	4.46	7.51
60	AO	4.76	8.05
80	AO	3.82	6.98
100	AO	4.02	6.39
120	AO	4.08	7.04
140	AO	4.41	7.48
160	AO	4.74	7.95
180	AO	4.39	7.53
200	AO	5.85	6.59

Table 8.7 Detected outliers in the RESEX series

Index	Date	Type	Size	U_0^*
29	5/68	AO	2.52	3.33
47	11/69	AO	−1.80	3.16
65	5/71	LvS	1.95	3.43
77	5/72	AO	4.78	5.64
83	11/72	AO	52.27	55.79
84	12/72	AO	27.63	27.16
89	5/73	AO	4.95	3.12

The classical procedure of Section 8.10.1 detects only two outliers: observations 120 and 160. The LS estimators of the parameters after removing the effect of these two outliers are $\widehat{\theta} = -0.48$ and $\widehat{\mu} = 0.36$, which are also far from the true values.

Example 8.8 *Continuation of Example 8.6. Tables and figures for this example are obtained with script* ***RESEX.R****.*

Table 8.7 shows the outliers and level shifts found by applying the robust procedure to the RESEX data. We observe two very large outliers corresponding to the last two months of 1972. The explanation is that November 1972 was a "bargain" month; that is, free installation of residential extensions, with a spillover effect, since not all November orders could be fulfilled that month.

8.10.3 REGARIMA models: estimation and outlier detection

A REGARIMA model is a regression model where the errors are an ARIMA time series. Suppose that we have T observations $(\mathbf{x}_1, y_1), \ldots, (\mathbf{x}_T, y_T)$, with $\mathbf{x}_i \in R^{k,}$, $y_i \in R$ satisfying

$$y_t = \boldsymbol{\beta}'\mathbf{x}_t + e_t$$

where $e_1, \ldots, e_T$ follow an ARIMA(p, d, q) model

$$\phi(B)(1 - B)^d e_t = \theta(B)u_t.$$

As in the preceding cases, we consider the situation when the actual observations are described by a REGARIMA model plus AO, IO and level shifts. In other words, instead of observing y_t we observe

$$y_t^* = y_t + \omega\,\xi_t^{(t_0)}, \tag{8.127}$$

where $\xi_t^{(t_0)}$ is as in the ARIMA model.

All the procedures for ARIMA models described in the preceding sections can be extended to REGARIMA models.

Define for each value of $\boldsymbol{\beta}$,

$$\widehat{e}_t(\boldsymbol{\beta}) = y_t^* - \boldsymbol{\beta}'\mathbf{x}_t, \ t = 1, \ldots, T$$

and put

$$\widehat{w}_t(\boldsymbol{\beta}) = (1 - B)^d \widehat{e}_t(\boldsymbol{\beta}), \quad t = d + 1, \ldots, T.$$

When $\boldsymbol{\beta}$ is the true parameter, $\widehat{w}_t(\boldsymbol{\beta})$ follows an ARMA(p, q) model with an AO, IO or level shift. Then it is natural to define for any $\boldsymbol{\beta}$ and $\boldsymbol{\lambda} = (\boldsymbol{\phi}, \boldsymbol{\theta})$ the residuals $\widehat{u}_t(\boldsymbol{\beta}, \boldsymbol{\lambda})$ as in (8.58), but replacing y_t by $\widehat{w}_t(\boldsymbol{\beta})$; that is,

$$\widehat{u}_t(\boldsymbol{\beta}, \boldsymbol{\lambda}) = \widehat{w}_t(\boldsymbol{\beta}) - \phi_1 \widehat{w}_{t-1}(\boldsymbol{\beta}) - \ldots - \phi_p \widehat{w}_{t-p}(\boldsymbol{\beta}) + \theta_1 \widehat{u}_{t-1}(\boldsymbol{\beta}, \boldsymbol{\lambda}) + \ldots$$
$$+ \theta_q \widehat{u}_{t-q}(\boldsymbol{\beta}, \boldsymbol{\lambda}) \quad (t = p + d + 1, \ldots, T).$$

Then the LS estimator is defined as $(\boldsymbol{\beta}, \boldsymbol{\lambda})$ minimizing

$$\sum_{t=p+d+1}^{T} \widehat{u}_t^2(\boldsymbol{\beta}, \boldsymbol{\lambda}),$$

and an M-estimator is defined as $(\boldsymbol{\beta}, \boldsymbol{\lambda})$ minimizing

$$\sum_{t=p+d+1}^{T} \rho\left(\frac{\widehat{u}_t(\boldsymbol{\beta}, \boldsymbol{\lambda})}{\widehat{\sigma}}\right),$$

where $\widehat{\sigma}$ is the scale estimator of the innovations u_t. As in the case of regression with independent errors, the LS estimator is very sensitive to outliers, and M-estimators with bounded ρ are robust when the u_ts are heavy tailed, but not for other types of outliers like AOs.

Let $\widetilde{u}_t(\boldsymbol{\beta}, \boldsymbol{\lambda})$ be the filtered residuals corresponding to the series $\widehat{e}_t(\boldsymbol{\beta})$ using the ARIMA(p, d, q) model with parameter $\boldsymbol{\lambda}$. Then we can define Fτ-estimators as in Section 8.8.3; that is,

$$(\hat{\boldsymbol{\beta}}, \hat{\boldsymbol{\lambda}}) = \arg\min_{\boldsymbol{\lambda}} \ Q^*(\hat{\boldsymbol{\beta}}, \boldsymbol{\lambda}),$$

where

$$Q^*(\hat{\boldsymbol{\beta}}, \boldsymbol{\lambda}) = \sum_{t=1}^{T} \log a_t^2(\boldsymbol{\lambda}) + T \log\left(\tau^2\left(\frac{\widetilde{u}_1(\hat{\boldsymbol{\beta}}, \boldsymbol{\lambda})}{a_1(\boldsymbol{\lambda})}, \ldots, \frac{\widetilde{u}_T(\hat{\boldsymbol{\beta}}, \boldsymbol{\lambda})}{a_T(\boldsymbol{\lambda})}\right)\right).$$

The robust procedure for detecting the outliers and level shifts of Section 8.10.2 can also easily be extended to REGARIMA models. For details on the Fτ-estimators and outliers and level shift detection procedures for REGARIMA models, see Bianco *et al.* (2001).

8.11 Robustness measures for time series

8.11.1 Influence function

In all situations considered so far, we have a finite-dimensional vector λ of unknown parameters (say, $\lambda = (\phi_1, \ldots, \phi_p, \theta_1, \ldots, \theta_q, \mu)'$ for ARMA models) and an estimator $\widehat{\lambda}_T = \widehat{\lambda}_T(y_1, \ldots, y_T)$. When y_t is a strictly stationary process, it holds under very general conditions that $\widehat{\lambda}_T$ converges in probability to a vector $\widehat{\lambda}_\infty$, which depends on the joint (infinite-dimensional) distribution F of $\{y_t : t = 1, 2, \ldots\}$.

Künsch (1984) extends Hampel's definition (3.4) of the influence function to time series in the case that $\widehat{\lambda}_T$ is defined by M-estimating equations that depend on a fixed number k of observations:

$$\sum_{t=k}^{n} \boldsymbol{\Psi}(\mathbf{y}_t, \widehat{\lambda}_T) = 0, \tag{8.128}$$

where $\mathbf{y}_t = (y_t, \ldots, y_{t-k+1})'$. Strict stationarity implies that the distribution F_k of $\mathbf{y}_t$ does not depend on t. Then, for a general class of stationary processes, $\widehat{\lambda}_\infty$ exists and depends only on F_k, and is the solution of the equation

$$\mathrm{E}_{F_k} \boldsymbol{\Psi}(\mathbf{y}_t, \lambda) = \mathbf{0}. \tag{8.129}$$

For this type of time series, the Hampel influence function could be defined as

$$\mathrm{IF_H}(\mathbf{y}; \widehat{\lambda}, F_k) = \lim_{\varepsilon \downarrow 0} \frac{\widehat{\lambda}_\infty[(1-\varepsilon)F_k + \varepsilon\delta_{\mathbf{y}}] - \widehat{\lambda}_\infty(F_k)}{\varepsilon}, \tag{8.130}$$

where $\mathbf{y} = (y_k, \ldots, y_1)'$, and the subscript H stands for the Hampel definition. Then, proceeding as in Section 5.11.1, it can be shown that for estimators of the form (8.128) the analog of (3.48) holds. Then, by analogy with (3.29), the gross-error sensitivity is defined as $\sup_{\mathbf{y}} \|\mathrm{IF_H}(\mathbf{y}; \widehat{\lambda}, F_k)\|$, where $\|.\|$ is a convenient norm.

If y_t is an AR(p) process, it is natural to generalize LS through M-estimators of the form (8.129) with $k = p + 1$. Künsch (1984) found the Hampel-optimal estimator for this situation, which turns out to be a GM-estimator of Schweppe form (5.86).

However, this definition has several drawbacks:

- This form of contamination is not the one we would like to represent. The intuitive idea of a contamination rate $\varepsilon = 0.05$ is that about 5% of the observations are altered. But in the definition (8.130), ε is the proportion of outliers in each k-dimensional marginal. In general, given ε and $\mathbf{y}$, there exists no process such that all its k-dimensional marginals are $(1-\varepsilon)F_k + \varepsilon\delta_{\mathbf{y}}$.
- The definition cannot be applied to processes such as ARMA, in which the natural estimating equations do not depend on finite-dimensional distributions.

An alternative approach was taken by Martin and Yohai (1986) who introduced a new definition of *influence functional* for time series, which we now briefly discuss.

We assume that observations y_t are generated by the general replacement outliers model

$$y_t^\varepsilon = (1 - z_t^\varepsilon)x_t + z_t^\varepsilon w_t, \tag{8.131}$$

where x_t is a stationary process (typically normally distributed) with joint distribution F_x, w_t is an outlier-generating process and z_t^ε is a 0–1 process with $\mathrm{P}(z_t^\varepsilon = 1) = \varepsilon$. This model encompasses the AO model through the choice $w_t = x_t + v_t$ with v_t independent of x_t, and provides a pure replacement model when w_t is independent of x_t. The model can generate both isolated and patch outliers of various lengths through appropriate choices of the process z_t^ε. Assume that $\widehat{\lambda}_\infty(F_y^\varepsilon)$ is well defined for the distribution F_y^ε of y_t^ε. Then the *time series influence function* $\mathrm{IF}(\{F_{x,z,w}^\varepsilon\}; \widehat{\lambda})$ is the directional derivative at F_x:

$$\mathrm{IF}(\{F_{x,z,w}^\varepsilon\}; \widehat{\lambda}) = \lim_{\varepsilon \downarrow 0} \frac{1}{\varepsilon}(\widehat{\lambda}_\infty(F_y^\varepsilon) - \widehat{\lambda}_\infty(F_x)) \tag{8.132}$$

where $F_{x,z,w}^\varepsilon$ is the joint distribution of the processes x_t, z_t^ε and w_t.

The first argument of IF is a distribution, and so in general the time series IF is a functional on a distribution space, which is to be contrasted with $\mathrm{IF_H}$ which is a function on a finite-dimensional space. However, in practice we often choose special forms of the outlier-generating process w_t such as constant amplitude outliers; for example, for AOs we may let $w_t \equiv x_t + v$ and for pure ROs we let $w_t = v$, where v is a constant.

Although the time series IF is similar in spirit to $\mathrm{IF_H}$, it coincides with the latter only in the very restricted case that $\widehat{\lambda}_T$ is permutation invariant and (8.131) is restricted to an i.i.d. pure RO model (see Corollary 4.1 in Martin and Yohai, 1986).

While IF is generally different from $\mathrm{IF_H}$, there is a close relationship between both: if $\widehat{\lambda}$ is defined by (8.128), then under regularity conditions:

$$\mathrm{IF}(\{F_{x,z,w}\}; \widehat{\lambda}) = \lim_{\varepsilon \downarrow 0} \frac{\mathrm{E}[\mathrm{IF_H}(\mathbf{y}_k; \widehat{\lambda}, F_{x,k})]}{\varepsilon} \tag{8.133}$$

where $F_{x,k}$ is the k-dimensional marginal of F_x and the distribution of $\mathbf{y}_k$ is the k-dimensional marginal of F_y^ε.

The above result is proved in Theorem 4.1 of Martin and Yohai (1986), where a number of other results concerning the time series IF are presented. In particular:

- Conditions are established which aid in the computation of time series IFs.
- IFs are computed for LS and robust estimators of AR(1) and MA(1) models, and the results reveal the differing behaviors of the estimators for both isolated and patchy outliers.
- It is shown that for MA models, bounded ψ-functions do not yield bounded IFs, whereas redescending ψ-functions do yield bounded IFs.
- Optimality properties are established for a class of estimators known as generalized RA estimators.

8.11.2 Maximum bias

In Chapter 3 we defined the maximum asymptotic bias of an estimator $\widehat{\theta}$ at a distribution F in an ε-contamination neighborhood of a parametric model. This definition made sense for i.i.d. observations, but cannot be extended in a straightforward manner to time series. A basic difficulty is that the simple mixture model $(1-\varepsilon)F_\theta + \varepsilon G$ that suffices for independent observations is not adequate for time series for the reasons given in the previous section.

As a simple case, consider estimation of ϕ in the AR(1) model $x_t = \phi x_{t-1} + u_t$, where u_t has $\mathrm{N}(0, \sigma_u^2)$ distribution. The asymptotic value of the LS estimator and of the M- and GM-estimators depends on the joint distribution $F_{2,y}$ of y_1 and y_2. Specification of $F_{2,y}$ is more involved than the two-term mixture distribution $(1-\varepsilon)$ $F_\theta + \varepsilon G$ used in the definition of bias given in Section 3.3. For example, suppose we have the AO model given by (8.5), where υ_t is an i.i.d. series independent of x_t with contaminated normal distribution (8.6). Denote by $\mathrm{N}_2(\mu_1, \mu_2, \sigma_1^2, \sigma_2^2, \gamma)$ the bivariate normal distribution with means μ_1 and μ_2, variances σ_1^2 and σ_2^2 and covariance γ, and call $\sigma_x^2 = \mathrm{Var}\,(x_t) = \sigma_u^2/(1-\phi^2)$. Then the joint distribution $F_{2,y}$ is a normal mixture distribution with four components:

$$\begin{aligned}&(1-\varepsilon)^2\mathrm{N}_2(0,0,1,1,\phi) + \varepsilon(1-\varepsilon)\mathrm{N}_2(0,0,1+\sigma_\upsilon^2,1,\phi)\\&+\varepsilon(1-\varepsilon)\mathrm{N}_2(0,0,1,1+\sigma_\upsilon^2,\phi)\\&+\varepsilon^2\mathrm{N}_2(0,0,1+\sigma_\upsilon^2,1+\sigma_\upsilon^2,\phi).\end{aligned} \tag{8.134}$$

The four terms correspond to the cases of no outliers, an outlier in y_1, an outlier in y_2, and outliers in both y_1 and y_2, respectively.

This distribution is even more complicated when modeling patch outliers in υ_t, and things get much more challenging for estimators that depend on joint distributions of order greater than two, such as AR(p), MA(q) and ARMA(p,q) models, where one must consider either p-dimensional joint distributions or joint distributions of all orders.

Martin and Jong (1977) took the above joint distribution modeling approach in computing maximum bias curves for a particular class of GM-estimators of an AR(1) parameter under both isolated and patch AO models. But it seems difficult to extend such calculations to higher-order models, and typically one has to resort to simulation methods to estimate maximum bias and BP (see Section 8.11.4 for an example of simulation computation of maximum bias curves).

A simple example of bias computation was given in Section 8.1.3. The asymptotic value of the LS estimator is the correlation between y_1 and y_2, and as such can be computed from the mixture expression (8.134), as the reader may verify (Problem 8.12).

Note that the maximum bias in (8.13) is $|\rho_\upsilon(1) - \phi|$, which depends upon the value of ϕ, and this feature holds in general for ARMA models.

8.11.3 Breakdown point

Extending the notion of BP given in Section 3.2 to the time series setting presents some difficulties.

The first is how "contamination" is defined. One could simply consider the finite BP for observations $y_1, \ldots, y_T$, as defined in Section 3.2.5, and then define the asymptotic BP by letting $T \to \infty$. The drawback of this approach is that it is intractable except in very simple cases. We are thus led to consider contamination by a process such as AO or RO, with the consequence that the results will depend on the type of contaminating process considered.

The second is how "breakdown" is defined. This difficulty is due to the fact that in time series models the parameter space is generally bounded, and moreover the effect of outliers is more complicated than with location, regression or scale.

This feature can be seen more easily in the AR(1) case. It was seen in Section 8.1.3 that the effect on the LS estimator of contaminating a process x_t with an AO process v_t is that the estimator may take on any value between the lag-1 autocorrelations of x_t and v_t. If v_t is arbitrary, then the asymptotic value of the estimator may be arbitrarily close to the boundary $\{-1, 1\}$ of the parameter space, and thus there would be breakdown according to the definitions of Section 3.2.

However, in some situations it is considered more reasonable to take only isolated (that is, i.i.d.) AOs into account. In this case the worst effect of the contamination is to shrink the estimator toward zero, and this could be considered as breakdown if the true parameter is not null. One could define breakdown as the estimator approaching ± 1 or 0, but it would be unsatisfactory to tailor the definition in an ad-hoc manner to each estimator and type of contamination.

A completely general definition that takes these problems into account was given by Genton and Lucas (2003). The intuitive idea is that breakdown occurs for some contamination rate ε_0 if further increasing the contamination rate does not further enlarge the range of values taken on by the estimator over the contamination neighborhood. In particular, for the case of AR(1) with independent AOs, it follows from the definition that breakdown occurs if the estimator can be taken to zero.

The details of the definition are very elaborate and are therefore omitted here.

8.11.4 Maximum bias curves for the AR(1) model

Here we present maximum bias curves from Martin and Yohai (1991) for three estimators of ϕ for a centered AR(1) model with RO:

$$y_t = x_t(1 - z_t) + w_t z_t, \quad x_t = \phi x_{t-1} + u_t$$

where z_t are i.i.d. with

$$\mathrm{P}(z_t = 1) = \gamma, \quad \mathrm{P}(w_t = c) = \mathrm{P}(w_t = -c) = 0.5.$$

The three considered estimators are

- the estimator obtained by modelling the outliers found using the procedure described in Section 8.10.1 (Chang *et al.*, 1988);
- the median of slopes estimator Med (y_t/y_{t-1}), which, as mentioned in Section 5.11.2, has bias-optimality properties;
- a filtered M-scale robust estimator, which is the same as the Fτ-estimator except that an M-scale was used by Martin and Yohai instead of the τ-scale, which is the approach recommended in this book.

The curves were computed by a Monte Carlo procedure. Let $\widehat{\phi}_T(\varepsilon, c)$ be the value of any of the three estimators for sample size T. For sufficiently large T, the value of $\widehat{\phi}_T(\varepsilon, c)$ will be negligibly different from its asymptotic value $\widehat{\phi}_\infty(\varepsilon, c)$; $T = 2000$ was used for the purpose of this approximation. Then the maximum asymptotic bias was approximated as

$$B(\varepsilon) = \sup_c |\widehat{\phi}_T(\varepsilon, c) - \phi| \tag{8.135}$$

by search on a grid of c values from 0 to 6 with a step size of 0.02. We plot the signed value of $B(\varepsilon)$ in Figure 8.14 for the case $\phi = 0.9$. The results clearly show the superiority of the robust filtered M-scale estimator, which has relatively small bias over the entire range of ε from 0 to 0.4, with estimator breakdown (not shown) occurring about $\varepsilon = 0.45$. Similar results would be expected for the Fτ-estimator. The estimator obtained using the classical outlier detection procedure of Chang *et al.* (1988) has quite poor global performance: while its maximum bias behaves similarly to that of the robust filtered M-scale estimator for small ε, the estimator breaks down in the presence of white-noise contamination, with a bias of essentially -0.9 for ε a little less than 0.1. The GM-estimator has a maximum bias behavior in between that of the other two estimators, with rapidly increasing maximum bias as ε increases

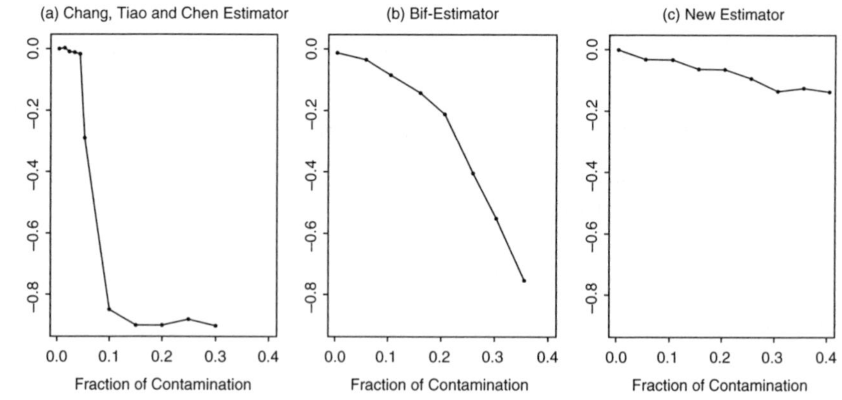

Figure 8.14 Maximum bias curves ("BIF" indicates the GM-estimator)

beyond roughly 0.1, but is not quite broken down at $\varepsilon = 0.35$. However, one can conjecture from the maximum bias curve that breakdown to zero occurs by $\varepsilon = 0.35$. We note that other types of bounded influence GM-estimators that use redescending functions can apparently achieve better maximum bias behavior than this particular GM-estimator (see Martin and Jong, 1977).

8.12 Other approaches for ARMA models

8.12.1 Estimators based on robust autocovariances

The class of robust estimators based on robust autocovariances (RA estimators) was proposed by Bustos and Yohai (1986). These estimators are based on a convenient robustification of the estimating LS equations.

Let $\boldsymbol{\lambda} = (\boldsymbol{\phi}, \boldsymbol{\theta}, \gamma)$. As a particular case of the results to be proved in Section 8.15, the equations for the LS estimator can be reexpressed as

$$\sum_{t=p+i+1}^{T} \widehat{u}_t(\boldsymbol{\lambda}) \sum_{j=0}^{t-p-i-1} \pi_j(\boldsymbol{\phi})\widehat{u}_{t-j-i}(\boldsymbol{\lambda}) = 0, \; i = 1, \ldots, p,$$

$$\sum_{t=p+i+1}^{T} \widehat{u}_t(\boldsymbol{\lambda}) \sum_{j=0}^{t-p-i-1} \zeta_j(\boldsymbol{\theta})\widehat{u}_{t-j-i}(\boldsymbol{\lambda}) = 0, \; i = 1, \ldots, q$$

and

$$\sum_{t=p+1}^{T} \widehat{u}_t(\boldsymbol{\lambda}) = 0, \tag{8.136}$$

where $\widehat{u}_t(\boldsymbol{\lambda})$ is defined in (8.58), and π_j and ζ_j are the coefficients of the inverses of the AR and MA polynomials; that is,

$$\phi^{-1}(B) = \sum_{j=0}^{\infty} \pi_j(\boldsymbol{\phi})B^j$$

and

$$\theta^{-1}(B) = \sum_{j=0}^{\infty} \zeta_j(\boldsymbol{\theta})B^j.$$

This system of equations can be written as

$$\sum_{j=0}^{T-p-i-1} \pi_j(\boldsymbol{\phi})M_{i+j}(\boldsymbol{\lambda}) = 0, \; i = 1, \ldots, p, \tag{8.137}$$

$$\sum_{j=0}^{T-p-i-1} \zeta_j(\boldsymbol{\theta})M_{i+j}(\boldsymbol{\lambda}) = 0, \; i = 1, \ldots, q \tag{8.138}$$

and (8.136), where

$$M_j(\lambda) = \sum_{t=p+j+1}^{T} \hat{u}_t(\lambda)\hat{u}_{t-j}(\lambda) = 0. \quad (8.139)$$

The RA estimators are obtained by replacing the term $M_j(\lambda)$ in (8.137) and (8.138) with

$$M_j^*(\lambda) = \sum_{t=p+j+1}^{T} \eta(\hat{u}_t(\lambda), \hat{u}_{t-j}(\lambda)), \quad (8.140)$$

and (8.136) with

$$\sum_{t=p+1}^{T} \psi(\hat{u}_t(\lambda)) = 0,$$

where ψ is a bounded ψ-function.

The name of this family of estimators comes from the fact that $M_j/(T-j-p)$ is an estimator of the autocovariance of the residuals $\hat{u}_t(\lambda)$, and $M_j^*/(T-j-p)$ is a robust version thereof.

Two types of η-functions are considered by Bustos and Yohai (1986): Mallows-type functions of the form $\eta(u,v) = \psi^*(u)\psi^*(v)$ and Schweppe-type functions of the form $\eta(u,v) = \psi^*(uv)$, where ψ^* is a bounded ψ-function. The functions ψ and ψ^* can be taken, for example, in the Huber or bisquare families. These estimators have good robustness properties for AR(p) models with small p. However, the fact that they use regular residuals makes them vulnerable to outliers when p is large or $q > 0$. They are consistent and asymptotically normal. A heuristic proof is given in Section 8.15. The asymptotic covariance matrix is of the form $b(\psi, F)V_{LS}$, where $b(\psi, F)$ is a scalar term (Bustos and Yohai, 1986).

8.12.2 Estimators based on memory-*m* prediction residuals

Suppose that we want to fit an ARMA(p, q) model using the series y_t, $1 \le t \le T$. It is possible to define M-estimators, GM-estimators and estimators based on the minimization of a residual scale using residuals based on a memory-m predictor, where $m \ge p + q$. This last condition is needed to ensure that the estimators are well defined.

Consider the memory-m best linear predictor when the true parameter is $\lambda = (\boldsymbol{\phi}, \boldsymbol{\theta}, \mu)$:

$$\hat{y}_{t,m}(\lambda) = \mu + \varphi_{m,1}(\boldsymbol{\phi}, \boldsymbol{\theta})(y_{t-1} - \mu) + \ldots + \varphi_{m,m}(\boldsymbol{\phi}, \boldsymbol{\theta})(y_{t-m} - \mu),$$

where $\varphi_{m,i}$ are the coefficients of the predictor defined in (8.31) (here we call them $\varphi_{m,i}$ rather than $\phi_{m,i}$ to avoid confusion with the parameter vector $\boldsymbol{\phi}$). Masarotto (1987) proposed estimating the parameters using memory-m residuals defined by

$$\hat{u}_{t,m}(\lambda) = y_t - \hat{y}_{t,m}(\lambda), \ t = m+1, \ldots, T.$$

Masarotto proposed this approach for GM-estimators, but actually it can be used with any robust estimator, including MM-estimators or estimators based on the minimization of a robust residual scale

$$\widehat{\sigma}(\widehat{u}_{p+1,m}(\lambda), \ldots, \widehat{u}_{T,m}(\lambda)),$$

where $\widehat{\sigma}$ is an M- or τ-scale. Since one outlier spoils $m+1$ memory-m residuals, the robustness of these procedures depends on how large the value of m is.

For AR(p) models, the memory-p residuals are the regular residuals $\widehat{u}_t(\lambda)$ given in (8.20), and therefore no new estimators are defined here.

One shortcoming of the estimators based on memory-m residuals is that the convergence to the true values holds only under the assumption that the process y_t is Gaussian.

8.13 High-efficiency robust location estimators

In Section 8.2 we described the AR(p) model in the two equivalent forms (8.14) and (8.18). We have been somewhat cavalier about which of these two forms to use in fitting the model, implicitly thinking that the location parameter μ is a nuisance parameter that is unimportant. That being the case, there is a temptation to use a simple robust location estimator $\widehat{\mu}$ for the centering, say for $\widehat{\mu}$ an ordinary location M-estimator, as described in Section 2.3. However, the location parameter may be of interest for its own sake, and there may be disadvantages in using an ordinary location M-estimator for the centering approach to fitting an AR model.

Use of the relationship (8.17) leads naturally to the location estimator

$$\widehat{\mu} = \frac{\widehat{\gamma}}{1 - \sum_{i=0}^{p} \widehat{\phi}_i}. \tag{8.141}$$

It is easy to check that the same form of location estimator is obtained for an ARMA(p, q) model in intercept form. In the context of M-estimators or GM-estimators of AR and ARMA models we call the estimator (8.141) a *proper* location M- (or GM-) estimator.

It turns out that use of an ordinary location M-estimator has two problems when applied to ARMA models. The first is that selection of the tuning constant to achieve a desired high efficiency when the innovations are normally distributed depends upon the model parameters, which are not known in advance. This problem is most severe for ARMA(p, q) models with $q > 0$. The second problem is that the efficiency of the ordinary M-estimator can be exceedingly low relative to the proper M-estimator. Details are provided by Lee and Martin (1986), who show that

- For an AR(1) model, the efficiency of the ordinary M-estimator relative to the proper M-estimator is between 10% and 20% for $\phi = \pm 0.9$ and approximately 60% for $\phi = \pm 0.5$.

- For an MA(1) model, the relative efficiency is above approximately 80% for positive θ but is around 50% for $\theta = -0.5$ and is arbitrarily low as θ approaches -1. The latter was shown by Grenander (1981) to be a point of super-efficiency.

The conclusion is that one should not use the ordinary location M-estimator for AR and ARMA processes when one is interested in location for its own sake. Furthermore, the severe loss of efficiency of the ordinary location M-estimator that is obtained for some parameter values raises doubts about its use for centering purposes, even when one is not interested in location for its own sake. It seems from the evidence at hand that it is prudent to fit the intercept form of AR and ARMA models, and when the location estimator is needed it can be computed from expression (8.141).

8.14 Robust spectral density estimation

8.14.1 Definition of the spectral density

Any second-order stationary process y_t defined for integer t has a spectral representation

$$y_t = \int_{-1/2}^{1/2} \exp(i2\pi tf)dZ(f) \tag{8.142}$$

where $Z(f)$ is a complex orthogonal increments process on $(-1/2, 1/2]$; that is, for any $f_1 < f_2 \le f_3 < f_4$

$$\mathrm{E}\{(Z(f_2) - Z(f_1))\overline{(Z(f_4) - Z(f_3))}\} = 0,$$

where $\bar{z}$ denotes the conjugate of the complex number z. See, for example, Brockwell and Davis (1991). This result says that any stationary time series can be interpreted as the limit of a sum of sinusoids $A_i \cos(2\pi f_i t + \Phi_i)$, with random amplitudes A_i and random phases Φ_i. The process $Z(f)$ defines an increasing function $G(f) = \mathrm{E}|Z(f)|^2$, with $G(-1/2) = 0$ and $G(1/2) = \sigma^2 = \mathrm{Var}(y_t)$. The function $G(f)$ is called the *spectral distribution* function, and when its derivative $S(f) = G'(f)$ exists it is called the *spectral density* function of y_t. Other commonly used terms for $S(f)$ are *power spectral density*, *spectrum* and *power spectrum*. We assume for purposes of this discussion that $S(f)$ exists, which implies that y_t has been centered by subtracting its mean. The more general case of a discrete time process on time intervals of length Δ is easily handled with slight modifications to the above (see, for example, Bloomfield, 1976 and Percival and Walden, 1993).

Using the orthogonal increments property of $Z(f)$, it is immediately found that the lag-k covariances of y_t are given by

$$C(k) = \int_{-1/2}^{1/2} \exp(i2\pi kf)S(f)df. \tag{8.143}$$

Therefore the $C(k)$ are the Fourier coefficients of $S(f)$ and so we have the Fourier series representation

$$S(f) = \sum_{k=-\infty}^{\infty} C(k) \exp(-i2\pi fk). \tag{8.144}$$

8.14.2 AR spectral density

It is easy to show that for a zero-mean AR(p) process with parameters $\phi_1, \ldots, \phi_p$ and innovations variance σ_u^2 the spectral density is given by

$$S_{AR,p}(f) = \frac{\sigma_u^2}{|H(f)|^2} \tag{8.145}$$

where

$$H(f) = 1 - \sum_{k=1}^{p} \phi_k \exp(i2\pi fk). \tag{8.146}$$

The importance of this result is that any continuous and nonzero spectral density $S(f)$ can be approximated arbitrarily closely and uniformly in f by an AR(p) spectral density $S_{AR,p}(f)$ for sufficiently large p (Grenander and Rosenblatt, 1957).

8.14.3 Classic spectral density estimation methods

The classic, most frequently used way to estimate spectral density is a nonparametric method based on smoothing the periodogram. The steps are as follows. Let y_t, $t = 1, \ldots, T$, be the observed data, let d_t, $t = 1, \ldots, T$, be a *data taper* that goes smoothly to zero at both ends, and form the modified data $\widetilde{y}_t = d_t y_t$. Then use the fast Fourier transform (FFT; Bloomfield, 1976) to compute the discrete Fourier transform

$$X(f_k) = \sum_{t=1}^{T} \widetilde{y}_t \exp(-i2\pi f_k t) \tag{8.147}$$

where $f_k = k/T$ for $k = 0, 1, \ldots, [T/2]$. Use the result to form the *periodogram*:

$$\widehat{S}(f_k) = \frac{1}{T} |X(f_k)|^2. \tag{8.148}$$

It is known that the periodogram is an approximately unbiased estimator of $S(f)$ for large T, but it is not a consistent estimator. For this reason, $\widehat{S}(f_k)$ is smoothed in the frequency domain to obtain an improved estimator of reduced variability, namely

$$\overline{S}(f_k) = \sum_{m=-M}^{M} w_m \widehat{S}(f_m), \tag{8.149}$$

where the smoothing weights w_m are symmetric with

$$w_m = w_{-m} \quad \text{and} \quad \sum_{m=-M}^{M} w_m = 1.$$

The purpose of the data taper is to reduce the so-called *leakage* effect of implicit truncation of the data with a rectangular window; originally, data tapers such as a cosine window or Parzen window were used. For details on this and other aspects of spectral density estimation, see Bloomfield (1976). A much preferred method is to use a prolate spheroidal taper, whose application in spectral analysis was pioneered by Thomson (1977). See also Percival and Walden (1993).

Given the result in Section 8.14.2 one can also use a parametric AR($\widehat{p}$) approximation approach to estimating the spectral density based on parameter estimators $\widehat{\phi}_1, \ldots, \widehat{\phi}_{\widehat{p}}$ and $\widehat{\sigma}_u^2$; here $\widehat{p}$ is an estimator of the order p, obtained through a selection criterion such as AIC, BIC or FPE which are discussed in Brockwell and Davis (1991). In this case we compute

$$\widehat{S}_{AR,\widehat{p}}(f) = \frac{\widehat{\sigma}_u^2}{\left|1 - \sum_{k=1}^{\widehat{p}} \widehat{\phi}_k \exp(i2\pi fk)\right|^2} \tag{8.150}$$

on a grid of frequency values $f = f_k$.

8.14.4 Prewhitening

Prewhitening is a filtering technique introduced by Blackman and Tukey (1958), in order to transform a time series into one whose spectrum is nearly flat. One then estimates the spectral density of the prewhitened series, with a greatly reduced impact of leakage bias, and then transforms the prewhitened spectral density back, using the frequency domain equivalent of inverse filtering, in order to obtain an estimator of the spectrum for the original series. Tukey (1967) says:

> If low frequencies are 10^3, 10^4, or 10^5 times as active as high ones, a not infrequent phenomenon in physical situations, even a fairly good window is too leaky for comfort. The cure is not to go in for fancier windows, but rather to preprocess the data toward a flatter spectrum, to analyze this *prewhitened* series, and then to adjust its estimatord spectrum for the easily computable effects of preprocessing.

The classic (nonrobust) way to accomplish the overall estimation method is to use the following modified form of the AR spectrum estimator (8.150):

$$\overline{S}_{AR,\widehat{p}}(f) = \frac{\overline{S}_{\widehat{u},\widehat{p}}(f)}{\left|1 - \sum_{k=1}^{\widehat{p}} \widehat{\phi}_k \exp(i2\pi fk)\right|^2} \tag{8.151}$$

where $\overline{S}_{\widehat{u},\widehat{p}}(f)$ is a smoothed periodogram estimator as described above, but applied to the fitted AR *residuals* $\widehat{u}_t = y_t - \widehat{\phi}_1 y_{t-1} - \ldots - \widehat{\phi}_{\widehat{p}} y_{t-\widehat{p}}$. The estimator $\overline{S}_{AR,\widehat{p}}(f)$ provides substantial improvement on the simpler estimator $\widehat{S}_{AR,\widehat{p}}(f)$ in (8.150) by replacing the numerator estimator $\widehat{\sigma}_u^2$ that is fixed, independent of frequency, with the frequency-varying estimator $\overline{S}_{r,\widehat{p}}(f)$. The order estimator $\widehat{p}$ may be obtained with an AIC or BIC order selection method (the latter is known to be preferable). Experience indicates that use of moderately small fixed orders p_o in the range from two to six will often suffice for effective prewhitening, suggesting that automatic order selection will often result in values of $\hat{p}$ in a similar range.

8.14.5 Influence of outliers on spectral density estimators

Suppose the AO model $y_t = x_t + \upsilon_t$ contains a single additive outlier υ_{t_0} of size A. Then the periodogram $\hat{S}_y(f_k)$ based on the observations y_t will have the form

$$\hat{S}_y(f_k) = \hat{S}_x(f_k) + \frac{A^2}{T} + 2\frac{A}{T}\mathrm{Re}[X(f_k)\exp(i2\pi f_k)] \tag{8.152}$$

where $\hat{S}_x(f_k)$ is the periodogram based on the outlier-free series x_t and Re denotes the real part. Thus the outlier causes the estimator to be raised by the constant amount A^2/T at all frequencies, plus the amount of the oscillatory term

$$\frac{2A}{T}\mathrm{Re}[X(f_k)\exp(i2\pi f_k)]$$

that varies with frequency. If the spectrum amplitude varies over a wide range with frequency, the effect of the outlier can be to obscure small but important peaks (corresponding to small-amplitude oscillations in the x_t series) in low-amplitude regions of the spectrum. It can be shown that a pair of outliers can generate an oscillation whose frequency is determined by the time separation of the outliers, and whose impact can also obscure features in the low-amplitude region of the spectrum (Problem 8.13).

To get an idea of the impact of AOs more generally, we focus on the mean and variance of the smoothed periodogram estimators $\overline{S}(f_k)$ under the assumption that x_t and υ_t are independent, and that the conditions of consistency and asymptotic normality of $\overline{S}(f_k)$ hold. Then for moderately large sample sizes, the mean and variance of $\overline{S}(f_k)$ are given approximately by

$$\mathrm{E}\overline{S}(f_k) = S_y(f_k) = S_x(f_k) + S_\upsilon(f_k) \tag{8.153}$$

and

$$\mathrm{Var}(\overline{S}(f_k)) = S_y(f_k)^2 = S_x(f_k)^2 + S_\upsilon(f_k)^2 + 2S_x(f_k)S_\upsilon(f_k). \tag{8.154}$$

Thus AOs cause both bias and inflated variability of the smoothed periodogram estimator. If υ_t is i.i.d. with variance σ_υ^2, the bias is just σ_υ^2 and the variance is inflated by the amount $\sigma_\upsilon^2 + 2S_x(f_k)\sigma_\upsilon^2$.

Striking examples of the influence that outliers can have on spectral density estimators were given by Kleiner *et al.* (1979) and Martin and Thomson (1982). The most dramatic and compelling of these examples is in the former paper, where the data consist of 1000 measurements of diameter distortions along a section of an advanced wave-guide designed to carry over 200,000 simultaneous telephone conversations. In this case, the data are a "space" series but can be treated in the same manner as a time series as far as spectrum analysis is concerned. Two relatively minor outliers due to a

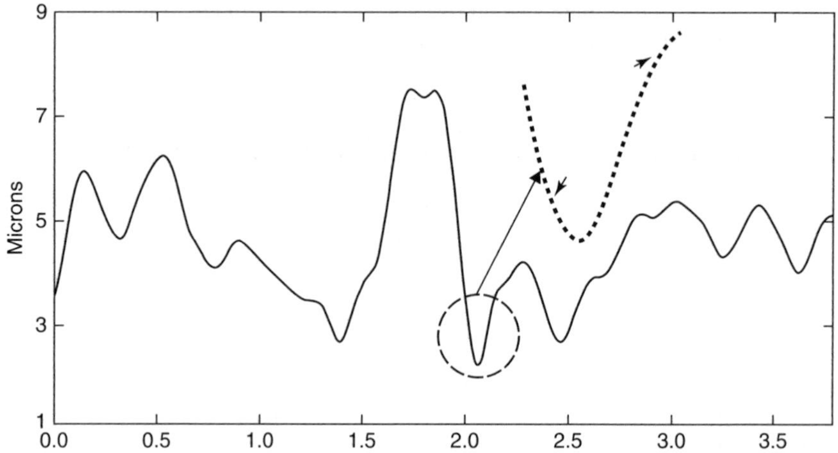

Figure 8.15 Wave-guide data: diameter distortion measurements against distance

malfunctioning of the recording instrument, and not noticeable in simple plots of the data, obscure important features of a spectrum having a very wide dynamic range (in this case the ratio of the prediction variance to the process variance of an AR(7) fit is approximately 10^{-6}!). Figure 8.15 (from Kleiner *et al.*, 1979) shows the diameter distortion measurements as a function of distance along the wave-guide, and points out that the two outliers are noticeable only in a considerably amplified local section of the data. Figure 8.16 shows the differenced series (a "poor man's prewhitening"), which clearly reveals the location of the two outliers as doublets; Figure 8.17 shows the classic periodogram-based estimator (dashed line) with the oscillatory artifact

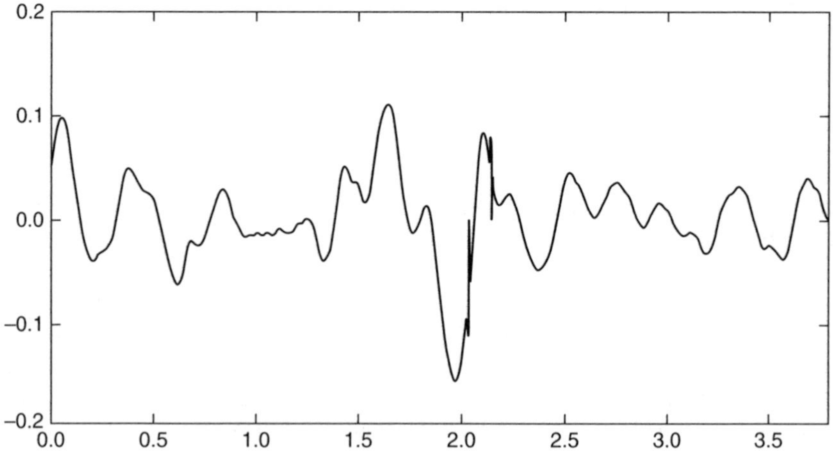

Figure 8.16 Wave-guide data: differenced series

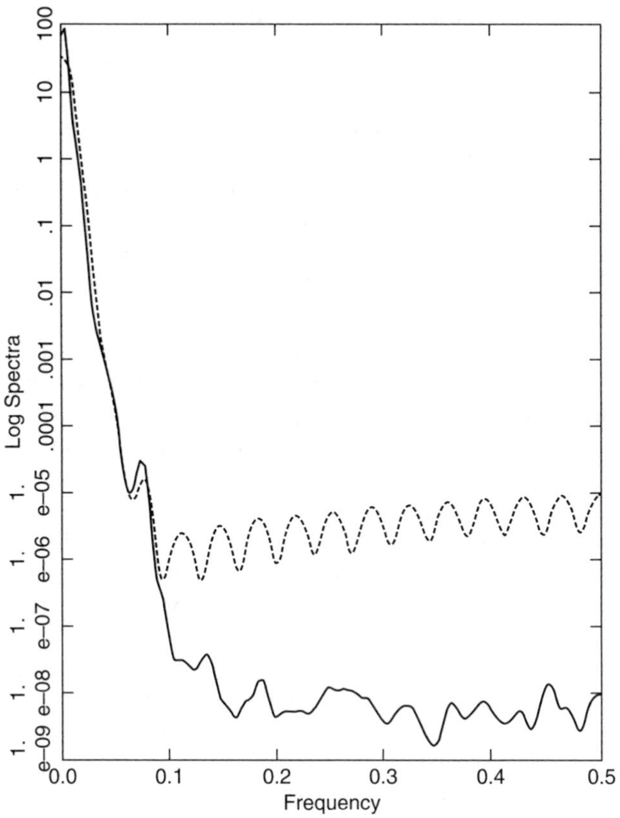

Figure 8.17 Wave-guide data: classical (- - - - - -) and robust (– -) spectra

caused by the outliers, along with a robust estimator (solid line) that we describe next. Note in the latter figure that the classic estimator has an eight-decade dynamic range, while the robust estimator has a substantially increased dynamic range of close to eleven decades, and reveals features that have known physical interpretations that are totally obscured in the classical estimator; see Kleiner *et al.* (1979) for details.

8.14.6 Robust spectral density estimation

Our recommendation is to compute robust spectral density estimators by robustifying the prewhitened spectral density (8.151) as follows. The AR parameter estimators $\widehat{\phi}_1, \widehat{\phi}_2, \ldots, \widehat{\phi}_{\widehat{p}}$ and $\widehat{\sigma}_u^2$ are computed using the Fτ-estimator, and $\widehat{p}$ is computed using the robust order selection method of Section 8.6.6. Then, to compute a robust smoothed spectral density estimator $\overline{S}_{\widetilde{u}*,\widehat{p}}(f)$, the nonrobust residual estimators

$$\widehat{u}_t = y_t - \widehat{\phi}_1 y_{t-1} - \ldots - \widehat{\phi}_{\widehat{p}} y_{t-\widehat{p}}$$

are replaced by the robust residual estimators, defined as

$$\widetilde{u}_t^* = \widehat{x}_{t|t} - \widehat{\phi}_1\widehat{x}_{t-1|t-1} - \widehat{\phi}_2\widehat{x}_{t-2|t-1} - \dots - \widehat{\phi}_{\widehat{p}}\widehat{x}_{t-\widehat{p}|t-1},$$

where $\widehat{x}_{t-i|t-1}$, $i = 0, 1, \dots, \widehat{p}$, are obtained from the robust filter. Note that these robust prediction residuals differ from the robust prediction residuals $\widetilde{u}_t$ (8.76) in Section 8.6.2 in that the latter have $y_t - \mu$ where we have $\widehat{x}_{t|t}$. We make this replacement because we do not want outliers to influence the smoothed periodogram estimator based on the robust residuals. Also, we do not bother with an estimator of μ because, as mentioned at the beginning of the section, one always works with de-meaned series in spectral analysis.

Note that our approach in this chapter – of using robust filtering – results in replacing outliers with one-sided predictions based on previous data. It is quite natural to think about improving this approach by using a *robust smoother*, as mentioned at the end of Section 8.6.2. See Martin and Thomson (1982) for the algorithm and its application to spectral density estimation. The authors show, using the wave-guide data, that it can be unsafe to use the robust filter algorithm if the AR order is not sufficiently large or the tuning parameters are changed somewhat, while the robust smoother algorithm results in a more reliable outlier interpolation and associated spectral density estimator; see Martin and Thomson (1982, Figs 24–27).

Kleiner *et al.* (1979) also show good results for some examples using a pure robust AR spectral density estimator; that is, the robust smoothed spectral density estimator $\overline{S}_{\widetilde{u}^*,\widehat{p}}(f)$ is replaced with a robust residuals variance estimator $\widehat{\sigma}_u^2$ and a sufficiently high AR order is used. Our feeling is that this approach is only suitable for spectrum analysis contexts where the user is confident that the dynamic range of the spectrum is not very large, at most two or three decades.

The reader interested in robust spectral density estimation can find more details and several examples in Kleiner *et al.* (1979) and Martin and Thomson (1982). Martin and Thomson (1982, Section III) point out that small outliers may not only obscure the lower part of the spectrum but also may inflate innovation variance estimators by orders of magnitude.

8.14.7 Robust time-average spectral density estimator

The classic approach to spectral density estimation described in Section 8.14.3 reduces the variability of the periodogram by averaging periodogram values in the frequency domain, as indicated in (8.149). In some applications with large amounts of data, it may be advantageous to reduce the variability by averaging the periodogram in the *time* domain, as originally described by Welch (1967). The idea is to break the time series data up into M equal-length contiguous segments of length N, compute the periodogram $\widehat{S}_m(f_k) = \frac{1}{T}|X_m(f_k)|^2$ at each frequency $f_k = k/N$

on the mth segment, and at each f_k form the smoothed periodogram estimator

$$\overline{S}(f_k) = \frac{1}{M}\sum_{m=1}^{M} \widehat{S}_m(f_k). \tag{8.155}$$

The problem with this estimator is that even a single outlier in the mth segment can spoil the estimator $\widehat{S}_m(f_k)$, as discussed previously. One way to robustify this estimator is to replace the sample mean in (8.155) with an appropriate robust estimator. One should not use a location M-estimator that assumes a symmetric nominal distribution for the following reason. Under normality, the periodogram may be represented by the approximation

$$2\widehat{S}_m(f_k) \approx s_k Y \tag{8.156}$$

where Y is a chi-squared random variable with two degrees of freedom and $s_k = \mathrm{E}\widehat{S}_m(f_k) \approx S(f_k)$ for large T. Thus estimation of $S(f_k)$ is equivalent to estimating the scale of an exponential distribution.

Under AO- or RO-type outlier contamination, a reasonable approximate model for the distribution of the periodogram $\widehat{S}_m(f_k)$ is the contaminated exponential distribution

$$(1-\varepsilon)\mathrm{Ex}(s_k) + \varepsilon \mathrm{Ex}(s_{c,k}), \tag{8.157}$$

where $\mathrm{Ex}(\alpha)$ is the exponential distribution with mean α. Here outliers may result in $s_{c,k} > s_k$, at least at some frequencies f_k. Thus the problem is to find a good robust estimator of s_k in the contaminated exponential model (8.157). It must be kept in mind that the overall data series can have a quite small fraction of contamination and still influence many of the segment estimators $\widehat{S}_m(f_k)$, and hence a high BP estimator of s_k is desirable. Consider a more general model of the form (8.157), in which the contaminating distribution $\mathrm{Ex}(s_{c,k})$ is replaced with the distribution of any positive random variable. As mentioned in Section 5.2.2, the min–max bias estimator of scale for this case is very well approximated by a scaled median (Martin and Zamar, 1989) with scaling constant $(0.693)^{-1}$ for Fisher-consistency for the nominal exponential distribution. Thus it is recommended to replace the nonrobust time-average estimator (8.155) with the scaled median estimator

$$\overline{S}(f_k) = \frac{1}{0.693}\,\mathrm{Med}\{\widehat{S}_m(f_k), m = 1, \ldots, M\}. \tag{8.158}$$

This estimator can be expected to work well in situations where less than half of the time segments of data contain influential outliers.

The idea of replacing the sample average in (8.155) with a robust estimator of the scale of an exponential distribution was considered by Thomson (1977) and discussed by Martin and Thomson (1982), with a focus on using an asymmetric truncated mean as the robust estimator. See also Chave *et al.* (1987) for an application.

8.15 Appendix A: Heuristic derivation of the asymptotic distribution of M-estimators for ARMA models

To simplify, we replace $\widehat{\sigma}$ in (8.65) by its asymptotic value σ. Generally, $\widehat{\sigma}$ is calibrated so that when u_t is normal, $\sigma^2 = \sigma_u^2 = \mathrm{E}u_t^2$. Differentiating (8.65) we obtain

$$\sum_{t=p+1}^{T} \psi\left(\frac{\widehat{u}_t(\widehat{\lambda})}{\sigma}\right) \frac{\partial \widehat{u}_t(\widehat{\lambda})}{\partial \lambda} = 0. \tag{8.159}$$

We leave it as an exercise (Problem 8.9) to show that

$$\frac{\partial \widehat{u}_t(\lambda)}{\partial \mu} = -\frac{1 - \phi_1 - \ldots - \phi_p}{1 - \theta_1 - \ldots - \theta_q}, \tag{8.160}$$

$$\frac{\partial \widehat{u}_t(\lambda)}{\partial \boldsymbol{\phi}_i} = -\phi^{-1}(B)\widehat{u}_{t-i}(\lambda) \tag{8.161}$$

and

$$\frac{\partial \widehat{u}_t(\lambda)}{\partial \boldsymbol{\theta}_j} = \theta^{-1}(B)\widehat{u}_{t-j}(\lambda). \tag{8.162}$$

Let

$$\mathbf{z}_t = \left.\frac{\partial \widehat{u}_t(\lambda)}{\partial \lambda}\right|_{\lambda=\lambda_0}, \quad \mathbf{W}_t = \left.\frac{\partial^2 \widehat{u}_t(\lambda)}{\partial \lambda^2}\right|_{\lambda=\lambda_0},$$

where λ_0 is the true value of the parameter. Observe that

$$\mathbf{z}_t = (\mathbf{c}_t, \mathbf{d}_t, \xi)'$$

with ξ defined in (8.63) and

$$\mathbf{c}_t = -(\phi^{-1}(B)u_{t-1}, \ldots, \phi^{-1}(B)u_{t-p})',$$
$$\mathbf{d}_t = (\theta^{-1}(B)u_{t-1}, \ldots, \theta^{-1}(B)u_{t-q})'.$$

Since $\widehat{u}_t(\lambda_0) = u_t$, a first-order Taylor expansion yields

$$\sum_{t=p+1}^{T} \psi\left(\frac{u_t}{\sigma}\right)\mathbf{z}_t + \left(\frac{1}{\sigma}\sum_{t=p+1}^{T} \psi'\left(\frac{u_t}{\sigma}\right)\mathbf{z}_t\mathbf{z}_t' + \sum_{t=p+1}^{T} \psi\left(\frac{u_t}{\sigma}\right)\mathbf{W}_t\right)(\widehat{\lambda} - \lambda_0) \simeq 0$$

and then

$$T^{1/2}(\widehat{\lambda} - \lambda_0) \simeq \mathbf{B}^{-1}\left(\frac{1}{T^{1/2}}\sum_{t=p+1}^{T} \psi\left(\frac{u_t}{\sigma}\right)\mathbf{z}_t\right) \tag{8.163}$$

with

$$\mathbf{B} = \frac{1}{\sigma T} \sum_{t=p+1}^{T} \psi'\left(\frac{u_t}{\sigma}\right) \mathbf{z}_t \mathbf{z}_t' + \frac{1}{T} \sum_{t=p+1}^{T} \psi\left(\frac{u_t}{\sigma}\right) \mathbf{W}_t.$$

We shall show that

$$\underset{T\to\infty}{\mathrm{p\,lim}}\ \beta = \frac{1}{\sigma} \mathrm{E}\psi'\left(\frac{u_t}{\sigma}\right) \mathrm{E}\mathbf{z}_t\mathbf{z}_t', \tag{8.164}$$

and that

$$\frac{1}{T^{1/2}} \sum_{t=p+1}^{T} \psi\left(\frac{u_t}{\sigma}\right) \mathbf{z}_t \to_d \mathrm{N}_{p+q+1}\left(\mathbf{0}, \mathrm{E}\psi\left(\frac{u_t}{\sigma}\right)^2 \mathrm{E}\mathbf{z}_t\mathbf{z}_t'\right). \tag{8.165}$$

From (8.163), (8.164) and (8.165), we get

$$T^{1/2}(\hat{\lambda} - \lambda_0) \to_d \mathrm{N}_{p+q+1}(\mathbf{0}, \mathbf{V}_M), \tag{8.166}$$

where

$$\mathbf{V}_M = \frac{\sigma^2 \mathrm{E}\psi(u_t/\sigma)^2}{(\mathrm{E}\psi'(u_t/\sigma))^2} (\mathrm{E}\mathbf{z}_t\mathbf{z}_t')^{-1}. \tag{8.167}$$

It is not difficult to show that the terms on the left-hand side of (8.164) are uncorrelated and have the same mean and variance. Hence it follows from the weak law of large numbers that

$$\underset{T\to\infty}{\mathrm{p\,lim}} \frac{1}{\sigma T} \sum_{t=p+1}^{T} \left(\psi'\left(\frac{u_t}{\sigma}\right) \mathbf{z}_t\mathbf{z}_t' + \psi\left(\frac{u_t}{\sigma}\right) \mathbf{W}_t\right)$$
$$= \left[\frac{1}{\sigma} \mathrm{E}\left(\psi'\left(\frac{u_t}{\sigma}\right) \mathbf{z}_t\mathbf{z}_t'\right) + \mathrm{E}\left(\psi\left(\frac{u_t}{\sigma}\right) \mathbf{W}_t\right)\right].$$

Then, (8.164) follows from the fact that u_t is independent of $\mathbf{z}_t$ and $\mathbf{W}_t$ and (8.66).

Recall that a sequence of random vectors $\mathbf{q}_n \in R^k$ converges in distribution to $\mathrm{N}_k(\mathbf{0}, \mathbf{A})$ if and only if each linear combination $\mathbf{a}'\mathbf{q}_n$ converges in distribution to $\mathrm{N}(0, \mathbf{a}'\mathbf{A}\mathbf{a})$ (Feller, 1971). Then, to prove (8.165) it is enough to show that for any $\mathbf{a} \in R^{p+q+1}$

$$\frac{1}{\sqrt{T}} \sum_{t=p+1}^{T} H_t \to_d \mathrm{N}(0, v_0), \tag{8.168}$$

where

$$H_t = \psi(u_t/\sigma)\mathbf{a}'\mathbf{z}_t$$

and

$$v_0 = \mathrm{E}(H_t)^2 = \mathbf{a}'\left(\mathrm{E}\psi\left(\frac{u_t}{\sigma}\right)^2 \mathrm{E}(\mathbf{z}_t\mathbf{z}_t')\right)\mathbf{a}.$$

Since the variables are not independent, the standard central limit theorem cannot be applied. However, it can be shown that the stationary process H_t satisfies

$$\mathrm{E}(H_t|H_{t-1},\ldots,H_1)=0 \quad \text{a.s.}$$

and is hence a so-called *martingale difference sequence*. Therefore, by the central limit theorem for martingales (see Theorem 23.1 of Billingsley, 1968) (8.168) holds, and hence (8.165) is proved.

We shall now find the form of the covariance matrix $\mathbf{V}_M$. Let

$$\phi^{-1}(B)u_t=\sum_{i=0}^{\infty}\pi_i u_{t-i}$$

and

$$\theta^{-1}(B)u_t=\sum_{i=0}^{\infty}\zeta_i u_{t-i},$$

where $\pi_0=\zeta_0=1$. We leave it as an exercise (Problem 8.10) to show that $\mathrm{E}(\mathbf{z}_t\mathbf{z}_t')$ has the following form:

$$\mathrm{E}(\mathbf{z}_t\mathbf{z}_t')=\begin{bmatrix}\sigma_u^2\mathbf{D} & \mathbf{0}\\ 0 & \xi^2\end{bmatrix}, \tag{8.169}$$

where $\mathbf{D}=\mathbf{D}(\phi,\theta)$ is a symmetric $(p+q)$ matrix with elements

$$D_{i,j}=\sum_{k=0}^{\infty}\pi_k\pi_{k+j-i} \text{ if } i\le j\le p$$

$$D_{i,p+j}=\sum_{k=0}^{\infty}\zeta_k\pi_{k+j-i} \text{ if } i\le p,\ j\le q,\ i\le j$$

$$D_{i,p+j}=\sum_{k=0}^{\infty}\pi_k\zeta_{k+i-j} \text{ if } i\le p,\ j\le q,\ j\le i$$

$$D_{p+i,p+j}=\sum_{k=0}^{\infty}\zeta_k\zeta_{k+j-i} \text{ if } i\le j\le q.$$

Therefore, the asymptotic covariance matrix of $\hat{\boldsymbol{\lambda}}$ is

$$\mathbf{V}_M=\frac{\sigma^2\mathrm{E}\psi(u_t/\sigma)^2}{(\mathrm{E}\psi'(u_t/\sigma))^2}\begin{bmatrix}\sigma_u^{-2}\mathbf{D}^{-1} & \mathbf{0}\\ \mathbf{0} & \xi^{-2}\end{bmatrix}.$$

In the case of the LS estimator, since $\psi(u)=2u$ and $\psi'(u)=2$, we have

$$\frac{\sigma^2\mathrm{E}\psi(u_t/\sigma)^2}{(\mathrm{E}\psi'(u_t/\sigma))^2}=\mathrm{E}u_t^2=\sigma_u^2,$$

and hence the asymptotic covariance matrix is

$$\mathbf{V}_{LS} = \begin{bmatrix} \mathbf{D}^{-1} & \mathbf{0} \\ \mathbf{0} & \sigma_u^2/\xi^2 \end{bmatrix}.$$

In consequence we have

$$\mathbf{V}_M = \frac{\sigma^2 \mathrm{E}\psi(u_t/\sigma)^2}{\sigma_u^2(\mathrm{E}\psi'(u_t/\sigma))^2}\mathbf{V}_{LS}.$$

In the AR(p) case, the matrix $\sigma_u^2\mathbf{D}$ coincides with the covariance matrix $\mathbf{C}$ of $(y_t, y_{t-1}, \ldots, y_{t-p+1})$ used in (8.49).

8.16 Appendix B: Robust filter covariance recursions

The vector $\mathbf{m}_t$ appearing in (8.82) is the first column of the covariance matrix of the state prediction error $\widehat{\mathbf{x}}_{t|t-1} - \mathbf{x}_t$:

$$\mathbf{M}_t = \mathrm{E}(\widehat{\mathbf{x}}_{t|t-1} - \mathbf{x}_t)(\widehat{\mathbf{x}}_{t|t-1} - \mathbf{x}_t)' \tag{8.170}$$

and

$$s_t = \sqrt{M_{t,11}} = \sqrt{m_{t,1}} \tag{8.171}$$

is the standard deviation of the observation prediction error $y_t - \hat{y}_{t|t-1} = y_t - \widehat{x}_{t|t-1}$. The recursion for $\mathbf{M}_t$ is

$$\mathbf{M}_t = \mathbf{\Phi} P_{t-1} \mathbf{\Phi} + \sigma_u^2 \mathbf{dd}', \tag{8.172}$$

where $\mathbf{P}_t$ is the covariance matrix of the state-filtering error $\widehat{\mathbf{x}}_{t|t} - \mathbf{x}_t$:

$$\mathbf{P}_t = \mathrm{E}(\widehat{\mathbf{x}}_{t|t} - \mathbf{x}_t)(\widehat{\mathbf{x}}_{t|t} - \mathbf{x}_t)'. \tag{8.173}$$

The recursion equation for $\mathbf{P}_t$ is

$$\mathbf{P}_t = \mathbf{M}_t - \frac{1}{s_t^2} W\left(\frac{\widetilde{u}_t}{s_t}\right)\mathbf{m}_t\mathbf{m}_t'$$

where $W(u) = \psi(u)/u$.

Reasonable initial conditions for the robust filter are $\widehat{\mathbf{x}}_{0|0} = (0, 0, \ldots, 0)'$, and $\mathbf{P}_0 = \hat{\mathbf{P}}_x$, where $\hat{\mathbf{P}}_x$ is a $p \times p$ robust estimator of the covariance matrix for $(y_{t-1}, y_{t-2}, \ldots, y_{t-p})$.

When applying the robust Durbin–Levinson algorithm to estimate an AR(p) model, the above recursions need to be computed for each of a sequence of AR orders $m = 1, \ldots, p$. Accordingly, we shall take $\sigma_u^2 = \sigma_{u,m}^2$, where $\sigma_{u,m}^2$ is the variance of the memory-m prediction error of x_t; that is,

$$\sigma_{u,m}^2 = \mathrm{E}(x_t - \widetilde{\phi}_{m,1}x_{t-1} - \cdots - \widetilde{\phi}_{m,m}x_{t-m})^2.$$

Then we need an estimator $\widehat{\sigma}^2_{u,m}$ of $\sigma^2_{u,m}$ for each m. This can be accomplished by using the following relationships:

$$\sigma^2_{u,1} = (1 - \widetilde{\phi}^2_{1,1})\sigma^2_x, \tag{8.174}$$

where σ^2_x is the variance of x_t, and

$$\sigma^2_{u,m} = (1 - \widetilde{\phi}^2_{m,m})\sigma^2_{u,m-1}. \tag{8.175}$$

In computing (8.91) for $m = 1$, we use the estimator $\widehat{\sigma}^2_{u,1}$ of $\sigma^2_{u,1}$, parameterized as a function of $\widetilde{\phi} = \widetilde{\phi}_{1,1}$ using (8.174)

$$\widehat{\sigma}^2_{u,1}(\widetilde{\phi}) = (1 - \widetilde{\phi}^2)\widehat{\sigma}^2_x \tag{8.176}$$

where $\widehat{\sigma}^2_x$ is a robust estimator of σ^2_x based on the observations y_t. For example, we might use an M- or τ-scale, or the simple estimator $\widehat{\sigma}_x = \text{MADN}(y_t)/0.6745$. Then when computing (8.91) for $m > 1$, we use the estimator $\widehat{\sigma}^2_{u,m}$ of $\sigma^2_{u,m}$ parameterized as a function of $\widetilde{\phi} = \widetilde{\phi}_{m,m}$ using (8.175)

$$\widehat{\sigma}^2_{u,m}(\phi) = (1 - \widetilde{\phi}^2)\widehat{\sigma}^2_{u,m-1} \tag{8.177}$$

where $\widehat{\sigma}_{u,m-1}$ is the minimized robust scale $\widehat{\sigma}$ in (8.91) for the order-$(m - 1)$ fit.

Since the function in (8.91) may have more than one local extremum, the minimization is performed by means of a grid search on $(-1, 1)$.

8.17 Appendix C: ARMA model state-space representation

Here we describe the state-space representation, (8.101) and (8.103), for an ARMA(p, q) model, and show how to extend it to ARIMA and SARIMA models. We note that a state-space representation of ARMA models is not unique, and the particular representation we chose was that by Ledolter (1979) and Harvey and Phillips (1979). For other representations see Akaike (1974b), Jones (1980) and Chapter 12 of Brockwell and Davis (1991).

Let

$$\phi(B)(x_t - \mu) = \theta(B)u_t.$$

Define $\boldsymbol{\alpha}_t = (\alpha_{1,t}, \ldots, \alpha_{p,t})$, where

$$\alpha_{1,t} = x_t - \mu,$$

$$\alpha_{j,t} = \phi_j(x_{t-1} - \mu) + \ldots + \phi_p(x_{t-p+j-1} - \mu) - \theta_{j-1}u_t - \ldots - \theta_q u_{t-q+j-1},$$
$$j = 2, \ldots, q + 1$$

and

$$\alpha_{j,t} = \phi_j(x_{t-1} - \mu) + \ldots + \phi_p(x_{t-p+j-1} - \mu),\ j = q+2, \ldots, p.$$

It is left as an exercise to show that the state-space representation (8.101) holds where $\mathbf{d}$ and Φ are given by (8.102) and (8.103) respectively. In the definition of $\mathbf{d}$, we take $\theta_i = 0$ for $i > q$.

The case $q \geq p$ is reduced to the above procedure on observing that y_t can be represented as an ARMA$(q+1, q)$ model where $\phi_i = 0$ for $i \geq p$. Thus, in general the dimension of $\boldsymbol{\alpha}$ is $k = \max(p, q+1)$.

The above state-space representation is easily extended to represent an ARIMA(p, d, q) model (8.108) by writing it as

$$\phi^*(B)(y_t - \mu) = \theta(B)u_t \tag{8.178}$$

where $\phi^*(B) = \phi(B)(1-B)^d$ has order $p^* = p + d$. Now we just proceed as above, with the ϕ_i replaced by the ϕ_i^* coefficients in the polynomial $\phi^*(B)$, resulting in the state-transition matrix Φ^*. For example, in the case of an ARIMA$(1, 1, q)$ model we have $\phi_1^* = 1 + \phi_1$ and $\phi_2^* = -\phi_1$. The order of Φ^* is $k^* = \max(p^*, q+1)$.

The above approach also easily handles the case of a SARIMA model (8.109). One just defines

$$\phi^*(B) = \phi(B)\Phi(B^s)(1-B)^d(1-B^s)^D, \tag{8.179}$$

$$\theta^*(B) = \theta(B)\Theta(B^s) \tag{8.180}$$

and specifies the state-transition matrix Φ^* and vector $\mathbf{d}^*$ based on the coefficients of polynomials $\phi^*(B)$ and $\theta^*(B)$ of order p^* and q^* respectively. The order of Φ^* is now $k = \max(p^*, q^* + 1)$.

8.18 Recommendations and software

For ARIMA and REGARIMA models we recommend the filtered τ-estimators (Section 8.8.3) computed with **arima.rob** (`robustarima`).

When used for ARIMA models without covariables, the formula should be of the form x ~ 1, where x is the name of the time series. In this case the intercept corresponds to the mean of the time series, or to the mean of the differenced time series if the specified order of differencing is positive.

8.19 Problems

8.1. Show that $|\widehat{\rho}(1)| \leq 1$ for $\widehat{\rho}(1)$ in (8.3). Also show that if the summation in the denominator in (8.3) ranges only from 1 to $T-1$, then $|\widehat{\rho}(1)|$ can be larger than one.

8.2. Show that for a "doublet" outlier at t_0 (i.e., $y_{t_0} = A = -y_{t_0+1}$) with $t_0 \in (1, T)$, the limiting value as $A \to \infty$ of $\widehat{\rho}(1)$ in (8.3) is -0.5.

8.3. Show that the limiting value as $A \to \infty$ of $\widehat{\rho}(1)$ defined in (8.4), when there is an isolated outlier of size A, is $-1/T + O(1/T^2)$.

8.4. Construct a probability model for additive outliers v_t that has non-overlapping patches of length $k > 0$, such that $v_t = A$ within each patch and $v_t = 0$ otherwise, and with $\mathrm{P}(v_t \neq 0) = \varepsilon$.

8.5. Verify the expression for the Yule–Walker equations given by (8.28).

8.6. Verify that for an AR(1) model with parameter ϕ we have $\rho(1) = \phi$.

8.7. Show that the LS estimator of the AR(p) parameters given by (8.26) is equivalent to solving the Yule–Walker equation(s) (8.28) with the true covariances replaced by the sample ones (8.30).

8.8. Prove the orthogonality condition (8.37).

8.9. Verify (8.160)–(8.162).

8.10. Prove (8.169).

8.11. Prove (8.40).

8.12. Verify (8.13) using (8.134).

8.13. Calculate the spectral density for the case that x_{t_0} and x_{t_0+k} are replaced by A and $-A$ respectively.

9

Numerical Algorithms

Computing M-estimators involves function minimization and/or solving nonlinear equations. General methods based on derivatives – like the Newton-Raphson procedure for solving equations – are widely available, but they are inadequate for this specific type of problem, for the reasons given in Section 2.10.5.1.

In this chapter we consider some details of the iterative algorithms used to compute M-estimators, as described in earlier chapters.

9.1 Regression M-estimators

We shall justify the algorithm in Section 4.5 for solving (4.39); this includes location as a special case. Consider the problem

$$h(\boldsymbol{\beta}) = \min,$$

where

$$h(\boldsymbol{\beta}) = \sum_{i=1}^{n} \rho\left(\frac{r_i(\boldsymbol{\beta})}{\sigma}\right),$$

where $r_i(\boldsymbol{\beta}) = y_i - \mathbf{x}_i'\boldsymbol{\beta}$ and σ is any positive constant.

It is assumed that the $\mathbf{x}_i$ are not collinear, otherwise there would be multiple solutions. It is assumed that $\rho(r)$ is a ρ-function, that the function $W(x)$ defined in (2.31) is nonincreasing in $|x|$, and that ψ is continuous. These conditions are easily verified for the Huber and the bisquare functions.

Robust Statistics: Theory and Methods (with R), Second Edition.
Ricardo A. Maronna, R. Douglas Martin, Victor J. Yohai and Matías Salibián-Barrera.

Companion website: www.wiley.com/go/maronna/robust

It will be proved that h does not increase at each iteration, and that if there is a single stationary point $\boldsymbol{\beta}_0$ of h; that is, a point satisfying

$$\sum_{i=1}^{n} \psi\left(\frac{r_i(\boldsymbol{\beta}_0)}{\sigma}\right)\mathbf{x}_i = \mathbf{0}, \tag{9.1}$$

then the algorithm converges to it.

For $r \geq 0$ let $g(r) = \rho(\sqrt{r})$. It follows from $\rho(r) = g(r^2)$ that

$$W(r) = 2g'(r^2) \tag{9.2}$$

and hence $W(r)$ is nonincreasing for $r \geq 0$ if and only if g' is nonincreasing.

We claim that

$$g(y) \leq g(x) + g'(x)(y - x); \tag{9.3}$$

that is, the graph of g lies below the tangent line. To show this, assume first that $y > x$ and note that by the intermediate value theorem,

$$g(y) - g(x) = (y - x)g'(\xi),$$

where $\xi \in [x, y]$. Since g' is nonincreasing, $g'(\xi) \leq g'(x)$. The case $y < x$ is dealt with likewise.

A function g with a nonincreasing derivative satisfies for all x, y and all $\alpha \in [0, 1]$

$$g(\alpha x + (1 - \alpha)y) \geq \alpha g(x) + (1 - \alpha)g(y); \tag{9.4}$$

that is, the graph of g lies above the secant line. Such functions are called *concave*. Conversely, a differentiable function is concave if and only if its derivative is nonincreasing. For twice differentiable functions, concavity is equivalent to having a nonpositive second derivative.

Define the matrix

$$\mathbf{U}(\boldsymbol{\beta}) = \sum_{i=1}^{n} W\left(\frac{r_i(\boldsymbol{\beta})}{\sigma}\right)\mathbf{x}_i\mathbf{x}_i',$$

which is nonnegative definite for all $\boldsymbol{\beta}$, and the function

$$f(\boldsymbol{\beta}) = \arg\min_{\boldsymbol{\gamma}} \sum_{i=1}^{n} W\left(\frac{r_i(\boldsymbol{\beta})}{\sigma}\right)(y_i - \mathbf{x}_i'\boldsymbol{\gamma})^2.$$

The algorithm can then be written as

$$\boldsymbol{\beta}_{k+1} = f(\boldsymbol{\beta}_k). \tag{9.5}$$

A fixed point $\boldsymbol{\beta}_0$ – that is, one satisfying $f(\boldsymbol{\beta}_0) = \boldsymbol{\beta}_0$ – is also a stationary point (9.1).

Given $\boldsymbol{\beta}_k$, put for simplicity $w_i = W(r_i(\boldsymbol{\beta}_k)/\sigma)$. Note that $\boldsymbol{\beta}_{k+1}$ satisfies

$$\sum_{i=1}^{n} w_i \mathbf{x}_i y_i = \sum_{i=1}^{n} w_i \mathbf{x}_i \mathbf{x}_i' \boldsymbol{\beta}_{k+1} = \mathbf{U}(\boldsymbol{\beta}_k)\boldsymbol{\beta}_{k+1}. \tag{9.6}$$

We shall show that

$$h(\boldsymbol{\beta}_{k+1}) \leq h(\boldsymbol{\beta}_k). \tag{9.7}$$

We have, using (9.3) and (9.2),

$$\begin{aligned} h(\boldsymbol{\beta}_{k+1}) - h(\boldsymbol{\beta}_k) &\leq \frac{1}{\sigma^2} \sum_{i=1}^{n} g'\left(\frac{r_i(\boldsymbol{\beta}_k)^2}{\sigma^2}\right)(r_i(\boldsymbol{\beta}_{k+1})^2 - r_i(\boldsymbol{\beta}_k)^2) \\ &= \frac{1}{2\sigma^2} \sum_{i=1}^{n} w_i (r_i(\boldsymbol{\beta}_{k+1}) - r_i(\boldsymbol{\beta}_k))(r_i(\boldsymbol{\beta}_{k+1}) + r_i(\boldsymbol{\beta}_k)). \end{aligned}$$

But since

$$\begin{aligned} &r_i(\boldsymbol{\beta}_{k+1}) - r_i(\boldsymbol{\beta}_k) = (\boldsymbol{\beta}_k - \boldsymbol{\beta}_{k+1})'\mathbf{x}_i \quad \text{and} \\ &r_i(\boldsymbol{\beta}_{k+1}) + r_i(\boldsymbol{\beta}_k) = 2y_i - \mathbf{x}_i'(\boldsymbol{\beta}_k + \boldsymbol{\beta}_{k+1}) \end{aligned}$$

we have, using (9.6),

$$\begin{aligned} h(\boldsymbol{\beta}_{k+1}) - h(\boldsymbol{\beta}_k) &\leq \frac{1}{2\sigma^2}(\boldsymbol{\beta}_k - \boldsymbol{\beta}_{k+1})' \sum_{i=1}^{n} w_i \mathbf{x}_i \mathbf{x}_i'(2\boldsymbol{\beta}_{k+1} - \boldsymbol{\beta}_k - \boldsymbol{\beta}_{k+1}) \\ &= \frac{1}{2\sigma^2}(\boldsymbol{\beta}_k - \boldsymbol{\beta}_{k+1})'\mathbf{U}(\boldsymbol{\beta}_k)(\boldsymbol{\beta}_{k+1} - \boldsymbol{\beta}_k) \leq 0 \end{aligned}$$

since $\mathbf{U}(\boldsymbol{\beta}_k)$ is nonnegative definite. This proves (9.7).

We shall now prove the convergence of $\boldsymbol{\beta}_k$ to $\boldsymbol{\beta}_0$ in (9.1). To simplify the proof, we make the stronger assumption that ρ is increasing and hence $W(r) > 0$ for all r. Since the sequence $h(\boldsymbol{\beta}_k)$ is nonincreasing and is bounded from below, it has a limit h_0. Hence the sequence $\boldsymbol{\beta}_k$ is bounded, otherwise there would be a subsequence $\boldsymbol{\beta}_{k_j}$ converging to infinity, and since ρ is increasing, so would $h(\boldsymbol{\beta}_{k_j})$.

Since $\boldsymbol{\beta}_k$ is bounded, it has a subsequence that has a limit $\boldsymbol{\beta}_0$, which by continuity satisfies (9.5) and is hence a stationary point. If it is unique, then $\boldsymbol{\beta}_k \to \boldsymbol{\beta}_0$; otherwise, there would exist a subsequence bounded away from $\boldsymbol{\beta}_0$, which in turn would have a convergent subsequence, which would have a limit different from $\boldsymbol{\beta}_0$, which would also be a stationary point. This concludes the proof of (9.1).

Another algorithm is based on *pseudo-observations*. Put

$$\tilde{y}_i(\boldsymbol{\beta}) = \mathbf{x}_i'\boldsymbol{\beta} + \hat{\sigma}\psi\left(\frac{r_i(\boldsymbol{\beta})}{\hat{\sigma}}\right).$$

Then (4.40) is clearly equivalent to

$$\sum_{i=1}^{n} \mathbf{x}_i(\tilde{y}_i(\hat{\boldsymbol{\beta}}) - \mathbf{x}_i'\hat{\boldsymbol{\beta}}) = 0.$$

Given $\boldsymbol{\beta}_k$, the next step of this algorithm is finding $\boldsymbol{\beta}_{k+1}$ such that

$$\sum_{i=1}^{n} \mathbf{x}_i(\tilde{y}_i(\boldsymbol{\beta}_k) - \mathbf{x}_i'\boldsymbol{\beta}_{k+1}) = 0,$$

which is an *ordinary* LS problem. The procedure can be shown to converge (Huber and Ronchetti, 2009, Sec. 7.8) but it is much slower than the reweighting algorithm.

9.2 Regression S-estimators

Here we deal with the descent algorithm described in Section 5.7.1.1. As explained there, the algorithm coincides with the one for M-estimators. The most important result is that, if W is nonincreasing, then at each step $\hat{\sigma}$ does not increase.

To see this, consider at step k the vector $\boldsymbol{\beta}_k$ and the respective residual scale σ_k, which satisfies

$$\frac{1}{n}\sum_{i=1}^{n} \rho\left(\frac{r_i(\boldsymbol{\beta}_k)}{\sigma_k}\right) = \delta.$$

The next vector $\boldsymbol{\beta}_{k+1}$ is obtained from (9.5) (with σ replaced by σ_k), and hence satisfies (9.7). Therefore

$$\frac{1}{n}\sum_{i=1}^{n} \rho\left(\frac{r_i(\boldsymbol{\beta}_{k+1})}{\sigma_k}\right) \le \frac{1}{n}\sum_{i=1}^{n} \rho\left(\frac{r_i(\boldsymbol{\beta}_k)}{\sigma_k}\right) = \delta. \tag{9.8}$$

Since σ_{k+1} satisfies

$$\frac{1}{n}\sum_{i=1}^{n} \rho\left(\frac{r_i(\boldsymbol{\beta}_{k+1})}{\sigma_{k+1}}\right) = \delta, \tag{9.9}$$

and ρ is nondecreasing, it follows from (9.9) and (9.8) that

$$\sigma_{k+1} \le \sigma_k. \tag{9.10}$$

9.3 The LTS-estimator

We shall justify the procedure in Section 5.7.1.2. Call $\hat{\sigma}_1$ and $\hat{\sigma}_2$ the scales corresponding to $\hat{\boldsymbol{\beta}}_1$ and $\hat{\boldsymbol{\beta}}_2$, respectively. For $k = 1, 2$ let $r_{ik} = y_i - \mathbf{x}_i'\hat{\boldsymbol{\beta}}_k$ be the respective residuals, and call $r^2_{(i)k}$ the ordered squared residuals. Let $I \subset \{1, \ldots, n\}$ be the set of indices corresponding to the h smallest r_{i1}^2. Then

$$\hat{\sigma}_2^2 = \sum_{i=1}^{h} r_{(i)2}^2 \leq \sum_{i \in I} r_{i2}^2 \leq \sum_{i \in I} r_{i1}^2 = \sum_{i=1}^{h} r_{(i)1}^2 = \hat{\sigma}_1^2.$$

9.4 Scale M-estimators

9.4.1 Convergence of the fixed-point algorithm

We shall show that the algorithm (2.80) given for solving (2.49) converges.

Define W as in (2.54). It is assumed again that $\rho(r)$ is a ρ-function of $|r|$. For $r \geq 0$ define

$$g(r) = \rho(\sqrt{r}). \tag{9.11}$$

It will be assumed that g is *concave* (see below (9.4)). To make things simpler, we assume that g is twice differentiable, and that $g'' < 0$.

The concavity of g implies that W is nonincreasing. In fact, it follows from $W(r) = g(r^2)/r^2$ that

$$W'(r) = \frac{2}{r^3}(r^2 g'(r^2) - g(r^2)) \leq 0,$$

since (9.3) implies for all t

$$0 = g(0) \leq g(t) + g'(t)(0 - t) = g(t) - tg'(t). \tag{9.12}$$

Put for simplicity $\theta = \sigma^2$ and $y_i = x_i^2$. Then (2.49) can be rewritten as

$$\frac{1}{n}\sum_{i=1}^{n} g\left(\frac{y_i}{\theta}\right) = \delta$$

and (2.80) can be rewritten as

$$\theta_{k+1} = h(\theta_k), \tag{9.13}$$

with

$$h(\theta) = \frac{1}{n\delta}\sum_{i=1}^{n} g\left(\frac{y_i}{\theta}\right)\theta. \tag{9.14}$$

It will be shown that h is nondecreasing and concave. It suffices to prove these properties for each term of (9.14). In fact, for all y,

$$\frac{d}{d\theta}\left(\theta g\left(\frac{y}{\theta}\right)\right) = g\left(\frac{y}{\theta}\right) - \frac{y}{\theta}g'\left(\frac{y}{\theta}\right) \geq 0 \tag{9.15}$$

because of (9.12); and

$$\frac{d^2}{d\theta^2}\left(\theta g\left(\frac{y}{\theta}\right)\right) = g''\left(\frac{y}{\theta}\right)\frac{y^2}{\theta^3} \leq 0 \tag{9.16}$$

because $g'' < 0$.

We shall now deal with the resolution of the equation

$$h(\theta) = \theta.$$

Assume it has a unique solution θ_0. We shall show that

$$\theta_k \rightarrow \theta_0.$$

Note first that $h'(\theta_0) < 1$. For

$$h(\theta_0) = \int_0^{\theta_0} h'(t)dt,$$

and if $h'(\theta_0) \geq 1$, then $h'(t) > 1$ for $t < \theta_0$, and hence $h(\theta_0) > \theta_0$. Assume first that $\theta_1 > \theta_0$. Since h is nondecreasing, $\theta_2 = h(\theta_1) \geq h(\theta_0) = \theta_0$. We shall prove that $\theta_2 < \theta_1$. In fact,

$$\theta_2 = h(\theta_1) \leq h(\theta_0) + h'(\theta_0)(\theta_1 - \theta_0) < \theta_0 + (\theta_1 - \theta_0) = \theta_1.$$

In the same way, it follows that $\theta_0 < \theta_{k+1} < \theta_k$. Hence the sequence θ_k decreases, and since it is bounded from below, it has a limit. The case $\theta_1 < \theta_0$ is treated likewise.

Actually, the procedure can be accelerated. Given three consecutive values θ_k, θ_{k+1} and θ_{k+2}, the straight line determined by the points (θ_k, θ_{k+1}) and $(\theta_{k+1}, \theta_{k+2})$ intersects the identity diagonal at the point (θ^*, θ^*) with

$$\theta^* = \frac{\theta_{k+1}^2 - \theta_k\theta_{k+2}}{2\theta_{k+1} - \theta_{k+2} - \theta_k}.$$

Then set $\theta_{k+3} = \theta^*$. The accelerated procedure also converges under the given assumptions.

9.4.2 Algorithms for the non-concave case

If the function g in (9.11) is not concave, the algorithm is not guaranteed to converge to the solution. In this case (2.49) has to be solved by using a general equation-solving procedure. For given $x_1, \ldots, x_n$, let

$$h(\sigma) = \frac{1}{n}\sum_{i=1}^{n} \rho\left(\frac{x_i}{\sigma}\right) - \delta. \tag{9.17}$$

Then we have to solve $h(\sigma) = 0$. Procedures using derivatives, like the Newton–Raphson, cannot be used, since the boundedness of ρ implies that h' is not bounded away from zero. Safe procedures without derivatives require locating the solution in an interval $[\sigma_1, \sigma_2]$ such that $\mathrm{sgn}(h(\sigma_1)) \neq \mathrm{sgn}(h(\sigma_2))$. The simplest is the bisection method, but faster ones exist and can be found, for example, in Brent (1973).

To find σ_1 and σ_2, recall that h is nonincreasing. Let $\sigma_0 = \text{Med}(|\mathbf{x}|)$ and set $\sigma_1 = \sigma_0$. If $h(\sigma_1) > 0$, we are done; else set $\sigma_1 = \sigma_1/2$ and continue halving σ_1 until $h(\sigma_1) > 0$. The same method yields σ_2.

9.5 Multivariate M-estimators

Location and covariance will be treated separately for the sake of simplicity. A very detailed treatment of the convergence of the iterative reweighting algorithm for simultaneous estimation was given by Arslan (2004).

Location involves solving

$$h(\boldsymbol{\mu}) = \min,$$

with

$$h(\boldsymbol{\mu}) = \sum_{i=1}^{n} \rho(d_i(\boldsymbol{\mu})),$$

where

$$d_i(\boldsymbol{\mu}) = (\mathbf{x}_i - \boldsymbol{\mu})'\boldsymbol{\Sigma}^{-1}(\mathbf{x}_i - \boldsymbol{\mu}),$$

which implies (6.11). The procedure is as follows. Given $\boldsymbol{\mu}_k$, let

$$\boldsymbol{\mu}_{k+1} = \frac{1}{\sum_{i=1}^{n} w_i} \sum_{i=1}^{n} w_i \mathbf{x}_i,$$

with $w_i = W(d_i(\boldsymbol{\mu}_k))$ and $W = \rho'$. Hence

$$\sum_{i=1}^{n} w_i \mathbf{x}_i = \boldsymbol{\mu}_{k+1} \sum_{i=1}^{n} w_i. \tag{9.18}$$

Assume that W is nonincreasing, which is equivalent to ρ being concave. It will be shown that $h(\boldsymbol{\mu}_{k+1}) \le h(\boldsymbol{\mu}_k)$. The proof is similar to that of Section 9.1. It is easy to show that the problem can be reduced to $\boldsymbol{\Sigma} = \boldsymbol{I}$, so that $\text{d}_i(\boldsymbol{\mu}) = \|\mathbf{x}_i - \boldsymbol{\mu}\|^2$. Using the concavity of ρ and then (9.18),

$$\begin{aligned} h(\boldsymbol{\mu}_{k+1}) - h(\boldsymbol{\mu}_k) &\le \sum_{i=1}^{n} w_i [\|\mathbf{x}_i - \boldsymbol{\mu}_{k+1}\|^2 - \|\mathbf{x}_i - \boldsymbol{\mu}_k\|^2] \\ &= (\boldsymbol{\mu}_k - \boldsymbol{\mu}_{k+1})' \sum_{i=1}^{n} w_i (2\mathbf{x}_i - \boldsymbol{\mu}_k - \boldsymbol{\mu}_{k+1}) \\ &= (\boldsymbol{\mu}_k - \boldsymbol{\mu}_{k+1})'(\boldsymbol{\mu}_{k+1} - \boldsymbol{\mu}_k) \sum_{i=1}^{n} w_i \le 0. \end{aligned}$$

The treatment of the covariance matrix is more difficult (Maronna, 1976).

9.6 Multivariate S-estimators

9.6.1 S-estimators with monotone weights

For the justification of the algorithm in Section 6.8.2, we shall show that if the weight function is nonincreasing, and hence ρ is concave, then

$$\hat{\sigma}_{k+1} \leq \hat{\sigma}_k. \tag{9.19}$$

Given $\boldsymbol{\mu}_k$ and $\boldsymbol{\Sigma}_k$, define $\hat{\sigma}_{k+1}$, $\boldsymbol{\mu}_{k+1}$ and $\boldsymbol{\Sigma}_{k+1}$ as in (6.59)–(6.60). It will be shown that

$$\sum_{i=1}^{n} \rho\left(\frac{d(\mathbf{x}_i, \boldsymbol{\mu}_{k+1}, \boldsymbol{\Sigma}_{k+1})}{\hat{\sigma}_k}\right) \leq \sum_{i=1}^{n} \rho\left(\frac{d(\mathbf{x}_i, \boldsymbol{\mu}_k, \boldsymbol{\Sigma}_k)}{\hat{\sigma}_k}\right). \tag{9.20}$$

In fact, the concavity of ρ yields (putting w_i for the w_{ki} of (6.59)):

$$\sum_{i=1}^{n} \rho\left(\frac{d(\mathbf{x}_i, \boldsymbol{\mu}_{k+1}, \boldsymbol{\Sigma}_{k+1})}{\hat{\sigma}_k}\right) - \sum_{i=1}^{n} \rho\left(\frac{d(\mathbf{x}_i, \boldsymbol{\mu}_k, \boldsymbol{\Sigma}_k)}{\hat{\sigma}_k}\right) \tag{9.21}$$

$$\leq \frac{1}{\hat{\sigma}_k} \sum_{i=1}^{n} w_i [d(\mathbf{x}_i, \boldsymbol{\mu}_{k+1}, \boldsymbol{\Sigma}_{k+1}) - d(\mathbf{x}_i, \boldsymbol{\mu}_k, \boldsymbol{\Sigma}_k)]. \tag{9.22}$$

Note that $\boldsymbol{\mu}_{k+1}$ is the weighted mean of the $\mathbf{x}_i$ with weights w_i, and hence it minimizes $\sum_{i=1}^{n} w_i(\mathbf{x}_i - \boldsymbol{\mu})'\mathbf{A}(\mathbf{x}_i - \boldsymbol{\mu})$ for any positive definite matrix $\mathbf{A}$. Therefore

$$\sum_{i=1}^{n} w_i d(\mathbf{x}_i, \boldsymbol{\mu}_{k+1}, \boldsymbol{\Sigma}_{k+1}) \leq \sum_{i=1}^{n} w_i d(\mathbf{x}_i, \boldsymbol{\mu}_k, \boldsymbol{\Sigma}_{k+1})$$

and hence the sum on the right-hand side of (9.21) is not larger than

$$\sum_{i=1}^{n} w_i d(\mathbf{x}_i, \boldsymbol{\mu}_k, \boldsymbol{\Sigma}_{k+1}) - \sum_{i=1}^{n} w_i d(\mathbf{x}_i, \boldsymbol{\mu}_k, \boldsymbol{\Sigma}_k) \tag{9.23}$$

$$= \sum_{i=1}^{n} \mathbf{y}_i \boldsymbol{\Sigma}_{k+1}^{-1} \mathbf{y}_i' - \sum_{i=1}^{n} \mathbf{y}_i \boldsymbol{\Sigma}_k^{-1} \mathbf{y}_i', \tag{9.24}$$

with $\mathbf{y}_i = \sqrt{w_i}(\mathbf{x}_i - \boldsymbol{\mu}_k)$. Since

$$\boldsymbol{\Sigma}_{k+1} = \frac{\mathbf{C}}{|\mathbf{C}|^{1/p}} \quad \text{with} \quad \mathbf{C} = \frac{1}{n}\sum_{i=1}^{n} \mathbf{y}_i \mathbf{y}_i',$$

we have that $\boldsymbol{\Sigma}_{k+1}$ is the sample covariance matrix of the $\mathbf{y}_i$ normalized to unit determinant, and by (6.35) it minimizes the sum of squared Mahalanobis distances among matrices with unit determinant. Since $|\boldsymbol{\Sigma}_k| = |\boldsymbol{\Sigma}_{k+1}| = 1$, it follows that (9.23) is ≤ 0, which proves (9.20).

Since

$$\frac{1}{n}\sum_{i=1}^{n}\rho\left(\frac{d(\mathbf{x}_i,\boldsymbol{\mu}_{k+1},\boldsymbol{\Sigma}_{k+1})}{\hat{\sigma}_{k+1}}\right)=\frac{1}{n}\sum_{i=1}^{n}\rho\left(\frac{d(\mathbf{x}_i,\boldsymbol{\mu}_k,\boldsymbol{\Sigma}_k)}{\hat{\sigma}_k}\right),$$

the proof of (9.19) follows like that of (9.10).

9.6.2 The MCD

The justification of the "concentration step" in Section 6.8.6 proceeds as in Section 9.3. Put for $k = 1, 2$: $d_{ik} = d(\mathbf{x}_i, \boldsymbol{\mu}_k, \boldsymbol{\Sigma}_k)$ and call $d_{(i)k}$ the respective ordered values and $\hat{\sigma}_1, \hat{\sigma}_2$ the respective scales. Let $I \subset \{1, \ldots, n\}$ be the set of indices corresponding to the smallest h values of d_{i1}. Then $\boldsymbol{\mu}_2$ and $\boldsymbol{\Sigma}_2$ are the mean and the normalized sample covariance matrix of the set $\{\mathbf{x}_i : i \in I\}$. Hence (6.35) applied to that set implies that

$$\sum_{i\in I} d_{i2} \leq \sum_{i\in I} d_{i1} = \sum_{i=1}^{h} d_{(i)1},$$

and hence

$$\sigma_2^2 = \sum_{i=1}^{h} d_{(i)2} \leq \sum_{i\in I} d_{i2} \leq \sigma_1^2.$$

9.6.3 S-estimators with non-monotone weights

Note first that if ρ is not concave, the algorithm (2.80) is not guaranteed to yield the scale σ, and hence the approach in Section 9.4.2 must be used to compute σ.

Now we describe the modification of the iterative algorithm for the S-estimator. Call $(\hat{\boldsymbol{\mu}}_N, \hat{\boldsymbol{\Sigma}}_N)$ the estimators at iteration N, and $\sigma(\hat{\boldsymbol{\mu}}_N, \hat{\boldsymbol{\Sigma}}_N)$ the respective scale. Call $(\tilde{\boldsymbol{\mu}}_{N+1}, \tilde{\boldsymbol{\Sigma}}_{N+1})$ the values given by a step of the reweighting algorithm.

If $\sigma(\tilde{\boldsymbol{\mu}}_{N+1}, \tilde{\boldsymbol{\Sigma}}_{N+1}) < \sigma(\hat{\boldsymbol{\mu}}_N, \hat{\boldsymbol{\Sigma}}_N)$, then we proceed as usual, setting

$$(\hat{\boldsymbol{\mu}}_{N+1}, \hat{\boldsymbol{\Sigma}}_{N+1}) = (\tilde{\boldsymbol{\mu}}_{N+1}, \tilde{\boldsymbol{\Sigma}}_{N+1}).$$

If instead

$$\sigma(\tilde{\boldsymbol{\mu}}_{N+1}, \tilde{\boldsymbol{\Sigma}}_{N+1}) \geq \sigma(\hat{\boldsymbol{\mu}}_N, \hat{\boldsymbol{\Sigma}}_N), \tag{9.25}$$

then for a given $\xi \in R$ put

$$(\hat{\boldsymbol{\mu}}_{N+1}, \hat{\boldsymbol{\Sigma}}_{N+1}) = (1-\xi)(\hat{\boldsymbol{\mu}}_N, \hat{\boldsymbol{\Sigma}}_N) + \xi(\tilde{\boldsymbol{\mu}}_{N+1}, \tilde{\boldsymbol{\Sigma}}_{N+1}). \tag{9.26}$$

Then it can be shown that there exists $\xi \in (0, 1)$ such that

$$\sigma(\hat{\boldsymbol{\mu}}_{N+1}, \hat{\boldsymbol{\Sigma}}_{N+1}) < \sigma(\hat{\boldsymbol{\mu}}_N, \hat{\boldsymbol{\Sigma}}_N). \tag{9.27}$$

The details are given below in Section 9.6.4.

If the situation in (9.25) occurs, then the algorithm proceeds as follows. Let $\xi_0 \in (0,1)$. Set $\xi = \xi_0$ and compute (9.26). If (9.27) occurs, we are done. Otherwise, set $\xi = \xi\xi_0$ and repeat the former steps, and so on. At some point we must have (9.27). In our programs we use $\xi = 0.7$.

A more refined method would be a line search; that is, to compute (9.26) for different values of ξ and choose the one yielding minimum σ. Our experiments do not show that this extra effort yields better results.

It must be noted that when the computation is near a local minimum, it may happen that because of rounding errors, no value of ξ yields a decrease in σ. Hence it is advisable to stop the search when ξ is less than a small prescribed constant and retain $(\hat{\boldsymbol{\mu}}_N, \hat{\boldsymbol{\Sigma}}_N)$ as the final result.

9.6.4 *Proof of (9.27)

Let $h(\mathbf{z}) : R^m \to R$ be a differentiable function, and call $\mathbf{g}$ its gradient at the point $\mathbf{z}$. Then for any $\mathbf{b} \in R^m$, $h(\mathbf{z} + \xi\mathbf{b}) = h(\mathbf{z}) + \xi\mathbf{g}'\mathbf{b} + o(\xi)$. Hence if $\mathbf{g}'\mathbf{b} < 0$, we have $h(\mathbf{z} + \xi\mathbf{b}) < h(\mathbf{z})$ for sufficiently small ξ.

We must show that we are indeed in this situation. To simplify the exposition, we deal only with $\boldsymbol{\mu}$; we assume Σ fixed, and without loss of generality we may take $\boldsymbol{\Sigma} = \boldsymbol{I}$. Then $d(\mathbf{x}, \boldsymbol{\mu}, \boldsymbol{\Sigma}) = \|\mathbf{x} - \boldsymbol{\mu}\|^2$. Call $\sigma(\boldsymbol{\mu})$ the solution of

$$\frac{1}{n}\sum_{i=1}^{n} \rho\left(\frac{\|\mathbf{x}_i - \boldsymbol{\mu}\|^2}{\sigma}\right) = \delta. \tag{9.28}$$

Call $\mathbf{g}$ the gradient of $\sigma(\boldsymbol{\mu})$ at a given $\boldsymbol{\mu}_1$. Then differentiating (9.28) with respect to $\boldsymbol{\mu}$ yields

$$\sum_{i=1}^{n} w_i[2\sigma(\mathbf{x} - \boldsymbol{\mu}_1) + \|\mathbf{x} - \boldsymbol{\mu}_1\|^2\mathbf{g}] = 0,$$

with

$$w_i = W\left(\frac{\|\mathbf{x}_i - \boldsymbol{\mu}_1\|^2}{\sigma}\right),$$

and hence

$$\mathbf{g} = -\frac{2\sigma}{\sum_{i=1}^{n} w_i\|\mathbf{x} - \boldsymbol{\mu}_1\|^2}\sum_{i=1}^{n} w_i(\mathbf{x}_i - \boldsymbol{\mu}_1). \tag{9.29}$$

Call $\boldsymbol{\mu}_2$ the result of an iteration of the reweighting algorithm; that is,

$$\boldsymbol{\mu}_2 = \frac{1}{\sum_{i=1}^{n} w_i}\sum_{i=1}^{n} w_i\mathbf{x}_i.$$

Then

$$\boldsymbol{\mu}_2 - \boldsymbol{\mu}_1 = \frac{1}{\sum_{i=1}^{n} w_i}\sum_{i=1}^{n} w_i(\mathbf{x}_i - \boldsymbol{\mu}_1), \tag{9.30}$$

and it follows from (9.30) and (9.29) that $(\boldsymbol{\mu}_2 - \boldsymbol{\mu}_1)'\mathbf{g} < 0$.

10

Asymptotic Theory of M-estimators

In order to compare the performances of different estimators, and also to obtain confidence intervals for the parameters, we need their distributions. Explicit expressions exist in some simple cases, such as sample quantiles, which include the median, but even these are in general intractable. It will be necessary to resort to approximating their distributions for large n, the so-called *asymptotic distribution*.

We shall begin with the case of a single real parameter, and we shall consider general M-estimators of a parameter θ defined by equations of the form

$$\sum_{i=1}^{n} \Psi(x_i, \theta) = 0. \tag{10.1}$$

For location, Ψ has the form $\Psi(x, \theta) = \psi(x - \theta)$ with $\theta \in R$; for scale, $\Psi(x, \theta) = \rho(|x|/\theta) - \delta$ with $\theta > 0$. If ψ (or ρ) is nondecreasing then Ψ is nonincreasing in θ.

This family contains maximum likelihood estimators (MLEs). Let $f_\theta(x)$ be a family of densities. The likelihood function for an i.i.d. sample $x_1, \ldots, x_n$ with density f_θ is

$$L = \prod_{i=1}^{n} f_\theta(x_i).$$

If f_θ is everywhere positive, and is differentiable with respect to θ with derivative $\dot{f}_\theta = \partial f_\theta / \partial \theta$, taking logs it is seen that the MLE is the solution of

$$\sum_{i=1}^{n} \Psi_0(x_i, \theta) = 0,$$

Robust Statistics: Theory and Methods (with R), Second Edition.
Ricardo A. Maronna, R. Douglas Martin, Victor J. Yohai and Matías Salibián-Barrera.

Companion website: www.wiley.com/go/maronna/robust

with

$$\Psi_0(x,\theta) = -\frac{\partial \log f_\theta(x)}{\partial \theta} = -\frac{\dot{f}_\theta(x)}{f_\theta(x)}. \tag{10.2}$$

10.1 Existence and uniqueness of solutions

We shall first consider the existence and uniqueness of solutions of (10.1). It is assumed that θ ranges in a finite or infinite interval (θ_1, θ_2). For location, $\theta_2 = -\theta_1 = \infty$; for scale, $\theta_2 = \infty$ and $\theta_1 = 0$. Henceforth the symbol ■ means "this is the end of the proof".

Theorem 10.1 *Assume that for each x, $\Psi(x,\theta)$ is nonincreasing in θ and*

$$\lim_{\theta \to \theta_1} \Psi(x,\theta) > 0 > \lim_{\theta \to \theta_2} \Psi(x,\theta) \tag{10.3}$$

(both limits may be infinite). Let

$$g(\theta) = \sum_{i=1}^{n} \Psi(x_i, \theta).$$

Then:

a) There is at least one point $\widehat{\theta} = \widehat{\theta}(x_1,..,x_n)$ at which g changes sign; that is,

$$g(\theta) \geq 0 \text{ for } \theta < \widehat{\theta} \text{ and } g(\theta) \leq 0 \text{ for } \theta > \widehat{\theta}$$

b) The set of such points is an interval.
c) If Ψ is continuous in θ, then $g(\widehat{\theta}) = 0$.
d) If Ψ is decreasing, then $\widehat{\theta}$ is unique.

Proof. It follows from (10.3) that

$$\lim_{\theta \to \theta_1} g(\theta) > 0 > \lim_{\theta \to \theta_2} g(\theta) \tag{10.4}$$

and the existence of $\widehat{\theta}$ follows from the monotonicity of g. If two values satisfy $g(\theta) = 0$, then the monotonicity of g implies that any value between them also does, which yields point (b). Statement (c) follows from the intermediate value theorem; and point (d) is immediate. ■

Example 10.1 *If $\Psi(x,\theta) = \text{sgn}(x - \theta)$ – which is neither continuous nor increasing – then*

$$g(\theta) = \#(x_i > \theta) - \#(x_i < \theta).$$

The reader can verify that for n odd, $n = 2m - 1$, g vanishes only at $\widehat{\theta} = x_{(m)}$, and for n even, $n = 2m$, it vanishes on the interval $(x_{(m)}, x_{(m+1)})$.

Example 10.2 *The equation for scale M-estimation (2.49) does not satisfy (10.3) since $\rho(0) = 0$ implies $\Psi(0, \theta) = -\delta < 0$ for all θ. But the same reasoning shows that (10.4) holds if*

$$\frac{\#(x_i = 0)}{n} < 1 - \frac{\delta}{\rho(\infty)}.$$

Uniqueness may hold without requiring the strict monotonicity of Ψ. For instance, Huber's ψ is not increasing, but the respective location estimator is unique unless there is a large gap in the middle of the data (Problem 10.7). A sufficient condition for the uniqueness of scale estimators is that $\rho(x)$ be increasing for all x such that $\rho(x) < \rho(\infty)$ (Problem 10.6).

10.1.1 Redescending location estimators

The above results do not cover the case of location estimators with a redescending ψ. In this case, uniqueness requires stronger assumptions than the case of monotone ψ. Uniqueness of the asymptotic value of the estimator requires that the distribution of x, besides being symmetric, is *unimodal*; that is, it has a density $f(x)$, which for some μ is increasing for $x < \mu$ and decreasing for $x > \mu$.

Theorem 10.2 *Let x have a density $f(x)$ which is a decreasing function of $|x|$, and let ρ be any ρ-function. Then $\lambda(\mu) = \mathrm{E}\rho(x - \mu)$ has a unique minimum at $\mu = 0$.*

Proof. Recall that ρ is even and hence its derivative ψ is odd. Hence the derivative of λ is

$$\lambda'(\mu) = -\int_{-\infty}^{\infty} f(x)\psi(x-\mu)dx$$
$$= \int_0^{\infty} \psi(x)\left[f(x-\mu) - f(x+\mu)\right] dx.$$

We shall show that $\lambda'(\mu) > 0$ if $\mu > 0$. It follows from the definition of ρ-function that $\psi(x) \geq 0$ for $x \geq 0$ and $\psi(x) > 0$ if $x \in (0, x_0)$ for some x_0. If x and μ are positive, then $|x - \mu| < |x + \mu|$ and hence $f(x - \mu) > f(x + \mu)$, which implies that the last integral above is positive. ■

If ψ is redescending and f is not unimodal, the minimum need not be unique. Let, for instance, f be a mixture: $f = 0.5f_1 + 0.5f_2$, where f_1 and f_2 are the densities of N $(k, 1)$ and N $(-k, 1)$, respectively. Then if k is large enough, $\lambda(\mu)$ has two minima, located near k and $-k$. The reason can be seen intuitively by noting that, if k is large, for $\mu > 0$, $\lambda(\mu)$ is approximately $0.5 \int \rho(x - \mu) f_1(x)\, dx$, which has a minimum at k.

Note that instead the asymptotic value of a monotone estimator is uniquely defined for this distribution.

10.2 Consistency

Let $x_1, \ldots, x_n$ now be i.i.d. with distribution F. We shall consider the behavior of the solution $\widehat{\theta}_n$ of (10.1) as a random variable. Recall that a sequence y_n of random variables *tends in probability* to y if $\mathrm{P}(|y_n - y| > \varepsilon) \to 0$ for all $\varepsilon > 0$; this will be denoted by $y_n \to_p y$ or $\mathrm{p}\lim y_n = y$. The sequence y_n tends *almost surely* (a.s.) or *with probability one* to y if $\mathrm{P}(\lim_{n\to\infty} y_n = y) = 1$. The expectation with respect to a distribution F will be denoted by E_F.

We shall need a general result.

Theorem 10.3 (Monotone convergence theorem) *Let y_n be a nondecreasing sequence of random variables such that $\mathrm{E}|y_n| < \infty$, and $y_n \to y$ with probability one. Then*

$$\mathrm{E}y_n \to \mathrm{E}y.$$

The proof can be found in Feller (1971).

Assume that $\mathrm{E}_F|\Psi(x, \theta)| < \infty$ for each θ, and define

$$\lambda_F(\theta) = \mathrm{E}_F\Psi(x, \theta). \tag{10.5}$$

Theorem 10.4 *Assume that $\mathrm{E}_F|\Psi(x, \theta)| < \infty$ for all θ. Under the assumptions of Theorem 10.1, there exists θ_F such that λ_F changes sign at θ_F.*

Proof. Proceed along the same lines as the proof of Theorem 10.1. The interchange of limits and expectations is justified by the monotone convergence theorem. ■

Note that if λ_F is continuous, then

$$\mathrm{E}_F\Psi(x, \theta_F) = 0. \tag{10.6}$$

Theorem 10.5 *If θ_F is unique, then $\widehat{\theta}_n$ tends in probability to θ_F.*

Proof. To simplify the proof, we shall assume $\widehat{\theta}_n$ unique. Then it will be shown that for any $\epsilon > 0$,

$$\lim_{n\to\infty} P(\widehat{\theta}_n < \theta_F - \epsilon) = 0.$$

Let

$$\widehat{\lambda}_n(\theta) = \frac{1}{n}\sum_{i=1}^{n} \Psi(x_i, \theta).$$

Since $\widehat{\lambda}_n$ is nonincreasing and $\widehat{\theta}_n$ is unique, $\widehat{\theta}_n < \theta_F - \epsilon$ implies $\widehat{\lambda}_n(\theta_F - \epsilon) < 0$. Since $\widehat{\lambda}_n(\theta_F - \epsilon)$ is the average of the i.i.d. variables $\Psi(x_i, \theta_F - \epsilon)$, and has expectation $\lambda(\theta_F - \epsilon)$ by (10.5), the law of large numbers implies that

$$\widehat{\lambda}_n(\theta_F - \epsilon) \to_p \lambda(\theta_F - \epsilon) > 0.$$

Hence

$$\lim_{n\to\infty} \mathrm{P}(\widehat{\theta}_n < \theta_F - \epsilon) \leq \lim_{n\to\infty} \mathrm{P}(\widehat{\lambda}_n(\theta_F - \epsilon) < 0) = 0.$$

The same method proves that $\mathrm{P}(\widehat{\theta}_n > \theta_F + \epsilon) \to 0$. ∎

Example 10.3 *For location $\Psi(x, \theta) = \psi(x - \theta)$. If $\psi(x) = x$, then Ψ is continuous and decreasing, the solution is $\widehat{\theta}_n = \overline{x}$ and $\lambda(\theta) = \mathrm{E}x - \theta$, so that $\theta_F = \mathrm{E}x$; convergence occurs only if the latter exists.*

If $\psi(x) = \mathrm{sgn}(x)$, we have

$$\lambda(\theta) = \mathrm{P}(x > \theta) - \mathrm{P}(x < \theta),$$

hence θ_F is a median of F, which is unique iff

$$F(\theta_F + \epsilon) > F(\theta_F - \epsilon)\ \forall\ \epsilon > 0. \tag{10.7}$$

In this case, for $n = 2m$ the interval $(x_{(m)}, x_{(m+1)})$ shrinks to a single point when $m \to \infty$. If 10.7 does not hold, the distribution of $\widehat{\theta}_n$ does not converge to a point-mass (Problem 10.2).

Note that for model (2.1), if ψ is odd and $\mathcal{D}(e)$ is symmetric about 0, then $\lambda(\theta) = 0$ so that $\theta_F = \theta$. For scale, Theorem 10.5 implies that estimators of the form (2.49) tend to the solution of (2.50) if it is unique.

10.3 Asymptotic normality

In Section 2.10.2, the asymptotic normality of M-estimators was proved heuristically, by replacing ψ with its first-order Taylor expansion. This procedure will now be made rigorous.

If the distribution of z_n tends to the distribution H of z, we shall say that z_n *tends in distribution* to z (or to H), and shall denote this by $z_n \to_d z$ (or $z_n \to_d H$). We shall need an auxiliary result.

Theorem 10.6 (Bounded convergence theorem) *Let y_n be a sequence of random variables such that $|y_n| \leq z$ where $\mathrm{E}z < \infty$ and $y_n \to y$ a.s.. Then $\mathrm{E}y_n \to \mathrm{E}y$.*

The proof can be found in (Feller, 1971).

Theorem 10.7 *Assume that* $A = \mathrm{E}\Psi(x, \theta_F)^2 < \infty$ *and that* $B = \lambda'(\theta_F)$ *exists and is nonnull. Then the distribution of* $\sqrt{n}\,(\widehat{\theta}_n - \theta_F)$ *tends to* $\mathrm{N}(0, \upsilon)$ *with*

$$\upsilon = \frac{A}{B^2}.$$

If $\dot{\Psi}(x, \theta) = \partial\Psi/\partial\theta$ *exists and verifies for all* x, θ

$$|\dot{\Psi}(x, \theta)| \leq K(x) \ \ with \ \ \mathrm{E}K(x) < \infty, \tag{10.8}$$

then $B = \mathrm{E}\dot{\Psi}(x, \theta_F)$.

Proof. To make things simpler, we shall make the extra (and unnecessary) assumptions that $\ddot{\Psi}(x, \theta) = \partial^2\Psi/\partial\theta^2$ exists and is bounded, and that $\dot{\Psi}$ verifies (10.8). A completely general proof may be found in Huber and Ronchetti (2009, Sec. 3.2).

Note first that the bounded convergence theorem implies $B = \mathrm{E}\dot{\Psi}(x, \theta_F)$. In fact,

$$B = \lim_{\delta\to 0} \mathrm{E}\frac{\Psi(x, \theta_F + \delta) - \Psi(x, \theta_F)}{\delta}.$$

The term in the expectation is $\leq K(x)$ by the mean value theorem, and for each x tends to $\dot{\Psi}(x, \theta_F)$.

A second-order Taylor expansion of Ψ at θ_F yields

$$\Psi(x_i, \widehat{\theta}_n) = \Psi(x_i, \theta_F) + (\widehat{\theta}_n - \theta_F)\dot{\Psi}(x_i, \theta_F) + \frac{1}{2}(\widehat{\theta}_n - \theta_F)^2\ddot{\Psi}(x_i, \theta_i)$$

where θ_i is some value (depending on x_i) between $\widehat{\theta}_n$ and θ_F. Summing over i yields

$$0 = A_n + (\widehat{\theta}_n - \theta_F)B_n + (\widehat{\theta}_n - \theta_F)^2 C_n,$$

where

$$A_n = \frac{1}{n}\sum_{i=1}^{n} \Psi(x_i, \theta_F),\ B_n = \frac{1}{n}\sum_{i=1}^{n} \dot{\Psi}(x_i, \theta_F),\ C_n = \frac{1}{2n}\sum_{i=1}^{n} \ddot{\Psi}(x_i, \theta_i)$$

and hence

$$\sqrt{n}(\widehat{\theta}_n - \theta_F) = -\frac{\sqrt{n}A_n}{B_n + (\widehat{\theta}_n - \theta_F)C_n}.$$

Since the i.i.d. variables $\Psi(x_i, \theta_F)$ have mean 0 (by (10.6)) and variance A, the central limit theorem implies that the numerator tends in distribution to $\mathrm{N}(0, A)$. The law of large numbers implies that $B_n \to_p B$; and since C_n is bounded and $(\widehat{\theta}_n - \theta_F) \to_p 0$ by the former theorem, Slutsky's lemma (Section 2.10.3) yields the desired result. ■

Example 10.4 (Location) *For the mean, the existence of A requires that of* $\mathrm{E}x^2$. *In general, if* ψ *is bounded, A always exists. If* ψ' *exists, then* $\lambda'(t) = \mathrm{E}\,\psi'(x - t)$. *For the median,* ψ *is discontinuous, but if F has a density f, explicit calculation yields*

$$\lambda(\theta) = P(x > \theta) - P(x < \theta) = 1 - 2F(\theta),$$

and hence $\lambda'(\theta_F) = -2f(\theta_F)$.

If $\lambda'(\theta_F)$ does not exist, $\widehat{\theta}_n$ tends to θ_F faster than $n^{-1/2}$, and there is no asymptotic normality. Consider, for instance, the median with F discontinuous. Let $\psi(x) = \text{sgn}\,(x)$, and assume that F is continuous except at zero, where it has its median and a point mass with $\text{P}\,(x = 0) = 2\delta$; that is,

$$\lim_{x\uparrow 0} F(x) = 0.5 - \delta,\ \lim_{x\downarrow 0} F(x) = 0.5 + \delta.$$

Then $\lambda(\theta) = 1 - 2F(\theta)$ has a jump at $\theta_F = 0$. We shall see that this entails $\text{P}(\widehat{\theta}_n = 0) \to 1$, and *a fortiori* $\sqrt{n}\,\widehat{\theta}_n \to_p 0$.

Let $N_n = \#(x_i < 0)$, which is binomial $\text{Bi}(n, p)$ with $p = 0.5 - \delta$. Then $\widehat{\theta}_n < 0$ implies $N_n > n/2$, and therefore

$$\text{P}(\widehat{\theta}_n < 0) \leq \text{P}(N_n/n > 0.5) \to 0$$

since the law of large numbers implies $N_n/n \to_p p < 0.5$. The same method yields $\text{P}(\widehat{\theta}_n > 0) \to 0$.

The fact that the distribution of $\widehat{\theta}_n$ tends to a normal $\text{N}(\theta_F, v)$ does *not* imply that the mean and variance of $\widehat{\theta}_n$ tend to θ_F and v (Problem 10.3). In fact, if F is heavy tailed, the distribution of $\widehat{\theta}_n$ will also be heavy tailed, with the consequence that its moments may not exist, or, if they do, they will give misleading information about $\mathcal{D}(\widehat{\theta}_n)$. In extreme cases, they may even not exist for any n. This shows that, as an evaluation criterion, the *asymptotic* variance may be better than the variance. Let $T_n = (\widehat{\theta}_n - \widehat{\theta}_\infty)\sqrt{n/v}$, where $\widehat{\theta}_n$ is the median and v its asymptotic variance under F, so that T_n should be approximately $\text{N}(0, 1)$. Figure 10.1 shows for the Cauchy distribution the normal Q–Q plot of T_n, that is, the comparison between the exact and the approximate quantiles of its distribution, for $n = 5$ and 11. It is seen that although the approximation improves in the middle when n increases, the tails remain heavy.

10.4 Convergence of the SC to the IF

In this section we prove (3.6) for general M-estimators (10.1). Call $\widehat{\theta}_n$ the solution of (10.1), and for a given x_0, call $\widehat{\theta}_{n+1}(x_0)$ the solution of

$$\sum_{i=0}^{n} \Psi(x_i, \theta) = 0. \tag{10.9}$$

The sensitivity curve is

$$\text{SC}_n(x_0) = (n + 1)\left(\widehat{\theta}_{n+1}(x_0) - \widehat{\theta}_n\right),$$

and

$$\text{IF}_{\widehat{\theta}}(x_0) = -\frac{\Psi(x_0, \theta_F)}{B},$$

with B and θ_F defined in Theorem 10.7 and (10.6) respectively.

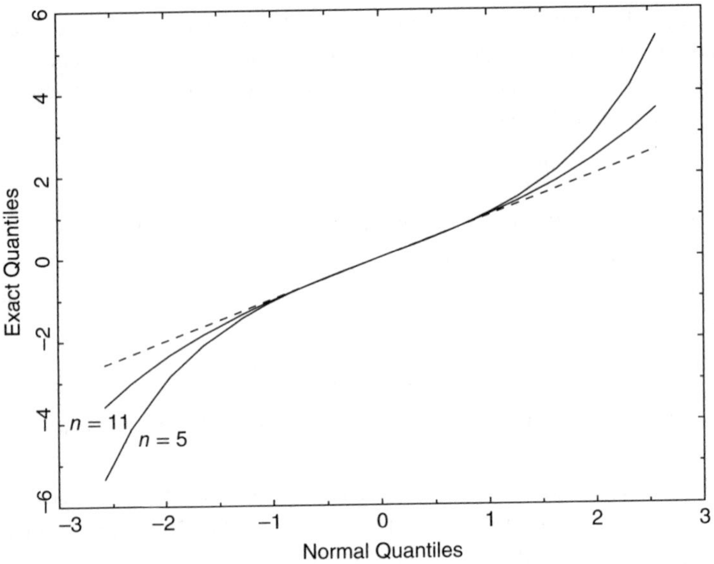

Figure 10.1 Q–Q plot of the sample median for Cauchy data. The dashed line is the identity diagonal.

Theorem 10.8 *Assume the same conditions as in Theorem 10.7. Then for each* x_0

$$SC_n(x_0) \to_p IF_{\hat{\theta}}(x_0).$$

Proof. Theorem 10.5 states that $\hat{\theta}_n \to_p \theta_F$. The same proof shows that also $\hat{\theta}_{n+1}(x_0) \to_p \theta_F$, since the effect of the term $\Psi(x_0, \theta)$ becomes negligible for large n. Hence

$$\Delta_n =: \hat{\theta}_{n+1}(x_0) - \hat{\theta}_n \to_p 0.$$

Using (10.1) and (10.9) and a Taylor expansion yields

$$0 = \Psi(x_0, \hat{\theta}_{n+1}(x_0)) + \sum_{i=1}^{n} \left[\Psi(x_i, \hat{\theta}_{n+1}(x_0)) - \Psi(x_i, \hat{\theta}_n) \right] \tag{10.10}$$

$$= \Psi(x_0, \hat{\theta}_{n+1}(x_0)) + \Delta_n \sum_{i=1}^{n} \dot{\Psi}(x_i, \hat{\theta}_n) + \frac{\Delta_n^2}{2} \sum_{i=1}^{n} \ddot{\Psi}(x_i, \theta_i),$$

where θ_i is some value between $\hat{\theta}_{n+1}(x_0)$ and $\hat{\theta}_n$. Put

$$B_n = \frac{1}{n} \sum_{i=1}^{n} \dot{\Psi}(x_i, \hat{\theta}_n), \quad C_n = \frac{1}{n} \sum_{i=1}^{n} \ddot{\Psi}(x_i, \theta_i).$$

Then C_n is bounded, and the consistency of $\widehat{\theta}_n$, plus a Taylor expansion, show that $B_n \to_p B$. It follows from (10.10) that

$$\text{SC}_n(x_0) = -\frac{\Psi(x_0, \widehat{\theta}_{n+1}(x_0))}{B_n + C_n\Delta_n/2}\frac{n+1}{n}.$$

And since $\Psi(x_0, \widehat{\theta}_{n+1}(x_0)) \to_p \Psi(x_0, \theta_F)$, the proof follows. ■

10.5 M-estimators of several parameters

We shall need the asymptotic distribution of M-estimators when there are several parameters. This happens in particular with the joint estimation of location and scale in Section 2.6.2, where we have two parameters, which satisfy a system of two equations. This situation also appears in regression (Chapter 4) and multivariate analysis (Chapter 6). Put $\boldsymbol{\theta} = (\mu, \sigma)$, and

$$\Psi_1(x, \boldsymbol{\theta}) = \psi\left(\frac{x-\mu}{\sigma}\right) \text{ and } \Psi_2(x, \boldsymbol{\theta}) = \rho_{\text{scale}}\left(\frac{x-\mu}{\sigma}\right) - \delta.$$

Then the simultaneous location–scale estimators satisfy

$$\sum_{i=1}^{n} \boldsymbol{\Psi}(x_i, \boldsymbol{\theta}) = \mathbf{0}, \tag{10.11}$$

with $\boldsymbol{\Psi} = (\Psi_1, \Psi_2)$. Here the observations x_i are univariate, but in general they may belong to any set $\mathcal{X} \subset R^q$, and we consider a vector $\boldsymbol{\theta} = (\theta_1, \ldots, \theta_p)'$ of unknown parameters, which ranges in a subset $\Theta \subset R^p$, which satisfies (10.11) where $\boldsymbol{\Psi} = (\Psi_1, \ldots, \Psi_p)$ is function of $\mathcal{X} \times \Theta \to R^p$. Existence of solutions must be dealt with in each situation. Uniqueness may be proved under conditions which generalize the monotonicity of Ψ in the case of a univariate parameter (as in part (d) of Theorem 10.1).

Theorem 10.9 *Assume that for all x and $\boldsymbol{\theta}$, $\boldsymbol{\Psi}(x, \boldsymbol{\theta})$ is differentiable and the matrix $\mathbf{D} = \mathbf{D}(x, \boldsymbol{\theta})$ with elements $\partial\Psi_i/\partial\theta_j$ is negative definite (i.e., $\mathbf{a}'\mathbf{D}\mathbf{a} < 0$ for all $\mathbf{a} \neq \mathbf{0}$). Put for given $x_1, \ldots, x_n$*

$$g(\boldsymbol{\theta}) = \sum_{i=1}^{n} \boldsymbol{\Psi}(x_i, \boldsymbol{\theta}).$$

If there exists a solution of $g(\boldsymbol{\theta}) = \mathbf{0}$, then this solution is unique.

Proof. We shall prove that

$$g(\boldsymbol{\theta}_1) \neq g(\boldsymbol{\theta}_2) \quad \text{if} \quad \boldsymbol{\theta}_1 \neq \boldsymbol{\theta}_2.$$

Let $\mathbf{a} = \boldsymbol{\theta}_2 - \boldsymbol{\theta}_1$, and define for $t \in R$ the function $h(t) = \mathbf{a}' g(\boldsymbol{\theta}_1 + t\mathbf{a})$, so that $h(0) = \mathbf{a}' g(\boldsymbol{\theta}_1)$ and $h(1) = \mathbf{a}' g(\boldsymbol{\theta}_2)$. Its derivative is

$$h'(t) = \sum_{i=1}^{n} \mathbf{a}'\mathbf{D}(x, \boldsymbol{\theta}_1 + t\mathbf{a})\mathbf{a} < 0 \ \forall \ t,$$

and hence $h(0) > h(1)$ which implies $g(\boldsymbol{\theta}_1) \neq g(\boldsymbol{\theta}_2)$. ■

To consider consistency, assume the x_i are i.i.d with distribution F, and put

$$\widehat{\lambda}_n(\boldsymbol{\theta}) = \frac{1}{n} \sum_{i=1}^{n} \boldsymbol{\Psi}(x_i, \boldsymbol{\theta}) \tag{10.12}$$

and

$$\lambda(\boldsymbol{\theta}) = \mathrm{E}_F \boldsymbol{\Psi}(x, \boldsymbol{\theta}). \tag{10.13}$$

Let $\widehat{\boldsymbol{\theta}}_n$ be any solution of (10.11); it is natural to conjecture that, if there is a unique solution θ_F of $\lambda(\boldsymbol{\theta}) = \mathbf{0}$, then as $n \to \infty$, $\widehat{\boldsymbol{\theta}}_n$ tends in probability to θ_F. General criteria are given in Huber and Ronchetti (2009, Sec. 6.2). However, their application to each situation must be dealt with separately.

For asymptotic normality, we can generalize Theorem 10.7. Assume that $\widehat{\theta}_n \to_p \theta_F$ and that λ is differentiable at θ_F, and call $\mathbf{B}$ the matrix of derivatives with elements

$$B_{jk} = \left. \frac{\partial \lambda_j}{\partial \theta_k} \right|_{\theta = \theta_F}. \tag{10.14}$$

Assume $\mathbf{B}$ is nonsingular. Then under general assumptions (Huber and Ronchetti, 2009, Sec. 6.3)

$$\sqrt{n}(\widehat{\boldsymbol{\theta}}_n - \theta_F) \to_d \mathrm{N}_p(\mathbf{0}, \mathbf{B}^{-1}\mathbf{A}\mathbf{B}^{-1'}) \tag{10.15}$$

where

$$\mathbf{A} = \mathrm{E}\boldsymbol{\Psi}(x, \theta_F)\boldsymbol{\Psi}(x, \theta_F)', \tag{10.16}$$

and $\mathrm{N}_p(\mathbf{t}, \mathbf{V})$ denotes the p-variate normal distribution with mean $\mathbf{t}$ and covariance matrix $\mathbf{V}$.

If $\dot{\Psi}_{jk} = \partial \Psi_j / \partial \theta_k$ exists and verifies, for all $x, \boldsymbol{\theta}$,

$$|\dot{\Psi}_{jk}(x, \boldsymbol{\theta})| \leq K(x) \text{ with } \mathrm{E}K(x) < \infty, \tag{10.17}$$

then $\mathbf{B} = \mathrm{E}\dot{\boldsymbol{\Psi}}(x, \theta_F)$, where $\dot{\boldsymbol{\Psi}}$ is the matrix with elements $\dot{\Psi}_{jk}$.

The intuitive idea behind the result is like that of (2.89)–(2.90): we take a first-order Taylor expansion of $\boldsymbol{\Psi}$ around θ_F and drop the higher-order terms. Before dealing with the proof of (10.15), let us see how it applies to simultaneous M-estimators of location–scale. Conditions for existence and uniqueness of solutions are given in Huber and Ronchetti (2009, Sec. 6.4) and Maronna and Yohai (1981). They may hold without requiring monotonicity of ψ. This holds in particular for the Student MLE. As can be expected, under suitable conditions they tend in probability to the solution (μ_0, σ_0) of the system of equations (2.74)–(2.75). The joint distribution of $\sqrt{n}(\widehat{\mu} - \mu_0, \widehat{\sigma} - \sigma_0)$ tends to the bivariate normal with mean 0 and covariance matrix

$$\mathbf{V} = \mathbf{B}^{-1}\mathbf{A}(\mathbf{B}^{-1})', \tag{10.18}$$

where

$$\mathbf{A} = \begin{bmatrix} a_{11} & a_{12} \\ a_{21} & a_{22} \end{bmatrix}, \ \mathbf{B} = \frac{1}{\sigma}\begin{bmatrix} b_{11} & b_{12} \\ b_{21} & b_{22} \end{bmatrix},$$

with

$$a_{11} = \mathrm{E}\psi(r)^2, \ a_{12} = a_{21} = \mathrm{E}(\rho_{\mathrm{scale}}(r) - \delta)\psi(r), \ a_{22} = \mathrm{E}(\rho_{\mathrm{scale}}(r) - \delta)^2,$$

where

$$r = \frac{x - \mu_0}{\sigma_0},$$

and

$$b_{11} = \mathrm{E}\psi'(r), \ b_{12} = \mathrm{E}r\psi'(r),$$

$$b_{21} = \mathrm{E}\rho'_{\mathrm{scale}}(r), \ b_{22} = \mathrm{E}r\rho'_{\mathrm{scale}}(r).$$

If ψ is odd, ρ_{scale} is even, and F is symmetric, the reader can verify (Problem 10.5) that $\mathbf{V}$ is diagonal,

$$\mathbf{V} = \begin{bmatrix} v_{11} & 0 \\ 0 & v_{22} \end{bmatrix},$$

so that $\widehat{\mu}$ and $\widehat{\sigma}$ are asymptotically independent, and their variances take on a simple form:

$$v_{jj} = \sigma_0^2 \frac{a_{jj}}{b_{jj}^2} \quad (j = 1, 2);$$

that is, the asymptotic variance of each estimator is calculated as if the other parameter were constant.

We shall now prove (10.15) under much more restricted assumptions. We shall need an auxiliary result.

Theorem 10.10 (Multivariate Slutsky lemma) *Let $\mathbf{u}_n$ and $\mathbf{v}_n$ be two sequences of random vectors and $\mathbf{W}_n$ a sequence of random matrices such that for some constant vector $\mathbf{u}$, random vector $\mathbf{v}$ and random matrix $\mathbf{W}$*

$$\mathbf{u}_n \to_p \mathbf{u}, \ \mathbf{v}_n \to_d \mathbf{V}, \ \mathbf{W}_n \to_p \mathbf{W}.$$

Then

$$\mathbf{u}_n + \mathbf{v}_n \to_d \mathbf{u} + \mathbf{v} \text{ and } \mathbf{W}_n\mathbf{v}_n \to_d \mathbf{W}\mathbf{v}.$$

The proof proceeds as in the univariate case.

Now we proceed with the proof of asymptotic normality under more restricted assumptions. Let $\widehat{\theta}_n$ be any solution of (10.11).

Theorem 10.11 *Assume that $\widehat{\boldsymbol{\theta}}_n \to_p \theta_F$, where θ_F is the unique solution of $\lambda_F(\boldsymbol{\theta}) = \mathbf{0}$. Let $\boldsymbol{\Psi}$ be twice differentiable with respect to $\boldsymbol{\theta}$ with bounded derivatives, and satisfying also (10.17). Then (10.15) holds.*

Proof. The proof follows that of Theorem 10.7. For each j, call $\ddot{\boldsymbol{\Psi}}_j$ the matrix with elements $\partial\Psi_j/\partial\theta_k\partial\theta_l$, and $\mathbf{C}_n(x, \boldsymbol{\theta})$ the matrix with jth row equal to $(\widehat{\boldsymbol{\theta}}_n - \theta_F)'\ddot{\boldsymbol{\Psi}}_j(x, \boldsymbol{\theta})$. By a Taylor expansion

$$\mathbf{0} = \widehat{\lambda}_n(\widehat{\boldsymbol{\theta}}_n) = \sum_{i=1}^{n} \left\{ \boldsymbol{\Psi}(x_i, \theta_F) + \dot{\Psi}(x_i, \theta_F) \left(\widehat{\boldsymbol{\theta}}_n - \theta_F\right) + \frac{1}{2}\mathbf{C}_n(x_i, \boldsymbol{\theta}_i) \left(\widehat{\boldsymbol{\theta}}_n - \theta_F\right) \right\}.$$

In other words

$$\mathbf{0} = \mathbf{A}_n + \left(\mathbf{B}_n + \overline{\mathbf{C}}_n\right)\left(\widehat{\boldsymbol{\theta}}_n - \theta_F\right),$$

with

$$\mathbf{A}_n = \frac{1}{n}\sum_{i=1}^{n} \boldsymbol{\Psi}(x_i, \theta_F),\ \mathbf{B}_n = \frac{1}{n}\sum_{i=1}^{n} \dot{\Psi}(x_i, \theta_F),\ \overline{\mathbf{C}}_n = \frac{1}{2n}\sum_{i=1}^{n} \mathbf{C}_n(x_i, \boldsymbol{\theta}_i);$$

that is, $\overline{\mathbf{C}}_n$ is the matrix with the jth row equal to $(\widehat{\boldsymbol{\theta}}_n - \boldsymbol{\theta}_F)'\ddot{\boldsymbol{\Psi}}_j^-$, where

$$\ddot{\boldsymbol{\Psi}}_j^- = \frac{1}{n}\sum_{i=1}^{n} \ddot{\boldsymbol{\Psi}}_j(x_i, \boldsymbol{\theta}_i),$$

which is bounded. Since $\widehat{\boldsymbol{\theta}}_n - \boldsymbol{\theta}_F \to_p \mathbf{0}$, this implies that also $\overline{\mathbf{C}}_n \to_p \mathbf{0}$. We have

$$\sqrt{n}(\widehat{\boldsymbol{\theta}}_n - \boldsymbol{\theta}_F) = -(\mathbf{B}_n + \overline{\mathbf{C}}_n)^{-1}\sqrt{n}\mathbf{A}_n.$$

Note that for $i = 1, 2, \ldots,$ the vectors $\boldsymbol{\Psi}(x_i, \theta_F)$ are i.i.d. with mean $\mathbf{0}$ (since $\lambda(\theta_F) = \mathbf{0}$) and covariance matrix $\mathbf{A}$, and the matrices $\dot{\boldsymbol{\Psi}}(x_i, \theta_F)$ are i.i.d. with mean $\mathbf{B}$. Hence when $n \to \infty$, the law of large numbers implies $\mathbf{B}_n \to_p \mathbf{B}$, which implies $\mathbf{B}_n + \overline{\mathbf{C}}_n \to_p \mathbf{B}$, which is nonsingular. The central limit theorem implies $\sqrt{n}\mathbf{A}_n \to_d \mathrm{N}_p(\mathbf{0}, \mathbf{A})$, and hence (10.15) follows by the multivariate version of Slutsky's lemma. ∎

10.6 Location M-estimators with preliminary scale

We shall consider the asymptotic behavior of solutions of (2.67). For each n let $\widehat{\sigma}_n$ be a dispersion estimator, and call $\widehat{\mu}_n$ the solution (assumed unique) of

$$\sum_{i=1}^{n} \psi\left(\frac{x_i - \mu}{\widehat{\sigma}_n}\right) = 0. \tag{10.19}$$

For consistency, it will be assumed that

A1 ψ is monotone and bounded with a bounded derivative;
A2 $\sigma = \text{p}\lim \widehat{\sigma}_n$ exists;
A3 the equation $\text{E}\psi((x-\mu)/\sigma) = 0$ has a unique solution μ_0.

Theorem 10.12 *If A1–A2–A3 hold, then* $\widehat{\mu}_n \to_p \mu_0$.

The proof follows the lines of Theorem 10.5, but the details require much more care, and are hence omitted.

Define now $u_i = x_i - \mu_0$ and

$$a = \text{E}\psi\left(\frac{u}{\sigma}\right)^2, \; b = \text{E}\psi'\left(\frac{u}{\sigma}\right), \; c = \text{E}\left(\frac{u}{\sigma}\right)\psi'\left(\frac{u}{\sigma}\right). \tag{10.20}$$

For asymptotic normality, assume

A4 the quantities defined in (10.20) exist and $b \neq 0$;
A5 $\sqrt{n}(\widehat{\sigma}_n - \sigma)$ converges to some distribution;
A6 $c = 0$.

Theorem 10.13 *Under A4–A5–A6, we have*

$$\sqrt{n}(\widehat{\mu}_n - \mu_0) \to_d \text{N}(0, v) \text{ with } v = \sigma^2 \frac{a}{b^2}. \tag{10.21}$$

Note that if x has a symmetric distribution, then μ_0 coincides with its center of symmetry, and hence the distribution of u is symmetric about zero, which implies (since ψ is odd) that $c = 0$.

Adding the assumption that ψ has a bounded second derivative, the theorem may be proved along the lines of Theorem 10.5, but the details are somewhat more involved. We shall content ourselves with a heuristic proof of (10.21) to exhibit the main ideas.

Put for brevity

$$\widehat{\Delta}_{1n} = \widehat{\mu}_n - \mu_0, \; \widehat{\Delta}_{2n} = \widehat{\sigma}_n - \sigma.$$

Then expanding ψ as in (2.89) yields

$$\psi\left(\frac{x_i - \widehat{\mu}_n}{\widehat{\sigma}_n}\right) = \psi\left(\frac{u_i - \widehat{\Delta}_{1n}}{\sigma + \widehat{\Delta}_{2n}}\right)$$
$$\approx \psi\left(\frac{u_i}{\sigma}\right) - \psi'\left(\frac{u_i}{\sigma}\right)\frac{\widehat{\Delta}_{1n} + \widehat{\Delta}_{2n}u_i/\sigma}{\sigma}.$$

Inserting the right-hand side of this expression into (2.67) and dividing by n yields

$$0 = A_n - \frac{1}{\sigma}\left(\widehat{\Delta}_{1n}B_n + \widehat{\Delta}_{2n}C_n\right),$$

where

$$A_n = \frac{1}{n}\sum_{i=1}^{n}\psi\left(\frac{u_i}{\sigma}\right),\ B_n = \frac{1}{n}\sum_{i=1}^{n}\psi'\left(\frac{u_i}{\sigma}\right),\ C_n = \frac{1}{n}\sum_{i=1}^{n}\left(\frac{u_i}{\sigma}\right)\psi'\left(\frac{u_i}{\sigma}\right),$$

and hence

$$\sqrt{n}\widehat{\Delta}_{1n} = \frac{\sigma\sqrt{n}A_n - \widehat{\Delta}_{2n}\sqrt{n}C_n}{B_n}. \tag{10.22}$$

Now A_n is the average of i.i.d. variables with mean 0 (by (10.19)) and variance a, and hence the central limit theorem implies that $\sqrt{n}A_n \to_d \mathrm{N}(0, a)$; the law of large numbers implies that $B_n \to_p b$. If $c = 0$, then $\sqrt{n}C_n$ tends to a normal by the central limit theorem, and since $\widehat{\Delta}_{2n} \to_p 0$ by hypothesis, Slutsky's lemma yields (10.21).

If $c \neq 0$, the term $\widehat{\Delta}_{2n}\sqrt{n}C_n$ does not tend to zero, and the asymptotic variance of $\widehat{\Delta}_{1n}$ will depend on that of $\widehat{\sigma}_n$ and also on the correlation between $\widehat{\sigma}_n$ and $\widehat{\mu}_n$.

10.7 Trimmed means

Although the numerical computing of trimmed means – and in general of L-estimators – is very simple, their asymptotic theory is much more complicated than that of M-estimators; even heuristic derivations are involved.

It is shown (see Huber and Ronchetti 2009, Sec. 3.3) that under suitable regularity conditions, $\widehat{\theta}_n$ converges in probability to

$$\theta_F = \frac{1}{1-2\alpha}\mathrm{E}_F x\mathrm{I}(k_1 \leq x \leq k_2), \tag{10.23}$$

where

$$k_1 = F^{-1}(\alpha),\ k_2 = F^{-1}(1-\alpha). \tag{10.24}$$

Let $F(x) = F_0(x-\mu)$, with F_0 symmetric about zero. Then $\theta_F = \mu$ (Problem 10.4).

If F is as above, then $\sqrt{n}(\widehat{\theta} - \mu) \to_d \mathrm{N}(0, v)$ with

$$v = \frac{1}{(1-2\alpha)^2}\mathrm{E}_F\ \psi_k(x-\mu)^2, \tag{10.25}$$

where ψ_k is Huber's function with $k = F_0^{-1}(1-\alpha)$, so that the asymptotic variance coincides with that of an M-estimator.

10.8 Optimality of the MLE

It can be shown that the MLE is "optimal", in the sense of minimizing the asymptotic variance, in a general class of asymptotically normal estimators (Shao, 2003). Here its optimality will be shown within the class of M-estimators of the form (10.1).

The MLE is an M-estimator, which under the conditions of Theorem 10.5 is Fisher-consistent; that is, it verifies (3.32). In fact, assume $\dot{f}_\theta = \partial f/\partial\theta$ is bounded. Then differentiating

$$\int_{-\infty}^{\infty} f_\theta(x)dx = 1$$

with respect to θ yields

$$0 = \int_{-\infty}^{\infty} \dot{f}_\theta(x)dx = -\int_{-\infty}^{\infty} \Psi_0(x,\theta)f_\theta(x)dx\ \forall\ \theta, \tag{10.26}$$

so that (10.6) holds (the interchange of integral and derivative is justified by the bounded convergence theorem).

Under the conditions of Theorem 10.7, the MLE has asymptotic variance

$$v_0 = \frac{A_0}{B_0^2},$$

with

$$A_0 = \int_{-\infty}^{\infty} \Psi_0^2(x,\theta)f_\theta(x)dx,\ B_0 = \int_{-\infty}^{\infty} \dot{\Psi}_0(x,\theta)f_\theta(x)dx,$$

with $\dot{\Psi}_0 = \partial\Psi_0/\partial\theta$. The quantity A_0 is called the *Fisher information.*

Now consider another M-estimator of the form (10.1), which is Fisher-consistent for θ; that is, such that

$$\int_{-\infty}^{\infty} \Psi(x,\theta)f_\theta(x)dx = 0\ \forall\ \theta, \tag{10.27}$$

and has asymptotic variance

$$v = \frac{A}{B^2},$$

with

$$A = \int_{-\infty}^{\infty} \Psi^2(x,\theta)f_\theta(x)dx,\ B = \int_{-\infty}^{\infty} \dot{\Psi}(x,\theta)f_\theta(x)dx.$$

It will be shown that

$$v_0 \leq v. \tag{10.28}$$

We shall show first that $B_0 = A_0$, which implies

$$v_0 = \frac{1}{A_0}. \tag{10.29}$$

In fact, differentiating the last member of (10.26) with respect to θ yields

$$0 = B_0 + \int_{-\infty}^{\infty} \Psi_0(x,\theta)\frac{\dot{f}_\theta(x)}{f_\theta(x)}f_\theta(x)dx = B_0 - A_0.$$

By (10.29), (10.28) is equivalent to

$$B^2 \leq A_0 A. \tag{10.30}$$

Differentiating (10.27) with respect to θ yields

$$B - \int_{-\infty}^{\infty} \Psi(x,\theta)\Psi_0(x,\theta) f_\theta(x)dx = 0.$$

The Cauchy–Schwarz inequality yields

$$\left(\int_{-\infty}^{\infty} \Psi(x,\theta)\Psi_0(x,\theta) f_\theta(x)dx\right)^2 \leq \left(\int_{-\infty}^{\infty} \Psi(x,\theta)^2 f_\theta(x)dx\right) \times \left(\int_{-\infty}^{\infty} \Psi_0^2(x,\theta) f_\theta(x)dx\right),$$

which proves (10.30).

10.9 Regression M-estimators: existence and uniqueness

From now on it will be assumed that $\mathbf{X}$ has full rank, and that $\widehat{\sigma}$ is fixed or estimated previously (i.e., it does not depend on $\widehat{\boldsymbol{\beta}}$). We first establish the existence of solutions of (4.39).

Theorem 10.14 *Let $\rho(r)$ be a continuous nondecreasing unbounded function of $|r|$. Then there exists a solution of (4.39).*

Proof. Here $\widehat{\sigma}$ plays no role, so that we may put $\widehat{\sigma} = 1$. Since ρ is bounded from below, so is the function

$$R(\boldsymbol{\beta}) = \sum_{i=1}^{n} \rho(y_i - \mathbf{x}_i'\boldsymbol{\beta}). \tag{10.31}$$

Call L its infimum; that is, the larger of its lower bounds. We must show the existence of $\boldsymbol{\beta}_0$ such that $R(\boldsymbol{\beta}_0) = L$. It will first be shown that $R(\boldsymbol{\beta})$ is bounded away from L if $||\boldsymbol{\beta}||$ is large enough. Let

$$a = \min_{||\boldsymbol{\beta}||=1} \max_{i=1,\ldots,n} |\mathbf{x}_i'\boldsymbol{\beta}|.$$

Then $a > 0$, since otherwise there would exist $\boldsymbol{\beta} \neq \mathbf{0}$ such that $\mathbf{x}_i'\boldsymbol{\beta} = 0$ for all i, which contradicts the full rank property. Let $b_0 > 0$ be such that $\rho(b_0) > 2L$, and b such that $ba - \max_i |y_i| \geq b_0$. Then $||\boldsymbol{\beta}|| > b$ implies $\max_i |y_i - \mathbf{x}_i'\boldsymbol{\beta}| \geq b_0$, and hence $R(\boldsymbol{\beta}) > 2L$. Thus minimizing R for $\boldsymbol{\beta} \in R^p$ is equivalent to minimizing it on the closed ball $\{||\boldsymbol{\beta}|| \leq b\}$. A well-known result of analysis states that a function that is continuous on a closed bounded set attains its minimum in it. Since R is continuous, the proof is completed. ■

Now we deal with the uniqueness of monotone M-estimators. Again we may take $\widehat{\sigma} = 1$.

Theorem 10.15 *Assume ψ nondecreasing. Put for given $(\mathbf{x}_i, y_i)$*

$$\mathbf{L}(\boldsymbol{\beta}) = \sum_{i=1}^{n} \psi\left(\frac{y_i - \mathbf{x}_i'\boldsymbol{\beta}}{\widehat{\sigma}}\right)\mathbf{x}_i.$$

Then (a) all solutions of $\mathbf{L}(\boldsymbol{\beta}) = \mathbf{0}$ minimize $R(\boldsymbol{\beta})$ defined in (10.31) and (b) if furthermore ψ has a positive derivative, then $\mathbf{L}(\boldsymbol{\beta}) = \mathbf{0}$ has a unique solution.

Proof.

(a) Without loss of generality we may assume that $\mathbf{L}(\mathbf{0}) = \mathbf{0}$. For a given $\boldsymbol{\beta}$ let $H(t) = R(t\boldsymbol{\beta})$ with R defined in (10.31). We must show that $H(1) \geq H(0)$. Since $dH(t)/dt = \boldsymbol{\beta}'\mathbf{L}(t\boldsymbol{\beta})$, we have

$$H(1) - H(0) = R(\boldsymbol{\beta}) - R(\mathbf{0}) = \sum_{i=1}^{n} \int_0^1 \psi(t\mathbf{x}_i'\boldsymbol{\beta} - y_i)(\mathbf{x}_i'\boldsymbol{\beta})dt.$$

If $\mathbf{x}_i'\boldsymbol{\beta} > 0$ (resp. < 0), then for $t > 0$ $\psi(t\mathbf{x}_i'\boldsymbol{\beta} - y_i)$ is greater (resp. smaller) than $\psi(-y_i)$. Hence

$$\psi(t\mathbf{x}_i'\boldsymbol{\beta} - y_i)(\mathbf{x}_i'\boldsymbol{\beta}) \geq \psi(-y_i)(\mathbf{x}_i'\boldsymbol{\beta}),$$

which implies that

$$R(\boldsymbol{\beta}) - R(\mathbf{0}) \geq \sum_{i=1}^{n} \psi(-y_i)(\mathbf{x}_i'\boldsymbol{\beta}) = \boldsymbol{\beta}'\mathbf{L}(\mathbf{0}).$$

(b) The matrix of derivatives of $\mathbf{L}$ with respect to $\boldsymbol{\beta}$ is

$$\mathbf{D} = -\frac{1}{\widehat{\sigma}}\sum_{i=1}^{n} \psi'\left(\frac{y_i - \mathbf{x}_i'\boldsymbol{\beta}}{\widehat{\sigma}}\right)\mathbf{x}_i\mathbf{x}_i',$$

which is negative definite; and the proof proceeds as that of Theorem 10.9. ∎

The former results do not cover MM- or S-estimators, since they are not monotonic. As was shown for location in Theorem 10.2, uniqueness holds for the asymptotic value of the estimator under the model $y_i = \mathbf{x}_i'\boldsymbol{\beta} + u_i$ if the u_is have a symmetric unimodal distribution.

10.10 Regression M-estimators: asymptotic normality

10.10.1 Fixed X

Now, to treat the asymptotic behavior of the estimator, we consider an infinite sequence $(\mathbf{x}_i, y_i)$ described by model (4.4). Call $\widehat{\boldsymbol{\beta}}_n$ the estimator (4.40) and $\widehat{\sigma}_n$ the scale estimator. Call $\mathbf{X}_n$ the matrix with rows $\mathbf{x}_i'$ $(i = 1, \ldots, n)$, which is assumed to have full rank. Then

$$\mathbf{X}_n'\mathbf{X}_n = \sum_{i=1}^{n} \mathbf{x}_i\mathbf{x}_i'$$

is positive definite, and hence it has a "square root"; that is, a (nonunique) $p \times p$ matrix $\mathbf{R}_n$ such that

$$\mathbf{R}_n'\mathbf{R}_n = \mathbf{X}_n'\mathbf{X}_n. \tag{10.32}$$

Call λ_n the smallest eigenvalue of $\mathbf{X}_n'\mathbf{X}_n$, and define

$$h_{in} = \mathbf{x}_i'(\mathbf{X}_n\mathbf{X}_n')^{-1}\mathbf{x}_i \tag{10.33}$$

and

$$M_n = \max\{h_{in} : i = 1, \ldots, n\}.$$

Define $\mathbf{R}_n$ as in (10.32). Assume

B1 $\lim_{n\to\infty}\lambda_n = \infty$;
B2 $\lim_{n\to\infty}M_n = 0$.

Then we have

Theorem 10.16 *Assume conditions A1–A2–A3 of Section 10.6. If B1 holds, then* $\widehat{\boldsymbol{\beta}}_n \to_p \boldsymbol{\beta}$. *If also B2 and A4–A5–A6 hold, then*

$$\mathbf{R}_n(\widehat{\boldsymbol{\beta}}_n - \boldsymbol{\beta}) \to_d \mathrm{N}_p(\mathbf{0}, \upsilon\mathbf{I}), \tag{10.34}$$

with υ *given by (10.21).*

The proof in a more general setting can be found in Yohai and Maronna (1979). For large n, the left-hand side of (10.34) has an approximate $\mathrm{N}_p(\mathbf{0}, \upsilon\mathbf{I})$ distribution, and from this, (4.43) follows since $\mathbf{R}_n^{-1}\mathbf{R}_n^{-1\prime} = (\mathbf{X}_n'\mathbf{X}_n)^{-1}$.

When $p = 1$ (fitting a straight line through the origin), condition B1 means that $\sum_{i=1}^n x_i^2 \to \infty$, which prevents the x_i from clustering around the origin. Condition B2 becomes

$$\lim_{n\to\infty} \frac{\max\left\{x_i^2 : i = 1, \ldots, n\right\}}{\sum_{i=1}^n x_i^2} = 0,$$

which means that none of the x_i^2 dominate the sum in the denominator; that is, there are no leverage points.

Now we consider a model with an intercept, namely (4.4). Let

$$\underline{\overline{\mathbf{x}}}_n = \mathrm{ave}_i(\underline{\mathbf{x}}_i) \text{ and } \mathbf{C}_n = \sum_{i=1}^{n} \left(\underline{\mathbf{x}}_i - \underline{\overline{\mathbf{x}}}_n\right)\left(\underline{\mathbf{x}}_i - \underline{\overline{\mathbf{x}}}_n\right)'.$$

Let $\mathbf{T}_n$ be any square root of $\mathbf{C}_n$, that is,

$$\mathbf{T}_n'\mathbf{T}_n = \mathbf{C}_n.$$

Theorem 10.17 *Assume the same conditions as in Theorem 10.16 excepting A3 and A6. Then*

$$\mathbf{T}_n(\widehat{\boldsymbol{\beta}}_{n1} - \boldsymbol{\beta}_1) \to_d \mathrm{N}_{p-1}(\mathbf{0}, v\mathbf{I}). \tag{10.35}$$

We shall give a heuristic proof for the case of a straight line; that is,

$$y_i = \beta_0 + \beta_1 x_i + u_i, \tag{10.36}$$

so that $\mathbf{x}_i = (1, x_i)$. Put

$$\overline{x}_n = \text{ave}_i(x_i),\ x_{in}^* = x_i - \overline{x}_n,\ \ C_n = \sum_{i=1}^n x_{in}^{*2}.$$

Then condition B1 is equivalent to

$$C_n \to \infty \text{ and } \frac{\overline{x}_n^2}{C_n} \to 0. \tag{10.37}$$

The first condition prevents the x_i from clustering around a point.

An auxiliary result will be needed. Let v_i, $i = 1, 2, \ldots$, be i.i.d. variables with a finite variance, and for each n let $a_{1n}, \ldots, a_{nn}$ be a set of constants. Let

$$V_n = \sum_{i=1}^n a_{in} v_i,\ \gamma_n = \mathrm{E}V_n,\ \tau_n^2 = \mathrm{Var}(V_n).$$

Then $W_n = (V_n - \gamma_n)/\tau_n$ has zero mean and unit variance, and the central limit theorem asserts that if for each n we have $a_{11} = \ldots = a_{nn}$, then $W_n \to_d \mathrm{N}(0, 1)$. It can be shown that this is still valid if the a_{in} are such that no term in V_n "dominates the sum", in the following sense:

Lemma 10.18 *If the a_{in}s are such that*

$$\lim_{n\to\infty} \frac{\max\left\{a_{in}^2 : i = 1, ..., n\right\}}{\sum_{i=1}^n a_{in}^2} = 0, \tag{10.38}$$

then $W_n \to_d \mathrm{N}(0, 1)$.

This result is a consequence of the so-called Lindeberg theorem (Feller, 1971). To see the need for condition (10.38), consider the v_i having a non-normal distribution G with unit variance. Take for each n: $a_{1n} = 1$ and $a_{in} = 0$ for $i > 1$. Then $V_n/\tau_n = v_1$, which has distribution G for all n, and hence does not tend to the normal.

To demonstrate Theorem 10.17 for the model (10.36), let

$$T_n = \sqrt{C_n}, \quad z_{in} = \frac{x_{in}^*}{T_n},$$

so that

$$\sum_{i=1}^{n} z_{in} = 0, \ \sum_{i=1}^{n} z_{in}^2 = 1. \tag{10.39}$$

Then (10.33) becomes

$$h_{in} = \frac{1}{n} + z_{in}^2,$$

so that condition B2 is equivalent to

$$\max\left\{z_{in}^2 \ : i = 1, \ldots, n\right\} \to 0. \tag{10.40}$$

We have to show that for large n, $\widehat{\beta}_{1n}$ is approximately normal, with variance v/C_n; that is, that $T_n(\widehat{\beta}_{1n} - \beta_1) \to_p \mathrm{N}(0, v)$.

The estimating equations are

$$\sum_{i=1}^{n} \psi\left(\frac{r_i}{\widehat{\sigma}_n}\right) = 0, \tag{10.41}$$

$$\sum_{i=1}^{n} \psi\left(\frac{r_i}{\widehat{\sigma}_n}\right) x_i = 0, \tag{10.42}$$

with $r_i = y_i - (\widehat{\beta}_{0n} + \widehat{\boldsymbol{\beta}}_{1n} x_i)$. Combining both equations yields

$$\sum_{i=1}^{n} \psi\left(\frac{r_i}{\widehat{\sigma}_n}\right) x_{in}^* = 0. \tag{10.43}$$

Put

$$\widehat{\Delta}_{1n} = \widehat{\boldsymbol{\beta}}_{1n} - \beta_1, \ \widehat{\Delta}_{0n} = \widehat{\boldsymbol{\beta}}_{0n} - \beta_0, \ \widehat{\Delta}_{2n} = \widehat{\sigma}_n - \sigma.$$

The Taylor expansion of ψ at t is

$$\psi(t + \varepsilon) = \psi(t) + \varepsilon\psi'(t) + o(\varepsilon), \tag{10.44}$$

where the last term is higher-order infinitesimal. Writing

$$r_i = u_i - \left(\widehat{\Delta}_{0n} + \widehat{\Delta}_{1n}(x_{in}^* + \overline{x}_n)\right),$$

expanding ψ at u_i/σ and dropping the last term in (10.44) yields

$$\psi\left(\frac{r_i}{\widehat{\sigma}_n}\right) = \psi\left(\frac{u_i - \left(\widehat{\Delta}_{0n} + \widehat{\Delta}_{1n}(x_{in}^* + \overline{x}_n)\right)}{\sigma + \widehat{\Delta}_{2n}}\right)$$

$$\approx \psi\left(\frac{u_i}{\sigma}\right) - \psi'\left(\frac{u_i}{\sigma}\right)\frac{\widehat{\Delta}_{0n} + \widehat{\Delta}_{1n}(x_{in}^* + \overline{x}_n) + \widehat{\Delta}_{2n}u_i/\sigma}{\sigma}. \tag{10.45}$$

Inserting (10.45) in (10.43), multiplying by σ and dividing by T_n, and recalling (10.39), yields

$$\sigma A_n = (T_n\widehat{\Delta}_{1n})\left(B_n + C_n\frac{\overline{x}_n}{T_n}\right) + \widehat{\Delta}_{0n}C_n + \widehat{\Delta}_{2n}D_n, \tag{10.46}$$

where

$$A_n = \sum_{i=1}^n \psi\left(\frac{u_i}{\sigma}\right)z_{in},\ B_n = \sum_{i=1}^n \psi'\left(\frac{u_i}{\sigma}\right)z_{in}^2,$$

$$C_n = \sum_{i=1}^n \psi'\left(\frac{u_i}{\sigma}\right)z_{in},\ D_n = \sum_{i=1}^n \psi'\left(\frac{u_i}{\sigma}\right)\frac{u_i}{\sigma}z_{in}.$$

Put

$$a = \mathrm{E}\psi\left(\frac{u}{\sigma}\right)^2,\ b = \mathrm{E}\psi'\left(\frac{u}{\sigma}\right),\ e = Var\left(\psi'\left(\frac{u}{\sigma}\right)\right).$$

Applying Lemma 10.18 to A_n (with $v_i = \psi(u_i/\sigma)$), and recalling (10.39) and (10.40), yields $A_n \to_d \mathrm{N}(0, a)$. The same procedure shows that C_n and D_n have normal limit distributions. Applying (10.39) to B_n yields

$$\mathrm{E}B_n = b,\ \mathrm{Var}(B_n) = e\sum_{i=1}^n z_{in}^4.$$

Now by (10.39) and (10.40)

$$\sum_{i=1}^n z_{in}^4 \le \max_{1\le i\le n} z_{in}^2 \to 0,$$

and then Tchebychev's inequality implies that $B_n \to_p b$. Recall that $\widehat{\Delta}_{2n} \to_p 0$ by hypothesis, and $\widehat{\Delta}_{0n} \to_p 0$ by Theorem 10.16. Since $\overline{x}_n/T_n \to 0$ by (10.37), application of Slutsky's lemma to (10.46) yields the desired result.

The asymptotic variance of $\widehat{\beta}_0$ may be derived by inserting (10.45) in (10.41). The situation is similar to that of Section 10.6: if $c = 0$ (with c defined in (10.20)) the proof can proceed, otherwise the asymptotic variance of $\widehat{\beta}_{0n}$ depends on that of $\widehat{\sigma}_n$.

10.10.2 Asymptotic normality: random X

Since the observations $\mathbf{z}_i = (\mathbf{x}_i, y_i)$ are i.i.d., this situation can be treated with the methods of Section 10.5. A regression M-estimator is a solution of

$$\sum_{i=1}^{n} \boldsymbol{\Psi}(z_i, \boldsymbol{\beta}) = \mathbf{0}$$

with

$$\boldsymbol{\Psi}(z, \beta) = x\psi(y - \mathbf{x}'\boldsymbol{\beta}).$$

We shall prove (5.14) for the case of σ known and equal to one. It follows from (10.15) that the asymptotic covariance matrix of $\widehat{\boldsymbol{\beta}}$ is $v\mathbf{V}_{\mathbf{x}}^{-1}$, with v given by (5.15) with $\sigma = 1$, and $\mathbf{V}_{\mathbf{x}} = \mathrm{E}\mathbf{x}\mathbf{x}'$. In fact, the matrices $\mathbf{A}$ and $\mathbf{B}$ in (10.16)–(10.14) are

$$\mathbf{A} = \mathrm{E}\psi(y - \mathbf{x}'\boldsymbol{\beta})^2\mathbf{x}\mathbf{x}', \quad \mathbf{B} = -\mathrm{E}\psi'(y - \mathbf{x}'\boldsymbol{\beta})\mathbf{x}\mathbf{x}',$$

and their existence is ensured by assuming that ψ and ψ' are bounded, and that $\mathrm{E}\ ||\mathbf{x}||^2 < \infty$. Under the model (5.1)–(5.2) we have

$$\mathbf{A} = \mathrm{E}\psi(u)^2\mathbf{V}_{\mathbf{x}}, \mathbf{B} = -\mathrm{E}\psi'(u)\mathbf{V}_{\mathbf{x}},$$

and the result follows immediately

10.11 Regression M estimators: Fisher-consistency

In this section we prove the Fisher-consistency (Section 3.5.3) of regression M estimators with random predictors. The case of monotone and redescending ψ will be considered separately. A general treatment of the consistency of regression M estimators is given in Fasano *et al.*, 2012.

10.11.1 Redescending estimators

Let the random elements $\mathbf{x} \in R^p$ and $y \in R$ satisfy the linear model

$$y = \mathbf{x}'\boldsymbol{\beta}_0 + u \tag{10.47}$$

with u independent of $\mathbf{x}$. Let $\widehat{\boldsymbol{\beta}} = \widehat{\boldsymbol{\beta}}(\mathbf{x}, y)$ be an M-estimating functional defined as $\widehat{\boldsymbol{\beta}} = \arg\min_{\beta} h(\boldsymbol{\beta})$ with

$$h(\boldsymbol{\beta}) = \mathrm{E}\rho(y - \mathbf{x}'\boldsymbol{\beta}), \tag{10.48}$$

where ρ is a given ρ-function (Definition 2.1). It will be proved that, under certain conditions on ρ and the distributions of $\mathbf{x}$ and u, $\widehat{\boldsymbol{\beta}}$ is Fisher-consistent; that is, $\widehat{\boldsymbol{\beta}}$ is the unique minimizer of h. It will be assumed that:

- ρ is a differentiable bounded ρ-function. It may be assumed that $\sup_t \rho(t) = 1$.
- The distribution F of u has a an even density f such that $f(t)$ is nondecreasing in $|t|$ and is decreasing in $|t|$ in a neighborhood of 0.
- The distribution G of $\mathbf{x}$ satisfies $\mathrm{P}(\boldsymbol{\beta}'\mathbf{x} = \mathbf{0}) < 1$ for all $\boldsymbol{\beta} \neq \mathbf{0}$.

Theorem 10.19 *$h(\boldsymbol{\beta}) > h(\boldsymbol{\beta}_0)$ for all $\boldsymbol{\beta} \neq \boldsymbol{\beta}_0$.*

The proof requires some auxiliary results.

Lemma 10.20 *Put $g(\lambda) = \mathrm{E}\rho(u - \lambda)$. Then:*

(i) *g is even*
(ii) *$g(\lambda)$ is nondecreasing for $\lambda \geq 0$*
(iii) *$g(0) < g(\lambda)$ for $\lambda \neq 0$.*

Proof of (i). Note that $\rho(u - \lambda)$ has the same distribution as $\rho(-u - \lambda) = \rho(u + \lambda)$, and therefore $g(\lambda) = g(-\lambda)$. ∎

Proof of (ii). For $\lambda \geq 0$ and $t \geq 0$ call $R_\lambda(t)$ the distribution function of $v = |u - \lambda|$:

$$R_\lambda(t) = P(|u - \lambda| \leq t) = \int_{\lambda - t}^{\lambda + t} f(s)ds.$$

Put $\dot{R}_\lambda(t) = \partial R_\lambda(t)/\partial\lambda$. Then

$$\dot{R}_\lambda(t) = f(\lambda + t) - f(\lambda - t) \leq 0. \tag{10.49}$$

Let λ_0 be such that for $t \in [0, \lambda_0]$, $f(t)$ is decreasing and $\psi(t) = \rho'(t) > 0$. Then

$$\dot{R}_\lambda(t) < 0 \text{ if } \lambda \leq \lambda_0 \text{ and } 0 < t \leq \lambda_0 \tag{10.50}$$

If follows from

$$R_\lambda(t) = \int_0^\lambda \dot{R}_\gamma(t)d\gamma$$

that $\lambda_1 \geq \lambda_2 \geq 0$ implies $R_{\lambda_1}(t) \geq R_{\lambda_2}(t)$, and therefore by (10.50)

$$R_\lambda(t) > R_0(t), \text{if } 0 < \lambda \leq \lambda_0 \text{ and } 0 < t < \lambda_0. \tag{10.51}$$

Integrating by parts yields

$$g(\lambda) = \mathrm{E}\rho(u - \lambda) = \int_0^\infty \rho(v)\frac{\partial R_\lambda(t)}{\partial t}dt$$

$$= R_\lambda(v)\rho(v)|_0^\infty - \int_0^\infty \rho'(v)R_\lambda(v)dv = 1 - \int_0^\infty \psi(t)R_\lambda(t)dt$$

and if $0 \le \lambda_1 \le \lambda_2$, (10.49) implies

$$g(\lambda_2) - g(\lambda_1) = \int_0^\infty \psi(t)(R_{\lambda_1}(t) - R_{\lambda_2}(t))dt \ge 0, \tag{10.52}$$

which proves (ii). ■

Proof of (iii). For $0 < \lambda \le \lambda_0$, we have

$$g(\lambda) - g(0) = \int_0^\lambda \psi(t)(R_0(t) - R_\lambda(t))dt + \int_\lambda^\infty \psi(t)(R_0(t) - R_\lambda(t))dt,$$

and since for $0 < t \le \lambda$ (10.51) implies $R_0(t) - R_\lambda(t) > 0$ and also $\psi(t) > 0$, it follows that

$$\int_0^\lambda \psi(t)(R_0(t) - R_\lambda(t))dt > 0 \ \text{ and } \ \int_\lambda^\infty \psi(t)(R_0(t) - R_\lambda(t))dt \ge 0,$$

and therefore

$$g(\lambda) > g(0) \text{ if } 0 < \lambda \le \lambda_0. \tag{10.53}$$

It follows from (10.52)–(10.53) that $g(\lambda) > g(0)$ if $0 < \lambda < \infty$. The result for $\lambda < 0$ follows from part (i) of the lemma. ■

Proof of the theorem. Let $\boldsymbol{\beta} \ne \boldsymbol{\beta}_0$ and put $\boldsymbol{\gamma} = \boldsymbol{\beta} - \boldsymbol{\beta}_0 \ne \mathbf{0}$. Then

$$h(\boldsymbol{\beta}) = \mathrm{E}\rho(y - \boldsymbol{\beta}'\mathbf{x}) = \mathrm{E}\rho(u - \boldsymbol{\gamma}'\mathbf{x}) = \mathrm{E}(\mathrm{E}[\rho(u - \boldsymbol{\gamma}'\mathbf{x})|\mathbf{x}]).$$

The independence of $\mathbf{x}$ and u implies that $\mathrm{E}[\rho(u - \boldsymbol{\gamma}'\mathbf{x})|\mathbf{x}] = g(\boldsymbol{\gamma}'\mathbf{x})$, and therefore $h(\boldsymbol{\beta}) = \mathrm{E}g(\boldsymbol{\gamma}'\mathbf{x}))$. Besides, since $h(\mathbf{0}) = g(0)$, we have

$$h(\boldsymbol{\beta}) - h(\mathbf{0}) = \mathrm{E}g(\boldsymbol{\gamma}'\mathbf{x})) - g(0).$$

Part (iii) of the lemma implies that $g(\boldsymbol{\gamma}'\mathbf{x}) - g(0) \ge 0$, and since $\mathrm{P}(\boldsymbol{\gamma}'\mathbf{x} > 0) > 0$, it follows that if we put $W = g(\boldsymbol{\gamma}'\mathbf{x}) - g(0)$, then $W \ge 0$ and $\mathrm{P}(W > 0) > 0$. Therefore

$$h(\boldsymbol{\beta}) - h(\mathbf{0}) = \mathrm{E}(W) > 0,$$

which completes the proof. ■

10.11.2 Monotone estimators

Let the random elements $\mathbf{x} \in R^p$ and $y \in R$ satisfy the linear model (10.47) with u independent of $\mathbf{x}$. Let $\widehat{\boldsymbol{\beta}} = \widehat{\boldsymbol{\beta}}(\mathbf{x}, y)$ be an M-estimating functional defined as the solution $\boldsymbol{\beta}$ of $\mathbf{h}(\boldsymbol{\beta}) = \mathbf{0}$ with

$$\mathbf{h}(\boldsymbol{\beta}) = \mathrm{E}\psi(y - \mathbf{x}'\boldsymbol{\beta})\mathbf{x} \tag{10.54}$$

where ψ is a given monotone ψ-function (Definition 2.2). It will be proved that under certain conditions on ψ and the distributions of $\mathbf{x}$ and u, $\widehat{\boldsymbol{\beta}}$ is Fisher-consistent; that is, $\boldsymbol{\beta} = \boldsymbol{\beta}_0$ is the only solution of $\mathbf{h}(\boldsymbol{\beta}) = \mathbf{0}$. It will be assumed that:

(A) ψ is odd, continuous and nondecreasing, and there exists a such that $\psi(t)$ is increasing on $[0, a]$.
(B) The distributon F of u has an even density f that is positive on $[-b, b]$ for some b.
(C) The distribution G of $\mathbf{x}$ satisfies $\mathrm{P}(\boldsymbol{\beta}'\mathbf{x} = \mathbf{0}) < 1$ for all $\boldsymbol{\beta} \neq \mathbf{0}$.

Theorem 10.21 $\mathbf{h}(\boldsymbol{\beta}) \neq \mathbf{0}$ *for all* $\boldsymbol{\beta} \neq \boldsymbol{\beta}_0$.

The proof requires some auxiliary results.

Lemma 10.22 *Let* $g(\lambda) = \mathrm{E}\psi(u - \lambda)$. *Then:*

(i) g is odd
(ii) g is nonincreasing
(iii) $g(\lambda) < 0$ *for* $\lambda > 0$.

Proof of (i). Since u and $-u$ have the same distribution, and ψ is odd, then $\psi(u - \lambda)$ has the same distribution as $\psi(-u - \lambda) = -\psi(u + \lambda)$, and hence $\mathrm{E}(\psi(u - \lambda)) = -\mathrm{E}(\psi(u + \lambda)$, which implies (i). ∎

Proof of (ii). Let $\lambda_1 < \lambda_2$. Then

$$g(\lambda_1) - g(\lambda_2) = \mathrm{E}(\psi(u - \lambda_1) - \psi(u - \lambda_2)) \tag{10.55}$$

Since ψ is nondecreasing and $u - \lambda_1 > u - \lambda_2$, we have $\psi(u - \lambda_1) - \psi(u - \lambda_2) \geq 0$, which implies $\mathrm{E}(\psi(u - \lambda_1) - \psi(u - \lambda_2)) \geq 0$, and hence $g(\lambda_1) - g(\lambda_2) \geq 0$. ∎

Proof of (iii). Note that by (A) and (B) there exists c such that for $|t| \leq c$, $\psi(t)$ is increasing and f is positive. Let $\lambda \in [0, c]$. Applying (10.49) to $\lambda_1 = 0$ and $\lambda_2 = c$, and recalling that $g(0) = 0$, yields $g(\lambda) = -\mathrm{E}(\psi(u) - \psi(u - \lambda))$ and hence

$$g(\lambda) = \mathrm{E}(\psi(u - \lambda) - \psi(u)) = \int_{-\infty}^{\infty} (\psi(t - \lambda) - \psi(t))f(t)dt. \tag{10.56}$$

Put $R(t, \lambda) = (\psi(t - \lambda) - \psi(t))f_0(t)$. Then

$$R(t, \lambda) \leq 0 \quad \text{for} \quad t \in R \quad \text{and} \quad \lambda \geq 0. \tag{10.57}$$

If $\lambda > 0$ and $0 \leq u \leq \lambda \leq c$ it follows that $|u - \lambda| \leq c$ which implies

$$R(t, \lambda) < 0 \text{ for } \lambda > 0 \text{ and } 0 \leq t \leq \lambda \leq a. \tag{10.58}$$

Using (10.56) decompose R as

$$\begin{aligned} g(\lambda) &= \int_{-\infty}^{0} R(t, \lambda)dt + \int_{0}^{\lambda} R(t, \lambda)du + \int_{\lambda}^{\infty} R(t, \lambda)du \\ &= I_1 + I_2 + I_3 \end{aligned}$$

Then (10.57) implies that $I_j \leq 0$ for $1 \leq j \leq 3$, and (10.58) implies $I_2 < 0$ and hence $g(\lambda) < 0$ for $\lambda \in [0, c]$. At the same time, (ii) implies that if $\lambda > a$ then $g(\lambda) \leq g(c) < 0$, which proves (iii). ■

Lemma 10.23 *Let $q(\lambda) = g(\lambda)\lambda$. Then $h(0) = 0$, $q(\lambda)$ is even and $q(\lambda) < 0$ for $\lambda \neq 0$.*

The proof is an immediate consequence of the previous lemma.

Proof of the theorem. We have

$$\mathbf{h}(\boldsymbol{\beta}_0) = \mathrm{E}\psi(y - \boldsymbol{\beta}_0'\mathbf{x})\mathbf{x} = \mathrm{E}\psi(u)\mathbf{x} = (\mathrm{E}\psi(u))(\mathrm{E}\mathbf{x}),$$

and since $\mathrm{E}\psi(u) = 0$ we have $\mathbf{h}(\boldsymbol{\beta}_0) = \mathbf{0}$.

Let $\boldsymbol{\beta} \neq \boldsymbol{\beta}_0$; then $\boldsymbol{\gamma} = \boldsymbol{\beta} - \boldsymbol{\beta}_0 \neq 0$. It will be shown that assuming $\mathbf{h}(\boldsymbol{\beta}) = \mathbf{0}$ yields a contradiction. For in that case

$$\boldsymbol{\gamma}'\mathbf{h}(\boldsymbol{\beta}) = \mathbf{0}, \tag{10.59}$$

and therefore

$$\boldsymbol{\gamma}'\mathbf{h}(\boldsymbol{\beta}) = \mathrm{E}\psi(y - \boldsymbol{\beta}'\mathbf{x})\boldsymbol{\gamma}'\mathbf{x} = \mathrm{E}\psi(u - \boldsymbol{\gamma}'\mathbf{x})\boldsymbol{\gamma}'\mathbf{x} = \mathrm{E}(\mathrm{E}[\psi(u - \boldsymbol{\gamma}'\mathbf{x})\boldsymbol{\gamma}'\mathbf{x}|\mathbf{x}]) = \mathrm{E}q(\boldsymbol{\gamma}'\mathbf{x})$$

But since $\mathrm{P}(\boldsymbol{\gamma}'\mathbf{x} \neq 0) > 0$, Lemma 10.23 implies that $\mathrm{E}q(\boldsymbol{\gamma}'\mathbf{x}) < 0$, and therefore $\mathbf{h}(\boldsymbol{\beta})'\boldsymbol{\beta} < 0$, which contradicts (10.59). ■

10.12 Nonexistence of moments of the sample median

We shall show that there are extremely heavy-tailed distributions for which the sample median has no finite moments of any order.

Let the sample $\{x_1, \ldots, x_n\}$ have a continuous distribution function F and an odd sample size $n = 2m + 1$. Then its median $\widehat{\theta}_n$ has distribution function G such that

$$\mathrm{P}(\widehat{\theta}_n > t) = 1 - G(t) = \sum_{j=0}^{m} \binom{n}{j} F(t)^j (1 - F(t))^{n-j}. \tag{10.60}$$

In fact, let $N = \#(x_i < t)$, which is binomial $\mathrm{Bi}(n, F(t))$. Then $\widehat{\theta}_n > t$ iff $N \leq m$, which yields (10.60).

It is easy to show, using integration by parts, that if T is a nonnegative variable with distribution function G, then

$$\mathrm{E}(T^k) = k \int_0^\infty t^{k-1}(1 - G(t))dt. \tag{10.61}$$

Now let

$$F(x) = \left(1 - \frac{1}{\log x}\right) \mathrm{I}(x \geq e).$$

Since for all positive r and s

$$\lim_{t\to\infty} \frac{t^r}{\log t^s} = \infty,$$

it follows from (10.61) that $\mathrm{E}\hat{\theta}_n^{\,k} = \infty$ for all positive k.

10.13 Problems

10.1. Let $x_1, ..., x_n$ be i.i.d. with continuous distribution function F. Show that the distribution function of the order statistic $x_{(m)}$ is

$$G(t) = \sum_{k=m}^{n} \binom{n}{k} F(t)^k (1 - F(t))^{n-k}$$

(Hint: for each t, the variable $N_t = \#\{x_i \leq t\}$ is binomial and verifies $x_{(m)} \leq t \Longleftrightarrow N_t \geq m$).

10.2. Let F be such that $F(a) = F(b) = 0.5$ for some $a < b$. If $x_1, .., x_{2m-1}$ are i.i.d. with distribution F, show that the distribution of $x_{(m)}$ tends to the average of the point masses at a and b.

10.3. Let $F_n = (1 - n^{-1})\mathrm{N}(0, 1) + n^{-1}\delta_{n^2}$, where δ_x is the point-mass at x. Verify that $F_n \to \mathrm{N}(0, 1)$, but its mean and variance tend to infinity.

10.4. Verify that if x is symmetric about μ, then (10.23) is equal to μ.

10.5. Verify that if ψ is odd, ρ is even, and F symmetric, then $\mathbf{V}$ in (10.18) is diagonal; and compute the asymptotic variances of $\hat{\mu}$ and $\hat{\sigma}$.

10.6. Show that scale M-estimators are uniquely defined if $\rho(x)$ is strictly increasing for all x such that $\rho(x) < \rho(\infty)$. (To make things easier, assume ρ differentiable).

10.7. Show that the location estimator with Huber's ψ_k and previous dispersion $\hat{\sigma}$ is uniquely defined unless there exists a solution $\hat{\mu}$ of $\sum_{i=1}^{n} \psi_k((x_i - \hat{\mu})/\hat{\sigma}) = 0$ such that $|x_i - \hat{\mu}| > k\hat{\sigma}$ for all i.

11

Description of Datasets

Here we describe the datasets used in the book

Alcohol

The solubility of alcohols in water is important in understanding alcohol transport in living organisms. This dataset from (Romanelli *et al.*, 2001) contains physicochemical characteristics of 44 aliphatic alcohols. The aim of the experiment was the prediction of the solubility on the basis of molecular descriptors. The columns are:

1. SAG: solvent accessible surface-bounded molecular volume
2. V: volume
3. Log PC: (octanol-water partitions coefficient)
4. P: polarizability
5. RM: molar refractivity
6. Mass
7. ln(solubility) (response)

Algae

This dataset is part of a larger one (http://kdd.ics.uci.edu/databases/coil/coil.html), which comes from a water quality study where samples were taken from sites on different European rivers over a period of approximately one year. These samples were

Robust Statistics: Theory and Methods (with R), Second Edition.
Ricardo A. Maronna, R. Douglas Martin, Victor J. Yohai and Matías Salibián-Barrera.

Companion website: www.wiley.com/go/maronna/robust

analyzed for various chemical substances. In parallel, algae samples were collected to determine the algal population distributions. The columns are:

1. Season (1,2,3,4 for winter, spring, summer and autumn)
2. River size (1,2,3 for small, medium and large)
3. Fluid velocity (1,2,3 for low, medium and high)

4-11. Content of nitrogen in the form of nitrates, nitrites and ammonia, and other chemical compounds

The response is the abundance of a type of algae (type 6 in the complete file). For simplicity we deleted the rows with missing values, or with null response values, and took the logarithm of the response.

Aptitude

There are three variables observed on 27 subjects:

- Score: numeric, represents scores on an aptitude test for a course
- Exp: numeric represents months of relevant previous experience
- Pass: binary response, 1 if the subject passed the exam at the end of the course and 0 otherwise.

The data may be downloaded as dataset 6.2 from the site http://www.jeremymiles.co .uk/regressionbook/data/.

Bus

This dataset from the Turing Institute, Glasgow, Scotland, contains measures of shape features extracted from vehicle silhouettes. The images were acquired by a camera looking downwards at the model vehicle from a fixed angle of elevation.

The following features were extracted from the silhouettes.

1. Compactness
2. Circularity
3. Distance circularity
4. Radius ratio
5. Principal axis aspect ratio
6. Maximum length aspect ratio
7. Scatter ratio
8. Elongatedness
9. Principal axis rectangularity

10. Maximum length rectangularity
11. Scaled variance along major axis
12. Scaled variance along minor axis
13. Scaled radius of gyration
14. Skewness about major axis
15. Skewness about minor axis
16. Kurtosis about minor axis
17. Kurtosis about major axis
18. Hollows ratio.

Glass

This is part of a file donated by Vina Speihler, describing the composition of glass pieces from cars. The columns are:

1. RI refractive index
2. Na_2O sodium oxide (unit measurement: weight percent in corresponding oxide, as are the rest of attributes)
3. MgO magnesium oxide
4. Al_2O_3 aluminum oxide
5. SiO_2 silcon oxide
6. K_2O potassium oxide
7. CaO calcium oxide.

Hearing

Prevalence rates in percent for men aged 55–64 with hearing levels 16 decibels or more above the audiometric zero. The rows correspond to different frequencies and to normal speech.

1. 500 Hz
2. 1000 Hz
3. 2000 Hz
4. 3000 Hz
5. 4000 Hz
6. 6000 Hz
7. Normal speech.

The columns classify the data into seven occupational groups:

1. Professional-managerial
2. Farm

3. Clerical sales
4. Craftsmen
5. Operatives
6. Service
7. Laborers.

Image

The data were supplied by A. Frery. They are a part of a synthetic aperture satellite radar image corresponding to a suburb of Munich.

Krafft

The Krafft point is an important physical characteristic of the compounds called surfactants, establishing the minimum temperature at which a surfactant can be used. The purpose of the experiment was to estimate the Krafft point of compounds as a function of their molecular structure.

The columns are:

1. Randiç index
2. Volume of tail of molecule
3. Dipole moment of molecule
4. Heat of formation
5. Krafft point (response).

Neuralgia

The data come from a study on the effect of iontophoretic treatment on elderly patients complaining of post-herpetic neuralgia. There were eighteen patients in the study, who were interviewed six weeks after the initial treatment and were asked if the pain had been reduced.

There are 18 observations on 5 variables:

- Pain: binary response: 1 if the pain eased, 0 otherwise.
- Treatment: binary variable: 1 if the patient underwent treatment, 0 otherwise.
- Age: the age of the patient in completed years.
- Gender: M (male) or F (female).
- Duration: pretreatment duration of symptoms (in months)

Oats

Yield of grain in grams per 16-foot row for each of eight varieties of oats in five replications in a randomized-block experiment.

Solid waste

The original data are the result of a study on production waste and land use by Golueke and McGauhey (1970), and contains nine variables. Here we consider the following six.

1. Industrial land (acres)
2. Fabricated metals (acres)
3. Trucking and wholesale trade (acres)
4. Retail trade (acres)
5. Restaurants and hotels (acres)
6. Solid waste (millions of tons), response.

Stack loss

The columns are:

1. Air flow
2. Cooling water inlet temperature (°C)
3. Acid concentration (%)
4. Stack loss, defined as the percentage of ingoing ammonia that escapes unabsorbed (response).

Toxicity

The aim of the experiment was to predict the toxicity of carboxylic acids on the basis of several molecular descriptors. The attributes for each acid are:

1. $\log(IGC_{50}^{-1})$: aquatic toxicity (response)
2. log Kow: partition coefficient
3. pKa: dissociation constant
4. ELUMO: energy of the lowest unoccupied molecular orbital
5. Ecarb: electrotopological state of the carboxylic group

6. Emet: electrotopological state of the methyl group
7. RM: molar refractivity
8. IR: refraction index
9. Ts: surface tension
10. P: polarizability.

Wine

This dataset, which is part of a larger one donated by Riccardo Leardi, gives the composition of several wines. The attributes are:

1. Alcohol
2. Malic acid
3. Ash
4. Alcalinity of ash
5. Magnesium
6. Total phenols
7. Flavanoids
8. Nonflavanoid phenols
9. Proanthocyanins
10. Color intensity
11. Hue
12. OD280/OD315 of diluted wines
13. Proline.

References

Abraham, B. and Chuang, A. (1989), Expectation–maximization algorithms and the estimation of time series models in the presence of outliers, *Journal of Time Series Analysis*, **14**, 221–234.

Adrover, J. and Yohai, V.J. (2002), Projection estimates of multivariate location. *The Annals of Statistics*, **30**, 1760–1781.

Adrover, J., Maronna, R.A. and Yohai, V.J. (2002), Relationships between maximum depth and projection regression estimates, *Journal of Statistical Planning and Inference*, **105**, 363–375.

Agostinelli, C. and Yohai, V.J. (2016), Composite robust estimators for linear mixed models. *Journal of American Statistical Association*, **111**, 1764–1774.

Agostinelli, C., Leung, A., Yohai, V.J. and Zamar, V.J. (2015), Robust estimation of multivariate location and scatter in the presence of cellwise and casewise contamination. *Test*, **24**, 441–461.

Agulló, J. (1996), Exact iterative computation of the multivariate minimum volume ellipsoid estimator with a branch and bound algorithm, *Proceedings of the 12th Symposium in Computational Statistics* (COMPSTAT 12), 175–180.

Agulló, J. (1997), Exact algorithms to compute the least median of squares estimate in multiple linear regression, in Y. Dodge (ed.), L_1-*Statistical Procedures and Related Topics*, Institute of Mathematical Statistics Lecture Notes – Monograph Series, vol. **31**, pp. 133–146.

Agulló, J. (2001), New algorithms for computing the least trimmed squares regression estimator, *Computational Statistics and Data Analysis*, **36**, 425–439.

Akaike, H. (1970), Statistical predictor identification, *Annals of the Institute of Statistical Mathematics*, **22**, 203–217.

Akaike, H. (1973), Information theory and an extension of the maximum likelihood principle, in B.N. Petran and F. Csaki (eds), *International Symposium on Information Theory* (2nd edn), pp. 267–281. Akademiai Kiadi.

Robust Statistics: Theory and Methods (with R), Second Edition.
Ricardo A. Maronna, R. Douglas Martin, Victor J. Yohai and Matías Salibián-Barrera.

Companion website: www.wiley.com/go/maronna/robust

Akaike, H. (1974a), A new look at the statistical model identification, *IEEE Transactions on Automatic Control*, **19**, 716–723.,

Akaike, H. (1974b), Markovian representation of stochastic processes and its application to the analysis of autoregressive moving average processes, *Annals of the Institute of Statistical Mathematics*, **26**, 363–387.

Albert, A. and Anderson, J.A. (1984), On the existence of maximum likelihohod estimates in logistic regression models, *Biometrika*,**71**, 1–10.

Alfons, A., Croux, C. and Gelper, S. (2013), Sparse least trimmed squares regression for analyzing high-dimensional data sets. *Annals of Applied Statistics*, **7**, 226–248.

Alqallaf, F.A., Konis, K.P., Martin, R.D. and Zamar, R.H. (2002), Scalable robust covariance and correlation estimates for data mining, in *Proceedings of SIGKDD 2002, Edmonton, Alberta, Canada*, Association of Computing Machinery (ACM).

Alqallaf, F.A., van Aelst, S., Yohai, V.J. and Zamar, R.H. (2009), Propagation of outliers in multivariate data. *Annals of Statistics*, **37**, 311–331.

Analytical Methods Committee (1989), Robust statistics – How not to reject outliers, *Analyst*, **114**, 1693–1702.

Anderson, D.K., Oti, R.S., Lord, C. and Welch, K (2009), Patterns of growth in adaptive social abilities among children with autism spectrum disorders.

Anderson, T.W. (1994), *The Statistical Analysis of Time Series*. John Wiley and Sons.

Journal of Abnormal Child Psychology, **37**, 1019–1034.

Arslan, O. (2004), Convergence behavior of an iterative reweighting algorithm to compute multivariate M-estimates for location and scatter, *Journal of Statistical Planning and Inference*, **118**, 115–128.

Bai, Z.D. and He, X. (1999), Asymptotic distributions of the maximal depth estimators for regression and multivariate location, *The Annals of Statistics*, **27**, 1616–1637.

Banerjee, O., Ghaoaui, L.E. and D'Aspremont, A. (2008), Model selection through sparse maximum likelihood estimation. *Journal of Machine Learning Research*, **9**, 485–516.

Barnett, V. and Lewis, T. (1998), *Outliers in Statistical Data* (3rd edn). John Wiley.

Barrodale, I. and Roberts, F.D.K. (1973), An improved algorithm for discrete l_1 linear approximation, *SIAM Journal of Numerical Analysis*, **10**, 839–848.

Belsley, D.A., Kuh, E. and Welsch, R.E. (1980), *Regression Diagnostics*. John Wiley.

Bergesio A. and Yohai, V.J. (2011), Projection estimators for generalized linear models. *Journal of the American Statistical Association*, **106**, 661–671.

Berrendero, J.R. and Zamar, R.H. (2001), The maxbias curve of robust regression estimates, *The Annals of Statistics*, **29**, 224–251.

Bianco, A.M. and Boente, G.L. (2002), On the asymptotic behavior of one-step estimates in heteroscedastic regression models. *Statistics and Probability Letters*, **60**, 33–47.

Bianco, A. and Yohai, V.J. (1996), Robust estimation in the logistic regression model, in H. Rieder (ed), *Robust Statistics, Data Analysis and Computer Intensive Methods*, Proceedings of the workshop in honor of Peter J. Huber, Lecture Notes in Statistics, vol. **109**, pp. 17–34. Springer-Verlag.

Bianco, A.M., García Ben, M., Martínez, E.J. and Yohai, V.J. (1996), Robust procedures for regression models with ARIMA errors, in A. Prat (ed.), *COMPSTAT 96, Proceedings in Computational Statistics*, pp. 27–38. Physica-Verlag.

Bianco, A.M., Boente, G.L. and Di Rienzo J. (2000), Some results of GM–based estimators in heteroscedastic regression models, *Journal of Statististical Planning and Inference*, **89**, 215–242.

Bianco, A.M., Garcia Ben, M., Martinez, E. and Yohai, V.J. (2001), Outlier detection in regression models with ARIMA errors using robust estimates, *Journal of Forecasting*, **20**, 565–579.

Bianco, A.M., Garcia Ben, M. and Yohai, V.J. (2005), Robust estimation for linear regression with asymmetric errors. *Canadian Journal of Statistics*, **33**, 511–528.

Bianco, A.M., Boente, G.L. and Rodrigues, I. (2013), Resistant estimators in Poisson and Gamma models with missing responses and an application to outlier detection. *Journal of Multivariate Analysis*, **114**, 209–226

Bickel, P.J. (1965), On some robust estimates of location. *The Annals of Mathematical Statistics*, **36**, 847–858.

Bickel, P.J. and Doksum, K.A. (2001), *Mathematical Statistics: Basic Ideas and Selected Topics*, vol. I (2nd edn). Prentice Hall.

Billingsley, P. (1968), *Convergence of Probability Measures*. John Wiley and Sons

Blackman, R.B. and Tukey, J.W. (1958), *The Measurement of Power Spectra*. Dover.

Bloomfield, P. (1976), *Fourier Analysis of Time Series: An Introduction*, John Wiley and Sons.

Bloomfield, P. and Staiger, W.L. (1983), *Least Absolute Deviations: Theory, Applications and Algorithms*, Birkhäuser.

Boente, G.L. (1983), Robust methods for principal components (in Spanish), PhD thesis, University of Buenos Aires.

Boente, G.L. (1987), Asymptotic theory for robust principal components, *Journal of Multivariate Analysis*, **21**, 67–78.

Boente, G.L. and Fraiman, R. (1999), Discussion of (Locantore *et al.*, 1999), *Test*, **8**,28–35.

Boente, G.L., Fraiman, R. and Yohai, V.J. (1987), Qualitative robustness for stochastic processes, *The Annals of Statistics*, **15**, 1293–1312.

Bond, N.W. (1979), Impairment of shuttlebox avoidance-learning following repeated alcohol withdrawal episodes in rats, *Pharmacology, Biochemistry and Behavior*, **11**, 589–591.

Bondell, H.D. and Stefanski, L.A. (2013), Efficient robust regression via two-stage generalized empirical likelihood, *Journal of the American Statistical Association*, **108**, 644–655.

Box, G.E.P., Hunter, W.G. and Hunter, J.S. (1978),*Statistics for Experimenters*. John Wiley and Sons.

Brandt, A. and Künsch, H.R. (1988), On the stability of robust filter-cleaners. *Stochastic Processes and their Applications*, **30**, 253–262.

Breiman, L., Friedman, J.H., Olshen, R.A. and Stone, C.J. (1984), *Classification and Regression Trees.*, Wadsworth.

Brent, R. (1973), *Algorithms for Minimisation Without Derivatives*. Prentice Hall.

Breslow, N.E. (1996), Generalized linear models: Checking assumptions and strengthening conclusions, *Statistica Applicata*, **8**, 23–41.

Brockwell, P.J. and Davis, R.A. (1991), *Introduction to Time Series and Forecasting*. Springer.

Brownlee, K.A. (1965), *Statistical Theory and Methodology en Science and Engineering* (2nd edn). John Wiley.

Brubacher. S.R. (1974), Time series outlier detection and modeling with interpolation, Bell Laboratories Technical Memo.

Bruce, A.G. and Martin, R.D. (1989), Leave–*k*–out diagnostics for time series (with discussion), *Journal of the Royal Statistical Society (B)*, **51**, 363–424.

Bühlmann, P. and van de Geer, S. (2011), Statistics for High-Dimensional Data. Springer-Verlag.

Bustos, O.H. (1982), General M-estimates for contaminated *p*th-order autoregressive processes: Consistency and asymptotic normality. Robustness in autoregressive processes. *Zeitschrift für Wahrscheinlichkeitstheorie und Verwandte Gebiete*, **59**, 491–504.

Bustos, O.H. and Yohai, V.J. (1986), Robust estimates for ARMA models, *Journal of the American Statistical Association*, **81**, 491–504.

Cai, T., Liu, W. and Luo, X. (2011), A constrained l1 minimization approach to sparse precision matrix estimation. *Journal of the American Statistical Association*, **106**, 594–607.

Campbell, N.A. (1980), Robust procedures in multivariate analysis I: Robust covariance estimation, *Applied Statistics*, **29**, 231–237.

Candés, E.J., Li, X. Ma, Y. and Wright, J. (2011), Robust principal component analysis?. *Journal of the ACM,* **58**, Article No. 11.

Cantoni, E. and Ronchetti, E. (2001), Robust inference for generalized linear models, *Journal of the American Statistical Association*, **96**, 1022–1030.

Carroll, R.J. and Pederson, S. (1993), On robustness in the logistic regression model, **55**, 693–706.

Carroll, R.J. and Ruppert, D. (1982), Robust estimation in heteroscedastic linear models, *The Annals of Statistics*, **10**, 429–441.

Chambers, J. (1977), *Computational Methods for Data Analysis*. John Wiley.

Chang, I., Tiao, G.C. and Chen, C. (1988), Estimation of time series parameters in the presence of outliers, *Technometrics*, **30**, 193–204

Chatterjee, S. and Hadi, A.S. (1988), *Sensitivity Analysis in Linear Regression.* John Wiley.

Chave, A.D., Thomson, D.J. and Ander, M.E. (1987), On the robust estimation of power spectra, coherence, and transfer functions, *Journal of Geophysical Research*, **92**, 633–648.

Chen, C. and Liu, L.M. (1993), Joint estimation of the model parameters and outlier effects in time series, *Journal of the American Statistical Association*, **88**, 284–297.

Chen, Z. and Tyler, D.E. (2002), The influence function and maximum bias of Tukey's median, *The Annals of Statistics*, **30**, 1737–1759.

Cheng, T.C. and Victoria-Feser, M.P. (2002), High-breakdown estimation of multivariate mean and covariance with missing observations. *British Journal of Mathematical & Statistical Psychology*, **55**, 317–335.

Chervoneva, I. and Vishnyakov, M. (2011), Constrained S-estimators for linear mixed effects models with covariance components. *Statistics in Medicine*, **30**, 1735–1750, 2011.

Clarke, B.R. (1983), Uniqueness and Fréchet diferentiability of functional solutions to maximum likelihood type equations, *The Annals of Statistics*, **11**, 1196–1205.

Cook, R.D. and Weisberg, S (1982), *Residuals and Influence in Regression,* Chapman and Hall.

Copt, S. and Victoria-Feser, M.P (2006), High breakdown inference for the mixed linear model. *Journal of American Statistical Association*, **101**, 292–300.

Croux. C., Dhaene, G. and Hoorelbeke, D. (2003), Robust standard errors for robust estimators, Discussion Papers Series 03.16, KU Leuven.

Croux, C. (1998), Limit behavior of the empirical influence function of the median, *Statistics and Probability Letters*, **37**, 331–340.

Croux, C. and Dehon, C. (2003), Estimators of the multiple correlation coefficient: local robustness and confidence intervals, *Statistical Papers*, **44**, 315–334.

Croux, C. and Haesbroeck, G. (2000), Principal component analysis based on robust estimators of the covariance or correlation matrix: influence functions and efficiencies, *Biometrika*, **87**, 603–618.

Croux, C. and Haesbroeck, G. (2003), Implementing the Bianco and Yohai estimator for logistic regression, *Computational Statistics and Data Analysis*, **44**, 273–295.

Croux, C. and Rousseeuw, P.J. (1992), Time-efficient algorithms for two highly robust estimators of scale, *Computational Statistics* **2**, 411–428.

Croux, C. and Ruiz-Gazen, A. (1996), A fast algorithm for robust principal components based on projection pursuit, in A. Prat (ed.), *Compstat: Proceedings in Computational Statistic*, pp. 211–216. Physica-Verlag.

Croux, C., Flandre, C. and Haesbroeck, G. (2002), The breakdown behavior of the maximum likelihood estimator in the logistic regression model, *Statistics and Probability Letters*, **60**, 377–386.

Daniel, C. (1978), Patterns in residuals in the two-way layout, *Technometrics*, **20**, 385–395.

Danilov, M. (2010), Robust estimation of multivariate scatter in non-affine equivariant scenarios. PhD thesis, University of British Columbia. Available at https://open.library.ubc.ca/cIRcle/collections/24/items/1.0069078.

Danilov, M., Yohai, V.J. and Zamar, R.H. (2012), Robust estimation of multivariate location and scatter in the presence of missing data. *Journal of the American Statistical Association.*, **107**, 1178–1186.

Davies, P.L. (1987), Asymptotic behavior of S-estimators of multivariate location parameters and dispersion matrices, *The Annals of Statistics*, **15**, 1269–1292.

Davies, P.L. (1990), The asymptotics of S-estimators in the linear regression model, *The Annals of Statistics,* **18**, 1651–1675.

Davies, P.L. (1992), The asymptotics of Rousseeuw's minimum volume ellipsoid estimator, *The Annals of Statistics*, **20**, 1828–1843.

Davis, R.A. (1996), Gauss-Newton and M-estimation for ARMA processes with infinite variance, *Stochastic Processes and their Applications*, **63**, 75–95.

Davis, R.A., Knight, K. and Liu, J. (1992), M-estimation for autoregression with infinite variance, *Stochastic Processes and their Applications*, **40**, 145–180.

Davison, A.C. and Hinkley, D.V. (1997), *Bootstrap Methods and their Application,* Cambridge University Press.

Dempster, A., Laird, N. and Rubin, D. (1977), Maximum likelihood from incomplete data via the EM algorithm. *Journal of the Royal Statistical Society, Series B*, **39**, 1–38.

Denby, L. and Martin, R.D. (1979), Robust estimation of the first-order autoregressive parameter, *Journal of the American Statistical Association*, **74**, 140–146.

de Vel, O., Aeberhard, S. and Coomans, D. (1993), Improvements to the classification performance of regularized discriminant analysis, *Journal of Chemometrics*, **7**, 99–115.

Devlin, S.J., Gnanadesikan, R. and Kettenring, J.R. (1981), Robust estimation of dispersion matrices and principal components, *Journal of the American Statistical Association*, **76**, 354–362.

Donoho, D.L. (1982), Breakdown properties of multivariate location estimators, PhD qualifying paper, Harvard University.

Donoho, D. and Gasko, M. (1992), Breakdown properties of location estimators based on halfspace depth and projected outlyingness, *The Annals of Statistics*, **20**, 1803–1827.

Donoho, D.L. and Huber, P.J. (1983), The notion of breakdown point, in P.J. Bickel, K.A. Doksum and J.L. Hodges (eds), *A Festshrift for Erich L. Lehmann*, pp. 157–184. Wadsworth.

Draper, R. and Smith, H. (1966), *Applied Regression Analysis*. John Wiley & Sons.

Draper, N.R. and Smith, H. (2001), *Applied Regression Analysis*, (3rd edn). John Wiley.

Durrett, R. (1996), *Probability Theory and Examples* (2nd edn). Duxbury Press.

Efron, B. and Tibshirani, R.J. (1993), *An Introduction to the Bootstrap*, Chapman and Hall.

Efron, B., Hastie, T., Johnstone, I. and Tibshirani, R. (2004), Least angle regression. *The Annals of Statistics*, **32**, 407–499.

Ellis, S.P. and Morgenthaler, S. (1992), Leverage and breakdown in L_1 regression, *Journal of the American Statistical Association*, **87**, 143–148.

Fan J. and Li R. (2001), Variable selection via nonconcave penalized likelihood and its oracle properties. *Journal of the American Statistical Association*, **96**, 1348–1360.

Farcomeni, A. (2014), Robust constrained clustering in presence of entry-wise outliers. *Technometrics*, **56**,102–111.

Fasano, M.V., Maronna, R.A., Sued, M. and Yohai, V.J. (2012), Continuity and differentiability of regression M functionals. *Bernouilli*, **8**, 1284–1309.

Feller, W. (1971), *An Introduction to Probability Theory and its Applications, vol. II* (2nd edn). John Wiley.

Fellner, W.H. (1986), Robust estimation of variance components. *Technometrics*, **28**, 51–60.

Fernholz, L.T. (1983), *Von Mises Calculus for Statistical Functionals*, Lecture Notes in Statistics No. 19. Springer,

Finney, D.J. (1947), The estimation from individual records of the relationship between dose and quantal response, *Biometrika*, **34**, 320–334.

Fox, A.J. (1972), Outliers in time series, *Journal of the Royal Statistical Society, Series B*, **34**, 350–363.

Frahm, G. and Jaekel, U. (2010), A generalization of Tyler's M-estimators to the case of incomplete data. *Computational Statistics & Data Analysis*, **54**, 374–393.

Fraiman, R., Yohai, V.J. and Zamar, R.H. (2001), Optimal M-estimates of location, *The Annals of Statistics*, **29**, 194–223.

Frank, I.E. and Friedman, J.H (1993), A statistical view of some chemometrics regression tools, *Technometrics*, **2**, 109–135.

Frery, A. (2005), Personal communication.

Friedman, J., Hastie, T. Höfling, H. and Tibshirani, R. (2007), Pathwise coordinate optimization. *Annals of Applied Statistics*, **1**, 302–332.

Friedman, H., Hastie, T. and Tibshirani R. (2008), Sparse inverse covariance estimation with the graphical lasso. *Biostatistics*, **9**, 432–441.

Gather, U. and Hilker, T. (1997), A note on Tyler's modification of the MAD for the Stahel–Donoho estimator, *The Annals of Statistics*, **25**, 2024–2026.

Genton, M.G. and Lucas, A. (2003), Comprehensive definitions of breakdown-points for independent and dependent observations, *Journal of the Royal Statistical Society Series B*, **65**, 81–94.

Genton, M.G. and Ma, Y. (1999), Robustness properties of dispersion estimators, *Statitics and Probability Letters*, **44**, 343–350.

Gervini, D. and Yohai, V.J. (2002), A class of robust and fully efficient regression estimators, *The Annals of Statistics*, **30**, 583–616.

Giltinan, D.M., Carroll, R.J. and Ruppert, D. (1986), Some new estimation methods for weighted regression when there are possible outliers, *Technometrics*, **28**, 219–230.

Gnanadesikan, R. and Kettenring, J.R. (1972), Robust estimates, residuals, and outlier detection with multiresponse data, *Biometrics*, **28**, 81–124.

Golueke, C.G. and McGauhey, P.H. (1970), *Comprehensive Studies of Solid Waste Management*, US Department of Health, Education and Welfare, Public Health Services Publication No. 2039.

Grenander, U. (1981), *Abstract Inference*. Wiley.

Grenander, U. and Rosenblatt, M. (1957), *Statistical Analysis of Stationary Time Series*, John Wiley and Sons.

Hampel, F.R. (1971), A general definition of qualitative robustness, *The Annals of Mathematical Statistics*, **42**, 1887–1896.

Hampel, F.R. (1974), The influence curve and its role in robust estimation., *The Annals of Statistics*, **69**, 383–393.

Hampel, F.R. (1975), Beyond location parameters: Robust concepts and methods, *Bulletin of the International Statistical Institute*,**46**, 375–382.

Hampel, F.R., Ronchetti, E.M., Rousseeuw, P.J. and Stahel, W.A. (1986), *Robust Statistics: The Approach Based on Influence Functions*. John Wiley.

Hannan, E.J. and Kanter, M. (1977), Autoregressive processes with infinite variance, *Journal of Applied Probability*, **14**, 411–415.

Harvey, A.C. and Phillips, G.D.A. (1979), Maximum likelihood estimation of regression models with autoregressive moving average disturbances, *Biometrika*, **66**, 49–58.

Hastie, T., Tibshirani, R. and Friedman, J. (2001), *The Elements of Statistical Learning*. Springer.

Hawkins, D.M. (1994), The feasible solution algorithm for least trimmed squares regression, *Computational Statistics and Data Analysis*, **17**, 185–196.

He, X. (1997), Quantile curves without crossing, *The American Statistician*, **51**, 186–192.

He, X and Portnoy, S. (1992), Reweighted LS estimators converge at the same rate as the initial estimator, *The Annals of Statistics*, **20**, 2161–2167.

Henderson, C., Kempthorne, O., Searle, S. and von Krosigk, C. (1959), The estimation of environmental and genetic trends from records subject to culling. *Biometrics*, **15**, 192–218.

Hennig, C. (1995), Efficient high-breakdown-point estimators in robust regression: Which function to choose?, *Statistics and Decisions*, **13**, 221–241.

Hettich, S. and Bay, S.D. (1999), The UCI KDD Archive, http://kdd.ics.uci.edu. University of California, Department of Information and Computer Science.

Hoerl, A.E. and Kennard, R.W. (1970), Ridge regression: biased estimation for nonorthogonal problems, *Technometrics*, **8**, 27–51.

Hössjer, O. (1992), On the optimality of S-estimators, *Statistics and Probability Letters*, **14**, 413–419.

Hsieh, C.J., Sustik, M.A., Dhillon, I.S. and Ravikumar, P. (2011), Sparse inverse covariance matrix estimation using quadratic approximation. *Advances in Neural Information Processing Systems,* **24**, 2330–2338.

Huber, P.J. (1964), Robust estimation of a location parameter, *The Annals of Mathematical Statistics*, **35**, 73–101.

Huber, P.J. (1965), A robust version of the probability ratio test, *The Annals of Mathematical Statistics*, **36,** 1753–1758.

Huber, P.J. (1967), The behavior of maximum likelihood estimates under nonstandard conditions, in *Proceedings of Fifth Berkeley Symposium of Mathematical Statistics and Probability*, vol. **1**, pp. 221–233. University of California Press.

Huber, P.J. (1968), Robust confidence limits, *Zeitschrift für Wahrscheinlichkeitstheorie und Verwandte Gebiete,* **10**, 269–278.

Huber, P.J. (1981). *Robust Statistics*. John Wiley.

Huber, P.J. (1984), Finite sample breakdown of M- and P-estimators, *The Annals of Statistics*, **12**, 119–126.

Huber, P.J. and Ronchetti, E.M. (2009), *Robust Statistics* (2nd edn). John Wiley.

Huber-Carol, C. (1970), Étude asymptotique des tests robustes, PhD thesis, Eidgenössische Technische Hochschule, Zürich.

Hubert, M. and Rousseeuw, P.J. (1996), Robust regression with a categorical covariable, in H. Rieder (ed.), *Robust Statistics, Data Analysis, and Computer Intensive Methods,* Lecture Notes in Statistics No. 109, pp. 215–224. Springer Verlag.

Hubert, M. and Rousseeuw, P.J. (1997), Robust regression with both continuous and binary regressors. *Journal of Statistical Planning and Inference*, **57**, 153–163.

Hubert, M., Rousseeuw, P.J. and Vanden Branden, K. (2012), ROBPCA: A new approach to robust principal component analysis. *Technometrics,* **47**, 64–79.

Johnson, W. (1985), Influence measures for logistic regression: Another point of view, *Biometrika*, **72**, 59–65.

Johnson, R.A. and Wichern, D.W. (1998), *Applied Multivariate Statistical Analysis,* Prentice Hall.

Jones, R.H. (1980), Maximum likelihood fitting of ARMA models to time series with missing observations, *Technometrics*, **22**, 389–396.

Khan, K., Van Aelst, S. and Zamar, R. (2007), Robust linear model selection based on least angle regression. *Journal of the American Statistical Association*, **102**, 1289–1299.

Kandel, R. (1991), *Our Changing Climate,* McGraw-Hill,

Kanter, M. and Steiger, W.L. (1974), Regression and autoregression with infinite variance, *Advances in Applied Probability*, **6**, 768–783.

Kent, J.T. and Tyler, D.E. (1996), Constrained M-estimation for multivariate location and scatter, *The Annals of Statistics*, **24**, 1346–1370.

Kim, J. and Pollard, D. (1990), Cube root asymptotics, *The Annals of Statistics,* **18**, 191–219.

Kleiner, B., Martin, R.D. and Thomson, D.J. (1979), Robust estimation of power spectra (with discussion), *Journal of the Royal Statistical Society Series B*, **41**, 313–351.

Knight, K. (1987), Rate of convergence of centered estimates of autoregressive parameters for infinite variance autoregressions, *Journal of Time Series Analysis*, **8**, 51–60.

Koenker, R. and Bassett, G.J. (1978), Regression quantiles, *Econometrica*, **46**, 33–50.

Koenker, R., Hammond, P. and Holly, A. (eds) (2005), *Quantile Regression*, Cambridge University Press.

Koller, M. (2012), Nonsingular subsampling for S-estimators with categorical predictors. Available at arXiv:1208.5595 [stat.CO].

Koller M. (2013), Robust estimation of linear mixed models. PhD thesis, ETH Zurich.

Koller, M. and Stahel, W.A. (2011), Sharpening Wald-type inference in robust regression for small samples. *Computational Statistics & Data Analysis*, **55**, 2504–2515.

Krasker, W.S. and Welsch, R.E. (1982), Efficient bounded influence regression estimation, *Journal of the American Statistical Association*, **77**, 595–604.

Künsch, H. (1984), Infinitesimal robustness for autoregressive processes, *The Annals of Statistics*, **12**, 843–863.

Künsch, H.R. Stefanski, L.A. and Carroll, R.J. (1991), Conditionally unbiased bounded-influence estimation in general regression models, with applications to generalized linear models, *Journal of the American Statistical Association*, **84**, 460–466.

Lauritzen, S.L. (1996), *Graphical Models.* Clarendon Press.

Ledolter, J. (1979), A recursive approach to parameter estimation in regression and time series models, *Communications in Statistics*, **A8**, 1227–1245.

Ledolter, J. (1991), Outliers in time series analysis: Some comments on their impact and their detection, in W. Stahel and S. Weisberg (eds.), *Directions in Robust Statistics and Diagnostics, Part I*, pp. 159–165. Springer–Verlag.

Lee, C.H. and Martin, R D. (1986), Ordinary and proper location M-estimates for autoregressive-moving average models, *Biometrika*, **73**, 679–686.

Lehmann, E.L. and Casella, G. (1998), *Theory of Point Estimation* (2nd edn), Springer Texts in Statistics.

Leung, A., Yohai, V.J. and Zamar, R.H. (2017) Multivariate Location and Scatter Matrix Estimation under Cellwise and Casewise Contamination. *Computational Statistics and Data Analysis*, **111**, 1–220.

Li, B. and Zamar, R.H. (1991), Min-max asymptotic variance when scale is unknown, *Statistics and Probability Letters*, **11**, 139–145.

Li, G. and Chen, Z. (1985), Projection-pursuit approach to robust dispersion matrices and principal components: primary theory and Monte Carlo. *Journal of the American Statistical Association*, **80**, 759–766.

Li, G., Peng, H. and Zhu, L. (2011), Nonconcave penalized M-estimation with a diverging number of parameters. *Statistica Sinica*, **21**, 391–419.

Lindsay, B.G (1988), Composite likelihood methods. *Contemporary Mathematics*, **80**, 221–239.

Little, R.J.A. (1988), Robust estimation of the mean and covariance matrix from data with missing values. *Journal of the Royal Statististical Society Series C*, **37**, 23–38.

Little, R.J.A. and Rubin, D.B. (2002), *Statistical Analysis with Missing Data* (2nd edn). Wiley.

Little, R.J.A. and Smith, P.J. (1987), Editing and imputation of quantitative survey data. *Journal of the American Statistical Association* **82**, 58–68.

Liu, R.Y. (1990), On a notion of data depth based on random simplices, *The Annals of Statistics*,**18**, 405–414.

Locantore, N., Marron, J.S., Simpson, D.G., Tripoli, N., Zhang, J.T. and Cohen, K.L. (1999), Robust principal components for functional data, *Test*, **8**, 1–28.

Lopuhaã, H.P. (1991), Multivariate τ-estimators for location and scatter, *Canadian Journal of Statistics*, **19**, 307–321.

Lopuhaã, H.P. (1992), Highly efficient estimators of multivariate location with high breakdown point, *The Annals of Statistics*, **20**, 398–413.

Ma, Y. and Genton, M.G. (2000), Highly robust estimation of the autocovariance function, *Journal of Time Series Analysis*, **21**, 663–684.

Mallows, C.L. (1975), *On some topics in robustness*, Unpublished memorandum, Bell Telephone Laboratories, Murray Hill, NJ.

Manku, G.S., Rajagopalan, S. and Lindsay, B. (1999), Random sampling techniques for space efficient online computation of order statistics of large data sets, ACM SGIMOD Record 28.

Marazzi, A., Paccaud F., Ruffieux, C. and Beguin, C. (1998), Fitting the distributions of length of stay by parametric models, *Medical Care*, **36**, 915–927.

Maronna, R.A. (1976), Robust M-estimators of multivariate location and scatter, *The Annals of Statistics*, **4**, 51–67.

Maronna, R.A. (2005), Principal components and orthogonal regression based on robust scales, *Technometrics*, **47**, 264–273.

Maronna, R.A (2011), Robust ridge regression for high-dimensional data. *Technometrics*, **53**, 44–53.

Maronna, R.A (2017), Improving the Peña-Prieto “KSD” procedure. arXiv:1708.03196 [stat.ME].

Maronna, R.A. and Yohai, V.J. (1981), Asymptotic behavior of general M–estimates for regression and scale with random carriers, *Zeitschrift für Wahrscheinlichkeitstheorie und Verwandte Gebiete*, **58**, 7–20.

Maronna, R.A. and Yohai, V.J. (1991), The breakdown point of simultaneous general M–estimates of regression and scale, *Journal of the American Statistical Association*, **86**, 699–703.

Maronna, R.A. and Yohai, V.J. (1993), Bias-robust estimates of regression based on projections, *The Annals of Statistics*, **21**, 965–990.

Maronna, R.A. and Yohai, V.J. (1995), The behavior of the Stahel–Donoho robust multivariate estimator, *Journal of the American Statistical Association*, **90**, 330–341.

Maronna, R.A. and Yohai, V.J. (1999), Robust regression with both continuous and categorical predictors. Technical Report, Faculty of Exact Sciences, University of Buenos Aires. Available by anonymous ftpat:ulises.ic.fcen.uba.ar.

Maronna, R.A. and Yohai, V.J. (2000), Robust regression with both continuous and categorical predictors, *Journal of Statistical Planning and Inference*, **89**, 197–214.

Maronna, R.A. and Yohai, V.J. (2008), Robust lower-rank approximation of data matrices with element-wise contamination, *Technometrics*, **50**, 295–304.

Maronna, R.A. and Yohai, V J. (2010), Correcting MM estimates for "fat" data sets. *Computational Statistics & Data Analysis*, **54**, 3168–3173.

Maronna, R.A. and Yohai, V J. (2013), Robust functional linear regression based on splines. *Computational Statistics and Data Analysis*, **65**, 46–55.

Maronna, R.A. and Yohai, V.J. (2014), High finite-sample efficiency and robustness based on distance-constrained maximum likelihood. *Computational Statistics and Data Analysis*, **83**, 262–274.

Maronna, R.A. and Yohai, V.J. (2017), Robust and efficient estimation of high dimensional scatter and location. *Computational Statistics and Data Analysis*, **109**, 64–75.

Maronna, R.A. and Zamar, R.H. (2002), Robust estimation of location and dispersion for high-dimensional data sets, *Technometrics*, **44**, 307–317.

Maronna, R.A., Bustos, O.H. and Yohai, V.J. (1979), Bias- and efficiency-robustness of general M-estimators for regression with random carriers, in T. Gasser and J.M. Rossenblat (eds.) *Smoothing Techniques for Curve Estimation*. Lecture Notes in Mathematics, vol. **757**, pp. 91–116. Springer Verlag,

Maronna, R.A., Stahel, W.A. and Yohai, V.J. (1992), Bias-robust estimators of multivariate scatter based on projections, *Journal of Multivariate Annalysis*, **42**, 141–161.

Martin, R.D. (1979), Approximate conditional mean type smoothers and interpolators, in T. Gasser and M. Rosenblatt (eds), *Smoothing Techniques for Curve Estimation*, pp. 117–143. Springer Verlag.

Martin, R.D. (1980), Robust estimation of autoregressive models, in D.R. Billinger and G.C. Tiao (eds), *Directions in Time Series*, pp. 228–254. Institute of Mathematical Statistics.

Martin, R.D. (1981), Robust methods for time series, in D.F. Findley (ed.), *Applied Time Series Analysis II*, pp. 683–659. Academic Press.

Martin, R.D. and Jong, J.M. (1977), Asymptotic properties of robust generalized M-estimates for the first-order autoregressive parameter, Bell Labs. Technical Memo, Murray Hill, NJ.

Martin, R.D. and Su, K.Y. (1985), Robust filters and smoothers: Definitions and design. Technical Report No. 58. Department of Statistics, University of Washington.

Martin, R.D. and Thomson, D.J. (1982), Robust-resistant spectrum estimation, *IEEE Proceedings*, **70**, 1097–1115.

Martin, R.D. and Yohai, V.J. (1985), Robustness in time series and estimating ARMA models, in E.J. Hannan, P.R. Krishnaiah and M.M. Rao (eds), *Handbook of Statistics, Volume 5: Time Series in the Time Domain*. Elsevier.

Martin, R.D. and Yohai, V.J. (1986), Influence functionals for time series (with discussion), *The Annals of Statistics*, **14**, 981–855.

Martin, R.D. and Yohai, V.J. (1991), Bias robust estimation of autoregression parameters, in W. Stahel and S. Weisberg (eds), *Directions in Robust Statistics and Diagnostics Part I*, IMA Volumes in Mathematics and its Applications, vol. 30. Springer-Verlag.

Martin, R.D. and Zamar, R.H. (1989), Asymptotically min-max bias-robust M-estimates of scale for positive random variables, *Journal of the American Statistical Association*, **84**, 494–501.

Martin, R.D. and Zamar, R.H. (1993a), Efficiency-constrained bias-robust estimation of location, *The Annals of Statistics*, **21**, 338–354.

Martin, R.D. and Zamar, R.H. (1993b), Bias robust estimates of scale, *The Annals of Statistics*, **21**, 991–1017.

Martin, R.D., Samarov, A. and Vandaele W. (1983), Robust methods for ARIMA models, in E. Zellner (ed.), *Applied Time Series Analysis of Economic Data*, pp. 153–177, Washington Bureau of the Census.

Martin, R.D., Yohai, V.J. and Zamar, R.H. (1989), Min–max bias robust regression, *The Annals of Statistics* **17**, 1608–1630.

Masarotto, G. (1987), Robust and consistent estimates of autoregressive-moving average parameters, *Biometrika*, **74**, 791–797.

Masreliez, C.J. (1975), Approximate non-Gaussian filtering with linear state and observation relations, *IEEE Transactions on Automatic Control*, **AC–20**, 107–110.

Masreliez, C.J. and Martin, R.D. (1977), Robust Bayesian estimation for the linear model and robustifying the Kalman filter, *IEEE Transactions on Automatic Control*, **AC-22**, 361–371.

Meinhold, R.J. and Singpurwalla, N.D. (1989), Robustification of Kalman filter models, *Journal of the American Statistical Association*, **84**, 479–88.

Mendes, B. and Tyler, D.E. (1996), Constrained M-estimation for regression, in *Robust Statistics, Data Analysis and Computer Intensive Methods (Schloss Thurnau, 1994)*, pp. 299–320, Lecture Notes in Statistics, vol. 109. Springer,

Menn, C. and Rachev S.T. (2005), A GARCH option pricing model with alpha-stable innovations, *European Journal of Operational Research*, **163**, 201–209.

Mikosch, T., Gadrich, T., Kluppelberg, C. and Adler, R.J. (1995), Parameter estimation for ARMA models with infinite variance innovations, *The Annals of Statistics*, **23**, 305–326.

Mili, L. and Coakley, C.W. (1996), Robust estimation in structured linear regression, *The Annals of Statistics*, **24**, 2593–2607.

Mili, L., Cheniae, M.G., Vichare, N.S. and Rousseeuw, P.J. (1996), Robust state estimation based on projection statistics, *IEEE Transactions on Power Systems*, **11**, 1118–1127.

Miller, A.J. (1990), *Subset Selection in Regression*, Chapman & Hall.

Montgomery, D.C., Peck, E.A. and Vining, G.G. (2001), *Introduction to Linear Regression Analysis* (3rd edn). John Wiley.

Muler, N. and Yohai, V.J. (2002), Robust estimates for ARCH processes, *Journal of Time Series Analysis*, **23**, 341–375.

Ollerer, V. and Croux, C. (2015), Robust high-dimensional precision matrix estimation. arXiv:1501.01219 [stat.ME].

Paindaveine, D. and van Bever, G. (2014), Inference on the shape of elliptical distributions based on the MCD. *Journal of Multivariate Analysis*, **129**, 1071–1083

Peña, D. (1987), Measuring the importance of outliers in ARIMA models, in M.L. Puri, J.P. Vilaplana and W. Wertz (eds.), *New Perspectives in Theoretical and Applied Statistics*, pp. 109–118. Wiley.

Peña, D. (1990), Influential observations in time series, *Journal of Business and Economic Statistics*, **8**, 235–241.

Peña, D. and Prieto, F.J. (2007), Combining random and specific directions for robust estimation of high-dimensional multivariate data. *Journal of Computational and Graphical Statistics,* **16**, 228–254.

Peña, D. and Yohai, V.J. (1999), A fast procedure for outlier diagnostics in large regression problems, *Journal of the American Statistical Association*, **94**, 434–445.

Percival, D.B. and Walden, A.T. (1993), *Spectral Analysis for Physical Applications: Multitaper and Conventional Univariate Techniques*, Cambridge University Press.

Piegorsch, W.W. (1992), Complementary log regression for generalized linear models, *The American Statistician*, **46**, 94–99.

Pires, R.C., Simões Costa, A. and Mili, L. (1999), Iteratively reweighted least squares state estimation through Givens rotations, *IEEE Transactions on Power Systems*, **14**, 1499–1505.

Portnoy, S. and Koenker, R. (1997), The Gaussian hare and the Laplacian tortoise: Computability of squared-error versus absolute-error estimators, *Statistical Science*, **12**, 299–300.

Pregibon, D. (1981), Logistic regression diagnostics, *The Annals of Statistics*, **9**, 705–724.

Pregibon, D. (1982), Resistant fits for some commonly used logistic models with medical applications, *Biometrics*, **38**, 485–498.

Qian, G. and Künsch, H.R. (1998), On model selection via stochastic complexity in robust linear regression, *Journal of Statistical Planning and Inference*, **75**, 91–116.

Rachev, S. and Mittnik, S. (2000), *Stable Paretian Models in Finance*. Wiley.

Richardson, A.M. and Welsh, A.H. (1995), Robust restricted maximum likelihood in mixed linear models. *Biometrics*, **51**, 1429–1439.

Rieder, H. (1978), A robust asymptotic testing model, *The Annals of Statistics*, **6**, 1080–1094.

Rieder, H. (1981), Robustness of one- and two-sample rank tests against gross errors, *The Annals of Statistics*, **9,** 245–265.

Roberts, J. and Cohrssen, J. (1968), Hearing levels of adults, US National Center for Health Statistics Publications, Series 11, No. 31.

Rocke, D.M. (1996), Robustness properties of *S*-estimators of multivariate location and shape in high dimension, *The Annals of Statistics*, **24,** 1327–1345.

Romanelli, G.P., Martino, C.M. and Castro, E.A. (2001), Modeling the solubility of aliphatic alcohols via molecular descriptors, *Journal of the Chemical Society of Pakistan*, **23**, 195–199.

Ronchetti, E. and Staudte, R.G. (1994), A robust version of Mallow's C_p, *Journal of the American Statistical Association*, **89**, 550–559.

Ronchetti, E., Field, C. and Blanchard, W. (1997), Robust linear model selection by cross-validation, *Journal of the American Statistical Association*, **92**, 1017–1023.

Rousseeuw, P.J. (1984), Least median of squares regression. *Journal of the American Statistical Association* **79**, 871–880.

Rousseeuw, P.J. (1985), Multivariate estimation with high breakdown point, in W. Grossmann, G. Pflug, I. Vincze and W. Wertz (eds), *Mathematical Statistics and its Applications* (vol. B), pp. 283–297. Reidel.

Rousseeuw, P.J. and Croux, C. (1993), Alternatives to the median absolute deviation, *Journal of the American Statistical Association*, **88**, 1273–1283.

Rousseeuw, P.J and Hubert, M. (1999), Regression depth. *Journal of the American Statististical Association*, **94,**388–402.

Rousseeuw, P.J. and Leroy, A.M. (1987), *Robust Regression and Outlier Detection*. John Wiley.

Rousseeuw, P.J. and van den Bossche, W. (2016), Detecting deviating data cells. arXiv:1601.07251 [stat.ME].

Rousseeuw, P.J. and van Driessen, K. (1999), A fast algorithm for the minimum covariance determinant estimator, *Technometrics*, **41**, 212–223.

Rousseeuw, P.J. and van Driessen, K. (2000), An algorithm for positive-breakdown regression based on concentration steps, in W. Gaul, O. Opitz and M. Schader (eds), *Data Analysis: Modeling and Practical Applications*, pp. 335–346. Springer Verlag,

Rousseeuw, P.J. and Wagner, J. (1994), Robust regression with a distributed intercept using least median of squares. *Computational Statististics and Data Analysis,* **17**, 65–76.

Rousseeuw, P.J and Yohai, V.J. (1984), Robust regression by means of S–estimators, In J. Franke, W. Härdley and R.D. Martin (eds), *Robust and Nonlinear Time Series*. Lectures Notes in Statistics vol. 26, pp. 256–272. Springer Verlag,

Ruppert, D. (1992), Computing S-estimators for regression and multivariate location/dispersion, *Journal of Computational and Graphical Statistics*, **1**, 253–270.

Salibian-Barrera, M. (2000), Contributions to the theory of robust inference. Unpublished PhD thesis. Department of Statistics, University of British Columbia, Vancouver, BC, Canada.

Salibian-Barrera, M. (2005), Estimating the p-values of robust tests for the linear model. *Journal of Statistical Planning and Inference*, **128**, 241–257.

Salibian-Barrera,M. and van Aelst, S. (2008), Robust model selection using fast and robust bootstrap, *Computational Statistics and Data Analysis*, **52**, 5121–5135.

Salibian-Barrera, M. and Yohai, V.J. (2006), A fast algorithm for S-regression estimates. *Journal of Computational and Graphical Statistics,* **15**, 414–427.

Salibian-Barrera, M. and Zamar, R.H. (2002), Bootstrapping robust estimates of regression. *The Annals of Statistics*, **30**, 556–582.

Salibian-Barrera, M., van Aelst, S., Willems, G. (2006), Principal components analysis based on multivariate MM estimators with fast and robust bootstrap. *Journal of the American Statistical Association*, **101**, 1198–1211.

Salibian-Barrera, M., Willems, G. and Zamar, R. (2008a), The fast-tau estimator for regression. *Journal of Computational and Graphical Statistics*, **17**, 659–682.

Salibian-Barrera, M., van Aelst, S. and Willems, G. (2008b), Fast and robust bootstrap. *Statistical Methods and Applications*, **17**, 41–71.

Salibian-Barrera, M., van Aelst, S. and Yohai, V.J. (2016), Robust tests for linear regression models based on tau-estimates. *Computational Statistics and Data Analysis*, **93**, 436–455.

Samarakoon, D.M. and Knight, K. (2005), A note on unit root tests with infinite variance noise. Unpublished manuscript.

Scheffé, H. (1959), *Analysis of Variance*. John Wiley.

Scherer, B. (2015), *Portfolio Construction and Risk Budgeting* (5th. edn). Risk Books.

Schweppe, F.C., Wildes, J. and Rom, D.B., (1970), Power system static-state estimation, Parts I, II and III, *IEEE Transactions on Power Apparatus and Systems*, **PAS-89**, 120–135.

Searle, S.R., Casella, G. and McCulloch, C.E. (1992), *Variance Components*. John Wiley and Sons.

Seber, G.A.F. (1984), *Multivariate Observations*. John Wiley.

Seber, G.A.F. and Lee, A.J. (2003), *Linear Regression Analysis* (2nd edn). John Wiley and Sons.

Shao, J. (2003), *Mathematical Statistics* (2nd edn). Springer.

Siebert, J.P. (1987), Vehicle recognition using rule based methods, Turing Institute Research Memorandum TIRM-87–018.

Silvapulle, M.J. (1991), Robust ridge regression based on an M–estimator. *Australian Journal of Statistics*, **33**, 319–333.

Simpson, J.R. and Montgomery, D.C. (1996), A biased-robust regression technique for the combined outlier-multicollinearity problem. *Journal of Statistical Computing and Simulation*, **56**, 1–20.

Simpson, D.G., Ruppert, D. and Carroll, R.J. (1992), On one-step GM-estimates and stability of inferences in linear regression, *Journal of the American Statistical Association*, **87**, 439–450.

Smith, R.E., Campbell, N.A. and Lichfield, A. (1984), Multivariate statistical techniques applied to pisolitic laterite geochemistry at Golden Grove, Western Australia, *Journal of Geochemical Exploration*, **22**, 193–216.

Smucler, E. and Yohai, V.J. (2015), Highly robust and highly finite sample efficient estimators for the linear model, in K. Nordhausen and S. Taskinen (eds), *Modern Nonparametric, Robust and Multivariate Methods, Festschrift in Honour of Hannu Oja*. Springer.

Smucler, E. and Yohai, V.J. (2017), Robust and sparse estimators for linear regession models. *Computational Statistics & Data Analysis,* **111**, 116–130

Sposito, V.A. (1987), On median polish and L_1 estimators, *Computational Statistics and Data Analysis,* **5**, 155–162.

Stahel, W.A. (1981), Breakdown of covariance estimators, Research report 31, Fachgruppe für Statistik, ETH, Zürich.

Stahel, W.A. and Welsh, A.H. (1997), Approaches to robust estimation in the simplest variance components model. *Journal of Statistical Planning and Inference*, **57**, 295–319.

Stapleton, J.H. (1995), *Linear Statistical Models.* John Wiley.

Staudte, R.G. and Sheather, S.J. (1990), *Robust Estimation and Testing*, John Wiley and Sons.

Stigler, S. (1973), Simon Newcomb, Percy Daniell and the history of robust estimation 1885–1920, *Journal of the American Statistics Association,* **68**, 872–879.

Stigler, S.M. (1977), Do robust estimators deal with *real* data?, *The Annals of Statistics*, **5**, 1055–1098.

Stigler, S.M. (1986), *The History of Statistics: The Measurement of Uncertainty before 1900*, Belkap Press of Harvard University Press.

Stromberg, A.J. (1993a), Computation of high breakdown nonlinear regression parameters, *Journal of the American Statistical Association*, **88**, 237–244.

Stromberg, A.J. (1993b), Computing the exact least median of squares estimate and stability diagnostics in multiple linear regression, *SIAM Journal of Scientific Computing*, **14**, 1289–1299.

Svarc, M., Yohai, V.J. and Zamar, R.H. (2002), Optimal bias-robust M-estimates of regression, in Y. Dodge (ed.), *Statistical Data Analysis Based on the L_1 Norm and Related Methods*, pp. 191–200. Birkhäuser.

Tarr, G., Muller, S. and Weber, N.C. (2015), Robust estimation of precision matrices under cellwise contamination. *Computational Statistics & Data Analysis*, **93**, 404–420

Tatsuoka, K.S. and Tyler, D.E. (2000), On the uniqueness of S-functionals and M-functionals under nonelliptical distributions, *The Annals of Statistics*, **28**, 1219–1243.

Thomson, D.J. (1977), Spectrum estimation techniques for characterization and development of WT4 waveguide, *Bell System Technical Journal*, **56**, 1769–1815 and 1983–2005.

Tibshirani, R. (1996), Regression shrinkage and selection via the lasso. *Journal of the Royal Statistical Society, Series B*, **58**, 267–288.

Tsay, R.S. (1988), Outliers, level shifts and variance changes in time series, *Journal of Forecasting*, **7**, 1–20.

Tukey, J.W. (1960), A survey of sampling from contaminated distributions, in I. Olkin (ed.) *Contributions to Probability and Statistics*. Stanford University Press.

Tukey, J.W. (1962), The future of data analysis, *The Annals of Mathematical Statistics* **33**, 1–67.

Tukey, J.W. (1967), An introduction to the calculations of numerical spectrum analysis, in B. Harris (ed.), *Proceedings of the Advanced Seminar on Spectral Analysis of Time Series*, pp. 25–46. John Wiley and Sons.

Tukey, J.W. (1975a), Useable resistant/robust techniques of analysis, in *Proceedings of First ERDA Symposium*, pp. 11–31, Los Alamos, New Mexico.

Tukey, J.W. (1975b), Comments on "Projection pursuit", *The Annals of Statistics*, **13**, 517–518.

Tukey, J.W. (1977), *Exploratory Data Analysis*, Addison-Wesley.

Tyler, D.E. (1983), Robustness and efficiency properties of scatter matrices, *Biometrika*, **70**, 411–420.

Tyler, D.E. (1987), A distribution-free M-estimator of multivariate scatter, *The Annals of Statistics*, **15**, 234–251.

Tyler, D.E. (1990), Breakdown properties of the M–estimators of multivariate scatter, Technical report, Department of Statistics, Rutgers University.

Tyler, D.E. (1994), Finite-sample breakdown points of projection-based multivariate location and scatter statistics, *The Annals of Statistics*, **22**, 1024–1044.

Valdora, M. and Yohai, V.J. (2014), Robust estimators for generalized linear models. *Journal of Statistical Planning and Inference*, **146**, 31–48.

van Aelst, S. and Willems, G. (2011), Robust and efficient one-way MANOVA tests. *Journal of the American Statistical Association*, **106**, 706–718.

Wang, H., Li, G. and Jiang, G. (2007), Robust regression shrinkage and consistent variable selection through the LAD-Lasso. *Journal of Business and Economic Statistics*, **25**, 347–355.

Wedderburn, R.W.M. (1974), Quasi-likelihood functions, generalized linear models, and the Gauss–Newton method, *Biometrika*, **61**, 439–447.

Weisberg, S. (1985), *Applied Linear Regression* (2nd edn).John Wiley.

Welch, P.D. (1967), The use of the fast Fourier transform for estimation of spectra: A method based on time averaging over short, modified periodograms, *IEEE Transactions on Audio and Electroacoustics*, **15**, 70–74.

West, M. and Harrison, P.J. (1997), *Bayesian Forecasting and Dynamic Models*, (2nd edn). Springer-Verlag.

Whittle, P. (1962), Gaussian estimation in stationary time series, *Bulletin of the International Statistical Institute*, **39**, 105–129.

Woodruff, D.L. and Rocke, D.M., (1994), Computable robust estimation of multivariate location and shape in high dimension using compound estimators, *Journal of the American Statistical Association*, **89**, 888–896.

Yohai, V.J. (1987), High breakdown–point and high efficiency estimates for regression, *The Annals of Statistics* **15**, 642–65.

Yohai, V.J. and Maronna, R.A. (1977), Asymptotic behavior of least-squares estimates for autoregressive processes with infinite variance, *The Annals of Statistics*, **5**, 554–560.

Yohai, V.J. and Maronna, R.A. (1979), Asymptotic behavior of M–estimates for the linear model, *The Annals of Statistics*, **7**, 258–268.

Yohai, V.J. and Zamar, R.H. (1988), High breakdown estimates of regression by means of the minimization of an efficient scale, *Journal of the American Statistical Association*, **83**, 406–413.

Yohai, V.J. and Zamar, R.H. (1997), Optimal locally robust M-estimates of regression, *Journal of Statistical Planning and Inference*, **57**, 73–92.

Yohai, V.J. and Zamar, R.H. (2004), Robust nonparametric inference for the median, *The Annals of Statistics,* **5**, 1841–1857.

Yohai, V.J., Stahel, W.A. and Zamar, R.H. (1991), A procedure for robust estimation and inference in linear regression, in W. Stahel and S. Weisberg (eds), *Directions in Robust Statistics and Diagnostics (Part II),* IMA Volumes in Mathematics and is Applications, pp. 365–374. Springer.

Yuan, Mand Lin, Y. (2006) Model selection and estimation in regression with grouped variables *Journal of the Royal Statistical Society Series B*, **68**, 49–67.

Yuan, M. and Lin, Y. (2007), Model selection and estimation in the gaussian graphical model. *Biometrika*, **94**, 19–35.

Zhao, Q. (2000), Restricted regression quantiles, *Journal of Multivariate Analysis*, **72**, 78–99.

Zou, H. (2006), The adaptive lasso and its oracle properties. *Journal of the American Statistical Association*, **101**, 1418–1429.

Zou, H. and Hastie, T. (2005), Regularization and variable selection via the elastic net. *Journal of the Royal Statistical Society Series B*, **67**, 301–320.

Zuo, Y., Cui, H. and He, X. (2004a), On the Stahel–Donoho estimator and depth-weighted means of multivariate data, *The Annals of Statistics*, **32**, 167–188.

Zuo, Y., Cui, H. and Young, D. (2004b), Influence function and maximum bias of projection depth based estimators, *The Annals of Statistics*, **32**, 189–218.

Zuo, Y. and Serfling, R. (2000), General notions of statistical depth function, *The Annals of Statistics,* **28**, 461–482.

Index

Robust Statistics: Theory and Methods (with R), Second Edition.
Ricardo A. Maronna, R. Douglas Martin, Victor J. Yohai and Matías Salibián-Barrera.

Companion website: www.wiley.com/go/maronna/robust